Modern Recording Techniques

Modern Recording Techniques

Eighth Edition

David Miles Huber
Robert E. Runstein

NEW YORK AND LONDON

First published 1974 by H.W. Sams

This edition published 2014 by Focal Press
70 Blanchard Road, Suite 402, Burlington, MA 01803

Simultaneously published in the UK by Focal Press
2 Park Square, Milton Park, Abingdon, Oxon OX14 4RN

Focal Press is an imprint of the Taylor & Francis Group, an informa business

Library of Congress Cataloging-in-Publication Data
Huber, David Miles.
Modern recording techniques / David Miles Huber, Robert E. Runstein.—Eighth edition.
pages cm
Includes index.
ISBN 978-0-240-82157-3 (paper back)
1. Sound—Recording and reproducing. 2. Magnetic recorders and recording. 3. Digital audiotape recorders and recording. I. Runstein, Robert E. II. Title.
TK7881.4.H783 2013
621.389′3—dc23 2013010598

ISBN: 978-0-240-82157-3 (pbk)
ISBN: 978-0-240-82464-2 (ebk)

Typeset in Giovanni Book
By Book Now Ltd, London

Printed and bound by Courier in Westford, MA

To those who came before ... To us ... and To those who will come after ... Enjoy the journey!

Contents

Acknowledgments

I'd like to thank my partner, Daniel Butler, for putting up with the general rantin', ravin' and all-round craziness that goes into writing a never-ending book project. I'd also like to express my thanks to all of my friends and family in the United States (Seattle, LA and Indiana), Europe (Berlin, Mol, Vienna and Amsterdam) and for my music collaborators who help me reach new heights, both in the studio and on-stage. I'd like to give special thanks to the folks at Easy Street Records in West Seattle, Drew, Yvonne and Uli (Berlin), Emiliano and Milber (Amsterdam), Sara and my buddies at Big Arts Labs (LA), Maurice Pastist (PMC Speakers, US), everyone at the Northwest Grammy Chapter and the main Grammy office (LA), as well as the folks at Steinberg North America (LA) and Ableton (Berlin).

A very special mention goes to my editor Terri Jadick, for truly being the best editor a guy could ever have, as well as to Meagan White and Anais Wheeler at focal for making my job go so smoothly.

Thank you all for your kindness and support!

David Miles Huber

www.davidmileshuber.com

CHAPTER 1

Introduction

The world of modern music and sound production is multifaceted. It's an exciting world of creative individuals: musicians, engineers, producers, manufacturers and business-people who are experts in such fields as music, acoustics, electronics, production, broadcast media, multimedia, marketing, graphics, law and the day-to-day workings of the business of music. The combined efforts of these talented people work together to create a single end product: music that can be marketed to the masses. The process of turning a creative spark into a final product takes commitment, talent, a creative production team, a marketing strategy and, often, money. Throughout the history of recorded sound, the process of capturing music and transforming it into a marketable product has always been driven by changes in technology and cultural tastes.

In the past, the process of turning one's own music into a final product required the use of a commercial recording studio, which was (and still is) equipped with specialized equipment and a professionally-skilled staff. With the introduction of the large-scale integrated (LSI) circuit, mass production and mass marketing—three of the most powerful forces in the Information Age—another option has arrived on the scene: the radical idea that musicians, engineers and/or producers can have their own project studio. Along with this concept comes the realization that almost anyone can afford, construct and learn to master their own personal audio production facility. In short, we're living in the midst of a techno-artistic revolution that puts more power, artistic control and knowledge directly into the hands of creative individuals from all walks of life ... a fact that assures that the industry will forever be a part of the creative life-force of CHANGE.

On the techno side, those who are new to the world of modern digital audio and multitrack production, musical instrument digital interface (MIDI), mixing, remixing and their production environments should be aware that years of dedicated practice are often required to develop the skills that are needed to successfully master the art and application of these technologies. In short, it takes time to master your craft. A person new to the recording or project studio

FIGURE 1.1
The historic (but newly-renovated) Capitol Records Studio A control room, LA. (Courtesy of Capitol Records, www.capitolrecords.com).

FIGURE 1.2
The API "Green Room" at Wisseloord, Hilversum, Netherlands. (Courtesy of Wisseloord, www.wisseloord.nl, Acoustics and design by Jochen Veith, jv-acoustics, www.jv-acoustics.de)

environment (Figures 1.1 and 1.2) might easily be overwhelmed by the amount and variety of equipment that's involved in the process; however, as you become familiar with the tools, toys and techniques of recording technology, a definite order to the studio's makeup will soon begin to emerge, with each piece of equipment and approach to production being designed to play a role in the overall scheme of making music and quality audio.

The goal of this book is to serve as a guide and reference tool to help you become familiar with the recording and production process. When used in conjunction with mentors, lots of hands-on experience, further reading, Web searching, soul searching and simple common sense, I hope this book will help you to understand the equipment and day-to-day practices of sound recording and production.

Although it's taken the modern music studio over a hundred years to evolve to its current level of technological sophistication, we have moved into an important evolutionary phase in the business of music and music production: the digital age. Truly, this is an amazing time in production history—we live in a time when we can choose between an amazing array of cost-effective and powerful tools for fully realizing our creative and human potential. As always, patience and a nose-to-the-grindstone attitude are needed in order to learn how to use them effectively, but today's technology can free you up for the really important stuff: making music and audio productions. In my opinion, these are definitely the good ol' days!

Tutorial: Diggin' Deep into the Web

This book, by its very nature, is an overview of recording technology and production. It's a very in-depth one, but there's absolutely no way that it can fully devote itself to all of the topics. However, we're lucky enough to have the Web at our disposal to help us dig deeper into a particular subject that we might not fully understand, or simply want to know more about. Giga-tons of sites can be found that are dedicated to even the most offbeat people, places, toys and things … and search engines can even help you find obscure information on how to fix a self-sealing stem-bolt on a 1905 sonic-driven nut cracker. As such, I strongly urge you to use the Web as an additional guide. For example, if there's a subject that you just don't get, look it up on www.wikipedia.org or simply Google it. Of course, there's a wealth of info that can be found by searching the innumerable www.youtube.com videos that relate to any number of hardware systems, software toys and production techniques. Further information relating to this book and the recording industry at large can also be found at www.modrec.com. Digging deeper into the Web will certainly provide you with a different viewpoint or another type of explanation … and having that "AH HA!" lightbulb go off (as well as the "hokey pokey") IS definitely what it's all about.

David Miles Huber (www.modrec.com)

THE RECORDING STUDIO

The commercial music studio (Figures 1.3 and 1.4) is made up of one or more acoustic spaces that are specially designed and tuned for the purpose of capturing the best possible sound on a recording medium. In addition, these facilities are often structurally isolated in order to keep outside sounds from entering the room and being recorded (as well as to keep inside sounds from leaking out and disturbing the surrounding neighbors). In effect, the most important characteristics that go into the making and everyday workings of such a facility include:

- A professional staff
- Professional equipment
- Professional, yet comfortable working environment

- Optimized acoustic and recording environment
- Optimized control room mixing environment

Recording studio spaces vary in size, shape and acoustic design (Figures 1.5 through 1.7) and usually reflect the personal taste of the owner or are designed to accommodate the music styles and production needs of clients, as shown by the following examples:

- A studio that records a wide variety of music (ranging from classical to rock) might have a large main room with smaller, isolated rooms off to the side for unusually loud or soft instruments, vocals, etc.

FIGURE 1.3
The historic Capitol Records Studio A, LA. (Courtesy of Capitol Records, www.capitolrecords.com).

FIGURE 1.4
Studio 1 at Wisseloord, Hilversum, Netherlands. (Courtesy of Wisseloord, www.wisseloord.nl, Acoustics and design by Jochen Veith, jv-acoustics, www.jv-acoustics.de).

- A studio designed for orchestral film scoring might be larger than other studio types. Such a studio will often have high ceilings to accommodate the large sound buildups that are often generated by a large the number of studio musicians.
- A studio used to produce audio for video, film dialogue, vocals and mixdown might consist of only a single, small recording space off the control room for overdub purposes.

In fact, there is no secret formula for determining the perfect studio design. Each studio design (Figures 1.8 and 1.9) has its own sonic character, layout, feel and

FIGURE 1.5
Studio Metronome's recording room, Brookline, NH. (Courtesy of Metronome Media Group, www.studiometronome.com, Photo by Bennett Chandler)

FIGURE 1.6
London Bridge Studios big tracking room, Seattle. (Courtesy of London Bridge Studios, www.londonbridgestudio.com, Photo by Christopher Nelson)

FIGURE 1.7
Classical scoring setup at Galaxy Hall, Galaxy Studios, Mol, Belgium. (Courtesy of Galaxy Studios, http://www.galaxy.be)

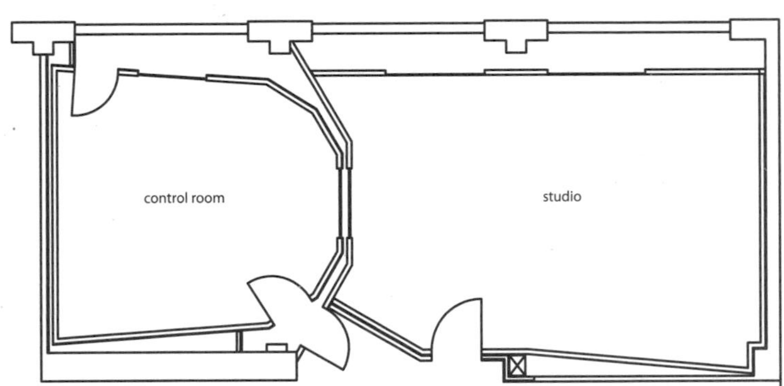

FIGURE 1.8
Basic floorplan for KMR Audio Berlin, Germany (Courtesy of KMR Audio, www.kmraudio.de; studio design by Fritz Fey, www.studioplan.de).

decor that are based on the personal tastes of its owners, the designer (if any), and the going studio rates (based on the studio's return on investment and the supporting market conditions).

During the 1970s, studios were generally small. Because of the advent of (and over-reliance on) artificial effects devices, they tended to be acoustically "dead" in that the absorptive materials tended to suck the life right out of the room. The basic concept was to eliminate as much of the original acoustic environment as possible and replace it with artificial ambience.

Fortunately, since the mid-1980s, a number of larger commercial studios that have the physical space have begun to move back to the original design concepts

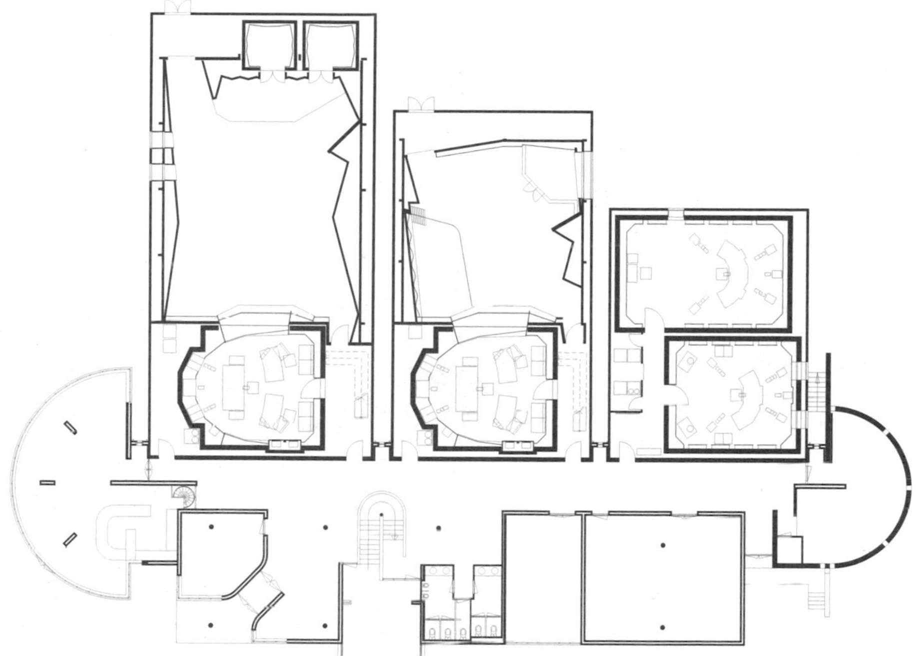

FIGURE 1.9 Floorplan for Wisseloord, Hilversum, Netherlands. (Courtesy of Wisseloord, www.wisseloord.nl, Acoustics and design by Jochen Veith, jv-acoustics, www.jv-acoustics.de)

of the 1930s and 40s, when studios were larger. This increase in size (along with the addition of one or more smaller iso-booths or rooms) has revived the art of capturing the room's original acoustic ambience along with the actual sound pickup. In fact, through improved studio design techniques, we have learned how to achieve the benefits of both earlier and modern-day recording eras by building a room that absorbs sound in a controlled manner thereby reducing unwanted leakage from an instrument to other mics in the room. This disperses reflections in a way that allows the studio to retain a well-developed reverberant and sonic personality of its own. This effect of combining direct and natural room acoustics is often used as a tool for "livening up" an instrument (when recorded at a distance), a technique that has become popular when recording live rock drums, string sections, electric guitars, choirs, etc.

THE CONTROL ROOM

A recording studio's *control room* (Figures 1.10 through 1.12) serves a number of purposes in the recording process. Ideally, the control room is acoustically isolated from the sounds that are produced in the studio, as well as from the surrounding, outer areas. It is optimized to act as a critical listening environment that uses carefully placed and balanced monitor speakers. This room also houses the majority of the studio's recording, control and effects-related equipment. At the heart of the control room is the recording console.

FIGURE 1.10 Patchwork Recordings, Atlanta, Georgia. (Courtesy of Russ Berger Design Group, Inc., www.rbdg.com)

FIGURE 1.11 London Bridge Studios big tracking room, Seattle. (Courtesy of London Bridge Studios, www.londonbridgestudio.com)

The *recording console* (also referred to as the *board* or *desk*) can be thought of as an artist's palette for the recording engineer, producer and musicians. The console allows the engineer to combine, control and distribute the input and output signals of most, if not all, of the devices found in the control room. The console's basic function is to allow for any combination of mixing (variable control over relative amplitude and signal blending between channels), spatial positioning (left/right or surround-sound control over front, center, rear and sub), routing (the ability to send any input signal from a source to a destination) and switching for the multitude of audio input/output signals that are commonly encountered in an audio production facility.

Tape machines might be located towards the rear of a control room, while digital audio workstations (DAWs) are often located at the side of the console

FIGURE 1.12
88D control room at Galaxy Studios, Mol, Belgium. (Courtesy of Galaxy Studios, http://www.galaxy.be)

or at the functional center if the DAW serves as the room's main recording/mixing device. Because of the added noise and heat generated by recorders, computers, power supplies, amplifiers and other devices, it's becoming more common for equipment to be housed in an isolated machine room that has a window and door adjoining the control room for easy access and visibility. In either case, DAW controller surfaces (which are used for computer-based remote control and mixing functions) and auto-locator devices (which are used for locating tape and media position cue points) are often situated in the control room near the engineer for easy access to all recording, mixing and transport functions. Effects devices (used to electronically alter and/or augment the character of a sound) and other signal processors are also placed nearby for easy accessibility (often being designed into an effects island or bay that's often located directly behind the console). In certain situations, a facility might not have a large recording space at all but simply have a small or midsized iso-room for recording overdubs (this is often the case for rooms that are used in audio-for-visual post-production and/or music remixing).

As with recording studio designs, every control room will usually have its own unique sound, feel, comfort factor and studio booking rate. Commercial control rooms often vary in design and amenities—from a room that's basic in form and function to one that is lavishly outfitted with the best toys and fully stocked kitchens in the business. Again, the style and layout are a matter of personal choice; however, as you'll see throughout this book, there are numerous guidelines that can help you make the most of a recording space. However, it's good to remember that as important as the layout, feel, and equipment are, the people (the staff, musicians and you) will almost always play the most prominent role in capturing the feel of a performance.

THE CHANGING FACES OF THE MUSIC STUDIO BUSINESS

As we've noted, the role of the professional recording studio has begun to change as a result of upsurges in project studios, audio for video and/or film, multimedia and Internet audio. These market forces have made it necessary for certain facilities to rethink their operational business strategies. Often, these changes have met with some degree of success, as is illustrated by the following examples:

- Personal production and home project studios have greatly reduced the need for an artist or producer to have constant and costly access to a professional facility. As a result, many pro studios now cater to artists and project studio owners who might have an occasional need for a larger space or better-equipped recording facility (e.g., for big drum sounds, string overdubs or an orchestral session). In addition, after an important project has been completed in a private studio, a professional facility might be needed to mix the production down into its final form. Most business-savvy studios are only too happy to capitalize on these new and constantly changing market demands.
- Upsurges in the need for audio for video and film postproduction have created new markets that allow professional recording studios to provide services to the local, national and international broadcast and visual production communities. Creative studios often enter into lasting relationships with audio-for-visual and broadcast production markets, so as to thrive in the tough business of music, when music production alone might not provide enough income to keep a studio afloat.
- Studios are also taking advantage of Internet audio distribution techniques by offering Web development, distribution and other audio-for-web services as an added incentive to their clients.
- A number of studios are also jumping directly into the business of music by offering advisory, business and networking services to artists and bands, sometimes signing the artists and funding tours in exchange for a piece of the business pie.

These and other aggressive marketing strategies (many of which may be unique to a particular area) are being widely adopted by commercial music and recording facilities to meet the changing market demands of new and changing media. No longer can a studio afford to place all of its eggs in one media basket. Tapping into changes in market forces and meeting them with new solutions are important for making it (or simply keeping afloat) in the business of music production and distribution. Make no mistake about it—starting, staffing and maintaining a production facility, as well as getting the clients' music heard, is serious work that requires dedication, stamina, innovation, guts and a definite dose of craziness.

OK ... now for the hard but often most interesting part. Let's take a moment to say that all-important word again: *business*. With the onset of more creative changes, opportunities and options, the only thing that stays constant is change, right? For example, there has been a steady onslaught of technological advances in audio production (such as those with software, portable recording and controller options that come with new generations of computer, laptop and pad technologies). However, beyond that, in reality many (if not most) of the technical aspects of music and audio production have stayed the same. What HAS drastically changed are the business aspects of the industry—most notably in how music is distributed, sold and marketed to the buying public.

I certainly don't have to remind you about how the traditional label distribution models have all changed in favor of the download era. Over the last several years, the game has been continuously evolving in a way that keeps even the most seasoned industry professionals on their toes ... and is this necessarily a bad thing? I think not! Innovation and ingenuity have always been part of the creative process—it's what keeps things new and exciting. However, innovation doesn't always come easily and is often at a price. Recording studios are constantly being challenged to ride the wave of innovation—some make it, some are simply unable to adapt to the new world of the personal studio, the internet and changing business models. In short, the reality is that the professional studio business is a tough one that requires its owners to constantly adapt to changing markets. If such a studio environment is where you'd like to be, you'll need to keep these thoughts in mind, so as to make yourself marketable to a business that will need your expertise – and make you a valuable member of the team.

THE PROJECT STUDIO

With the advent of affordable digital audio and analog recording systems, it's a foregone conclusion that the vast majority of music and audio recording/production systems are being built and designed for personal use. The rise of the *project studio* (Figures 1.13 and 1.14) has brought about monumental changes in the business of music and professional audio, in a way that has affected and altered almost every facet of the audio production community.

One of the greatest benefits of a project or portable production system centers around the idea that an artist can choose from a wide range of tools and toys to get the particular sounds that he or she likes. This technology is often extremely powerful, as the components combine to create a vast palette of sounds and handle a wide range of task-specific functions. Such systems often include a DAW (digital audio workstation computer for recording, MIDI sequencing, mixing and just about anything that relates to modern audio production), soft and hardware electronic instruments, effects devices ... as well as speaker monitoring.

FIGURE 1.13
Happiness is being able to record your own band in the basement. (Courtesy of Brooks Callison, callisonic.com)

FIGURE 1.14
Tony Sheppard's studio. (Courtesy of Tony Sheppard, www.tonysound.com, photo by Ed Colver)

Systems like these are constantly being installed in the homes of working and aspiring musicians, audio enthusiasts and DJs. Their sizes range from a corner in an artist's bedroom to a larger system that has been installed in a dedicated project studio room. All of these system types can be designed to handle a wide range of applications and have the important advantage of letting the artist produce his or her music in a comfortable, cost-effective, at-home environment whenever the creative mood hits. Such production luxuries, which would have literally cost a fortune 30 years ago, are now within the reach of almost all working and aspiring musicians. This revolution has been carried out under the motto "You don't have to have a million-dollar studio to make good music." Truly, the modern-day project and portable studio systems offer such a degree of cost-effective power and audio fidelity that they can often match the production quality of a professional recording facility …all you need to supply is knowledge, care, dedication and patience.

THE PORTABLE STUDIO

Of course, as laptops have grown in power, it has become a simple matter to load them with your favorite DAW software and audio interface, grab your favorite mics and headphones, put the entire system in your backpack and hit the road running. These systems have actually gotten powerful enough that they equal and sometimes surpass large tower-based studio systems, allowing you to record, edit, mix and produce on-the-go or in the studio without any compromises whatsoever. In the bedroom, on the beach or at a remote seaside island under battery or solar power (Figure 1.15)—there are literally no limits to the capability of these ever-growing studio systems.

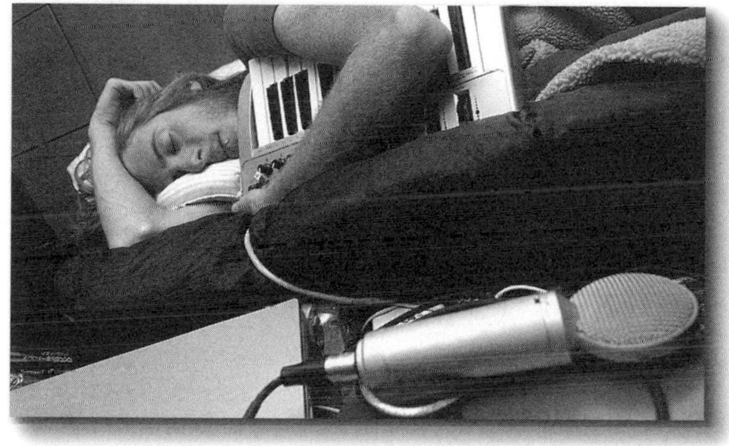

FIGURE 1.15
Studio on the go. ... (Courtesy of M-Audio, a division of Avid, www.m-audio.com)

THE iREVOLUTION

In recent years an even smaller studio revolution has taken place—the pad revolution. In fact, some of the greatest changes in audio production today are coming about as a direct result of the introduction of palm-held devices like the iPad and Android pad devices. Of course, the strengths of these devices are that they are extremely portable, wireless and offer an ever-increasing amount of processing power. A by-product of all this is overall cost-effectiveness. As an example, years ago, a remote controller device for a DAW would set you back $1200 or so...now, it's a simple matter of getting out your pad, downloading a controller app from the "store" and you'll have as many (or more) of the functions of its "wired" hardware counterpart at a ridiculously small cost ... or even for free (Figure 1.16).

As new apps (applications) come onto the market - literally on a daily basis - the pad revolution continues to make its mark on all forms of media production.

FIGURE 1.16
A pad can be used as a wireless control surface for a DAW in a way that simulates the functions of a larger controller at a meager fraction of the cost.

Within the fields of music and audio production, a pad can be used for such applications as:

- Audio recording and mixing:
- DAW and systems control
- Electronic instruments
- Composition

Obviously, as these devices become more powerful, they can be used for any number of on-the-go purposes that previously required a larger computer or laptop … this is truly a technological revolution in the making.

Studio in the Palm of your Hand

To take these ever-shrinking analogies to the *n*th degree, newer handheld recording systems can actually fit in your pocket to sample and record sounds with professional results, using either their internal high-quality mics or, in some cases, using external professional mics under phantom power. Truly, it is a small world after all (Figure 1.17)!

KNOWLEDGE IS POWER!

With all of these amazing tools that are at our disposal, it totally makes sense that artist, producers and aspiring recording professionals will create art using the toys, tools and techniques that are affordably available to them. However, technology isn't enough to create great art … a personal sense of drive, passion and ingenuity is required. One more ingredient is necessary to finish off this

FIGURE 1.17
Capturing sound on the go!

artistic "mix" … namely, a personal and never-ending search for knowledge to improve your craft. This all-important ingredient can be gained by:

- Reading about the equipment choices that are available to you on the Web or in the trade magazines
- Visiting and talking with others of like (and dissimilar) minds about their equipment, techniques and personal working styles (social networks can be a really effective learning tool)
- Enrolling in a recording course that best fits your needs, working style and budget
- Researching the type of equipment and room design before you make any purchases and, if possible, getting your hands on equipment before you make any final purchases (i.e. checking them out at your favorite music store)
- Experience and time … always the best teacher

The more you take the time to familiarize yourself with the options and possibilities that are available to you, the less likely you are to be unhappy about how you spent your hard-earned bucks. It is also important to point out that *having* the right equipment for the job isn't enough—it's important to *take the time* to learn how to use your tools to their fullest potential. Whenever possible, read the manual and get your feet wet by taking the various settings, functions and options for a test spin … long before you're under the time and emotional constraints of being in a session.

WHATEVER WORKS FOR YOU

As you begin to research the various types of recording and supporting systems that can be put to use in a project studio, you'll find that a wide variety of options are available. There are indeed hundreds, if not thousands, of choices for recording media, hardware types, software systems, speakers, effects devices … the list goes on. This should automatically tell us that no one tool is right for the job. As with everything in art (even the business of an art), there are many personal choices that can combine into the making of a working environment that's right for you. Whether you:

- Work with a hard-disk or tape-based system
- Choose to use analog or digital equipment (or both)
- Are a Mac or PC kind of person (pretty much a nonissue these days)
- Use this type of software or that …

Basically, it all comes down to the bottom line of … how does it sound? How does it move the audience? How can it be sold? In truth, no prospective buyer will turn down a song because it wasn't recorded on such-n-such a machine using speakers made by so-n-so—it's the feel, the emotion and the art that always seals the deal.

MAKING THE PROJECT STUDIO PAY FOR ITSELF

Beyond the obvious advantage of being able to record when, where and how you want to in your own project studio, there are several additional benefits to working in a personal environment. Here are ways that a project studio can help subsidize itself, at any number of levels:

- Setting your own schedule and saving money while you're at it! An obvious advantage of a project studio revolves around the idea that you can create your own music on your own schedule. Part of the expense of using a professional studio comes from having to be practiced and ready to roll on a specific date or range of dates. Having your own project studio frees you up to lay down practice tracks and/or record when the mood hits, without having to worry about punching the studio's time clock.
- For those who are in the business of music, media production or the related arts business, the equipment, building and utility payments can be written off as a tax-deductible expense (see Appendix B, "Tax Tips for Musicians").
- An individual artist or group might consider pre-producing a project in their own studio … allowing the time and expense billings to be tax-deductible.
- The same artists might consider recording part or all of their production at their own project studio. The money saved (and deducted) could be spent on a better mixdown facility, professional freelance engineer, solving legal issues (such as copyright) and/or marketing.
- The "signed artist/superstar approach" refers to the mega-artist who, instead of blowing the advance royalties on lavish parties in the studio (a sure way never to see any money from your hard work), will spend the bucks on building their own professional-grade project studio. After the project has been recorded, the artist will still have a tax-deductible facility

that can be operated as a business enterprise. When the next project comes along, the artist will still have a personal facility where they can record and put the saved advance bucks in the bank.

LIVE/ON-LOCATION RECORDING: A DIFFERENT ANIMAL

Unlike the traditional multitrack recording environment, where overdubs are often used to build up a song over time, *live/on-location recordings* are created on the spot, in real time, often during a single on-stage or in-the-studio performance, with little or no studio postproduction other than mixdown. A live recording might be very simple, possibly being recorded using only a few mics that are mixed directly to two or more tracks. Or, a more elaborate gig might call for a full-fledged multitrack setup, requiring the use of a temporary control room or fully equipped mobile recording van or truck (Figure 1.18). A more involved setup will obviously require a great deal of preparation and expertise, including a knowledge of the sound reinforcement combined with the live recording techniques that are necessary to capture instruments in a manner that provides enough isolation between the tracks so as to yield a high degree of control over the individual instruments during the mixdown phase, yet still have a live performance feel.

Although the equipment and system setups will be familiar to any studio engineer, live recording differs from its more controlled studio counterpart in that it happens in a world where the motto is "you only get one chance." When you're recording an event where the artist is spilling his or her guts to hundreds or even thousands of fans, it's critical for everything to run smoothly. Live recording usually requires a unique degree of preparedness, redundancy, system setup skills, patience, and, above all, experience. It's all about capturing the sound and feel of the moment, the first time … and for a brave few, that's a very exciting thing.

FIGURE 1.18
Studio Metronome's live recording audio truck, Brookline, NH. (Courtesy of Metronome Media Group, www.studiometronome.com, Photo by Bennett Chandler)

FIGURE 1.19
Skywalker Sound scoring stage with orchestra, Marin County, CA. (Courtesy of Skywalker Sound, www.skysound.com).

AUDIO FOR VIDEO AND FILM

In this day and age, it's simply impossible to overlook the importance that quality audio plays in the production of film and video. With the introduction of complex surround playback formats, high-budget music scores and special effects production, audio-for-film has long been an established and specialized industry that has a dramatic effect on the movie-goer's experience. Most definitely, audio-for film is an artform that has touched and helped shape our world culture (Figure 1.19).

Prior to the advent of the DVD, Blu-ray and home theater surround sound; broadcast audio and the home theater experience was almost an afterthought in a TV tube's eye. However, with the introduction of these new technologies, audio has matured to being a highly respected part of video media production. With the common use of surround sound in the creation of movie sound tracks, along with the growing popularity of surround-sound in-home and computer entertainment systems, the public has come to expect high levels of audio quality from their entertainment experience.

In modern-day production, MIDI, hard-disk recording, timecode and synchronization, automated mixdown and advanced effects have become everyday components of the audio-for-visual environment, requiring that professionals be highly specialized and skilled in order to meet the demanding schedules and production complexities.

AUDIO FOR GAMES

Most of the robot-zappin', daredevil-flyin', hufflepuff-boppin' addicts who are reading this book are very aware that one of the largest and most lucrative areas of multimedia audio production is the field of scoring, designing and producing audio for computer games (Figure 1.20)... Zaaaaaaapppppppp! Like most

FIGURE 1.20
Not your typical multimedia studio. Electronic Arts, Vancouver, Canada. (Courtesy of Walters-Storyk Design Group – designed by Beth Walters and John Storyk, www.wsdg.com. Photo credit – Robert Wolsch)

subcategories within audio production, this field of expertise has its own set of technical and scheduling requirements that center more around spreadsheets and databases than they do audio equipment. With the tens of thousands of voice and music cues that are commonly required to make a game into a fully interactive experience, an entirely different skillset is often required of a sound technician.

A rather interesting connection between orchestral recording and game audio has been on the upsurge. Just as film makes use of heavy orchestral scoring for dramatic effect, game audio has also begun to score using big-budget and big-name orchestras to create a bigger-than-life storyline.

THE TIMES, THEY'VE ALREADY CHANGED: MULTIMEDIA & THE WEB

With the integration of text, graphics, MIDI, digital audio and digitized video into almost every facet of the personal computer environment, the field of *multimedia* audio has become a fast-growing, established industry that represents an important and lucrative source of income for both creative individuals and production facilities alike. Of course, the use of audio-for-the-web can take any number of forms:

- A record label might decide to offer music and audio for books in a download-only format, requiring that the projects be mastered in a way that best suits the medium.
- An online language dictionary might require that all of the translations be recorded, so they can be easily heard and pronounced.
- An online download site might offer any number of music loops that can be downloaded and made into a personal remix that can be shared over the web.
- … The list is absolutely endless!

For decades, the industry has been crying foul over the breakup of the traditional record industry as we know it. In the early days of the web, a new kid on the block came onto the scene: "The MP3." This "ripping" and playback medium made it possible for entire song libraries to be compressed (data-wise) into devices that could be downloaded and streamed with relative ease. Such a simple beastie then progressed into a medium that would bring an entire industry to its virtual knees. With media sharing came the eventual revolution of the social network … allowing people to connect with each other in totally new ways never before thought possible. With the sharing of information, came the sharing of music and visual media … allowing you to see that DMH is currently listening to Mouse on Mars on his phone, whilst walking on a bridge in Belgrade, Serbia … it's a brave, new world after all … and we're all part of the big change!

THE PEOPLE WHO MAKE IT ALL HAPPEN

> "One of the most satisfying things about being in the professional audio [and music] industry is the sense that you are part of a community."
>
> **Frank Wells, editor, Pro Sound News**

When you get right down to the important stuff, the recording field is built around pools of talented individuals and service industries who work together toward a common goal: producing, selling and enjoying music. As such, it's the *people* in the recording industry who make the business of music happen. Recording studios and other businesses in the industry aren't only known for the equipment that they have but are often judged by the quality, knowledge, vision and combined personalities of their staff. The following sections describe but a few of the ways in which a person can be involved in this multifaceted industry. In reality, the types and descriptions of a job in this techno-artistic industry are limited only by the imagination. New ways of expressing a passion for music production and sales are being created every day … and if you see a new opportunity, the best way to make it happen is to roll up your sleeves and "just do it."

The artist

The strength of a recorded performance begins and ends with the artist. All of the technology in the world is of little use without the existence of the central ingredients of human creativity, emotion and individual technique. Just as the overall sonic quality of a recording is no better than its weakest link, it is the performer's job to see that music's main ingredient—its inner soul—is laid out for all to experience and hear. After all is said and done, a carefully planned and well-produced recording project is simply a gilded framework for the music's original drive, intention and emotion.

Studio musicians and arrangers

A project will often require additional musicians to add extra spice and depth to the artist's recorded performance. For example:

- An entire group of studio musicians might be called on to provide the best possible musical support for a high-profile artist or vocalist.
- A project might require musical ensembles (such as a choir, string section or background vocals) for a particular part or to give a piece a fuller sound.
- If a large ensemble is required, it might be necessary to call in a professional music contractor to coordinate all of the musicians and make the financial arrangements. The project might also require a music arranger, who can notate and possibly conduct the various musical parts.
- A member of a group might not be available or be up to the overall musical standards that are required by a project. In such situations, it's not uncommon for a professional studio musician to be called in to fit the bill.

In situations like these, a project that's been recorded in a private studio might benefit from the expertise of a professional studio that has a larger recording room, an analog multitrack for that certain sound and/or an engineer that knows how to better deal with a complicated situation.

The producer

Beyond the scheduling and budgetary aspects of coordinating a recording project, it is the job of a producer to help the artist and record company create the best possible recorded performance and final product that reflects the artist's vision. A producer can be hired for a project to fulfill a number of specific duties or might be given full, creative reign to help with any and all parts of the creative and business side of the process to get the project out to the buying public. More likely, however, a producer will act collaboratively with an artist or group to guide them through the recording process to get the best possible final product. This type of producer will often:

- Help the artist (and/or record label) create the best possible recorded performance and final product that reflects the artist's vision. This will often include a large dose of musical input, creative insight and mastery of the recording process.
- Assist in the selection of songs.
- Help to focus the artistic goals and performance in a way that best conveys the music to the targeted audience.
- Help to translate that performance into a final, salable product (with the technical and artistic help of an engineer and mastering engineer).

It's interesting to note that because engineers spend much of their working time with musicians and industry professionals with the intention of making their clients sound good, it's not uncommon for an engineer to take on the role of producer or co-producer (by default or by mutual agreement). Conversely, as producers and artists alike become increasingly more knowledgeable about recording technology, it's increasingly common to find them on the other side of the glass, sitting behind the controls of a console.

In addition, a producer might also be chosen for his or her ability to understand the process of selling a final recorded project from a business perspective to a label, film licensing entity or to the buying public. This type of producer can help with artist gain insights into the world of business, business law, budgeting and sales … always an important ingredient in the process.

Of course, in certain circumstances, a project producer might be chosen for his or her reputation alone and/or for giving a certain cache to a project that can help put a personal "brand" on the project, thereby adding to the project's stature and hopefully help grab the public's attention.

One final thing's for certain, the artist and/or label should take time to study what type of producer is needed (if any) and then agree upon his or her creative and financial role in the project *before* entering into the creative process.

The engineer

The role of an engineer can best be described as an interpreter in a techno-artistic field. He or she must be able to express the artist's music and the producer's concepts and intent through the medium of recording technology. This job is actually best classified as a techno-artform, because both music and recording are totally subjective and artistic in nature and rely on the tastes, experiences and feelings of those involved. During a recording session, one or more engineers can be used on a project to:

- Conceptualize the best technological approach for capturing a performance or music experience
- Translate the needs and desires of the artists and producer into a technological approach to capturing the music
- Document the process for other engineers or future production use.
- Place the musicians in the desired studio positions
- Choose and place the microphones or pickup connections
- Set levels and balances on the recording console or DAW mixing interface
- Capture the performance (onto hard disk or tape) in the best way possible
- Overdub additional musical parts into the session that might be needed at a later time
- Mix the project into a final master recording in any number of media formats (mono, stereo, and surround-sound)
- Help in meeting the needs for archiving and/or storing the project

In short, engineers use their talent and artful knowledge of recording media technology to convey the best possible finished sound for the intended media, the client and the buying public.

Assistant engineer

Many studios often train future engineers (or build up a low-wage staff) by allowing them to work as assistants or interns who can offer help to staff and visiting freelance engineers. The assistant engineer might do microphone and headphone setups, run DAW setup or tape machines, help with session documentation, do session breakdowns and (in certain cases) perform rough mixes and balance settings for the engineer on the console. With the proliferation of freelance engineers (engineers who are not employed by the studio but are retained by the artist, producer or record company to work on a particular project), the role of the assistant engineer has become even more important. It's often his or her role to guide freelance engineers through the technical aspects and quirks that are peculiar to the studio … and to generally babysit the technical and physical aspects of the place.

Traditionally, this has been a no- or low-wage job that can expose a "newbie" to a wide range of experiences and situations. With hard work and luck, many assistants have worked their way into the hot seat whenever an engineer quits or is unexpectedly ill. As in life, there are no guarantees in this position—you just never know what surprises are waiting around the next corner for those who rise to the occasion.

Maintenance engineer

The maintenance engineer's job is to see that the equipment in the studio is maintained in top condition and regularly aligned and repaired when necessary. Of course, with the proliferation of project studios, cheaper mass-produced equipment, shrinking project budgets and smaller staffs, most studios will not have a maintenance engineer on staff. Larger organizations (those with more than one studio) might employ a full-time staff maintenance engineer, whereas outside freelance maintenance engineers and technical service companies are often called in to service smaller commercial studios in both major and non-major markets.

Mastering engineer

Often a final master recording will need to be tweaked in terms of level, equalization (EQ) and dynamics so as to present the final "master" recording in the best possible sonic and marketable light. If the project calls for it, this job will fall to a mastering engineer, whose job it is to listen to and process the recording in a specialized, fine-tuned monitoring environment. Of course, mastering is a techno-artistic field in its own right. Beauty is definitely in the ear of the beholding client and one mastering engineer might easily have a completely different approach to the sound and overall feel to a project than the next bloke. However, make no mistake about it—the mastering of a project can have a profound impact on the final sound of a project, and the task of finding the right mastering engineer for the job should never be taken lightly. Further info on mastering can be found in Chapter 19.

The DJ

Let's not forget one of the more important links for getting the musical word out to the buying public: the disc jockey (DJ). Actually, the role of disc jockey can take on many modern-day forms:

- *On the air*: The DJ of the airwaves is still a very powerful voice for getting the word out about a particular musical product.
- *On the floor*: This DJ form often reinforces the music from the airwaves or the sub-culture and helps to promote music to the mainstream and countercultural party animals.
- *On the Web*: Probably one of the more up-and-coming voices for getting the promotional word of the latest music out to a large number of specially targeted online audiences.

The VJ

With the integration of video into the online and on-stage scene, the *video jockey* (VJ) has begun to cast a shadow across the virtual canvas of the music scene (Figure 1.21). Mixing videos and visual imagery in real time on stage and producing music videos for the Web is a relative must for many an aspiring artist or band.

FIGURE 1.21
"Mike" The VJ… outstanding in his field.

STUDIO MANAGEMENT

Running a business in the field of music and audio production requires the special talents of businesspeople who are knowledgeable about the inner workings of promotion, the music studio, the music business and, above all, the people. It requires constant attention to quirky details that would probably be totally foreign to someone outside the "biz." Studio management tasks include:

- *Management*: The studio manager, who might or might not be the owner, is responsible for managerial and marketing decisions for all of the inner workings of the facility and its business.
- *Bookings*: This staff person keeps track of most of the details relating to studio usage and billing.
- *Competent administration staff*: These assistants keep everyone happy and everything running as smoothly as possible.

Note, however, that some or all of these functions often vary from studio to studio. These and other equally important staff are necessary in order to successfully operate a commercial production facility on a day-to-day basis.

MUSIC LAW

It's never good for an artist, band or production facility to underestimate the importance of a good music lawyer. When entering into important business relationships, it's always a good idea to have a professional ally who can help you, your band or your company navigate the potentially treacherous waters of a poorly or vaguely written contract. Such a professional can serve a wide range of purposes, ranging from the primary duties of looking after their clients' interests and ensuring that they don't sign their careers away by entering into a life of indentured, nonprofit servitude … all the way to introducing an artist to the best possible music label or distribution network.

Music lawyers, like many in this business, can be involved in the working of a business or career in many ways; hence, various fee scales are used. For example, a new artist might meet up with a friend who knows about a bright, young, freshly-graduated music lawyer who has just passed the bar exam. By developing a relationship early on, there are any number of potentials for building trust, making special deals that are beneficial to both parties, etc. On the other hand, a more established lawyer could help solicit and shop a song, band or artist more effectively in a major music, TV or film market. As with most facets of the biz, answers to these questions are often situational and require intuition, careful reference checking and the building of trust over time. Again, it's important to remember that a good music lawyer is often a must and is the unsung hero of many a successful career.

WOMEN AND MINORITIES IN THE INDUSTRY

Ever since its inception, males have dominated the recording industry. I remember many a session in which the only women on the scene were female artists, secretaries or studio groupies in short dresses. Fortunately, over the years, women have begun to play a more prominent role, both in front of and behind the glass … and in every facet of studio production and the business of music as a whole (Figure 1.22). In recent decades, most of the resistance to including new and fresh blood into the biz has greatly reduced.

FIGURE 1.22 Women's Audio Mission, an organization formed to assist women in the industry. (Courtesy of the Women's Audio Mission, www.womensaudiomission.org.)

> No matter who you are, where you're from or what your race, gender, sexual or planetary orientation is, remember this universal truth: If your heart is in it and you're willing to work hard enough, you'll make it (whatever you perceive "it" to be).

BEHIND THE SCENES

In addition to the positions listed earlier, there are scores of professionals who serve as a backbone for keeping the business of music alive and functioning. Without the many different facets of music business—technology, production, distribution and law—the biz of music would be very, very different. A small sampling of the additional professional fields that help make it happen includes:

- Artist management
- Artist booking agents
- A&R (artist and repertoire)
- Manufacturing
- Music and print publishing
- Distribution
- Web development
- Graphic arts and layout
- Equipment design
- Audio company marketing
- Studio management
- Live sound
- Live sound tour management
- Acoustics
- Audio instruction
- Club management
- Sound system installation for nightclubs, airports, homes, etc.
- … and a lot more!

This incomplete listing serves as a reminder that the business of making music is full of diverse possibilities and extends far beyond the notion that in order to make it in the biz you'll have to sell your soul or be someone you're not. In short, there are many paths that can be taken in this techno-artistic business. Once you've found the one that best suits your own personal style, you can then begin the lifelong task of gaining knowledge and experience and pulling together a network of those who are currently working in the field.

It's also important to realize that finding the career niche that's right for you might not happen overnight. You might try your hand at one aspect of production, only to find that your passion is in another field. This isn't a bad thing. As the saying goes, "Wherever you may be, there you are!" Finding the path that's best for you is a lifelong ongoing quest ... the general idea is to work hard, learn and enjoy the ride.

CAREER DEVELOPMENT

It's a sure bet that those who are interested in getting into the business of audio will quickly find out that it can be a tough nut to crack. For every person who makes it, a large number won't. In short, there are a lot of people who are waiting in line to get into what is perceived by many to be a glamorous biz. So, how *do* you get to the front of the line? Well, folks, here's the key:

Self-motivation

The business of art (the techno-art of recording and music being no exception) is one that's generally reserved for self-starters and self-motivated people. Even if you get a degree from XYZ college or recording school, there's absolutely no guarantee that your dream studio will be knocking on the door with an offer in hand (in fact, they most certainly won't). It takes a large dose of perseverance, talent and personality to make it.

This may sound strange, but one of the best ways to get into the biz is to simply jump in and *start*. In fact, you might try this little trick ... find a stick (or one of those Scottish swords, if you have to) and get down on your knees, then "knight" yourself on the shoulder with the "sword" (figuratively or literally) and say: "I am now a ________!" Fill in the blank with whatever you want to be (engineer, artist, producer ... whatever) and arise Sir or Dame ________ ... you are now a ________! Simply become it ... right there on the spot! Now, make up a business card, start a business and begin contacting artists to work with (or make the first step toward becoming the creative person you want to be). All you have to do is work hard, believe in yourself and follow the Golden Rule.

In fact, there are many ways to get to the top of your own personal mountain. For example, you could get a diploma from a school of education or from the school of hard knocks (it usually ends up being from a bit of both)—but the goals and the paths are up to you. Like a mentor of mine always said: "Failure isn't a bad thing … but not trying is!"

Networking: "Showing up is HUGE"

The other half of the success equation lies in your ability to network with other people. Like the venerable expression says: "It's not [only] what you know, it's who you know." Maybe you have an uncle or a friend in the business or a friend who has an uncle … you just never know where help might come from next. This idea of getting to know someone who knows someone else is what makes the business world go around. So, don't be afraid to put your best face forward and start meeting people. If you want to work at XYZ Studios, hang out without being in the way. You never know, the engineer might need some help or might know someone who can help get you into the proverbial door. The longer you stick with it, the more people you'll meet, and you'll have a bigger and stronger network than you thought could ever be possible.

ANCIENT PROVERB

Being "in the right place at the right time" means being in the wrong (or right) place at the wrong time a thousand times! … In short, "Showing up is HUGE"!

So what are some good ways to get started?

- Join an industry association … such as The Recording Academy (Grammys), Grammy U(niversity), AES, etc.
- Attend conventions and industry business functions (both nationally and in your area)
- Visit a favorite studio and get to know them (and make it easy for them to get to know you)
- Online social Networking

Of course, I've been assuming that you want to get into the production side of the recording biz …. But for those who just want to learn the tools and toys of recording technology from an artist standpoint, the above networking tools apply even more to us blokes who want and need to get our names out to the music-consuming public. For business professionals, networking is essential—for the artist, networking is the driving force of your life.

So, when do you start this grand adventure? When do you start building your career? The obvious answer is *RIGHT NOW*. If you're in school, you have already started the process. If you're just hanging out with like-minded biz folks and or joined a local or national organization … that, too, is an equally strong start. Whatever you do, don't wait until you graduate or until some magic date in the future, because putting it off will just put you that much further behind.

In short, make yourself visible. Try not to be afraid when sending out a resume, demo link or CD of your work or when asking for a job. The worst thing they can do is say "No." You might also keep in mind that "No" could actually mean "No, not right now." You might actually ask if this is the case. If so, they might take your persistence into account before saying "No" two or three times. By picking a market and particular area, blanketing that area with resumes or EPKs and knocking on doors, you just never know what might happen. I know it's not easy, but pick yourself up (again), reevaluate your strategies and start pounding the streets (again). Just remember the self-motivation rule, "failing at something isn't a bad thing … not trying is!"

Here are a few additional networking and job placement tips to get you started:

- Make a personal Web page (Wordpress is an easy, free and powerful way to get started) and/ or a personal or band Facebook page.
- Send out *lots* of resumes or, better yet, make an online bio/resume link on your page.
- Choose a mentor who you can rely on and talk to (sometimes they fall out of the sky, sometimes you have to develop the relationship over time).
- Pick the areas you want to live in (if that is a factor).
- Pick the companies in that area and target them.
- Contact studios or companies in this area that might be looking for interns.
- Visit these companies just to hang out and see what they are like.
- Use your school counselors for intern placement.
- Always to remember to follow up.

THE RECORDING PROCESS

In this age of pro studios, project studios, digital audio workstations, groove tools and personal choices, it's easy to understand how the "different strokes for different folks" adage applies to recording, in that the differences between people and the tools they use allow the process of recording to be approached in many different ways. The cost-effective environment of the project studio has also brought music and audio production to a much wider audience, thus making the process much more personal. If we momentarily set aside the monumental process of creating music in its various styles and forms, the process of capturing sound onto a recorded medium will generally occur in seven distinct steps:

1. Preparation
2. Recording
3. Overdubbing
4. Mixdown
5. Mastering and song sequence editing
6. Product manufacturing
7. Marketing and sales

1 Preparation

One of the most important aspects of the recording process occurs *before* the artist and production team step into any type of studio: preparation. Questions like the following must be addressed long before you start:

- What is the goal?
- What is the budget?
- What are the estimated studio costs?
- Will there be enough time to work on tracking, vocals, mixing, and other important issues before running out of money?
- Will the project be released on physical media or download only?
- How much will it cost to manufacture the CDs and/or records?
- How will the music be distributed and sold? And to whom?
- Who will do the artwork and what is its budget?
- What will the advertising costs be?
- Is the group practiced enough?
- If the project doesn't have a producer, who will speak for the group when the going gets rough?
- Are the instruments, voices and attitudes ready for the task ahead?
- Will additional musicians be needed?
- Are there any legal issues to consider (an important question for any project)?
- How and when will the website be up and running?

These questions and a whole lot more will have to be addressed before it comes time to press the big red record button.

2 Recording

The first phase in multitrack production is the recording process. In this phase, one or more sound sources are picked up by a microphone or an electrical signal is directly recorded to a track or tracks (as often occurs when recording electric or electronic instruments) of a recording system.

Of course, hard-disk and multitrack recording technologies have the added flexibility of allowing multiple sound sources to be captured onto and played back from separate tracks in a disk- or tape-based environment. Because the recorded tracks are isolated from each other, any number of instruments can be recorded and rerecorded without affecting other instruments (with disk-based DAWs offering an almost unlimited track count and analog tape counts usually being offered in track groups of eight (e.g., 8, 16, 24). Of course, the biggest advantage to having individual tracks is that they can be altered, added, combined and edited at any time and in any order to best suit the production.

By recording a single instrument to a dedicated track (or group of tracks), it's possible to vary the level, spatial positioning (such as left/right or surround panning), EQ and signal processing and routing (in mixdown) without affecting the level or tonal qualities of other instruments that are recorded onto an adjacent track or tracks (Figure 1.23). This isolation allows leakage from nearby instruments or mic pickups to be reduced to such an insignificant level that individual tracks can be rerecorded and/or processed at a later time, without affecting the overall mix.

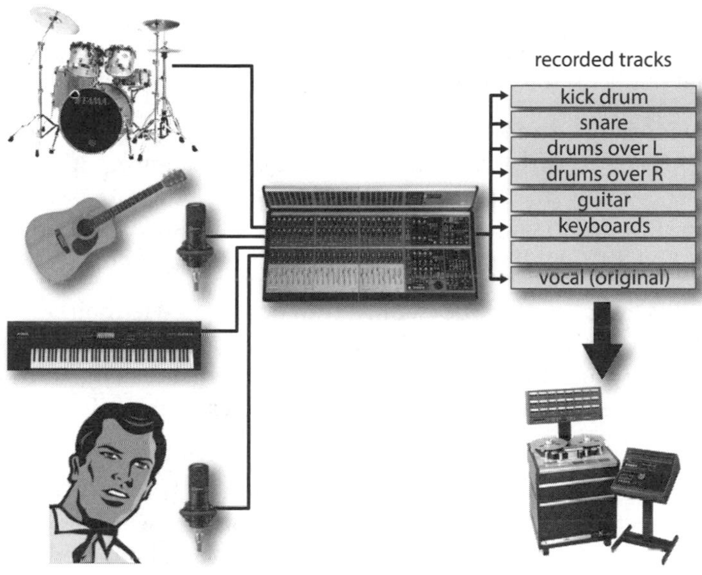

FIGURE 1.23
Basic representation of how isolated sound sources can be recorded to a DAW or multitrack recorder.

The basic tracks of a session can, of course, be built up in any number of ways. For example, the foundation of a session might be recorded all at once in a traditional fashion, or it can be built up using basic tracks that involve such acoustic instruments as drums, guitar, piano and a scratch vocal (used as a rough guide throughout the session until the final vocals can be laid down). Alternatively, these basic tracks might be made up of basic electronic music loops, samples or MIDI tracks that have been composed using a digital audio workstation (DAW). The various combinations of working styles, studio miking, isolation and instrument arrangements are literally limitless and, whenever possible, are best discussed or worked out in the early preparation stages.

From a technical standpoint, the microphones for each instrument are selected either by experience or by experimentation and are then connected to the desired console or audio interface inputs. Once done, the mic type and track selection should be noted within the DAW track notepads and/or a track sheet or piece of paper for easy input and track assignment in the studio or for choosing the same mic/placement during a subsequent overdub session (taking a photo or screenshot of any complicated setups and saving it within the session directory can also save your butt at a later time).

Some engineers find it convenient to standardize on a system that uses the same console mic input and tape/DAW track number for an instrument type at every session. For example, an engineer might consistently plug their favorite kick drum mic into input #1 and record it onto track #1, the snare mic onto #2, and so on. This way, the engineer instinctively knows which track belongs to a particular instrument without having to think too much about it. When recording to a DAW, track names, groupings and favorite identifying track colors can also be selected so as to easily identify the instrument or grouped type (an approach that lends itself to saving your favorite DAW setup as session template).

Once the instruments, baffles (an optional sound-isolating panel) and mics have been roughly placed and headphones that are equipped with enough extra cord to allow free movement have been distributed to each player, the engineer can now get down to business by using a setup sheet to label each input strip on the console with the name of the corresponding instrument.

At this time, the mic/line channels can be assigned to their respective tracks. Make sure you fully document the assignments and other session info on the song or project's track sheet (Figure 1.24). If a DAW is used, make sure each input track in the session is properly named for easy reference and track identification.

If the console is digital in nature, labeling tracks is often as simple as naming the track on the DAW which will then be shown on the console's channel readout display. If you're working in the analog world, label strips (which are often provided just below each channel input fader) can be marked with an erasable felt marker, or the age-old tactic of rolling out and marking a piece of paper masking tape could be used. (Ideally, the masking tape should be the type that doesn't leave a tacky residue on the console surface.)

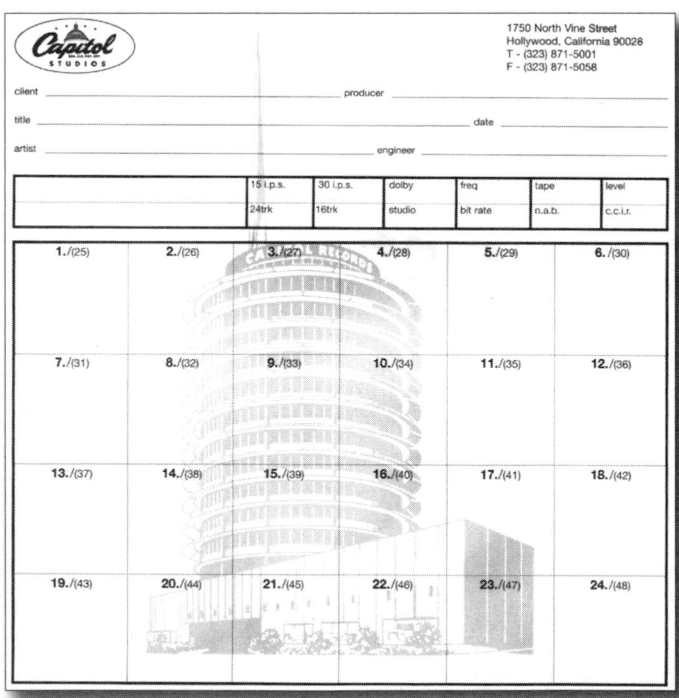

Capitol STUDIOS

1750 North Vine Street
Hollywood, California 90028
T - (323) 871-5001
F - (323) 871-5058

client ______ producer ______

title ______ date ______

artist ______ engineer ______

	15 i.p.s.	30 i.p.s.	dolby	freq	tape	level
	24trk	16trk	studio	bit rate	n.a.b.	c.c.i.r.

1./(25)	2./(26)	3./(27)	4./(28)	5./(29)	6./(30)
7./(31)	8./(32)	9./(33)	10./(34)	11./(35)	12./(36)
13./(37)	14./(38)	15./(39)	16./(40)	17./(41)	18./(42)
19./(43)	20./(44)	21./(45)	22./(46)	23./(47)	24./(48)

FIGURE 1.24
Example of a studio track log that can be used for instrument/track assignments. (Courtesy of Capitol Records, www.capitolrecords.com).

After all of the assignments and labeling have been completed, the engineer then can begin the process of setting levels for each instrument and mic input by asking each musician to play solo or by asking for a complete run-through of the song and listening one input at a time (using the solo function). By placing each of the channel and master output faders at their unity (0 dB) setting and starting with the EQ settings at the flat position, the engineer can then check each of the track meter readings and adjust the mic preamp gains to their optimum level while listening for potential preamp overload. If necessary, a gain pad can be inserted into the path in order to help eliminate distortion.

After these levels have been set, a rough headphone mix can be made so that the musicians can hear themselves. Mic choice and/or placements can be changed or EQ settings can be adjusted, if necessary, to obtain the best sound for each instrument, and dynamic limiting or compression could be carefully inserted and adjusted for any channel that requires dynamic attention. It's important to keep in mind that it's easier to change the dynamics of a track later during mixdown (particularly if the session is being recorded digitally) than to undo any changes that have been made during the recording phase.

Once this is done, the engineer and producer can again listen for tuning and extraneous sounds (such as buzzes or hum from guitar amplifiers or squeaks from drum pedals) and adjust or eliminate them. Soloing the individual tracks can ease the process of selectively listening for such unwanted sounds and for getting the best sound from an instrument without any distractions from the other tracks. If several mics are to be grouped into one or more tracks, the balance between them should be *carefully* set at this time.

After this procedure has been followed for all the instruments, the musicians should do a couple of practice *rundown* songs so that the engineer and producer can listen to how the instruments sound together before being recorded. (If disk or tape space isn't a major concern, you might consider recording these tracks, because they might turn out to be your best takes—you just never know!) During the rundown, you might also consider soloing the various instruments and instrument combinations as a final check and, finally, monitor all of the instruments together. Careful changes in EQ can be made at this time, making sure to note these changes in the DAW notepad or track sheet for future reference. These changes should be made sparingly, because final compensations are probably better made during the final mixdown phase.

While the song is being run down, the engineer can also make final adjustments to the recording levels and the headphone monitor mix. He or she can then check the headphone mix either by putting on a pair of phones connected to the cue system or by routing the mix to the monitor loudspeakers. If the musicians can't hear themselves properly, the mix should be changed to satisfy their monitoring needs (fortunately, this can usually be done without regard to the recorded track levels). If several cue systems are available, multiple headphone mixes can be built up to satisfy those with different balance needs. During a loud session, the musicians might ask you to turn up their level (or the overall headphone mix), so they can hear themselves above the ambient room leakage. It's important to note that high sound-pressure levels can cause the pitch of instruments to sound flat, so musicians might have trouble tuning or even singing with their headphones on. To avoid these problems, tuning shouldn't be done while listening through phones. The musicians should play their instruments at levels that they're accustomed to and adjust their headphone levels accordingly. For example, they might put only one cup over an ear, leaving the other ear free to hear the natural room sound.

The importance of proper headphone levels and a good cue balance can't be stressed enough, because they can either help or hinder a musician's overall performance. The same situation exists in the control room with respect to high monitor-speaker levels: Some instruments might sound out of tune, even when they aren't, and ear fatigue can easily impair your ability to properly judge sounds and their relative balance.

During the practice rundown, it's also a good idea to ask the musician(s) to play through the entire song so you'll know where the breaks, bridges and any other point of particular importance might be. Making notes and even writing down or entering the timing numbers (into a DAW using session markers or an analog recorder's auto-locator) can help speed up the process of finding a section during a take or overdub. You can also pinpoint the loud sections at this time, so as to avoid any overloads. If compression or limiting is used, you might keep an ear open to ensure that the instruments don't trigger an undue amount of gain reduction. (If the tracks are recorded digitally, you might consider applying gain reduction later during mixdown.) Even though an engineer might ask each musician to play as loudly as possible, they'll often play even louder when performing together. This fact may require further changes in the mic preamp gain, recording level and compression/limiting thresholds. Soloing each mic and listening for leakage can also help to check for separation and leakage problems between the instruments. If necessary, the relative positions of mics, instruments and baffles can be changed one last time.

3 Overdubbing

Once the basic tracks have been laid down, additional instrument and/or vocal parts can be added in a process known as *overdubbing*. During this phase, additional tracks are added by monitoring the previously recorded tape tracks (usually over headphones) while simultaneously recording new, doubled or augmented instruments and/or vocals onto one or more available tracks of a DAW or recorder (Figure 1.25). During the overdub (OD) phase, individual parts are added to an existing project until the song or soundtrack is complete. If the artist makes a mistake, no problem! Simply recue the DAW or rewind the tape back to the where the instrument begins and repeat the process until you've captured the best possible take. If a take goes *almost* perfectly except for a bad line or a few flubbed notes, it's possible to go back and rerecord the offending

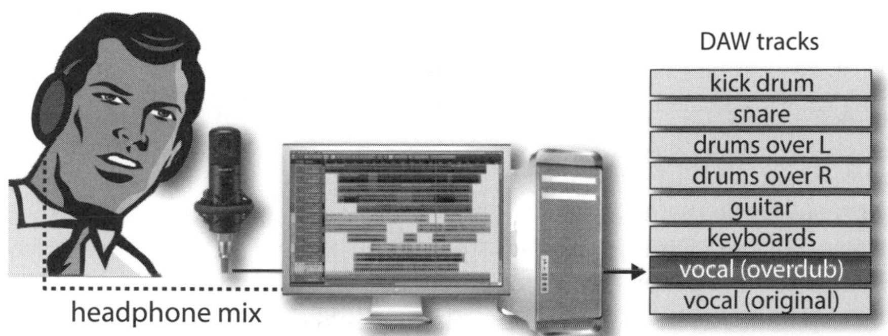

FIGURE 1.25
Overdubbing allows instruments and/or vocals to be added at a later time to existing tracks on a multitrack recording medium.

segment onto the same or a different track in a process known as *punching in*. If the musician lays down his or her part properly and the engineer dropped in and out of record at the correct times (either manually or under automation), the listener won't even know that the part was recorded in multiple takes … such is the magic of the recording process!

While recording onto the same track might make sense when an analog recorder is used, a DAW, on the other hand, doesn't have track limitations. Therefore, it often makes sense to record the overdub onto another track and then combine the edited tracks into a single track or set of separate grouped tracks. If multiple takes are needed to get the best performance, it's possible to record the different takes to a set of new tracks either manually or under a looped automation function. Once done, the engineer and producer can sit down and pick the best parts from the various takes and combine them into a final performance in a process that's known as *comping*.

In an overdub session, the same procedure is followed for mic selection, placement, EQ, and level … just as they occurred during a recording session (now you're beginning to see the need for good documentation). If only one instrument is to be overdubbed, the problem of having other instrument tracks leak into the new track won't exist. However, care should be taken to ensure that the headphones aren't too loud or are improperly placed on the artist's head, because excessive leakage from the headphone mix into the mic can occur.

Of course, the natural ambience of the session should be taken into account during an overdub. If the original tracks were made from a natural, roomy ensemble, it could be distracting to hear an added track that was obviously laid down in a different (usually deader) room environment.

If an analog recorder is to be used, it should be placed in the master sync mode to ensure that the previously recorded tracks will play back in sync from the record head (see Chapter 5 for more info). The master sync mode is set either at the recorder or using its auto-locator/remote control. Usually, the tape machine can automatically switch between monitoring the source (signals being fed to the recorder or console) and tape/sync (signals coming from the playback or record/sync heads). When recording to a DAW or tape, the control room monitor mix should prominently feature the instruments that are being recorded, so mistakes can be easily heard. During the initial rundown, the headphone cue mix can be adjusted to fit the musician's personal taste.

OPTIONS IN OVERDUBBING

At first glance, the process of overdubbing seems pretty cut and dried … put a mic out in the studio, put headphones on the performer and start recording. In fact, the process can present as many unique challenges and opportunities as the recording process. For starters, getting the right sound during OD is extremely important. The "Good Principle" applies as much here as within any phase. Asking these and other questions can cause lots of headaches later in the game:

- Is the performer in the right headspace to do the OD?
- Is the instrument properly tuned to the track?
- Is the mic's placement and distance appropriate to the track?
- If an overdub is being made over a previous overdub track, do all of the above match the previous session?

One other option that's available to the artist, producer and engineer during an overdub is the wide number of pickup options that can be used in the session to add effect and various mixdown options to the OD. I'm referring to the various ways that an instrument can be recorded to additional, separate tracks to add dramatic options during mixdown. For example:

- A standard close-mic pickup can be used to record the overdub.
- If the instrument is electric or electronic, it can be recorded "direct" to capture the pure instrument sound.
- Another mic could be placed at a 6′ to 10′ distance from the instrument (i.e. electric guitar amp) to get a fuller, more distant sound.
- It's also possible to place an additional distant (or stereo) pickup mic further back in the room to capture the room's overall reverberant sound.
- As always, if the instrument is electronic and has a MIDI out, it's wise to capture the MIDI performance to a track on the DAW.

So why go through all this additional work? Well, let's say that, at a later time, the mix engineer calls up your additional distant mics and adds it to the track and the producer goes wild … she just loves it! … or the track gets picked up as the soundtrack to a feature film and needs to be remixed in surround. By placing the distant room sounds in the rear speakers, the whole guitar sound just got super-huge, in a way that fills out the track amazingly well. On the flip-side, let's say that the producer likes the OD, but decides that amp isn't right for the track at all … all you'd have to do is play the direct signal back through that new Vox stack in the big studio (with new a whole new set of room mics, as well) and everybody's happy … or, you captured the MIDI tracks from a synth, which made it possible to change the patch to get a whole new sound, without having to re-perform the track. Are you starting to get the idea that the name of the game is versatility in production and/or mixdown?

④ Mixdown

When all of the tracks of a project have been recorded, assembled and edited, the time has come to individually *mix* the songs into their final media form (Figure 1.26). The mixdown process occurs by routing the various tracks of a DAW or tape-based recorder through a hardware or DAW virtual mixing console to alter the overall session's program with respect to:

- Relative level
- Spatial positioning (the physical placement of a sound within a stereo or surround field)
- Equalization (affecting the relative frequency balance of a track)
- Dynamics processing (altering the dynamic range of a track, group or output bus to optimize levels or to alter the dynamics of a track so that it "fits" better within a mix)
- Effects processing (adding reverb-, delay- or pitch-related effects to a mix in order to augment or alter the piece in a way that is natural, unnatural or just plain interesting)

If a DAW is being used to create the final mix "in-the-box," the channel faders can be adjusted for overall level, panning can be used and a basic array of effects can be programmed into the session mixer and, if available, a controller surface can be used to facilitate the mix by giving you hands-on control.

The first job at hand is to set up a rough mix of the song by adjusting the levels and the spatial pan positions. The artist and/or producer then listens to this mix and might ask the engineer to make specific changes. The instruments are often soloed one by one or in groups, allowing for any necessary EQ changes that might need to be made. The engineer, producer, and possibly the group can then

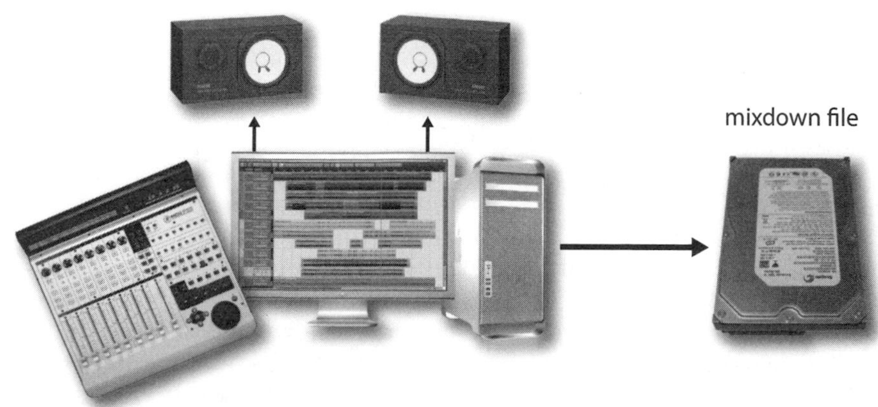

FIGURE 1.26
Basic representation of the DAW mixdown process.

begin the cooperative process of "building" the mix into its final form. Compression and limiting can be used on individual instruments as required, either to make them sound fuller, more consistent in level, or to prevent them from overloading the mix when raised to the desired mix output level. Once the mix begins to take shape, reverb and other effects types can be added to shape and add ambience in order to give close-miked sounds a more "live," spacious feeling, as well as to help blend the instruments. Once a mix has been made, it's then possible to virtually mix down "export or bounce" the overall mix to a specified mono, stereo or surround soundfile without the need for an external hardware recording device. It's also wise to note that various versions of the mix can be saved to the session directory, allowing you to go back to any preferred mix version at any time.

If an actual recording console is used in the process, it can be placed into the mixdown mode (or each input module can be switched to the line or tape position) and the fader label strips can be labeled with their respective instrument names. Channel and group faders should be set to unity gain (0 dB) and the master output faders should likewise be set with the monitor section being switched to feed the mixdown signal to the appropriate speakers. At this point, the console's automation features, if available, can be used.

If the hardware recording console isn't automation assisted, the fader settings will have to be changed during the mix in real time. This means that the engineer will have to memorize the various fader moves (often noting the transport counter to keep track of transition times). If more changes are needed than the engineer can handle alone, the assistant, producer or artist (who probably knows the transition times better than anyone) can help by controlling certain faders or letting you know when a transition is coming up. It's usually best, however, if the producer is given as few tasks as possible, so that he or she can concentrate fully on the music rather than the physical mechanics of the mix. The engineer then listens to the mix from a technical standpoint to detect any sounds or noises that shouldn't be present. If noises are recorded on tracks that aren't used during a section of a song, these tracks can be muted until needed. After the engineer practices the song enough to determine and learn all the changes, the mix can be recorded and faded at the end. The engineer might not want to fade the song during mixdown, because it will usually be performed after being transferred to a DAW (which can perform a fade much more smoothly than even the smoothest hand). Of course, if automation is available or if the mix is performed using a DAW, all of these moves can be performed much more easily and with full repeatability by using the software's own mix automation.

It's usually important for levels to be as consistent as possible between the various take versions and songs, and it's often wise to monitor at consistent, moderate listening levels. This is due to the variations in our ear's frequency response at different sound-pressure levels, which could result in inconsistencies between song balances. Ideally, the control room monitor level should be the same as might be heard at home, over the radio or in the car (between 70 and 90 dB

SPL), although certain music styles will simply "want" to be listened to at higher levels. Once the final mix or completed project master is made, you'll undoubtedly want to listen to the mix over different speaker systems (ranging from the smallest to the biggest/baddest, and crappiest to the most accurate you can find). It's usually wise to run off a few copies for the producer and band members to listen to in familiar environments (i.e. at home and in their cars). In addition, the mix should be tested for mono–stereo/surround compatibility to see if any changes in instrumental balances occur. If there are any changes in frequency balances or if phase becomes a problem when the mix is played in mono, the original mix might have to be modified.

⑤ Mastering

Once done, the final, edited mixdown of a project might be sent to a mastering engineer (Figure 1.27), so that fine-tuning adjustments can be made to the overall recording with respect to:

- Relative level balancing between songs within the project
- Dynamic level (altering the dynamics of a song so as to maximize its level for the intended media or to tighten up the dynamic balance, overall or within certain frequency bands)
- Equalization
- Overall level
- Song sequence editing

In essence, it's the job of a qualified mastering engineer to smooth over any level and spectral imbalances within a project and to present the final, recorded product in the best possible light for its intended media form. Commonly, the producer and/or artist will be faced with the question of whether to hire a qualified or well-known mastering engineer to put on the finishing touches, or to master the project themselves into a final product (using the producer's, engineer's or artist's talents). These questions should be thoroughly discussed in the preplanning phase, allowing for any on-the-spot change of plans. Who knows—you just might try mastering the project yourself (a process that takes patience, objectivity and a willingness to accept criticism). If it doesn't work out, you can

FIGURE 1.27
Future Disc Mastering McMinnville, Oregon. (Courtesy of Future Disc Systems, www.futurediscsystems.com).

always work with a trusted mastering engineer to present your masterpiece into its best possible form.

6 Product manufacturing

Last but never least in the production chain is the process of manufacturing the master recording into a final, salable product. Whether the final product is a compact disc or digital download, this process should be carefully overseen to ensure that the final product doesn't compromise all of the blood, sweat, tears and bucks that have gone into the creation of a project. These manufacturing phases should be carefully scrutinized, checked, and rechecked:

- Creating a manufacture master
- Art layout and printing
- Product packaging

Whenever possible, ask for a proof copy of the final duplicated product and artwork *before* it is mass produced. Receiving 1000 or more copies of your hard-earned project that has a flaw is as bad as being handed an accordion in hell. Further info on this topic can found in Chapter 19.

7 Marketing and sales

Although this section is mentioned last, it is by far one of the most important areas to be dealt with when contemplating the time, talent and financial effort involved in creating a recorded product. For starters, the following questions (and more) should all be answered long before the Record button is pressed and the first downbeat is played:

- Who is my audience?
- Will the project be distributed by a record company, or will I try to sell it myself?
- What should the final product look and sound like?
- What's my budget and how much is this going to cost me?
- How will it be marketed?

In this short section, I won't even attempt to cover this extremely important and complex topic, because these subjects have been fully discussed in a number of well-crafted books and searchable online articles.

THE TRANSDUCER

Because this chapter is an introduction to sound and recording technology, I'd like to take a quick moment to look at an important concept that is central to all music, sound, electronics and the art of sound recording: the *transducer*. If any conceptual tool can help you to understand the technological underpinnings of the art and process of recording, this is probably it!

FIGURE 1.28
The guitar and microphone as transducers.

Quite simply, a *transducer* is any device that changes one form of energy into another, corresponding form of energy. For example, a guitar is a transducer in that it takes the vibrations of picked or strummed strings (the medium), amplifies them through a body of wood, and converts these vibrations into corresponding sound-pressure waves, which are then perceived as sound (Figure 1.28).

A microphone is another example of a transducer. Here, sound-pressure waves (the medium) act on the mic's diaphragm and are converted into corresponding electrical voltages. The electrical signal from the microphone can then be amplified (not a process of transduction because the medium stays in its electrical form) and fed to a recording device. The recorder is a device that changes electrical voltages into analogous magnetic flux signals on magnetic tape or into representative digital data that can be encoded onto tape, hard disk or other type of disc. On playback, the stored magnetic signals or digital data are converted back to their original electrical form, amplified, and then fed to a speaker system. The speakers convert the electrical signal back into a mechanical motion (by way of magnetic induction), which, in turn, recreates the original air-pressure variations that were picked up by the microphone, and … ta-da … we have sound!

As can be gathered from Table 1.1, transducers can be found practically everywhere in the audio environment. In general, transducers (and the media they use) tend to be the weakest link in the audio system chain. Given our current technology, the process of changing the energy in one medium into a corresponding form of energy in another medium can't be accomplished perfectly (although hi-res digital coding gets very close). Noise, distortion and (often)

Table 1.1 Media Used by Transducers in the Studio to Transfer Energy

Transducer	From	To
Ear	Sound waves in air	Nerve impulses in the brain
Microphone	Sound waves in air	Electrical signals in wires
Record head	Electrical signals in wires	Magnetic flux on tape
Playback head	Magnetic flux on tape	Electrical signals in wires
Phonograph cartridge	Grooves cut in disk surface	Electrical signals in wires
Speaker	Electrical signals in wires	Sound waves in air

coloration of the sound are introduced to some degree and, unfortunately, these effects can only be minimized, not eliminated. Differences in design are another major factor that can affect sound quality. Even a slight design variation between two microphones, speaker systems, digital audio converters, guitar pickups or other transducers can cause them to sound quite different. This factor, combined with the complexity of music and acoustics, helps makes the field of recording the subjective and personal art form that it is.

It's interesting to note that fewer transducers are used in an all or largely digital recording system. In this situation, the acoustic waveforms that are picked up by a microphone are converted into electrical signals and then converted into digital form by an analog-to-digital (A/D) converter. The A/D converter changes these continuous electrical waveforms into corresponding discrete numeric values that represent the waveform's instantaneous, analogous voltage levels. Arguably, digital information has the distinct advantage over analog in that data can be transferred between electrical, magnetic and optical media with little or no degradation in quality. Because the information continues to be stored in its original, discrete binary form, no transduction process is involved (i.e., only the medium changes, while the data representing the actual information doesn't change). Does this mean that digital's better? Well, it's just another way of expressing sound through a medium, which, in the end, is just one of the many possible artistic and technological choices in the making and recording of sound and music. Beauty is, indeed, in the ear of the beholder.

CHAPTER 2

Sound and Hearing

When we make a recording, in effect we're actually capturing and storing sound into a memory media so that an original event can be re-created at a later date. If we start with the idea that *sound* is actually a concept that corresponds to the brain's perception and interpretation of a physical auditory stimulus, the study of sound can be divided into four areas:

- The basics of sound
- The characteristics of the ear
- How the ear is stimulated by sound
- The psychoacoustics of hearing

By understanding the physical nature of sound and the basics of how the ears change a physical phenomenon into a sensory one, we can discover how to best convey this science into the subjective art forms of music, sound recording and production.

THE BASICS OF SOUND

Sound arrives at the ear in the form of periodic variations in atmospheric pressure called *sound-pressure waves*. This is the same atmospheric pressure that's measured by the weather service with a barometer; however, the changes in pressure heard by the ear are too small in magnitude and fluctuate too rapidly to be observed on a barometer. An analogy of how sound waves travel in air can be demonstrated by bursting a balloon in a silent room. Before we stick it with a pin, the molecular motion of the room's atmosphere is at a normal resting pressure. The pressure inside the balloon is much higher, though, and the molecules are compressed much more tightly together—like people packed into a crowded subway car (Figure 2.1a). When the balloon is popped… KAPOW! (Figure 2.1b), the tightly compressed molecules under high pressure begin to exert an outward force on their neighbors in an effort to move toward areas of lower pressure. When the neighboring set of molecules has been compressed, they will continue to exert an outward force on the next set of lower-pressured

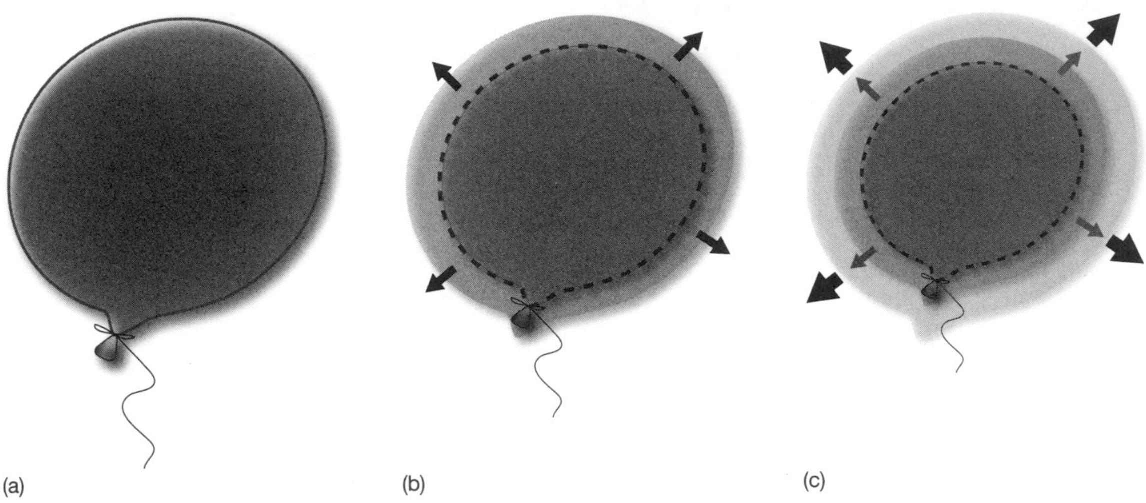

FIGURE 2.1
Wave movement in air as it moves away from its point of origin. (a) An intact balloon contains pressurized air. (b) When the balloon is popped, the compressed molecules exert a force on outer neighbors in an effort to move to areas of lower pressure. (c) The outer neighbors then exert a force on the next set of molecules in an effort to move to areas of lower pressure… and the process continues.

neighbors (Figure 2.1c) in an ongoing outward motion that continues until the molecules have used up all their energy in the form of heat.

Likewise, as a vibrating mass (such as a guitar string, a person's vocal chords or a loudspeaker) moves outward from its normal resting state, it squeezes air molecules into a compressed area, away from the sound source. This causes the area being acted on to have a greater than normal atmospheric pressure, a process called *compression* (Figure 2.2a). As the vibrating mass moves inward from its normal resting state, an area with a lower-than-normal atmospheric pressure will be created, in a process called *rarefaction* (Figure 2.2b). As the vibrating body cycles through its inward and outward motions, areas of higher and lower compression states are generated. These areas of high pressure will cause the wave to move outward from the sound source in the same way waves moved outward from the burst balloon. It's interesting (and important) to note that the molecules themselves don't move through air at the velocity of sound—only the sound wave itself moves through the atmosphere in the form of high-pressure compression waves that continue to push against areas of lower pressure (in an outward direction). This outward pressure motion is known as *wave propagation*.

WAVEFORM CHARACTERISTICS

A *waveform* is essentially the graphic representation of a sound-pressure level or voltage level as it moves through a medium over time. In short, a waveform lets

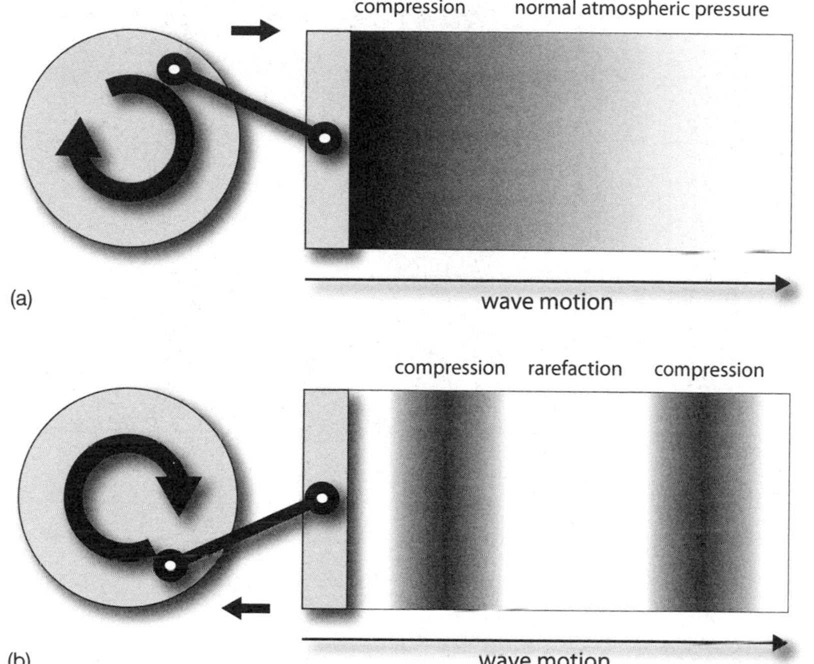

FIGURE 2.2
Effects of a vibrating mass on air molecules and their propagation.
(a) Compression—air molecules are forced together to form a compression wave.
(b) Rarefaction—as the vibrating mass moves inward, an area of lower atmospheric pressure is created.

us see and explain the actual phenomenon of wave propagation in our physical environment and will generally have the following fundamental characteristics:

- Amplitude
- Frequency
- Velocity
- Wavelength
- Phase
- Harmonic content
- Envelope

These characteristics allow one waveform to be distinguished from another. The most fundamental of these are amplitude and frequency (Figure 2.3). The following sections describe each of these characteristics. Although several math formulas have been included, it is by no means important that you memorize or worry about them. It's far more important that you grasp the basic principles of acoustics rather than fret over the underlying math.

Amplitude

The distance above or below the centerline of a waveform (such as a pure sine wave) represents the *amplitude* level of that signal. The greater the distance or displacement from that centerline, the more intense the pressure variation, electrical signal level, or physical displacement will be within a medium. Waveform

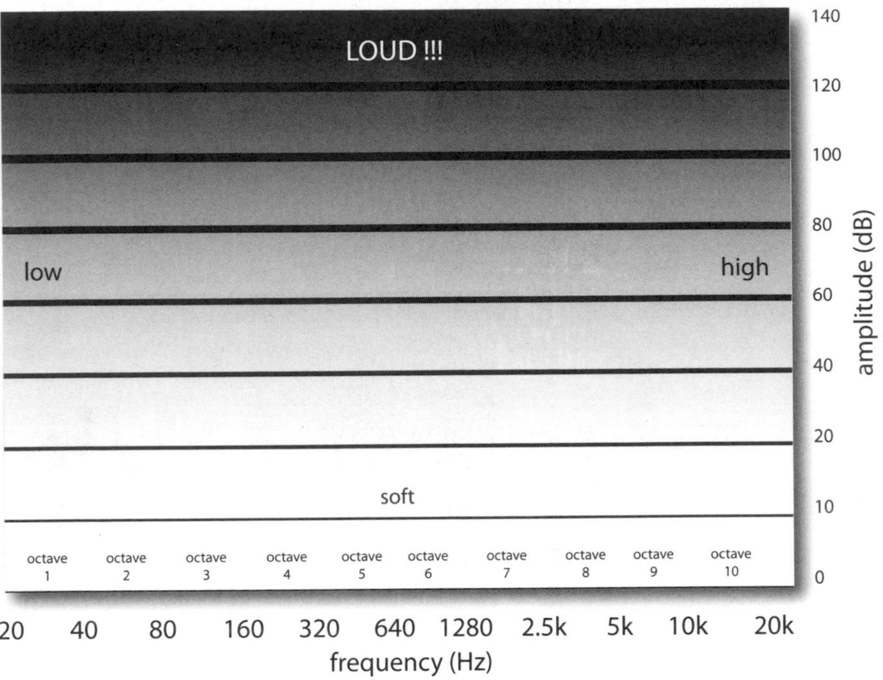

FIGURE 2.3
Amplitude and frequency ranges of human hearing.

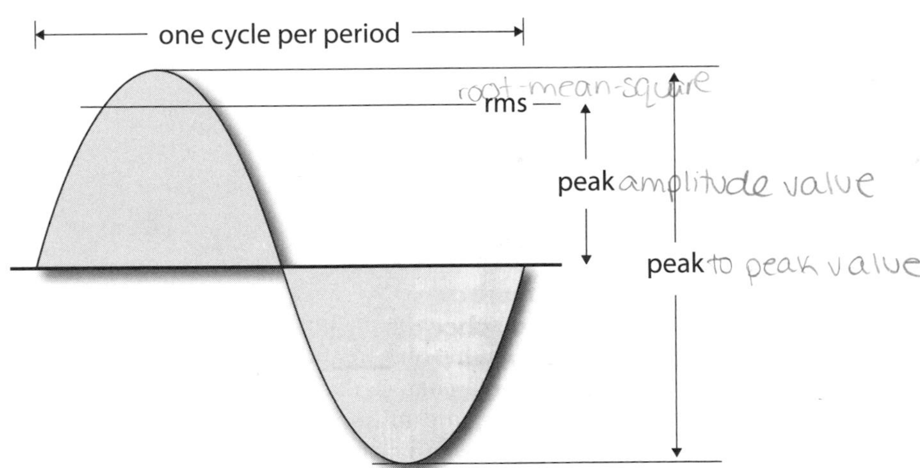

FIGURE 2.4
Graph of a sine wave showing the various ways to measure amplitude.

amplitudes can be measured in several ways (Figure 2.4). For example, the measurement of either the maximum positive or negative signal level of a wave is called its *peak amplitude value* (or peak level). The total measurement of the positive and negative peak signal levels is called the *peak-to-peak value*. The *root-mean-square (rms) value* was developed to determine a meaningful average level of a waveform over time (one that more closely approximates the level that's actually perceived by our ears and gives a better real-world measurement of overall signal amplitudes). The rms value of a sine wave can be calculated by squaring the amplitudes at points along the waveform and then taking the mathematical average of the combined results. The math isn't as important as the basic concept that the rms value of a perfect sine wave is equal to 0.707 times its instantaneous peak amplitude level. Because the square of a positive or negative value is always positive, the rms value will always be positive. The following simple equations show the relationship between a waveform's peak and rms values:

rms voltage = 0.707 x peak voltage

peak voltage = 1.414 x rms voltage

Frequency

The rate at which an acoustic generator, electrical signal or vibrating mass repeats within a cycle of positive and negative amplitude is known as the *frequency* of that signal. As the rate of repeated vibration increases within a given time period, the frequency (and thus the perceived pitch) will likewise increase ... and vice versa. One completed excursion of a wave (which is plotted over the 360° axis of a circle) is known as a *cycle* (Figure 2.5). The number of cycles that occur within a second (frequency) is measured in hertz (Hz). The diagram in Figure 2.6 shows the value of a waveform as starting at zero (0°). At time

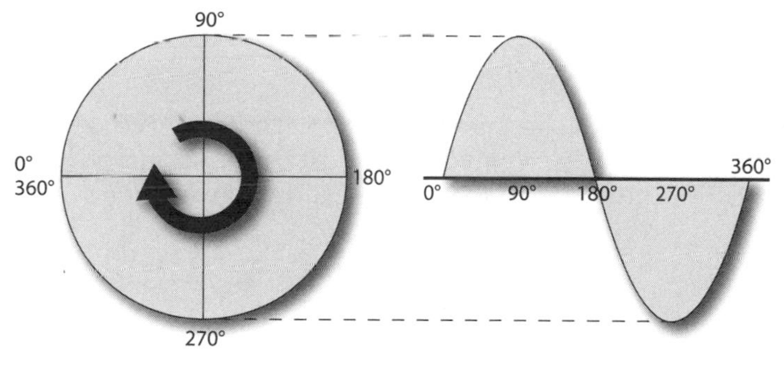

FIGURE 2.5
Cycle divided into the 360° of a circle.

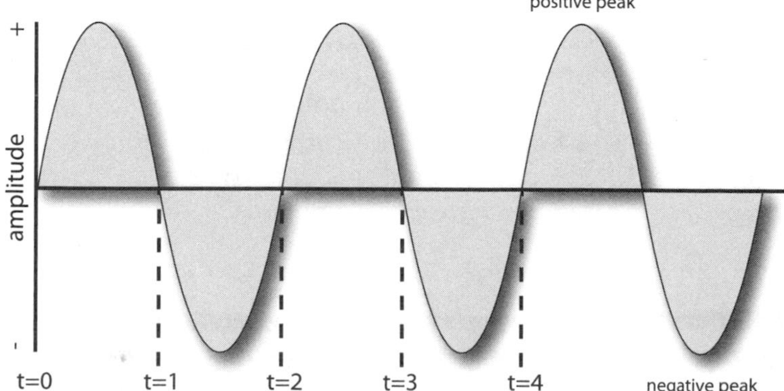

FIGURE 2.6
Graph of waveform amplitude over time.

$t = 0$, this value increases to a positive maximum value and then decreases back through zero, where the process begins all over again in a repetitive fashion. A cycle can begin at any angular degree point on the waveform; however, to be complete, it must pass through a single 360° rotation and end at the same point as its starting value. For example, the waveform that starts at $t = 0$ and ends at $t = 2$ constitutes a cycle, as does the waveform that begins at $t = 1$ and ends at $t = 3$.

Velocity

The *velocity* of a sound wave as it travels through air at 68°F (20°C) is approximately 1130 feet per second (ft/sec) or 344 meters per second (m/sec). This speed is temperature dependent and increases at a rate of 1.1 ft/sec for each Fahrenheit degree increase in temperature (2 ft/sec per Celsius degree).

Wavelength

The *wavelength* of a waveform (frequently represented by the Greek letter lambda, λ) is the physical distance in a medium between the beginning and the end of a cycle. The physical length of a wave can be calculated using:

$$\lambda = V/f$$

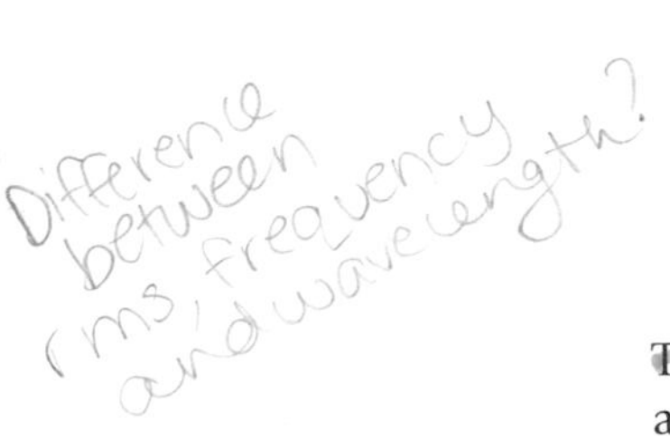

where λ is the wavelength in the medium

V is the velocity in the medium

f is the frequency (in hertz)

The time it takes to complete 1 cycle is called the *period* of the wave. To illustrate, a 30-Hz sound wave completes 30 cycles each second or 1 cycle every 1/30th of a second. The period of the wave is expressed using the symbol T:

$$T = 1/f$$

where T is the number of seconds per cycle.

Assuming that sound propagates at the rate of 1130 ft/sec, all that's needed is to divide this figure by the desired frequency. For example, the simple math for calculating the wavelength of a 30-Hz waveform would be 1130/30 = 37.6 feet long, whereas a waveform having a frequency of 300 Hz would be 1130/300 = 3.76 feet long (Figure 2.7). Likewise, a 1000-Hz waveform would work out as being 1130/1000 = 1.13 feet long, and a 10,000-Hz waveform would be 1130/10,000 = 0.113 feet long. From these calculations, you can see that whenever the frequency is increased, the wavelength decreases.

REFLECTION OF SOUND

Much like a light wave, sound reflects off a surface boundary at an angle that's equal to (and in an opposite direction of) its initial angle of incidence. This basic property is one of the cornerstones of the complex study of acoustics.

30 Hz
37.66 ft
300 Hz
3.76 ft

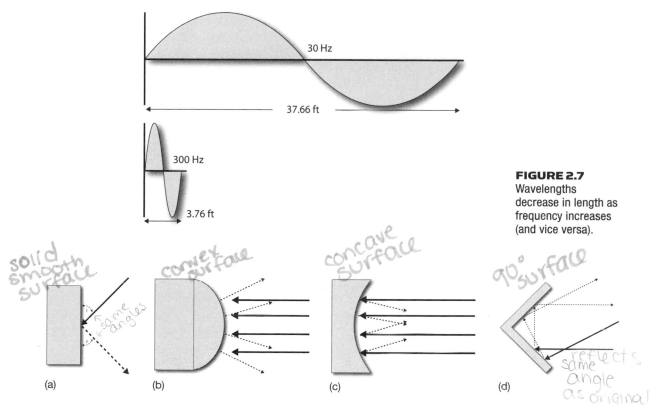

FIGURE 2.7
Wavelengths decrease in length as frequency increases (and vice versa).

(a) (b) (c) (d)

FIGURE 2.8
Incident sound waves striking surfaces with varying shapes: (a) single-planed, solid, smooth surface; (b) convex surface; (c) concave surface; (d) 90° corner reflection.

For example, Figure 2.8a shows how a sound wave reflects off a solid smooth surface in a simple and straightforward manner (at an equal and opposite angle). Figure 2.8b shows how a convex surface will splay the sound outward from its surface, radiating the sound outward in a wide dispersion pattern. In Figure 2.8c, a concave surface is used to focus a sound inward toward a single point, while a 90° corner (as shown in Figure 2.8d) reflects patterns back at angles that are equal to their original incident direction. This holds true both for the 90° corners of a wall and for intersections where the wall and floor meet. These corner reflections help to provide insights into how volume levels often build up in the corners of a room (particularly at wall-to-floor corner intersections).

DIFFRACTION OF SOUND

Sound has the inherent ability to diffract around or through a physical acoustic barrier. In other words, sound can bend around an object in a manner that reconstructs the signal back to its original form in both frequency and amplitude. For

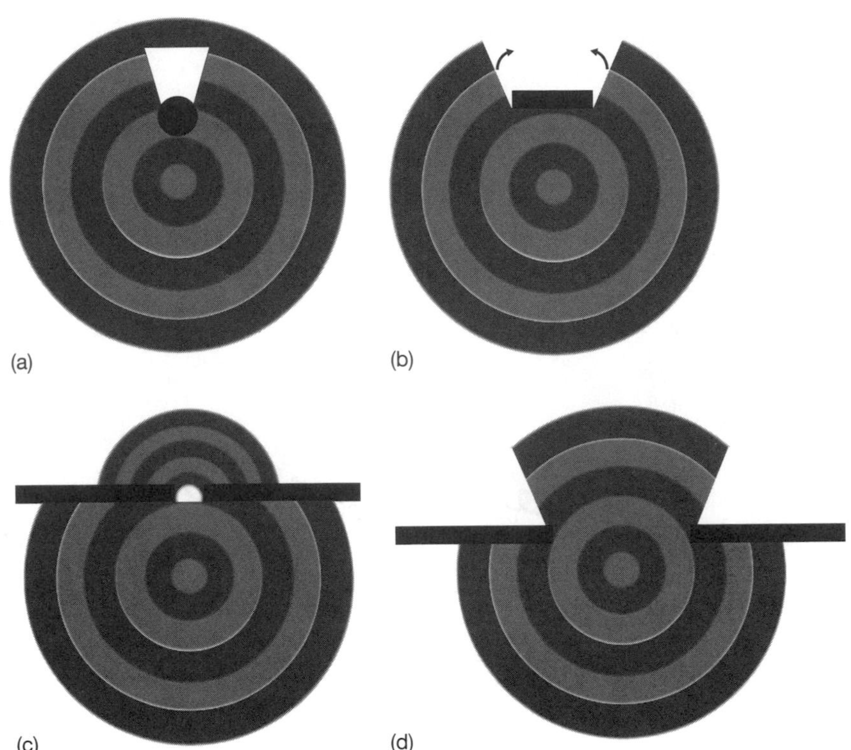

FIGURE 2.9 The effects of obstacles on sound radiation and diffraction. (a) A small obstacle will scarcely impede a longer wavelength signal. (b) A larger obstacle will obstruct the signal to a greater extent; the waveform will also reconstruct itself in the barrier's wake. (c) A small opening in a barrier will greatly impede a signal; the waveform will emanate from the opening and reconstruct itself as a new source point. (d) A larger opening allows sound to pass unimpeded, allowing it to quickly diffract back into its original shape.

example, in Figure 2.9a, we can see how a small obstacle will scarcely impede a larger acoustic waveform. Figure 2.9b shows how a larger obstacle can obstruct a larger portion of the waveform; however, past the obstruction, the signal bends around the area in the barrier's wake and begins to reconstruct itself. Figure 2.9c shows how the signal is able to radiate through an opening in a large barrier. Although the signal is greatly impeded (relative to the size of the opening), it nevertheless begins to reconstruct itself in wavelength and relative amplitude and begins to radiate outward as though it were a new point of origin. Finally, Figure 2.9d shows how a large opening in a barrier lets much of the waveform pass through relatively unimpeded.

FREQUENCY RESPONSE

The charted output of an audio device is known as its *frequency response curve* (when supplied with a reference input of equal level over the 20 to 20,000-Hz range of human hearing). This curve is used to graphically represent how a device will respond to the audio spectrum and, thus, how it will affect a signal's overall sound. As an example, Figure 2.10 shows the frequency response of several unidentified devices. In these and all cases, the *x*-axis represents the signal's measured frequency, while the *y*-axis represents the device's measured output

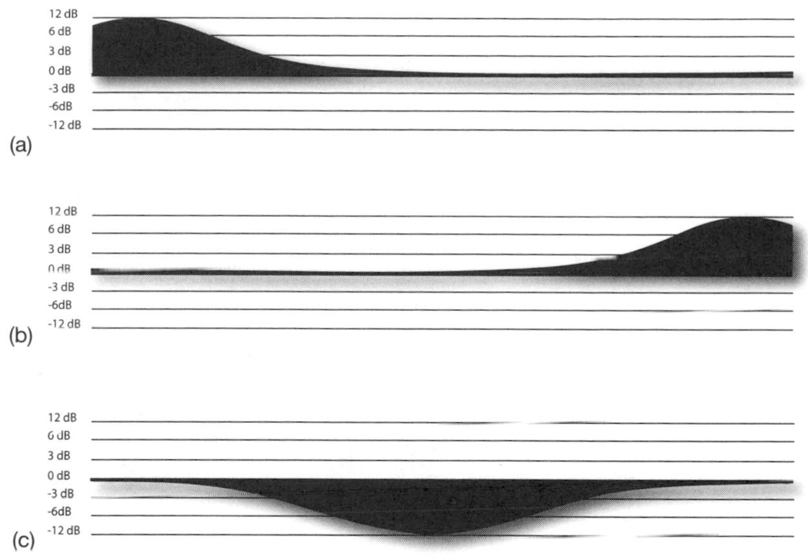

FIGURE 2.10
Frequency response curves: (a) curve showing a bass boost; (b) curve showing a boost at the upper end; (c) curve showing a dip in the midrange.

signal. These curves are created by feeding the input of an acoustic or electrical device with a constant-amplitude reference signal that sweeps over the entire frequency spectrum. The results are then charted on an amplitude vs. frequency graph that can be easily read at a glance. If the measured signal is the same level at all frequencies, the curve will be drawn as a flat, straight line from left to right (known as a *flat frequency response curve*). This indicates that the device passes all frequencies equally (with no frequency being emphasized or de-emphasized). If the output lowers or increases at certain frequencies, these changes will easily show up as dips or peaks in the chart.

Phase

Because we know that a cycle can begin at any point on a waveform, it follows that whenever two or more waveforms are involved in producing a sound, their relative amplitudes can (and most often will) be different at any one point in time. For simplicity's sake, let's limit our example to two pure tone waveforms (sine waves) that have equal amplitudes and frequency, but start their cyclic periods at different times. Such waveforms are said to be *out of phase* with respect to each other. Variations in *phase*, which are measured in degrees (°), can be described as a time delay between two or more waveforms. These delays are often said to have differences in relative phase degree angles (over the full rotation of a cycle, e.g., 90°, 180°, 270° or any angle between 0° and 360°). The sine wave (so named because its amplitude follows a trigonometric sine function) is usually considered to begin at 0° with an amplitude of zero; the waveform then increases to a positive maximum at 90°, decreases back to a zero

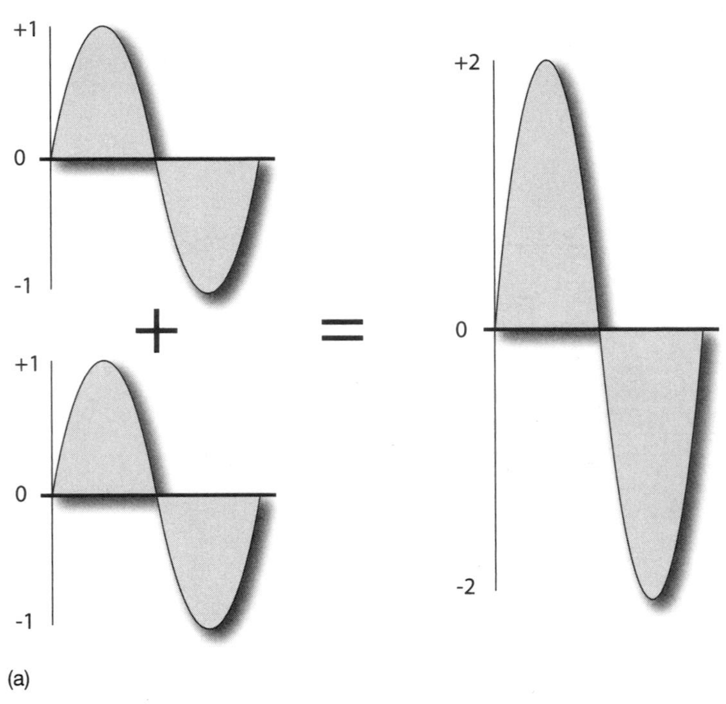

(a)

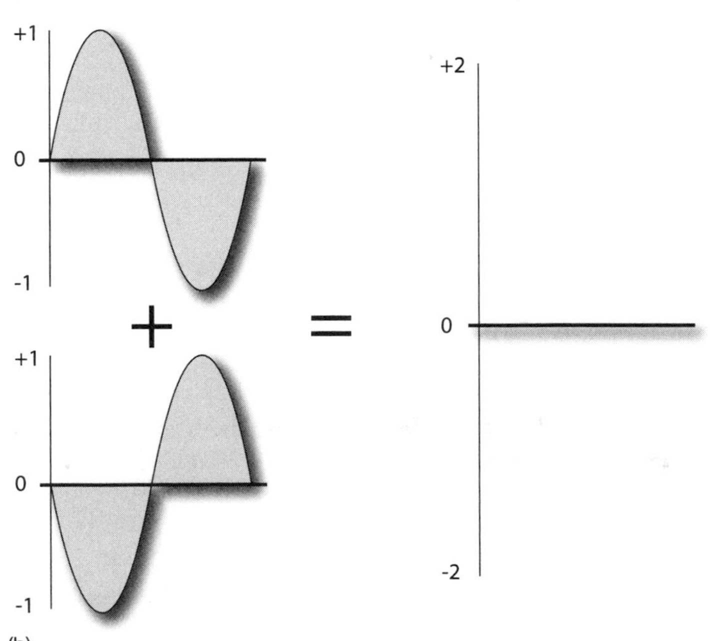

(b)

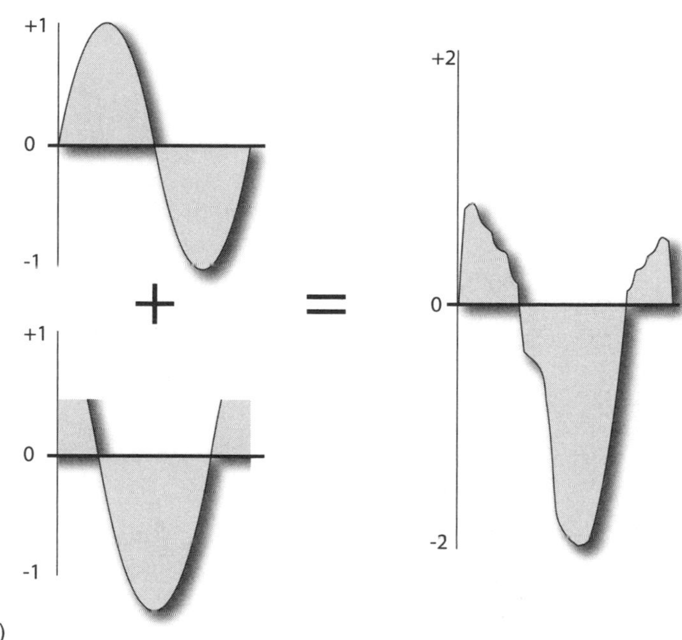

FIGURE 2.11
Combining sine waves of various phase relationships. (a) The amplitudes of in-phase waves increase in level when mixed together. (b) Waves of equal amplitude cancel completely when mixed 180° out of phase. (c) When partial phase angles are mixed, the combined signals will add in certain places and subtract in others.

amplitude at 180°, increases to a negative maximum value at 270°, and finally returns back to its original level at 360°, simply to begin all over again.

Whenever two or more waveforms arrive at a single location out of phase, their relative signal levels will be added together to create a combined amplitude level at that one point in time. Whenever two waveforms having the same frequency, shape and peak amplitude are completely *in phase* (meaning that they have no relative time difference), the newly combined waveform will have the same frequency, phase and shape, but will be double in amplitude (Figure 2.11a). If the same two waves are combined completely out of phase (having a phase difference of 180°), they will cancel each other out when added, which results in a relative value of zero amplitude (Figure 2.11b). If the second wave is only partially out of phase (by a degree other than 180°), the levels will be added at points where the combined amplitudes are positive and reduced in level where the combined result is negative (Figure 2.11c).

PHASE SHIFT

Phase shift is a term that describes one waveform's lead or lag in time with respect to another. Basically, it results from a time delay between two (or more) waveforms (with differences in acoustic distance being the most common source of this type of delay). For example, a 500 Hz wave completes one cycle every 0.002 sec. If you start with two in-phase, 500 Hz waves and delay one of them

Tutorial: Phase

1. Go to the Tutorial section of www.modrec.com, click on Phase Tutorial and download the 0° and 180° soundfiles.
2. Load the 0° file onto track 1 of the digital audio workstation (DAW) of your choice, making sure to place the file at the beginning of the track, with the signal panned center.
3. Load the same 0° file again into track 2.
4. Load the 180° file into track 3.
5. Play tracks 1 and 2 (by muting track 3) and listen to the results. The result should be a summed signal that is 3 dB louder.
6. Play tracks 1 and 3 (by muting track 2) and listen to the results. It should cancel, producing no output.
7. Offsetting track 3 (relative to track 1) should produce varying degrees of cancellation.
8. Feel free to zoom in on the waveforms, mix them down, and view the results. Cool, huh?

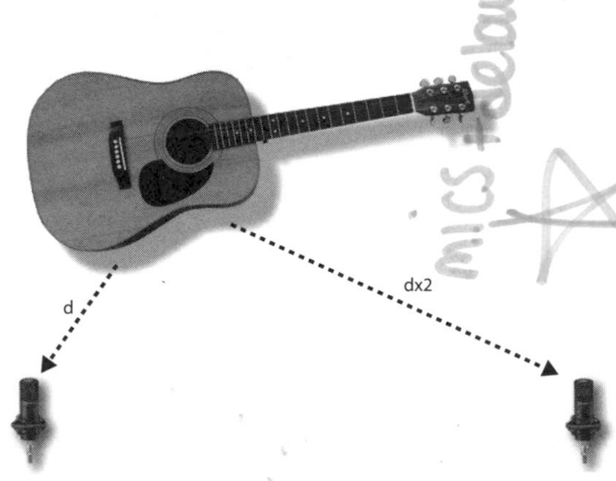

FIGURE 2.12 Cancellations can occur when a single source is picked up by two microphones.

by 0.001 sec (half the wave's period), the delayed wave will lag the other by one-half a cycle, or 180°. Another example might include a single source that's being picked up by two microphones that have been placed at different distances (Figure 2.12), thereby creating a corresponding time delay when the mics are mixed together. Such a delay can also occur when a single microphone picks up direct sounds as well as those that are reflected off of a nearby boundary. These signals will be in phase at frequencies where the path-length difference is equal to the signal's wavelength, and out of phase at those frequencies where the multiples fall at or near the half-wavelength distance. In all the above situations, these boosts and cancellations combine to alter the signal's overall frequency response at the pickup. For this and other reasons, acoustic leakage between microphones and reflections from nearby boundaries should be kept to a minimum whenever possible.

Harmonic content

Up to this point, our discussion has centered on the sine wave, which is composed of a single frequency that produces a pure sound at a specific pitch. Fortunately, musical instruments rarely produce pure sine waves. If they did, all of the instruments would basically sound the same, and music would be pretty boring. The factor that helps us differentiate between instrumental "voicings" is the presence of frequencies (called *partials*) that exist in addition to the fundamental pitch that's being played. Partials that are higher than the fundamental frequency are called *upper partials* or *overtones*. Overtone frequencies that are whole-number multiples of the fundamental frequency are called *harmonics*. For example, the frequency that corresponds to concert A is 440 Hz (Figure 2.13a). An 880 Hz

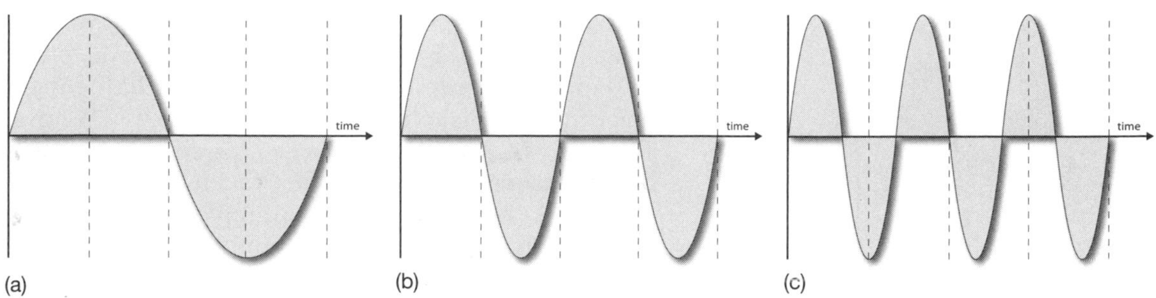

FIGURE 2.13
An illustration of harmonics: (a) first harmonic "fundamental waveform"; (b) second harmonic; (c) third harmonic.

wave is a harmonic of the 440-Hz fundamental because it is twice the frequency (Figure 2.13b). In this case, the 440 Hz fundamental is technically the first harmonic because it is 1 times the fundamental frequency, and the 880 Hz wave is called the second harmonic because it is 2 times the fundamental. The third harmonic would be 3 times 440 Hz, or 1320 Hz (Figure 2.13c). Some instruments, such as bells, xylophones and other percussion instruments, will often contain overtone partials that aren't harmonically related to the fundamental at all.

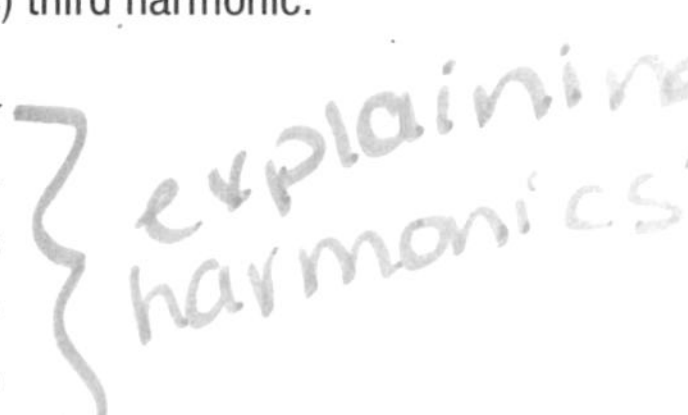

The ear perceives frequencies that are whole, doubled multiples of the fundamental as being related in a special way (a phenomenon known as the *musical octave*). For example, as concert A is 440 Hz (A3), the ear hears 880 Hz (A4) as being the next highest frequency that sounds most like concert A. The next related octave above that will be 1760 Hz (A5). Therefore, 880 Hz is said to be one octave above 440 Hz, and 1760 Hz is said to be two octaves above 440 Hz, etc. Because these frequencies are even multiples of the fundamental, they're known as *even harmonics*. Not surprisingly, frequencies that are odd multiples of the fundamental are called *odd harmonics*. In general, even harmonics are perceived as creating a sound that is pleasing to the ear, while odd harmonics will create a dissonant, harsher tone.

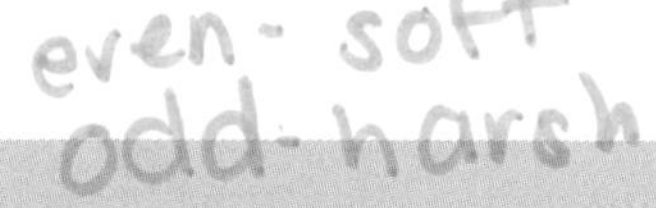

Tutorial: Harmonics

1. Go to the Tutorial section of www.modrec.com, click on Harmonics Tutorial and download all of the soundfiles.
2. Load the first-harmonic A440 file onto track 1 of the digital audio workstation (DAW) of your choice, making sure to place the file at the beginning of the track, with the signal panned center.
3. Load the second-, third-, fourth- and fifth-harmonic files into the next set of consecutive tracks.
4. Solo the first-harmonic track, then solo the first- and second-harmonic tracks. Do they sound related in nature?
5. Solo the first-harmonic track, then solo the first- and third-harmonic tracks. Do they sound more dissonant?
6. Solo the first-, second- and third-harmonic tracks. Do they sound related?
7. Solo the first-, third- and fifth-harmonic tracks. Do they sound more dissonant?

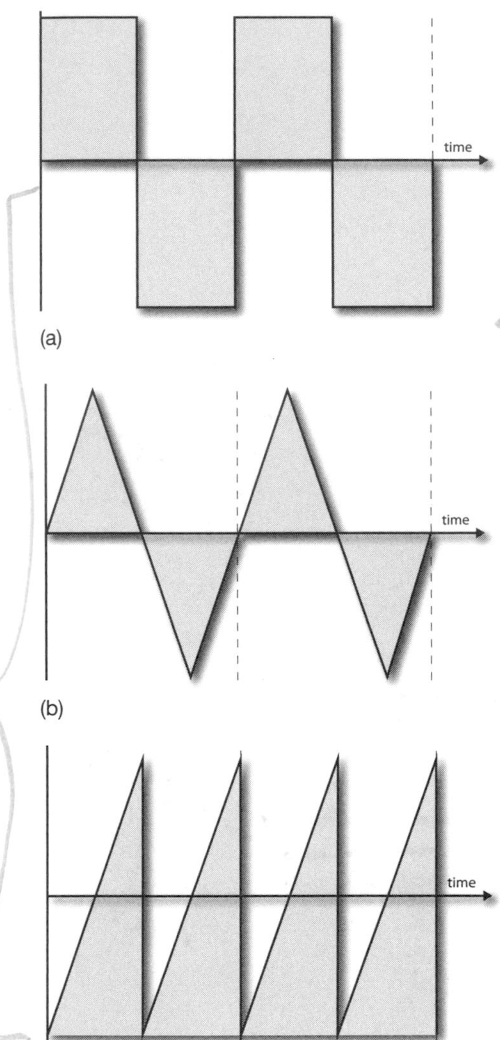

FIGURE 2.14
Simple waveforms: (a) square waves; (b) triangle waves; (c) sawtooth waves.

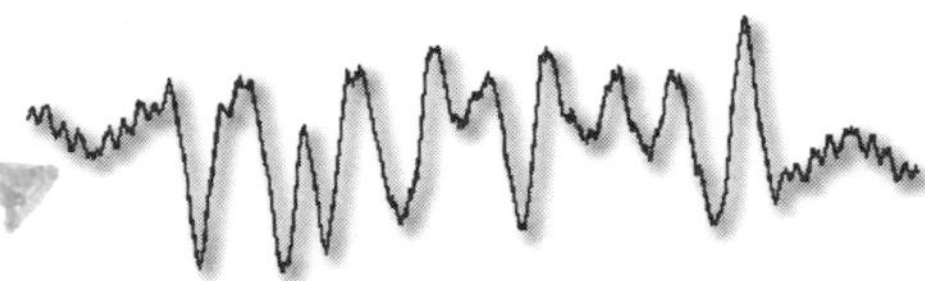

FIGURE 2.15
Example of a complex waveform.

Because musical instruments produce sound waves that contain harmonics with various amplitude and phase relationships, the resulting waveforms bear little resemblance to the shape of the single-frequency sine wave. Therefore, musical waveforms can be divided into two categories: simple and complex. Square waves, triangle waves and sawtooth waves are examples of *simple waves* that contain a consistent harmonic structure (Figure 2.14). They are said to be simple because they're continuous and repetitive in nature. One cycle of a square wave looks exactly like the next, and they are symmetrical about the zero line.

Complex waves, on the other hand, don't necessarily repeat and often are not symmetrical about the zero line. An example of a complex waveform (Figure 2.15) is one that's created by any naturally occurring sound (such as music or speech). Although complex waves are rarely repetitive in nature, all sounds can be mathematically broken down into a series of ever-changing combination of individual sine waves (or re-synthesized through a complex process known as Fourier analysis).

Regardless of the shape or complexity of a waveform that reaches the eardrum, the inner ear is able to perceive these component waveforms and then transmit the stimulus to the brain. This can be illustrated by passing a square wave through a bandpass filter that's set to pass only a narrow band of frequencies at any one time. Doing this would show that the square wave is composed of a fundamental frequency plus a number of harmonics that are made up of odd-number multiple frequencies (whose amplitudes decrease as the frequency increases). In Figure 2.16, we see how individual sine-wave harmonics can be combined to form a square wave.

If we were to analyze the harmonic content of sound waves that are produced by a violin and compare them to the content of the waves that are produced by a viola (with both playing concert A440 Hz), we would come up with results like those shown in Figure 2.17. Notice that the violin's harmonics differ in both degree and intensity from those of the viola. The harmonics and their relative intensities (which determine an instrument's characteristic sound) are called the *timbre* of an instrument. If we changed an instrument's harmonic balance, the sonic character of the instrument would also be changed.

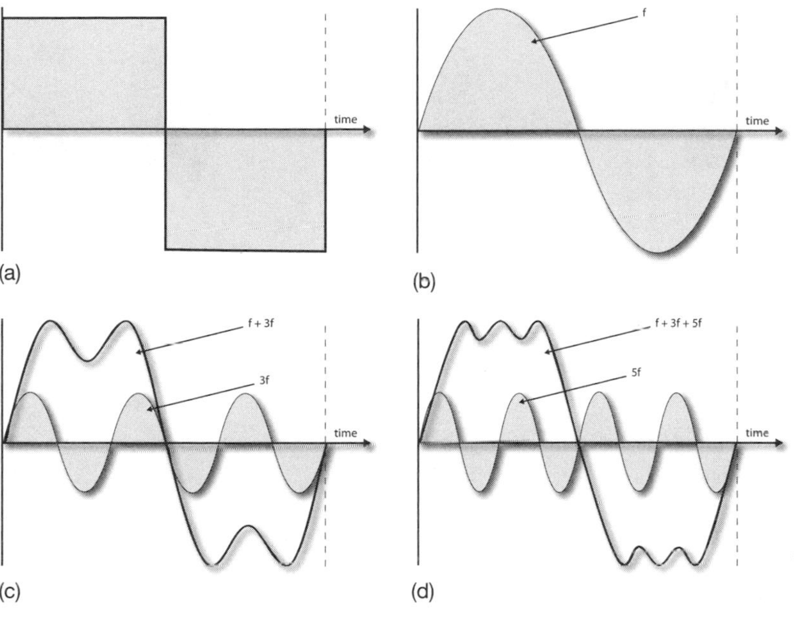

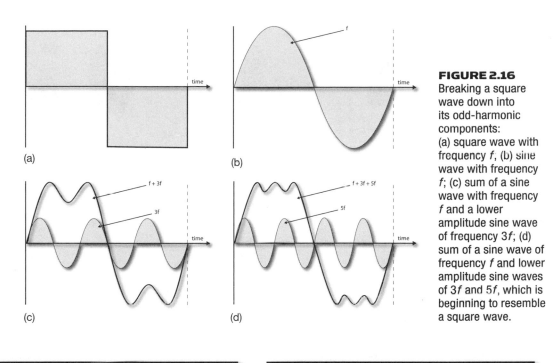

FIGURE 2.16
Breaking a square wave down into its odd-harmonic components: (a) square wave with frequency *f*, (b) sine wave with frequency *f*; (c) sum of a sine wave with frequency *f* and a lower amplitude sine wave of frequency 3*f*; (d) sum of a sine wave of frequency *f* and lower amplitude sine waves of 3*f* and 5*f*, which is beginning to resemble a square wave.

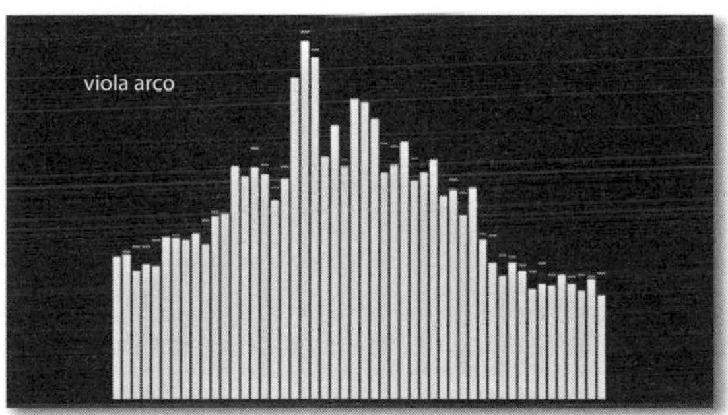

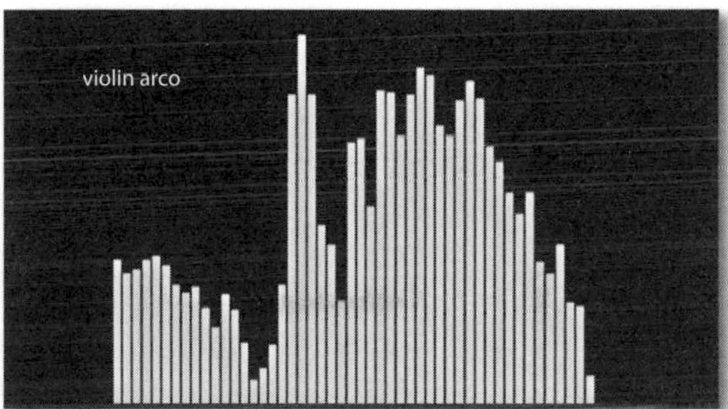

FIGURE 2.17
Harmonic structure of concert A440: (a) played on a viola; (b) played on a violin.

For example, if the violin's upper harmonics were reduced, the violin would sound more like a viola.

Because the relative harmonic balance is so important to an instrument's sound, the frequency response of a microphone, amplifier, speaker and all other elements in the signal path can have an effect on the timbral (tonal) balance of a sound. If the frequency response isn't flat, the timbre of the sound will be changed. For example, if the high frequencies are amplified less than the low and middle frequencies, then the sound will be duller than it should be. For this reason, a specific mic, equalizer or mic placement can be used as tools to vary the timbre of an instrument, thereby changing its subjective sound.

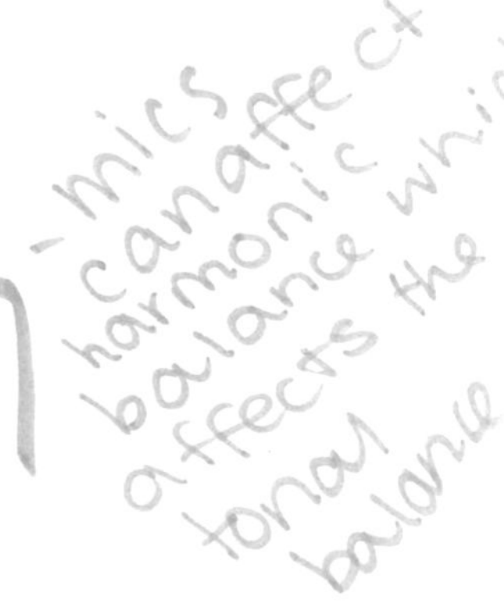

-mics can affect harmonic balance which affects the tonal balance

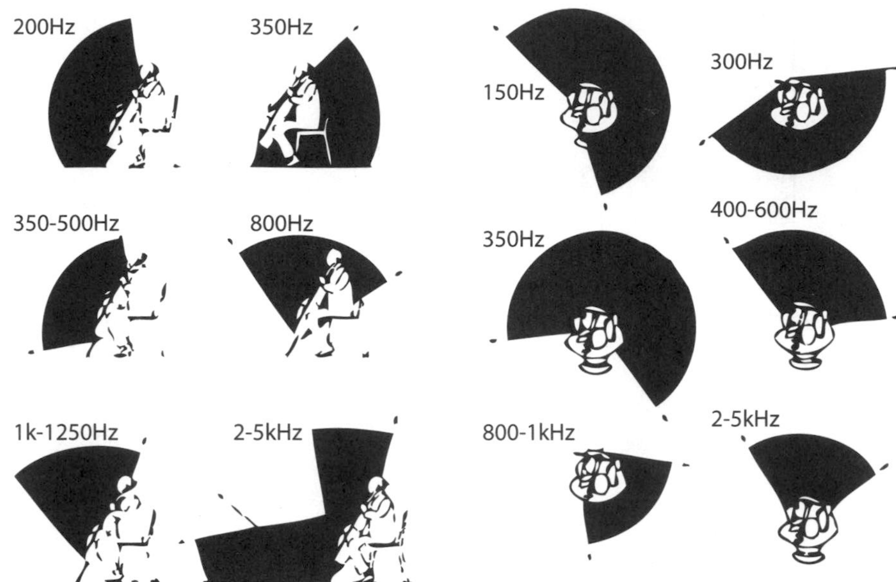

FIGURE 2.18
Radiation patterns of a cello as viewed from the side (left) and top (right).

In addition to the variations in harmonic balance that can exist between instruments and their families, it is common for the harmonic balance to vary with respect to the direction that a sound wave radiates from an instrument. Figure 2.18 shows the principal radiation patterns as they emanate from a cello (as seen from both the side and top views).

Envelope

Timbre isn't the only characteristic that helps us differentiate between instruments. Each one produces a sonic amplitude *envelope* that works in combination with timbre to determine its unique and subjective sound. The envelope (ADSR) of a waveform can be described as characteristic variations in level that occur in time over the duration of a played note. The envelope of an acoustic or electronically generated signal is composed of four sections that vary in amplitude over time:

- *Attack* (A) refers to the time taken for a sound to build up to its full volume when a note is initially sounded.
- *Decay* (D) refers to how quickly the sound levels off to a sustain level after the initial attack peak.
- *Sustain* (S) refers to the duration of the ongoing sound that's generated following the initial attack decay.
- *Release* (R) relates to how quickly the sound will decay once the note is released.

Figure 2.19a illustrates the envelope of a trombone note. The attack, decay times and internal dynamics produce a smooth, sustaining sound. A cymbal crash

(Figure 2.19b) combines a high-level, fast attack with a longer sustain and decay that creates a smooth, lingering shimmer. Figure 2.19c illustrates the envelope of a snare drum. Notice that the initial attack is much louder than the internal dynamics, while the final decay trails off very quickly, resulting in a sharp, percussive sound.

It's important to note that the concept of an envelope often relies on peak waveform values, while the human perception of loudness is proportional to the average wave intensity over a period of time (rms value). Therefore, high-amplitude portions of the envelope won't make an instrument sound loud unless the amplitude is maintained for a sustained period. Short high-amplitude sections tend to contribute to a sound's overall character, rather than to its loudness. By using a compressor or limiter, an instrument's character can often be modified by changing the dynamics of its envelope without changing its overall timbre.

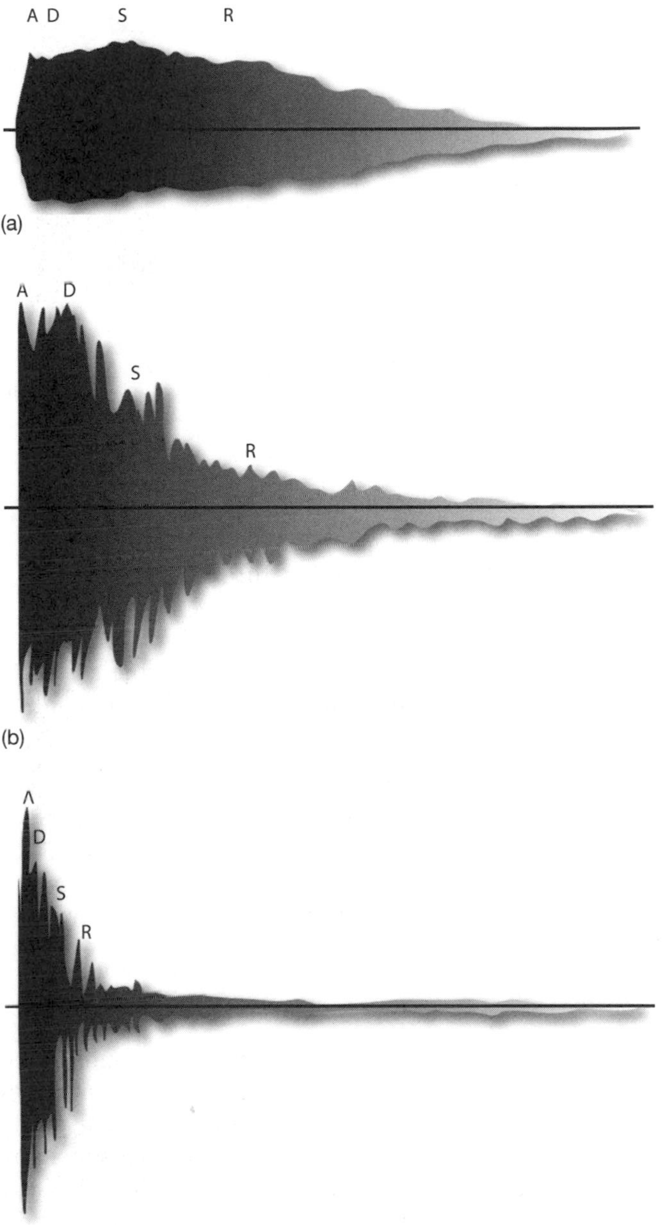

FIGURE 2.19
Various musical waveform envelopes: (a) trombone, (b) cymbal crash, and (c) snare drum, where A = attack, D = decay, S = sustain, and R = release.

LOUDNESS LEVELS: THE DECIBEL

The human ear operates over an energy range of approximately 10^{13}:1 (10,000,000,000,000:1), which is an extremely wide range. Since it's difficult for us to conceptualize number ranges that are this large, a logarithmic scale has been adopted to compress the measurements into figures that are more manageable. The unit used for measuring sound-pressure level (SPL), signal level and relative changes in signal level is the *decibel* (*dB*), a term that literally means 1/10th of a Bell (an older telephone transmission loss measurement unit that was named after Alexander Graham Bell, inventor of the telephone). In order to develop an understanding of the decibel, we first need to examine logarithms

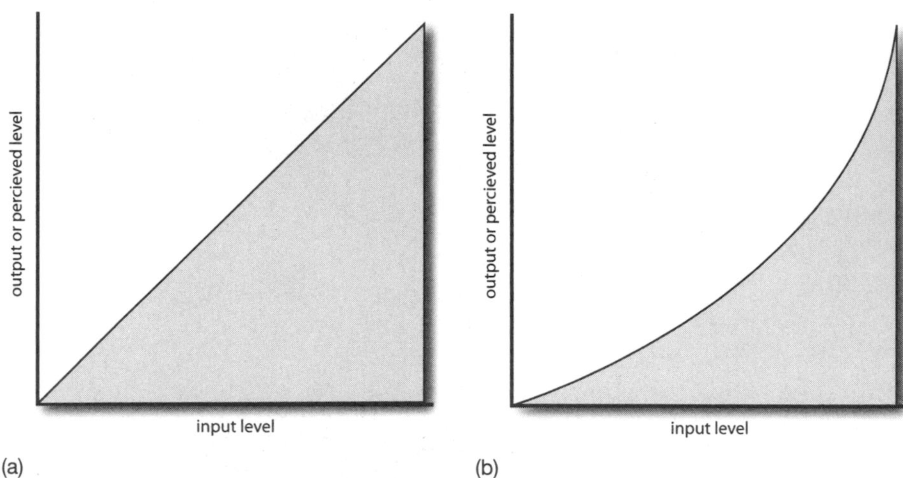

FIGURE 2.20
Linear and logarithmic curves: (a) linear; (b) logarithmic.

and the logarithmic scale (Figure 2.20). The *logarithm* (*log*) is a mathematical function that reduces large numeric values into smaller, more manageable numbers. Because logarithmic numbers increase exponentially in a way that's similar to how we perceive loudness (e.g., 1, 2, 4, 16, 128, 256, 65,536 ...), it expresses our perceived sense of volume more precisely than a linear curve can.

Before we delve into a deeper study of this important concept and how it deals with our perceptual senses, let's take a moment to understand the basic concepts and building block ideas behind the log scale, so as to get a better understanding of what examples such as "+3 dB at 10,000 Hz" really mean. Be patient with yourself! Over time, the concept of the decibel will become as much a part of your working vocabulary as ounces, gallons and miles per hour (or KG, liters and kilometers).

Logarithmic basics

In audio, we use logarithmic values to express the differences in intensities between two levels (often, but not always, comparing a measured level to a standard reference level). Because the differences between these two levels can be really, really big, a simpler system makes use of expressed values that are mathematical exponents of 10. To begin, finding the log of a number such as 17,386 without a calculator is not only difficult ... it's unnecessary! All that's really important to help you along are three simple guidelines:

- The log of the number 2 is 0.3.
- When a number is an integral power of 10 (e.g., 100, 1000, 10,000), the log can be found simply by adding up the number of zeros.
- Numbers that are greater than 1 will have a positive log value, while those less than 1 will have a negative log value.

Again, the first one is an easy fact to remember: The log of 2 is 0.3 … this will make sense shortly. The second one is even easier: The logs of numbers such as 100, 1000 or 10,000,000,000,000 can be arrived at by simply counting up the zeros. The last guideline relates to the fact that if the measured value is less than the reference value, the resulting log value will be negative. For example:

$$\log 2 = 0.3$$

$$\log \tfrac{1}{2} - \log 0.5 = -0.3$$

$$\log 10{,}000{,}000{,}000{,}000 = 13$$

$$\log 1000 = 3$$

$$\log 100 = 2$$

$$\log 10 = 1$$

$$\log 0.1 = -1$$

$$\log 0.01 = -2$$

$$\log 0.01 = -3$$

All other numbers can be arrived at by using a scientific calculator (most computers and cell phones have one built in); however, it's unlikely that you will ever need to know any log values beyond understanding the basic concepts that are listed above.

THE DECIBEL

Now that we've gotten past the absolute bare basics, I'd like to break with tradition again and attempt an explanation of the decibel in a way that's less complex and relates more to our day-to-day needs in the sound biz. First off, the decibel is a logarithmic value that "expresses differences in intensities between two levels." From this, we can infer that these levels are expressed by several units of measure, the most common being sound-pressure level (SPL), voltage (V) and power (wattage, or W). Now, let's look at the basic math behind these three measurements.

SOUND-PRESSURE LEVEL

Sound-pressure level is the acoustic pressure that's built up within a defined atmospheric area (usually a square centimeter, or cm^2). Quite simply, the higher the SPL, the louder the sound (Figure 2.21). In this instance, our measured reference (SPL_{ref}) is the threshold of hearing, which is defined as being the softest sound that an average person can hear. Most conversations will have an SPL of about 70 dB, while average home stereos are played at volumes ranging between 80 and 90 dB SPL. Sounds that are so loud as to be painful have SPLs of about

Typical A-Weighted Sound Levels dB re: $20uN/m^2$

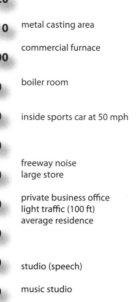

	dB	
	140	
50 Hp siren (100 ft)		
	130	
jet takoff (200 ft)	120	
riveting machine	110	metal casting area
		commercial furnace
cut off saw pnuematic hammer	100	
textive weaving plant subway train (20 ft)	90	boiler room
pneumatic drill (50 ft)	80	inside sports car at 50 mph
freight train (100 ft) vacuum cleaner (10 ft) speach (1 ft)	70	
	60	freeway noise large store
large transformer (200 ft)	50	private business office light traffic (100 ft) average residence
	40	
soft wisper (5 ft)	30	studio (speech)
	20	music studio
	10	
threshold of hearing youth 1k - 4kHz	0	

FIGURE 2.21
Chart of sound-pressure levels. (Courtesy of General Radio Company.)

130 to 140 dB (10,000,000,000,000 or more times louder than the 0 dB reference). We can arrive at an SPL rating by using the formula:

$$dB\ SPL = 20\log SPL/SPL_{ref}$$

where SPL is the measured sound pressure (in dyne/cm^2). SPL_{ref} is a reference sound pressure (the threshold limit of human hearing, 0.02 millipascals = 2 ten-billionths of our atmospheric pressure).

From this, I feel that the major concept that needs to be understood is the idea that SPL levels change with the square of the distance (hence, the 20 log part of the equation). This means that whenever a source/pickup distance is doubled, the SPL level will be reduced by 6 dB (20 log 1/2 = 20 x -0.3 = –6 dB SPL); as the distance is halved, it will increase by 6 dB (20 log 2/1 = 20 x 0.3 = 6 dB SPL), as shown in Figure 2.22.

VOLTAGE

Voltage can be thought of as the pressure behind electrons within a wire. As with acoustic energy, comparing one voltage level to another level (or reference level) can be expressed as dBv using the equation:

$$dBv = 20\log V/V_{ref}$$

where V is the measured voltage, and V_{ref} is a reference voltage (0.775 volts).

POWER

Power is usually a measure of wattage or current and can be thought of as the flow of electrons through a wire over time. Power is generally associated with audio

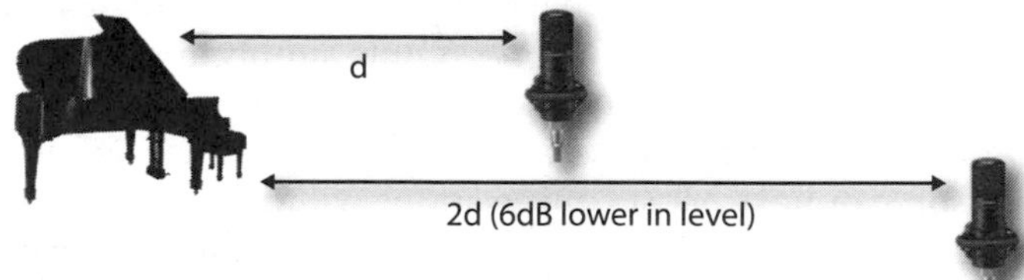

FIGURE 2.22
Doubling the distance of a pickup will lower the perceived direct signal level by 6 dB SPL.

signals that are carried throughout an audio production system. Unlike SPL and voltage, the equation for signal level (which is often expressed in dBm) is:

$$dBm = 10\log P/P_{ref}$$

where P is the measured wattage, and P_{ref} is referenced to 1 milliwatt (0.001 watt).

The simple heart of the matter

I am going to stick my neck out and state that, when dealing with decibels, it's far more common for working professionals to deal with the concept of power. The dBm equation expresses the spirit of the decibel term when dealing with the marking and measurements on an audio device or the numeric values in a computer dialog box. This is due to the fact that power is the unit of measure that's most often expressed when dealing with audio equipment controls; therefore, it's my personal opinion that the average working stiff only needs to grasp the following basic concepts:

- A 1 dB change is noticeable to most ears (but not by much).
- Turning something up by 3 dB will double the signal's level (believe it or not, doubling the signal level won't increase the perceived loudness as much as you might think).
- Turning something down by 3 dB will halve the signal's level (likewise, halving the signal level won't decrease the perceived loudness as much as you might think).
- The log of an exponent of 10 can be easily figured by simply counting the zeros (e.g., the log of 1,000 is 3). Given that this figure is multiplied by 10 ($10 \log P/P_{ref}$), turning something up by 10 dB will increase the signal's level 10-fold, 20 dB will yield a 100-fold increase, 30 dB will yield a 1,000-fold increase, etc.

Most pros know that turning a level fader up by 3 dB will effectively double its energy output (and vice versa). Beyond this, it's unlikely that anyone will ever ask, "Would you please turn that up a thousand times?" It just won't happen! However, when a pro asks his or her assistant to turn the gain up by 20 dB, that assistant will often instinctively know what 20 dB is… and what it sounds like. I guess I'm saying that the math really isn't nearly as important as the ongoing process of getting an instinctive feel for the decibel and how it relates to relative levels within audio production.

THE EAR

A sound source produces acoustic waves by alternately compressing and rarefying the air molecules between it and the listener, causing fluctuations that fall above and below normal atmospheric pressure. The human ear is a sensitive transducer that responds to these pressure variations by way of a series

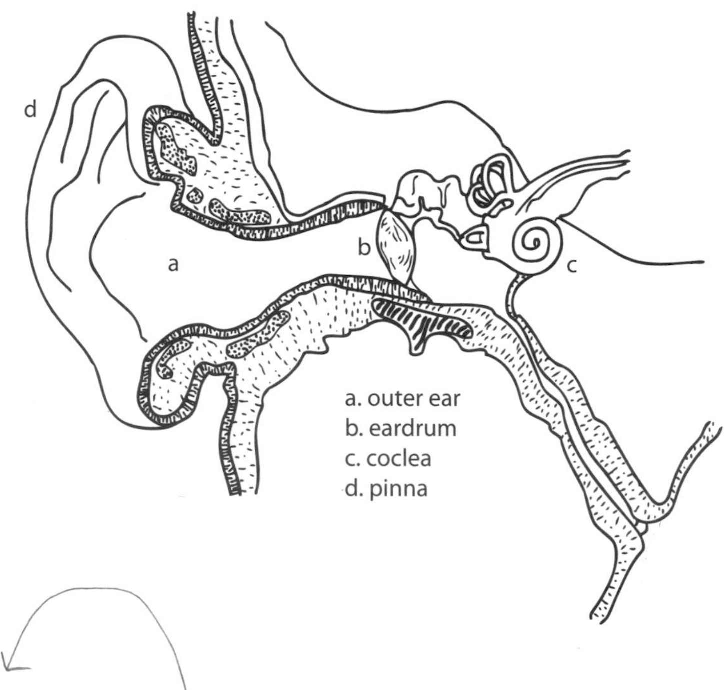

FIGURE 2.23
Outer, middle and inner ear.

of related processes that occur within the auditory organs … *our ears*. When these variations arrive at the listener, sound-pressure waves are collected in the aural canal by way of the outer ear's pinna. These are then directed to the eardrum, a stretched drum-like membrane (Figure 2.23), where the sound waves are changed into mechanical vibrations, which are transferred to the inner ear by way of three bones known as the hammer, anvil and stirrup. These bones act both as an amplifier (by significantly increasing the vibrations that are transmitted from the eardrum) and as a limiting protection device (by reducing the level of loud, transient sounds such as thunder or fireworks explosions). The vibrations are then applied to the inner ear (cochlea) – a tubular, snail-like organ that contains two fluid-filled chambers. Within these chambers are tiny hair receptors that are lined up in a row along the length of the cochlea. These hairs respond to certain frequencies depending on their placement along the organ, which results in the neural stimulation that gives us the sensation of hearing. Permanent hearing loss generally occurs when these hair/nerve combinations are damaged or as they deteriorate with age.

Threshold of hearing

In the case of SPL, a convenient pressure-level reference is the *threshold of hearing*, which is the minimum sound pressure that produces the phenomenon of hearing in most people and is equal to 0.0002 microbar. One microbar is equal to 1 millionth of normal atmospheric pressure, so you can see that the ear is an amazingly sensitive instrument. In fact, if the ear were any more sensitive, the thermal motion of molecules in the air would be audible! When referencing SPLs to 0.0002 microbar, this threshold level usually is denoted as 0 dB SPL, which is defined as the level that an average person can hear at a specific frequency only 50% of the time.

Threshold of feeling

An SPL that causes discomfort in a listener 50% of the time is called the *threshold of feeling*. It occurs at a level of about 118 dB SPL between the frequencies of 200 Hz and 10 kHz.

Threshold of pain

The SPL that causes pain in a listener 50% of the time is called the *threshold of pain* and corresponds to an SPL of 140 dB in the frequency range between 200 Hz and 10 kHz.

Taking care of your hearing

During the 1970s and early 1980s, recording studio monitoring levels were often turned up so high as to be truly painful. In the mid-1990s, a small band of powerful producers and record executives banded together to successfully reduce these average volumes down to tolerable levels (85 to 95 dB) … a practice that often continues to this day. Live sound venues and acts often continue the practice of raising house and stage volumes to chest-thumping levels. Although these levels are exciting, long-term exposure can lead to temporary or permanent hearing loss. So what types of hearing loss are there?

- *Acoustic trauma:* This happens when the ear is exposed to a sudden, loud noise in excess of 140 dB. Such a shock could lead to permanent hearing loss.
- *Temporary threshold shift:* The ear can experience temporary hearing loss when exposed to long-term, loud noise.
- *Permanent threshold shift:* Extended exposure to loud noises in a specific or broad hearing range can lead to permanent hearing loss in that range. In short, the ear becomes less sensitive to sounds in the damaged frequency range leading to a reduction in perceived volume… What?

Here are a few hearing conservation tips (courtesy of the House Ear Institute, www.hei.org) that can help reduce hearing loss due to long-term exposure of sounds over 115 dB:

- Avoid hazardous sound environments; if they're not avoidable, wear hearing protection devices, such as foam earplugs, custom-molded earplugs, or in-ear monitors.

A simple fact to remember: Once your hearing is gone … it's gone … for good!

- Monitor sound-pressure levels at or around 85 dB. The general rule to follow is if you're in an environment where you must raise your voice to be heard, then you're monitoring too loudly and should limit your exposure times.
- Take 15-minute "quiet breaks" every few hours if you're being exposed to levels above 85 dB.
- Musicians and other live entertainment professionals should avoid practicing at concert-hall levels whenever possible.
- Have your hearing checked periodically by a licensed audiologist.

PSYCHOACOUSTICS

The area of *psychoacoustics* deals with how and why the brain interprets a particular sound stimulus in a certain way. Although a great deal of study has been devoted to this subject, the primary device in psychoacoustics is the all-elusive brain … which is still largely unknown to present-day science.

AUDITORY PERCEPTION

From the outset, it's important to realize that the ear is a nonlinear device (what's received at your ears isn't always what you'll hear). It's also important to note that the ear's frequency response (its perception of timbre) changes with the loudness of the perceived signal. The "loudness" compensation switch found on many hi-fi preamplifiers is an attempt to compensate for this decrease in the ear's sensitivity to low- and high-frequency sounds at low listening levels.

The *Fletcher-Munson equal-loudness contour curves* (Figure 2.24) indicate the ear's average sensitivity to different frequencies at various levels. These indicate the sound-pressure levels that are required for our ears to hear frequencies along the curve as being equal in level to a 1000-Hz reference level (measured in phons). Thus, to equal the loudness of a 1-kHz tone at 110 dB SPL (a level typically created by a trumpet-type car horn at a distance of 3 feet), a 40-Hz tone has to be about 6 dB louder, whereas a 10-kHz tone must be 4 dB louder in order to be perceived as being equally loud. At 50 dB SPL (the noise level present in the average private business office), the level of a 40-Hz tone must be 30 dB louder and a 10-kHz tone 13 dB louder than a 1-kHz tone to be perceived as having the same volume. Thus, if a piece of music is mixed to sound great at a level of 85 to 95 dB, its bass and treble balance will actually be boosted when turned up (often a good thing). If the same piece were mixed at 110 dB SPL, it would sound both bass and treble shy when played back at lower levels … because no compensation for the ear's response was added to the mix. Over the years, it has generally been found that changes in apparent frequency balance are less apparent when monitoring at levels of 85 dB SPL.

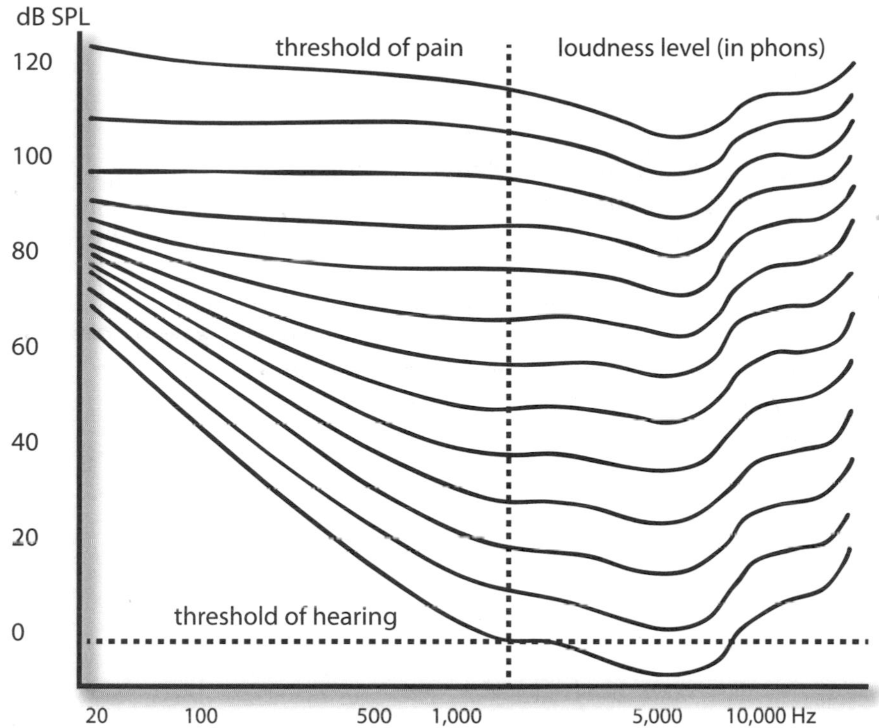

FIGURE 2.24
The Fletcher–Munson curve shows an equal loudness contour for pure tones as perceived by humans having an average hearing acuity. These perceived loudness levels are charted relative to sound-pressure levels at 1000 Hz.

In addition to the above, whenever it is subjected to sound waves that are above a certain loudness level, the ear can produce harmonic distortion that doesn't exist in the original signal. For example, the ear can cause a loud 1-kHz sine wave to be perceived as being a combination of 1-, 2-, 3-kHz waves, and so on. Although the ear might hear the overtone structure of a violin (if the listening level is loud enough), it might also perceive additional harmonics (thus changing the timbre of the instrument). This is one of several factors that implies that sound monitored at very loud levels could sound quite different when played back at lower levels.

The loudness of a tone can also affect our ear's perception of pitch. For example, if the intensity of a 100-Hz tone is increased from 40 to 100 dB SPL, the ear will hear a pitch decrease of about 10%. At 500 Hz, the pitch will change about 2% for the same increase in sound-pressure level. This is one reason why musicians find it difficult to tune their instruments when listening through loud headphones.

As a result of the nonlinearities in the ear's response, tones will often interact with each other rather than being perceived as being separate. Three types of interaction effects can occur:

- Beats
- Combination tones
- Masking

Beats

Two tones that differ only slightly in frequency and have approximately the same amplitude will produce an effect known as *beats*. This effect sounds like repetitive volume surges that are equal in frequency to the difference between these two tones. The phenomenon is often used as an aid for tuning instruments, because the beats slow down as the two notes approach the same pitch and finally stop when the pitches match. In reality, beats are a result of the ear's inability to separate closely pitched notes. This results in a third frequency that's created from the phase sum and difference values between the two notes.

Tutorial: Beats

1. Go to the Tutorial section of www.modrec.com, click on Beats Tutorial and download all of the soundfiles.
2. Load the 440Hz file onto track 1 of the digital audio workstation (DAW) of your choice, making sure to place the file at the beginning of the track, with the signal panned center.
3. Load the 445 and 450 Hz files into the next two consecutive tracks.
4. Solo and play the 440 Hz tone.
5. Solo both the 440 and 445 Hz tones and listen to their combined results. Can you hear the 5-Hz beat tone? (445 Hz – 440 Hz = 5 Hz)
6. Solo both the 445 and 450 Hz tones and listen to their combined results. Can you hear the 5-Hz beat tone? (450 Hz – 445 Hz = 5 Hz)
7. Now, solo both the 440 and 450 Hz tones and listen to their combined results. Can you hear the 10 Hz beat tone? (450 Hz – 440 Hz = 10 Hz)

Combination tones

Combination tones result when two loud tones differ by more than 50 Hz. In this case, the ear perceives an additional set of tones that are equal to both the sum and the difference between the two original tones … as well as being equal to the sum and difference between their harmonics. The simple formulas for computing the fundamental tones are:

$$\text{Sum tone} = f_1 + f_2$$

$$\text{Difference tone} = f_1 - f_2$$

Difference tones can be easily heard when they are below the frequency of both tones' fundamentals. For example, the combination of 2000 and 2500 Hz produces a difference tone of 500 Hz.

Masking

Masking is the phenomenon by which loud signals prevent the ear from hearing softer sounds. The greatest masking effect occurs when the frequency of the sound and the frequency of the masking noise are close to each other. For example, a 4k-Hz tone will mask a softer 3.5-kHz tone but has little effect on the audibility of a quiet 1000-Hz tone. Masking can also be caused by harmonics of the masking tone (e.g., a 1-kHz tone with a strong 2-kHz harmonic might mask a 1900-Hz tone). This phenomenon is one of the main reasons why stereo placement and equalization are so important to the mixdown process. An instrument that sounds fine by itself can be completely hidden or changed in character by louder instruments that have a similar timbre. Equalization, mic choice or mic placement might have to be altered to make the instruments sound different enough to overcome any masking effect.

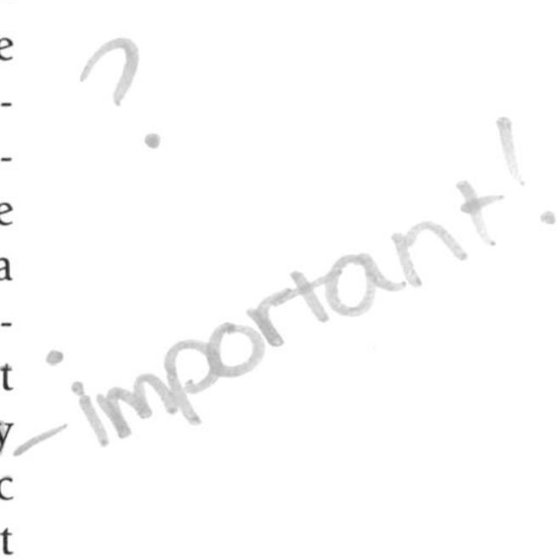

Tutorial: Masking

1. Go to the Tutorial section of www.modrec.com, click on Masking Tutorial and download all of the soundfiles.
2. Load the 1000 Hz file onto track 1 of the digital audio workstation (DAW) of your choice, making sure to place the file at the beginning of the track, with the signal panned center.
3. Load the 3800- and 4000-Hz files into the next two consecutive tracks.
4. Solo and play the 1000-Hz tone.
5. Solo both the 1000- and the 4000-Hz tones and listen to their combined results. Can you hear both of the tones clearly?
6. Solo and play the 3800-Hz tone.
7. Solo both the 3800- and the 4000-Hz tones and listen to their combined results. Can you hear both of the tones clearly?

PERCEPTION OF DIRECTION

Although one ear can't discern the direction of a sound's origin, two ears can. This capability of two ears to localize a sound source within an acoustic space is called *spatial* or *binaural localization*. This effect is the result of three acoustic cues that are received by the ears:

- Interaural intensity differences
- Interaural arrival-time differences
- The effects of the pinnae (outer ears)

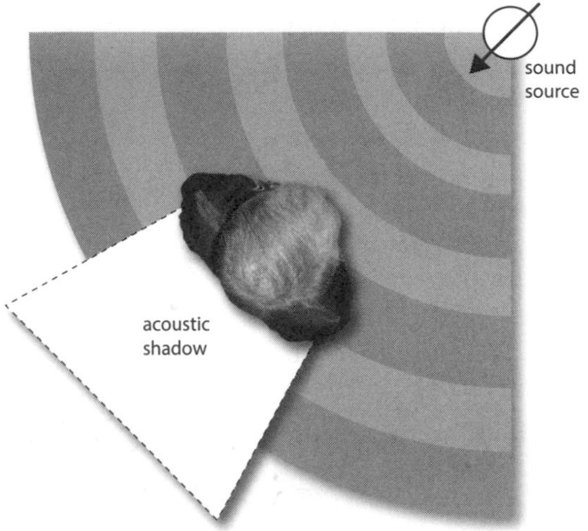

FIGURE 2.25
The head casts an acoustic shadow that helps with localization at middle to upper frequencies.

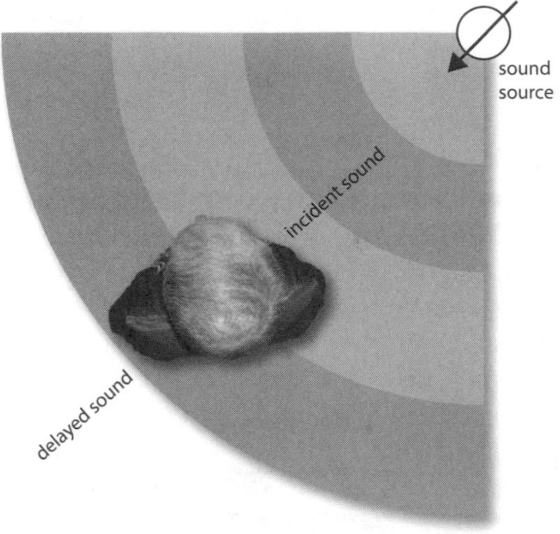

FIGURE 2.26
Interaural arrival-time differences occurring at lower frequencies.

Middle to higher frequency sounds originating from the right side will reach the right ear at a higher intensity level than the left ear, causing an interaural intensity difference. This volume difference occurs because the head casts an acoustic block or shadow, allowing only reflected sounds from surrounding surfaces to reach the opposite ear (Figure 2.25). Because the reflected sound travels farther and loses energy at each reflection - in our example the intensity of sound perceived by the left ear will be greatly reduced, resulting in a signal that's perceived as originating from the right.

This effect is relatively insignificant at lower frequencies, where wavelengths are large compared to the head's diameter, allowing the wave to easily bend around its acoustic shadow. For this reason, a different method of localization (known as *interaural arrival-time differences*) is employed at lower frequencies (Figure 2.26). In both Figures 2.25 and 2.26, small time differences occur because the acoustic path length to the left ear is slightly longer than the path to the right ear. The sound pressure therefore arrives at the left ear at a later time than the right. This method of localization (in combination with interaural intensity differences) helps to give us lateral localization cues over the entire frequency spectrum.

Intensity and delay cues allow us to perceive the direction of a sound's origin but not whether the sound originates from the front, behind or below. The pinna (Figure 2.27), however, makes use of two ridges that reflect sound into the ear. These ridges introduce minute time delays between the direct sound (which reaches the entrance of the ear canal) and the sound that's reflected from the ridges (which varies according to source location). It's interesting to note that beyond 130° from the front of our face, the pinna is able to reflect and delay sounds by 0 and 80 microseconds (µsec), making rear localization possible. Ridge 2 (see Figure 2.27) has been reported to produce delays of between 100 and 330 µsec that help us to locate sources in the vertical plane. The delayed reflections from both ridges are then combined with the direct sound to produce frequency–response colorations that are compared within the brain to determine source location. Small movements of the head can also provide additional position information.

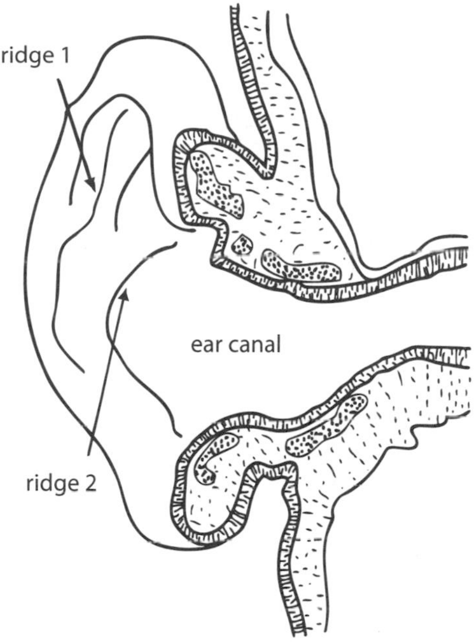

FIGURE 2.27
The pinna and its reflective ridges for determining vertical location information.

If there are no differences between what the left and right ears hear, the brain assumes that the source is the same distance from each ear. This phenomenon allows us to position sound not only in the left and right loudspeakers but also monophonically between them. If the same signal is fed to both loudspeakers, the brain perceives the sound identically in both ears and deduces that the source must be originating from directly in the center. By changing the proportion that's sent to each speaker, the engineer changes the relative interaural intensity differences and thus creates the illusion of physical positioning between the speakers. This placement technique is known as *panning* (Figure 2.28).

PERCEPTION OF SPACE

In addition to perceiving the direction of sound, the ear and brain combine to help us perceive the size and physical characteristics of the acoustic space in which a sound occurs. When a sound is generated, a percentage reaches the listener directly, without encountering any obstacles. A larger portion, however, is propagated to the many surfaces of an acoustic enclosure. If these surfaces are reflective, the sound is bounced back into the room and toward the listener. If the surfaces are absorptive, less energy will be reflected back to the listener. Three types of reflections are commonly generated within an enclosed space (Figure 2.29):

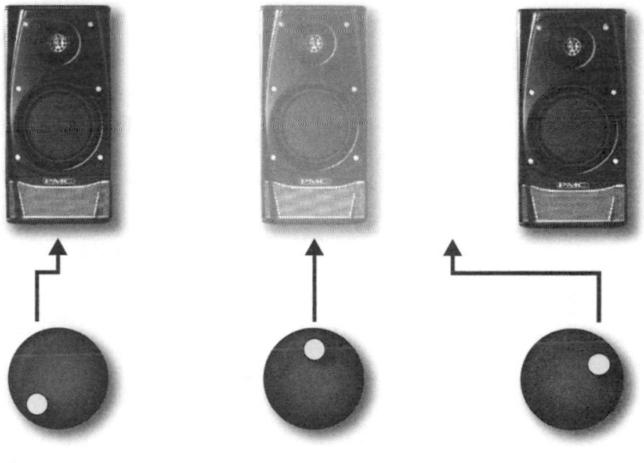

- Direct sound
- Early reflections
- Reverberation

FIGURE 2.28
Pan pot settings and their relative spatial positions.

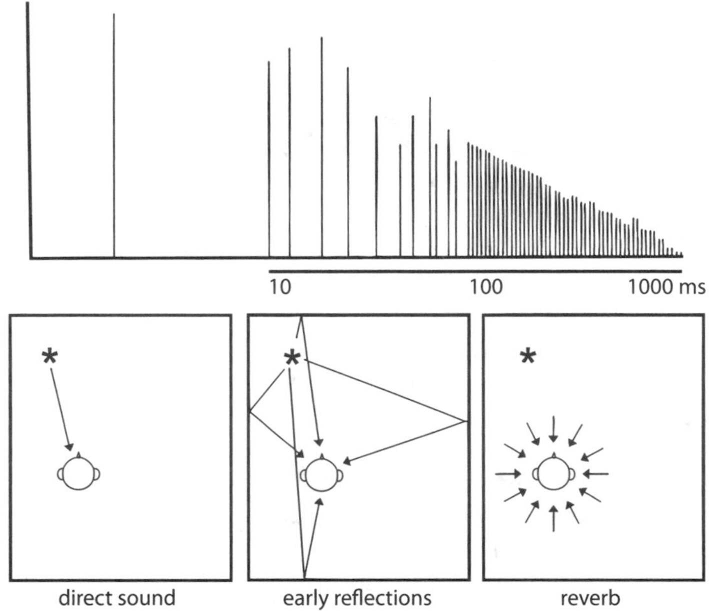

FIGURE 2.29
The three distinct soundfield types that are generated within an enclosed space.

Direct sound

In air, sound travels at a constant speed of about 1130 feet per second, so a wave that travels from the source to the listener will follow the shortest path and arrive at the listener's ear first. This is called the *direct sound*. Direct sounds determine our perception of a sound source's location and size and conveys the true timbre of the source.

Early reflections

Waves that bounce off of surrounding surfaces in a room must travel further than direct sound to reach the listener and therefore arrive after the direct sound and from a multitude of directions. These waves form what are called *early reflections*. Early reflections give us clues as to the reflectivity, size and general nature of an acoustic space. These sounds generally arrive at the ears less than 50 msec after the brain perceives the direct sound and are the result of reflections off of the largest, most prominent boundaries within a room. The time elapsed between hearing the direct sound and the beginning of the early reflections helps to provide information about the size of the performance room. Basically, the farther the boundaries are from the source and listener, the longer the delay before it's reflected back to the listener.

Another aspect that occurs with early reflections is called *temporal fusion*. Early reflections arriving at the listener within 30 msec of the direct sound are not only audibly suppressed, but are also fused with the direct sound. In effect, the ear can't distinguish the closely occurring reflections and considers them to be part of the direct sound. The 30-msec time limit for temporal fusion isn't absolute; rather, it depends on the sound's envelope. Fusion breaks down at 4 msec for transient clicks, whereas it can extend beyond 80 msec for slowly evolving sounds (such as a sustained organ note or legato violin passage). Despite the fact that the early reflections are suppressed and fused with the direct sound, they still modify our perception of the sound, making it both louder and fuller.

Reverberation

Whenever room reflections continue to bounce off of room boundaries, a randomly decaying set of sounds can often be heard after the source stops in the form of *reverberation*. A highly reflective surface absorbs less of the wave energy at each reflection and allows the sound to persist longer after the initial sound stops (and vice versa). Sounds reaching the listener 50-msec later in time are perceived as a random and continuous stream of reflections that arrive from all directions. These densely spaced reflections gradually decrease in amplitude and add a sense of warmth and body to a sound. Because it has undergone multiple reflections, the timbre of the reverberation is often quite different from the direct sound (with the most notable difference being a roll-off of high frequencies and a slight bass emphasis).

The time it takes for a reverberant sound to decrease to 60 dB below its original level is called its *decay time* or *reverb time* and is determined by the room's absorption characteristics. The brain is able to perceive the reverb time and timbre of the reverberation and uses this information to form an opinion on the hardness or softness of the surrounding surfaces. The loudness of the perceived direct sound increases rapidly as the listener moves closer to the source, while the reverberation levels will often remain the same, because the diffusion is roughly constant throughout the room. This ratio of the direct sound's loudness to the reflected sound's level helps listeners judge their distance from the sound source.

Whenever artificial reverb and delay units are used, the engineer can generate the necessary cues to convince the brain that a sound was recorded in a huge, stone-walled cathedral – when, in fact, it was recorded in a small, absorptive room. To do this, the engineer programs the device to mix the original unreverberated signal with the necessary early delays and random reflections. Adjusting the number and amount of delays on an effects processor gives the engineer control over all of the necessary parameters to determine the perceived room size, while decay time and frequency balance can help to determine the room's perceived surfaces. By changing the proportional mix of direct-to-processed sound, the engineer/producer can place the sound source at either the front or rear of the artificially created space.

CHAPTER 3

Studio Acoustics and Design

The *Audio Cyclopedia* defines *acoustics* as "a science dealing with the production, effects and transmission of sound waves; the transmission of sound waves through various mediums, including reflection, refraction, diffraction, absorption and interference; the characteristics of auditoriums, theaters and studios, as well as their design." We can see from this description that the proper acoustic design of music recording, project and audio-for-visual or broadcast studios is no simple matter. A wide range of complex variables and interrelationships often come into play in the creation of a successful acoustic and monitoring design. When designing or redesigning an acoustic space, the following basic requirements should be considered:

- *Acoustic isolation:* This prevents external noises from transmitting into the studio environment through the air, ground or building structure. It can also prevent feuds that can arise when excessive volume levels leak out into the surrounding neighborhood.
- *Frequency balance:* The frequency components of a room shouldn't adversely affect the acoustic balance of instruments and/or speakers. Simply stated, the acoustic environment shouldn't alter the sound quality of the original or recorded performance.
- *Acoustic separation:* The acoustic environment should not interfere with intelligibility and should offer the highest possible degree of acoustic separation within the room (often a requirement for ensuring that sounds from one instrument aren't unduly picked up by another instrument's microphone).
- *Reverberation:* The control of sonic reflections within a space is an important factor for maximizing the intelligibility of music and speech. No matter how short the early reflections and reverb times are, they will add an important psychoacoustic sense of "space" in the sense that they can give our brain subconscious cues as to a room's size, number of reflective boundaries, distance between the source and listener, and so forth.
- *Cost factors:* Not the least of all design and construction factors is cost. Multimillion-dollar facilities often employ studio designers and

> construction teams to create a plush decor that has been acoustically tuned to fit the needs of both the owners and their clients. Owners of project studios and budget-minded production facilities, however, can also take full advantage of the same basic acoustic principles and construction techniques and apply them in cost-effective ways.

This chapter will discuss many of the basic acoustic principles and construction techniques that should be considered in the design of a music or sound production facility. I'd like to emphasize that any or all of these acoustical topics can be applied to any type of audio production facility and aren't only limited to professional music studio designs. For example, owners of modest project and bedroom studios should know the importance of designing a control room that's symmetrical and, hopefully, sounds good. It doesn't cost anything to know that if one speaker is in a corner and the other is on a wall, the perceived center image will be off balance. As with many techno-artistic endeavors, studio acoustics and design are a mixture of fundamental physics (in this case, mostly dimensional mathematics) and an equally large dose of common sense and dumb luck. More often than not, acoustics is an artistic science that melds physics with the art of intuition and experience.

STUDIO TYPES

Although the acoustical fundamentals are the same for most studio design types, differences will often follow the form, function and budgets required by the tasks at hand. Some of the more common studio types include:

- Professional music studios
- Audio-for-visual production environments
- Project studios

The professional recording studio

The *professional recording studio* (Figures 3.1 and 3.2) is first and foremost a commercial business, so its design, decor, and acoustical construction requirements are often much more demanding than those of a privately owned project studio. In some cases, an acoustical designer and experienced construction team are placed in charge of the overall building phase of a professional facility. In others, the studio's budget is just too low to hire such professionals, which places the studio owners and staff squarely in charge of designing and constructing the entire facility. Whether you happen to have the luxury of building a new facility from the ground up or are renovating a studio within an existing shell, you would probably benefit from a professional studio designer's experience and skills. Such expert advice often proves to be cost effective in the long run, because errors in design judgment can lead to cost overruns, lost business due to unexpected delays or the unfortunate state of living with mistakes that could have been easily avoided.

FIGURE 3.1
Blade Studios Control Room A, Shreveport, Louisiana. (Courtesy of Blade Studios, www.bladestudios.com)

FIGURE 3.2
Blade Studios Studio A, Shreveport, Louisiana. (Courtesy of Blade Studios, www.bladestudios.com)

The audio-for-visual production environment

An audio-for-visual production facility is used for video, film and game postproduction (often simply called "post") and includes such facets as music recording for film or other media (scoring), score mixdown, automatic dialog replacement (ADR—the replacement of on- and off-screen dialog to visual media) and Foley (the replacement and creation of on- and off-screen sound effects). As with music studios, audio-for-visual production facilities can range from high-end facilities that can accommodate the posting needs of network video or feature film productions (Figure 3.3) to a simple, budget-minded project studio that's equipped with video and a digital audio workstation. As with the music studio,

FIGURE 3.3
Paragon Studios Control Room A setup for post-production, Franklin, Tennessee. (Courtesy of Solid State Logic, www.solid-state-logic.com).

FIGURE 3.4
Gettin' it all going in the bedroom studio. (Courtesy of Yamaha Corporation of America, www.yamaha.com)

audio-for-visual construction and design techniques often span a wide range of styles and scope in order to fit the budget needs at hand.

The project studio

It goes without saying that the vast majority of audio production studios fall into the project studio category. This basic definition of such a facility is open to interpretation. It's usually intended as a personal production resource for recording music, audio-for-visual production, multimedia production, voiceovers … you name it. Project studios can range from being fully commercial in nature to smaller setups that are both personal and private (Figure 3.4). All of these possible studio types have been designed with the idea of giving artists the flexibility of making their art in a personal, off-the-clock environment that's both cost and time effective. The design and construction considerations for creating a privately owned project studio often differ from the design considerations for a professional music facility in two fundamental ways:

- Building constraints
- Cost

Generally, a project studio's room (or series of rooms) is built into an artist's home or a rented space where the construction and dimensional details are already defined. This fact (combined with inherent cost considerations) often leads the owner/artist to employ cost-effective techniques for sonically treating a room. Even if the room has little or no treatment, keep in mind that a basic knowledge of acoustical physics and room design can be a handy and cost-effective tool as your experience, production needs and business abilities grow.

Modern-day digital audio workstations (DAWs) have squarely placed the Mac and PC within the ergonomics and functionality of the pro and home project studio (Figure 3.5). In fact, in many cases, the DAW *is* the project studio. With the advent of self-powered speaker monitors, cost-effective microphones and hardware DAW controllers, it's a relatively simple matter to design a powerful production system into almost any existing space.

FIGURE 3.5
785 Records & Publishing/Denise Rich Songs, New York. (Courtesy of Solid State Logic, www.solid-state-logic.com).

With regard to setting up any production/monitoring environment, I'd like to first draw your attention to the need for symmetry in any critical monitoring environment. A symmetrical acoustic environment around the central mixing axis can work wonders toward creating a balanced left/right and surround image. Fortunately, this generally isn't a difficult goal to achieve. An acoustical and speaker placement environment that isn't balanced between the left-hand and right-hand sides will allow for differing reflections, absorption coefficients and variations in frequency response that can adversely affect the imaging and balance of your final mix. Further information on this important subject can be found later in this chapter … consider this your first heads-up on an important topic.

PRIMARY FACTORS GOVERNING STUDIO AND CONTROL ROOM ACOUSTICS

Regardless of which type of studio facility is being designed, built and used, a number of primary concerns should be addressed in order to achieve the best possible acoustic results. In this section, we'll take a close look at such important and relevant aspects of acoustics as:

- Acoustic isolation
- Symmetry in control room and monitoring design
- Frequency balance
- Absorption
- Reflection
- Reverberation

Although several mathematical formulas have been included in the following sections, it's by no means necessary that you memorize or worry about them. By far, I feel that it's more important that you grasp the basic principles of acoustics rather than worry about the underlying math. *Remember:* More often than not, acoustics is an artistic science that blends math with the art of intuition and experience.

Acoustic isolation

Because most commercial and project studio environments make use of an acoustic space to record sound, it's often wise and necessary to employ effective isolation techniques into their design in order to keep external noises to a minimum. Whether that noise is transmitted through the medium of air (e.g., from nearby auto, train, or jet traffic) or through solids (e.g., from air-conditioner rumbling, underground subways, or nearby businesses), special construction techniques will often be required to dampen these extraneous sounds (Figure 3.6).

FIGURE 3.6
Various isolation, absorption, and reflective acoustical treatments for the construction of a recording/monitoring environment. (Courtesy of Auralex Acoustics, www.auralex.com)

If you happen to have the luxury of building a studio facility from the ground up, a great deal of thought should be put into selecting the studio's location. If a location has considerable neighborhood noise, you might have to resort to extensive (and expensive) construction techniques that can "float" the rooms (a process that effectively isolates and uncouples the inner rooms from the building's outer foundations). If there's absolutely no choice of studio location and the studio happens to be located next to a factory, just under the airport's main landing path or over the subway's uptown line … you'll simply have to give in to destiny and build acoustical barriers to these outside interferences.

The reduction in the sound-pressure level (SPL) of a sound source as it passes through an acoustic barrier of a certain physical mass (Figure 3.7) is termed the *transmission loss* (TL) of a signal. This attenuation can be expressed (in dB) as:

$$TL = 14.5 \log M + 23$$

where TL is the transmission loss in decibels, and M is the surface density (or combined surface densities) of a barrier in pounds per square foot (lb/ft^2).

Because transmission loss is frequency dependent, the following equation can be used to calculate transmission loss at various frequencies with some degree of accuracy:

$$TL = 14.5 \log Mf - 16$$

where f is the frequency (in hertz).

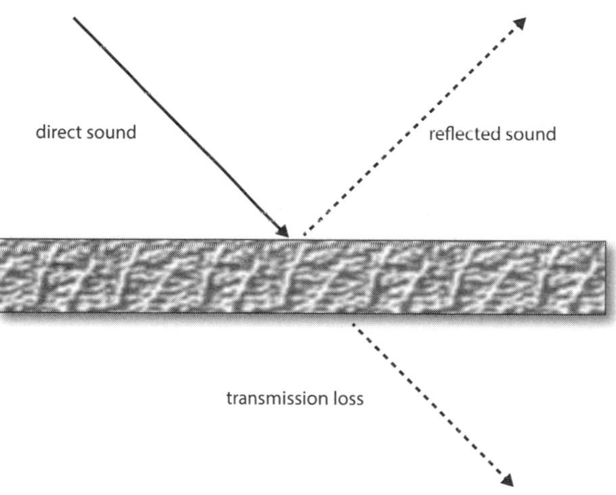

FIGURE 3.7
Transmission loss refers to the reduction of a sound signal (in dB) as it passes through an acoustic barrier.

Both common sense and the preceding two equations tell us that heavier acoustic barriers will yield a higher transmission loss. For example, Table 3.1 tells us that a 12-inch-thick wall of dense concrete (yielding a surface density of 150 lb/ft^2) offers a much greater resistance to the transmission of sound than can a 4-inch cavity filled with sand (which yields a surface density of 32.3 lb/ft^2). From the second equation ($TL = 14.5 \log Mf - 16$), we can also draw the conclusion that, for a given acoustic barrier, transmission losses will increase as the frequency rises. This can be easily illustrated by closing the door of a car that has its sound system turned up, or by shutting a single door to a music studio's control room. In both instances, the high frequencies will be greatly reduced in level, while the bass frequencies will be impeded to a much lesser extent. From this, the goal would seem to be to build a studio wall, floor, ceiling, window or door out of the thickest and most dense material that's available; however, expense and physical space often play roles in determining just how much of a barrier can be built to achieve the desired isolation. As such, a balance must usually be struck when using both space- and cost-effective building materials.

Table 3.1 Surface densities of common building materials.

Material	Thickness (inches)	Surface Density (lb/ft^2)
Brick	4	40.0
	8	80.0
Concrete (lightweight)	4	33.0
	12	100.0
Concrete (dense)	4	50.0
	12	150.0
Glass	—	3.8
	—	7.5
	—	11.3
Gypsum wallboard	—	2.1
	—	2.6
Lead	1/16	3.6
Particleboard	—	1.7
Plywood	—	2.3
Sand	1	8.1
	4	32.3
Steel	—	10.0
Wood	1	2.4

WALLS

When building a studio wall or reinforcing an existing structure, the primary goal is to reduce leakage (increase the transmission loss) through a wall as much as possible over the audible frequency range. This is generally done by:

- Building a wall structure that is as massive as is practically possible (both in terms of cubic and square foot density)
- Eliminating open joints that can easily transmit sound through the barrier
- Dampening structures, so that they are well supported by reinforcement structures and are free of resonances

The following guidelines can be helpful in the construction of framed walls that have high transmission losses:

- If at all possible, the inner and outer wallboards should not be directly attached to the same wall studs. The best way to avoid this is to alternately stagger the studs along the floor and ceiling frame, so that the front/back facing walls aren't in physical contact with each other (Figure 3.8).

- Each wall facing should have a different density to reduce the likelihood of increased transmission due to resonant frequencies that might be sympathetic to both sides. For example, one wall might be constructed of two 5/8-inch gypsum wallboards, while the other wall might be composed of soft fiberboard that's surfaced with two 1/2-inch gypsum wallboards.
- If you're going to attach gypsum wallboards to a single wall face, you can increase transmission loss by mounting the additional layers (not the first layer) with adhesive caulking rather than using screws or nails.
- Spacing the studs 24 inches on center instead of using the traditional 16-inch spacing yields a slight increase in transmission loss.
- To reduce leakage that might make it through the cracks, apply a bead of non-hardening caulk sealant to the inner gypsum wallboard layer at the wall-to-floor, wall-to-ceiling and corner junctions.

Generally, the same amount of isolation is required between the studio and the control room as is required between the studio's interior and exterior environments. The proper building of this wall is important, so that an accurate tonal balance can be heard over the control-room monitors without promoting leakage between the rooms or producing resonances within the wall that would audibly color the signal. Optionally, a specially designed cavity, called a *soffit*, can be designed into the front-facing wall of the control room to house the larger studio monitors. This superstructure allows the main, farfield studio monitors to be mounted directly into the wall to reduce reflections and resonances in the monitoring environment.

It's important for the soffit to be constructed to high standards, using a multiple-wall or high-mass design that maximizes the density with acoustically tight construction techniques in order to reduce leakage between the two rooms. Cutting corners by using substandard (and even standard) construction techniques in

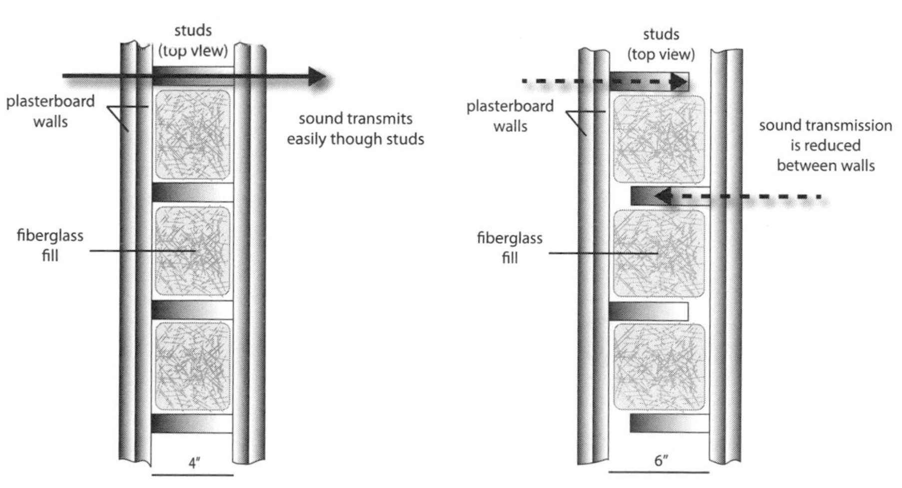

FIGURE 3.8
Double, staggered stud construction greatly reduces leakage by decoupling the two wall surfaces from each other: (a) top view showing walls that are directly tied to wall studs (allowing sound to easily pass through). (b) top view showing walls with offset, non-touching studs (so that sound doesn't easily pass from wall to wall).

the building of a studio soffit can lead to unfortunate side effects, such as wall resonances, rattles, and increased leakage. Typical wall construction materials include:

- *Concrete:* This is the best and most solid material, but it is often expensive and it's not always possible to pour cement into an existing design.
- *Bricks (hollow-form or solid-facing):* This excellent material is often easier to place into an existing room than concrete.
- *Gypsum plasterboard:* Building multiple layers of plasterboard onto a double-walled stud frame is often the most cost- and design-efficient approach to reducing resonances and maximizing transmission loss. It's often a good idea to reduce these resonances by filling the wall cavities with rockwool or fiberglass, while bracing the internal structure to add an extra degree of stiffness.

Studio monitors can be designed into the soffit in a number of ways. In one expensive approach, the speakers' inner enclosures are cavities designed into walls that are made from a single concrete pour. Under these conditions, resonances are completely eliminated. Another less expensive approach has the studio monitors resting on poured concrete pedestals; in this situation, inserts can be cast into the pedestals that can accept threaded rebar rods (known as all-thread). By filing the rods to a chamfer or a sharper point, it's possible to adjust the position, slant and height of the monitors for final positioning into the soffit's wall framing. The most common and affordable approach uses traditional wood framing in order to design a cavity into which the speaker enclosures can be designed and positioned. Extra bracing and heavy construction should be used to reduce resonances.

FLOORS

For many recording facilities, the isolation of floor-borne noises from room and building exteriors is an important consideration. For example, a building that's located on a busy street and whose concrete floor is tied to the building's ground foundation might experience severe low-frequency rumble from nearby traffic. Alternatively, a second-floor facility might experience undue leakage from a noisy downstairs neighbor or, more likely, might interfere with a quieter neighbor's business. In each of these situations, increasing the isolation to reduce floor-borne leakage and/or transmission is essential. One of the most common ways to isolate floor-related noise is to construct a "floating" floor that is structurally decoupled from its subfloor foundation.

Common construction methods for floating a professional facility's floor uses either neoprene "hockey puck" isolation mounts, U-Boat floor floaters (Figure 3.9), or a continuous underlay, such as a rubberized floor mat. In these cases, the underlay is spread over the existing floor foundation and then covered with an overlaid plywood floor structure. In more extreme situations, this superstructure could be covered with reinforcing wire mesh and finally topped with a 4-inch layer of concrete (Figure 3.10). In either case, the isolated floor is then ready for carpeting, wood finishing, painting or any other desired surface.

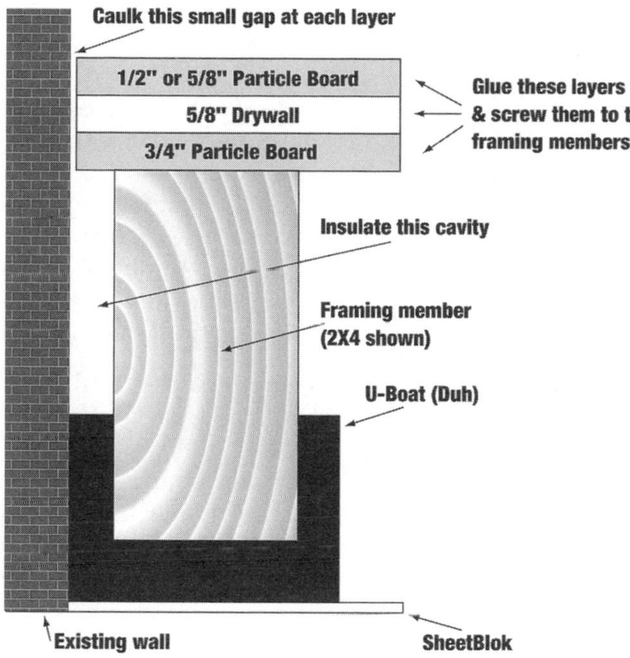

FIGURE 3.9
U-Boat™ floor beam float channels can be placed under a standard 2 × 4 floor frame to increase isolation. Floor floaters should be placed every 16 inches under a 2 × floor joist. (Courtesy of Auralex Acoustics, www.auralex.com)

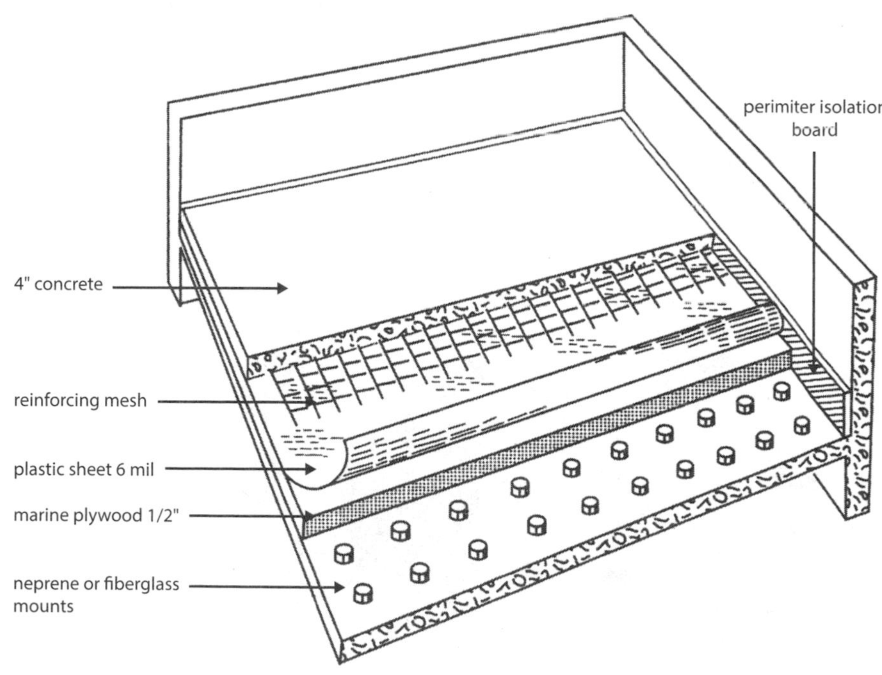

FIGURE 3.10
Basic guidelines for building a concrete floating floor using neoprene mounts.

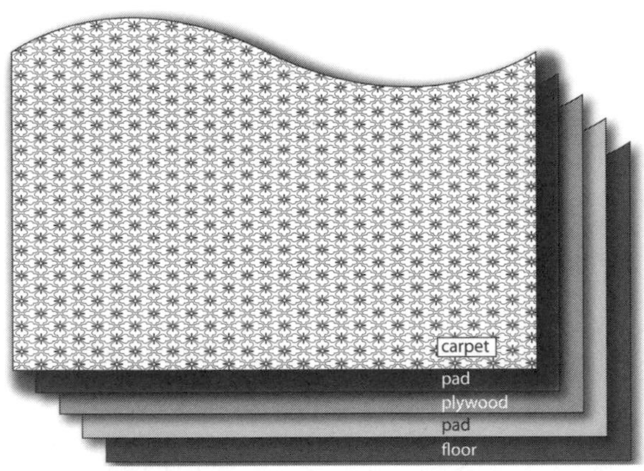

FIGURE 3.11
An alternative, cost-effective way to float an existing floor by layering relatively inexpensive materials.

An even more cost- and space-effective way to decouple a floor involves layering the original floor with a rubberized or carpet foam pad. A 1/2- or 5/8-inch layer of tongue-and-groove plywood or oriented strand board (OSB) is then laid on top of the pad. These should not be nailed to the subfloor; instead, they can be stabilized by glue or by locking the pieces together with thin, metal braces. Another foam pad can then be laid over this structure and topped with carpeting or any other desired finishing material (Figure 3.11).

It is important for the floating superstructure to be isolated from both the under-flooring and the outer wall. Failing to isolate these allows floor-borne sounds to be transmitted through the walls to the subfloor, and vice versa (often defeating the whole purpose of floating the floor). These wall perimeter isolation gaps can be sealed with pliable decoupling materials such as widths of soft mineral fiberboard, neoprene, silicone or other pliable materials.

RISERS

As we saw from the equation TL = 14.5 log Mf – 16, low-frequency sound travels through barriers much more easily than does high-frequency sound. It stands to reason that strong, low-frequency energy is transmitted more easily than high-frequency energy between studio rooms, from the studio to the control room or to outside locations. In general, the drum set is most likely to be the biggest leakage offender. By decoupling much of a drum set's low-frequency energy from a studio floor, many of the low-frequency leakage problems can be reduced. In most cases, the problem can be fixed by using a drum riser. Drum risers are available commercially (Figure 3.12), or they can be easily constructed. In order to reduce unwanted resonances, drum risers should be constructed using

FIGURE 3.12
HoverDeck™ 88 isolation riser. (Courtesy of Auralex Acoustics, www.auralex.com)

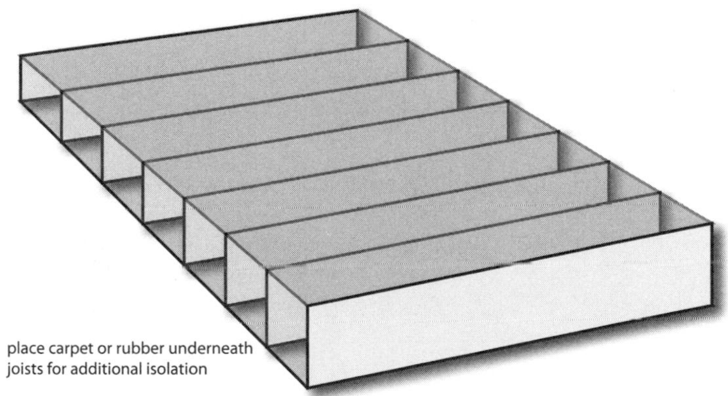

FIGURE 3.13 General construction details for a homemade drum riser.

2 × 6-inch or 2 × 8-inch beams for both the frame and the supporting joists (spaced at 16 or 12 inches on center, as shown in Figure 3.13). Sturdy 1/2- or 5/8-inch tongue-and-groove plywood panels should be glued to the supporting frames with carpenter's glue (or a similar wood glue) and then nailed or screwed down (using heavy-duty, galvanized fasteners). When the frame has dried, rubber coaster float channels or (at the very least) strips of carpeting should be attached to the bottom of the frame … and the riser will be ready for action.

CEILINGS

Foot traffic and other noises from above a sound studio or production room are another common source of external leakage. Ceiling noise can be isolated in a number of ways. If foot traffic is your problem and you're fortunate enough to own the floors above you, you can reduce this noise by simply carpeting the overhead hallway or by floating the upper floor. If you don't have that luxury, one approach to isolating ceiling-borne sounds is to hang a false structure from the existing ceiling or from the overhead joists (as is often done when a new room is being constructed). This technique can be fairly cost effective when spring or "Z" suspension channels are used (Figure 3.14). Z channels are often screwed to the ceiling joists to provide a flexible, yet strong support to which a hanging wallboard ceiling can be attached. If necessary, fiberglass or other sound-deadening materials can be placed into the cavities between the overhead structures. Other more expensive methods use spring support systems to hang false ceilings from an existing structure.

WINDOWS AND DOORS

Access to and from a studio or production room area (in the form of windows and doors) can also be a potential source of sound leakage. For this reason, strict attention needs to be given to window and door design and construction. Visibility in a studio is extremely important within a music production environment. For example, when multiple rooms are involved, good visibility serves to promote effective communication between the producer or engineer and the studio musician (as well as among the musicians themselves). For this

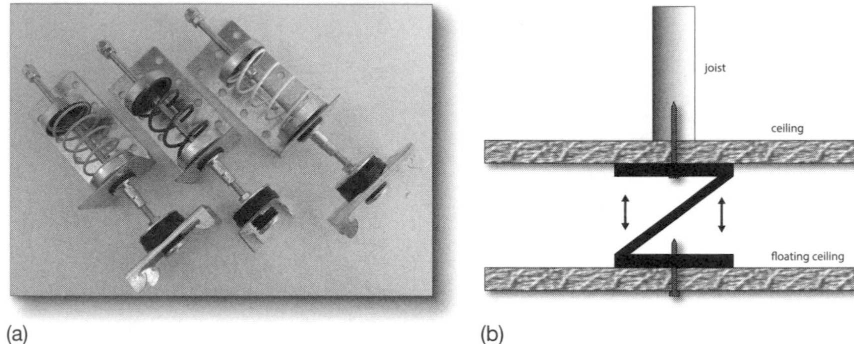

FIGURE 3.14
Ceiling isolator systems. (a) RSIC-SI-1 (Resilient Sound Isolation Clips), (courtesy of PAC International, Inc; www.pac-intl.com). (b) Z channels can be used to hang a floating ceiling from an existing overhead structure.

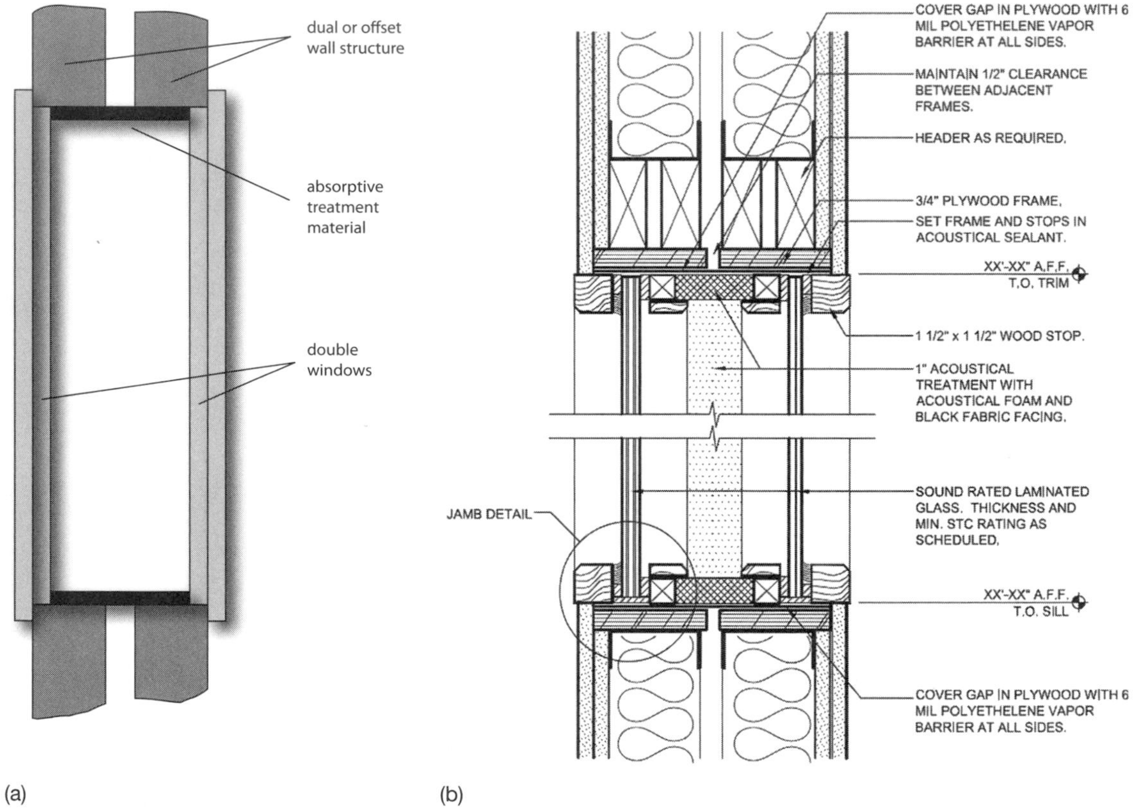

dual or offset wall structure
absorptive treatment material
double windows
(a)

COVER GAP IN PLYWOOD WITH 6 MIL POLYETHELENE VAPOR BARRIER AT ALL SIDES.
MAINTAIN 1/2" CLEARANCE BETWEEN ADJACENT FRAMES.
HEADER AS REQUIRED.
3/4" PLYWOOD FRAME.
SET FRAME AND STOPS IN ACOUSTICAL SEALANT.
XX'-XX" A.F.F. T.O. TRIM
1 1/2" x 1 1/2" WOOD STOP.
1" ACOUSTICAL TREATMENT WITH ACOUSTICAL FOAM AND BLACK FABRIC FACING.
SOUND RATED LAMINATED GLASS. THICKNESS AND MIN. STC RATING AS SCHEDULED.
JAMB DETAIL
XX'-XX" A.F.F. T.O. SILL
COVER GAP IN PLYWOOD WITH 6 MIL POLYETHELENE VAPOR BARRIER AT ALL SIDES.
(b)

FIGURE 3.15
Detail for a practical window construction between the control room and studio. (a) simplified drawing (b) detailed drawing. (Courtesy of Russ Berger Design Group, Inc., www.rbdg.com)

reason, windows have been an important factor in studio design since the beginning. The design and construction details for a window often vary with studio needs and budget requirements and can range from being deep, double-plate cavities that are built into double-wall constructions (Figure 3.15) to more

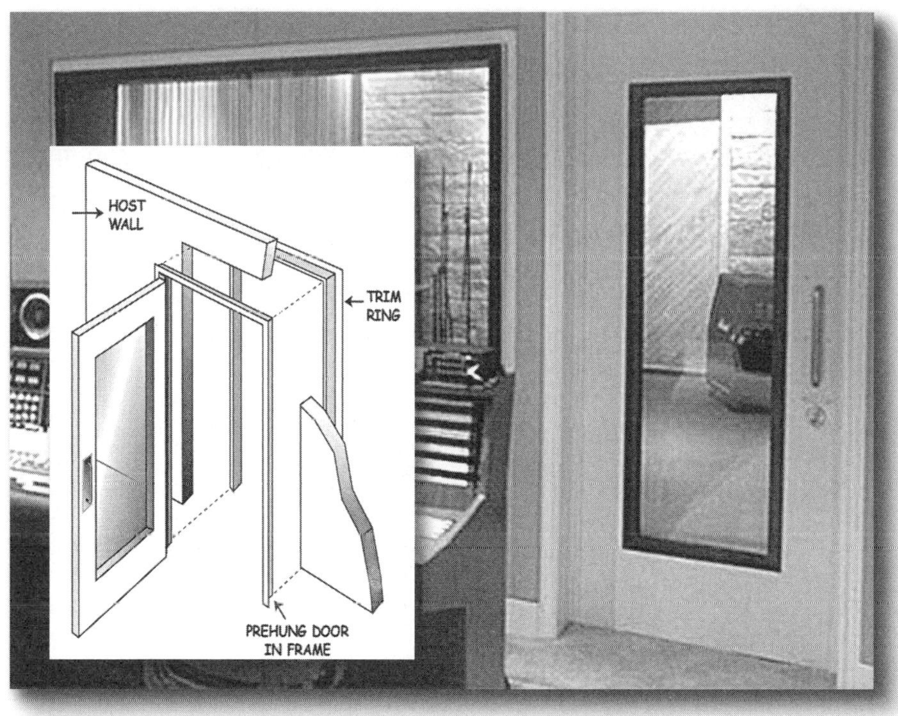

FIGURE 3.16
A SoundSecure™ studio door. (Courtesy of ETS-Lingren, www.ets-lindgren.com)

modest prefab designs that are built into a single wall. Other more expensive designs include floor-to-ceiling windows that create a virtual "glass wall," as well as those that offer sweeping vistas, which are designed into poured concrete soffit walls.

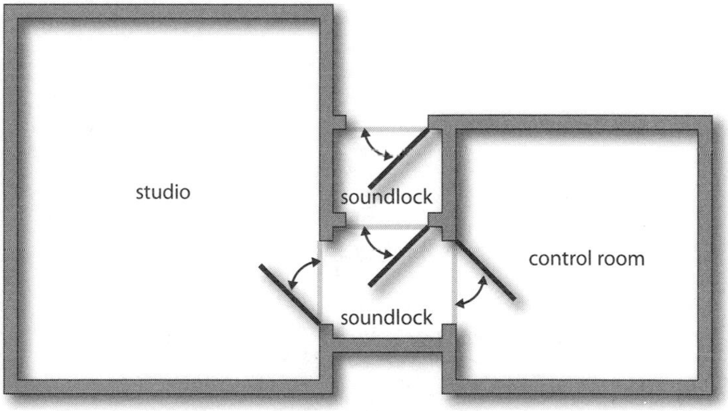

Access doors to and from the studio, control room, and exterior areas should be constructed of solid wood or high-quality acoustical materials (Figure 3.16), as solid doors generally offer higher TL values than their cheaper, hollow counterparts. No matter which door type is used, the appropriate seals, weatherstripping, and doorjambs should be used throughout so as to reduce leakage through the cracks. Whenever possible, double-door designs should be used to form an acoustical *sound lock* (Figure 3.17). This construction technique dramatically reduces leakage because the air trapped between the two solid barriers offers up high TL values.

FIGURE 3.17
Example of a sound lock door system design.

ISO-ROOMS AND ISO-BOOTHS

Isolation rooms (*iso-rooms*) are acoustically isolated or sealed areas that are built into a music studio or just off of a control room (Figure 3.18). These recording areas can be used to separate louder instruments from softer ones (and vice versa) in order to reduce leakage and to separate instrument types by volume to maintain control over the overall ensemble balance. For example:

- To eliminate leakage when recording scratch vocals (a guide vocal track that's laid down as a session reference), a vocalist might be placed in a small room while the rhythm ensemble is placed in the larger studio area.
- A piano or other instrument could be isolated from the larger area that's housing a full string ensemble.
- A B3 organ could be blaring away in an iso-room while backing vocals are being laid down in the main room. … The possibilities are endless.

An iso-room can be designed to have any number of acoustical properties. By having multiple rooms and/or iso-room designs in a studio, several acoustical environments can be offered that range from being more reflective (live) to absorptive (dead) … or a specific room can be designed to better fit the acoustical needs of a particular instrument (e.g., drums, piano or vocals). These rooms can be designed as totally separate areas that can be accessed from the main studio or control room, or they might be directly tied to the main studio by way of sliding walls or glass sliding doors. In short, their form and function can be put to use to fit the needs and personality of the session.

Isolation booths (*iso-booths*) provide the same type of isolation as an iso-room, but are often much smaller (Figure 3.19). Often called *vocal booths*, these mini-studios are perfect for isolating vocals and single instruments from the larger

FIGURE 3.18
Iso-room design (located at right) at Studio Records, LLC, Ft. Worth, TX. (Courtesy of Russ Berger Design Group, Inc., www.rbdg.com)

FIGURE 3.19
Example of an iso-booth in action.

studio. In fact, rooms that have been designed and built for the express purpose of mixing down a recording will often only have an iso-booth … and no other recording room. Using this space-saving option, vocals or single instruments can be easily overdubbed on site, and should more space be needed a larger studio can be booked to fit the bill.

NOISE ISOLATION WITHIN THE CONTROL ROOM

Isolation between rooms and the great outdoors isn't the only noise-related issue in the modern-day recording or project studio. The proliferation of computers, multitrack tape machines and cooling systems has created issues that present their own Grinch-like types of noise, Noise, NOISE, NOISE!!! This usually manifests itself in the form of system fan noise, transport tape noise and computer-related sounds from CPUs, case fans, hard drives and the like.

When it comes to isolating tape transport and system fan sounds, should budget and size constraints permit, it is often wise to build an iso-room or iso-closet that's been specifically designed and ventilated for containing such equipment. An equipment room that has easy-access doors that provide for current/future wiring needs can add a degree of peace-'n'-quiet and an overall professionalism that will make both you and your clients happy.

Within a smaller studio or project studio space, such a room isn't always possible; however, with care and forethought the whizzes and whirrs of the digital era can be turned into a nonissue that you'll be proud of.

Here are a few examples of the most common problems and their solutions:

- Place the computer(s) in an isolated case, alcove or room (care needs to be taken to monitor the CPU/case temperatures so as not to harm your system).
- Connect the studio computers via a high-speed network to a remote server location.
- Replace fans with quieter ones. By doing some careful Web searching or by talking to your favorite computer salesperson, it's often possible to install CPU and case fans that are quieter than most off-the-shelf models.

ACOUSTIC PARTITIONS

Movable acoustic *partitions* (also known as *flats* or *gobos*) are commonly used in studios to provide on-the-spot barriers to sound leakage. By partitioning a musician and/or instrument on one or more sides and then placing the mic inside the temporary enclosure, isolation can be greatly improved in a flexible way that can be easily changed as new situations arise. Acoustic partitions are currently available on the commercial market in various design styles and types for use in a wide range of studio applications (Figure 3.20). For those on a budget, or who have particular isolation needs, it's relatively simple to get out the workshop tools and make your own flats that are based around wood frames, fiberglass or other acoustically absorptive materials with your favorite colored fabric coverings—and ingenious craftsmanship (Figure 3.21).

If you can't get a flat when you need one, you can often improvise using common studio and household items. For example, a simple partition can be easily made on the spot by grabbing a mic/boom stand combination and retracting the boom halfway at a 90° angle to make a T-shape. Simply drape a blanket or heavy coat over the T-bar and voilà—you've built a quick-'n'-dirty dividing flat.

When using a partition, it's important to be aware that musicians' need to be able to see each other, the conductor and the producer. Musicality and human connectivity almost always take precedence over technical issues.

Symmetry in control room design

While many professional studios are built from the ground up using standard acoustic and architectural guidelines, most budget-minded production and project studios are often limited by their own unique sets of building, space and acoustic constraints. Even though the design of a budget, project or bedroom control room might not be acoustically perfect, if speakers are to be used in the monitoring environment, certain ground rules of acoustical physics must be followed in order to create a proper listening environment.

One of the most important acoustic design rules in a monitoring environment is the need for symmetrical reflections on all axes within the design of a control room or single-room project studio. In short, the center and acoustic *imaging* (ability to discriminate placement and balance in a stereo or surround field)

(a)

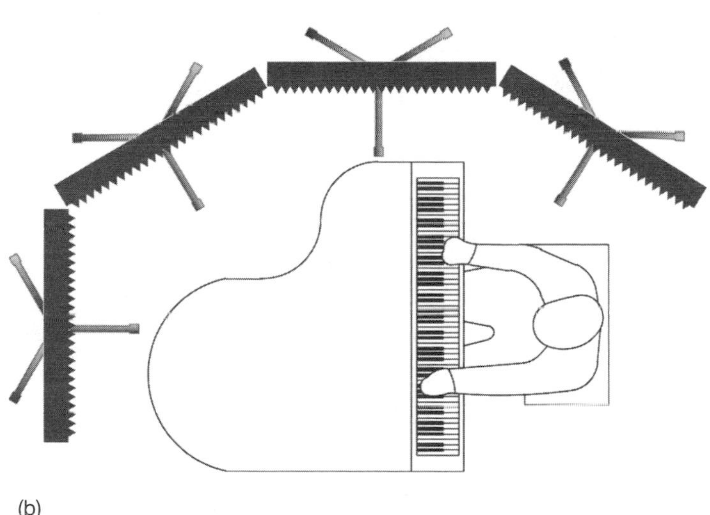

(b)

FIGURE 3.20
Acoustic partition flat examples: (a) S5–2L "Sorber" baffle system (courtesy of ClearSonic Mfg., Inc., www.clearsonic.com); (b) piano panel setup (courtesy of Auralex Acoustics, www.auralex.com)

is best when the listener, speakers, walls and other acoustical boundaries are symmetrically centered about the listener's position (often in an equilateral triangle). In a rectangular room, the best low-end response can be obtained by orienting the console and loudspeakers into the room's long dimension (Figure 3.22a). Should space or other room considerations come into play, centering the listener/monitoring position at a 45° angle within a symmetrical corner

(a)

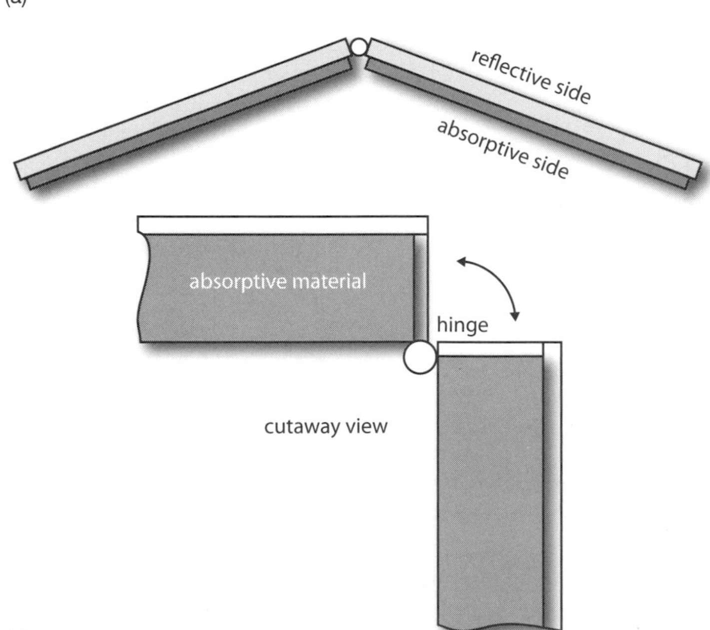

(b)

FIGURE 3.21
Examples of a homemade flat: (a) the "blanket and a boom" trick; (b) homemade flat design.

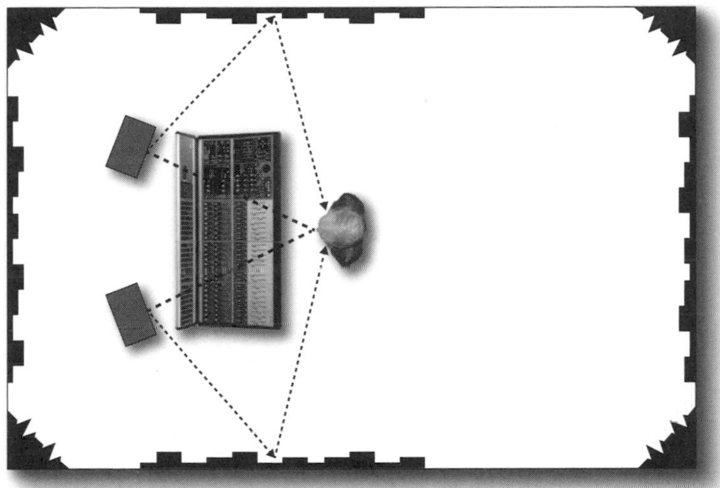

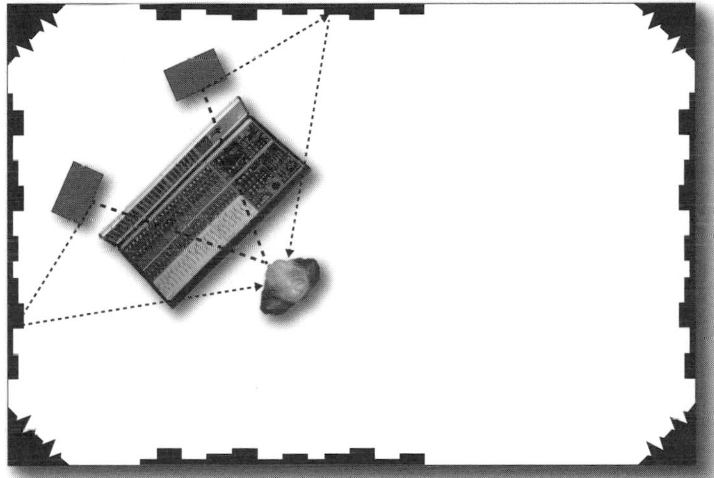

FIGURE 3.22
Various acceptable symmetries in a monitoring environment: (a) Acoustic reflections must be symmetrical about the listener's position. In addition, orienting a control room along the long dimension can extend the room's low-end response. (b) Placing the listening environment symmetrically in a corner is another example of how the left/right imagery can be improved over an off-center placement.

(Figure 3.22b) is another example of how the left/right imagery can be reasonably maintained.

Should any primary boundaries of a control room (especially wall or ceiling boundaries near the mixing position) be asymmetrical from side to side, sounds heard by one ear will receive one combination of direct and reflected sounds, while the other ear will hear a different acoustic balance (Figure 3.23). This condition can drastically alter the sound's center image characteristics, so that when a sound is actually panned between the two monitor speakers the sound will appear to be centered; however, when the sound is heard in another studio or standard listening environment the imaging may be off center. To avoid this

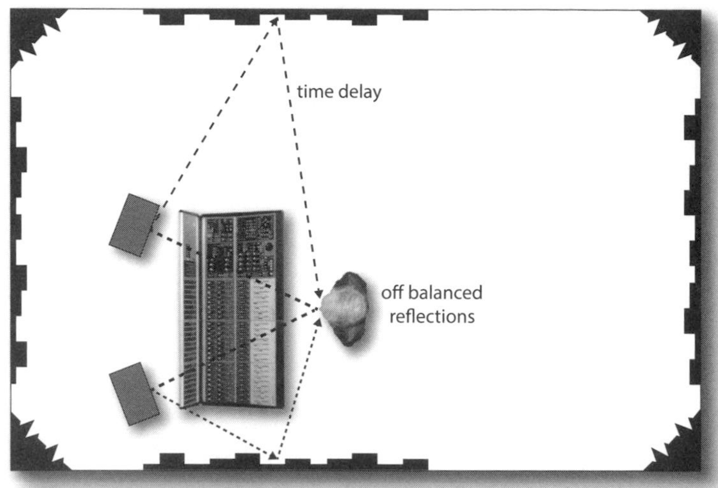

FIGURE 3.23
Placing the monitoring environment off-center and in a corner will affect the audible center image, and placing one speaker in a 90° corner can cause an off-center bass buildup and adversely affect the mix's imagery. Shifting the listener/monitoring position into the center will greatly improve the left/right imagery.

problem, care should be taken to ensure that both the side and ceiling boundaries are largely symmetrical with respect to each other and that all of the speaker level balances are properly set.

While we're on the subject of the relationship between the room's acoustic layout and speaker placement, it's always wise to place nearfield and all other speaker enclosures at points that are equidistant to the listener in the stereo and surround field. Whenever possible, speaker enclosures should be placed 1 to 2 feet away from the nearest wall and/or corner, which helps to avoid bass buildups that acoustically occur at boundary and corner locations. In addition to strategic speaker placement, homemade or commercially available isolation pads can be used to reduce resonances that often occur whenever enclosures are placed directly onto a table or flat surface.

Frequency balance

Another important factor in room design is the need for maintaining the original *frequency balance* of an acoustic signal. In other words, the room should exhibit a relatively flat frequency response over the entire audio range without adding its own particular sound coloration. The most common way to control the tonal character of a room is to use materials and design techniques that govern the acoustical reflection and absorption factors.

REFLECTIONS

One of the most important characteristics of sound as it travels through air is its ability to reflect off a boundary's surface at an angle that's equal to (and opposite of) its original angle of incidence (Figure 3.24). Just as light bounces off a mirrored surface or multiple reflections can appear within a mirrored room, sound reflects throughout room surfaces in ways that are often amazingly complex. Through careful control of these reflections, a room can be altered to improve its frequency response and sonic character.

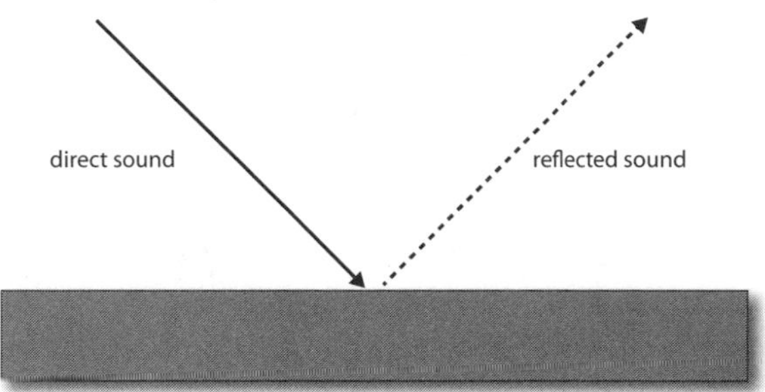

FIGURE 3.24
Sound reflects off a surface at an angle equal (and opposite) to its original angle of incidence, much as light will reflect off a mirror.

In Chapter 2, we learned that sonic reflections can be controlled in ways that disperse the sound outward in a wide-angled pattern (through the use of a convex surface) or focus them on a specific point (through the use of a concave surface). Other surface shapes, on the other hand, can reflect sound back at various other angles. For example, a 90° corner will reflect sound back in the same direction as its incident source (a fact that accounts for the additive acoustic buildups at various frequencies at or near a wall-to-corner or corner-to-floor intersection).

The all-time winner of the "avoid this at all possible cost" award goes to constructions that include opposing parallel walls in its design. Such conditions give rise to a phenomenon known as *standing waves*. Standing waves (also known as room modes) occur when sound is reflected off of parallel surfaces and travels back on its own path, thereby causing phase differences to interfere with a room's amplitude response (Figure 3.25). Room modes are expressed as integer multiples of the length, width and depth of the room and indicate which multiple is being referred to for a particular reflection.

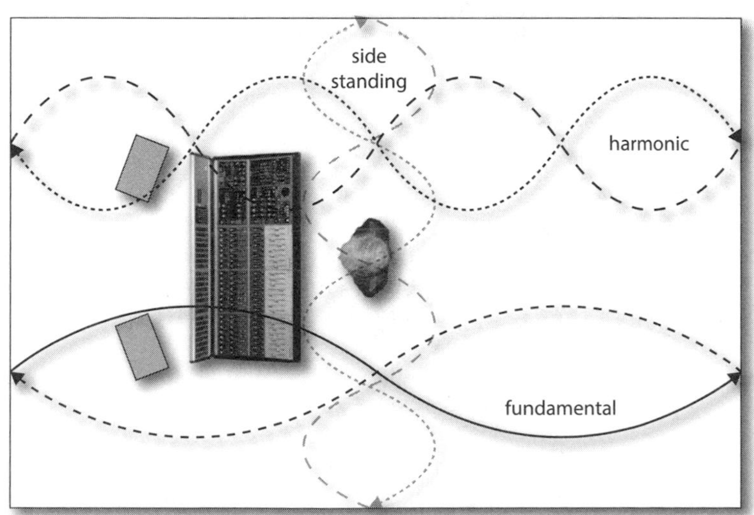

FIGURE 3.25
Standing waves within a room with reflective parallel surfaces can potentially cancel and reinforce frequencies within the audible spectrum, causing changes in its response.

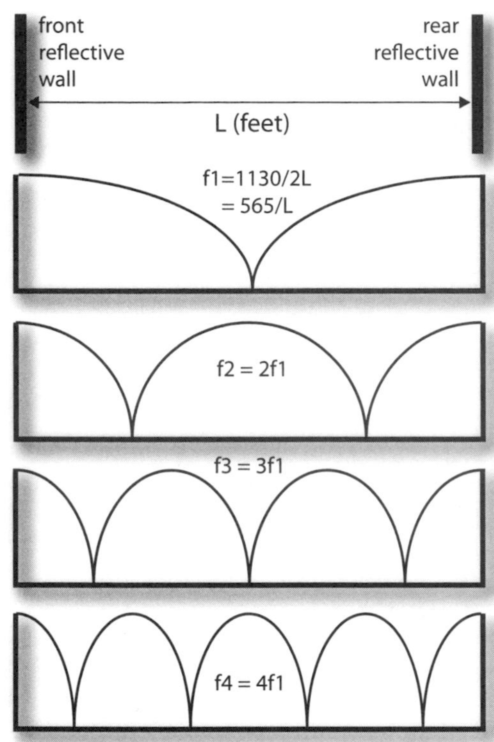

FIGURE 3.26
The reflective, parallel walls create an undue number of standing waves, which occur at various frequency intervals (f_1, f_2, f_3, f_4, and so on).

Walking around a room with moderate to severe mode problems produces the sensation of increasing and/or decreasing volume levels at various frequencies throughout the area. These perceived volume changes are due to amplitude (phase) cancellations and reinforcements of the combined reflected waveforms at the listener's position. The distance between parallel surfaces and the signal's wavelength determines the nodal points that can potentially cause sharp peaks or dips at various points in the response curve (up to or beyond 19 dB) at the affected fundamental frequency (or frequencies) and upper harmonic intervals (Figure 3.26). This condition exists not only for opposing parallel walls but also for all parallel surfaces (such as between the floor and ceiling or between two reflective flats). From this discussion, it's obvious that the most effective way to prevent standing waves is to construct walls, boundaries and ceilings that are nonparallel.

If the room in question is rectangular or if further sound-wave dispersion is desired, diffusers can be attached to the wall and/or ceiling boundaries to help break up standing waves. Diffusers (Figure 3.27) are acoustical boundaries that reflect the sound wave back at various angles that are wider than the original incident angle (thereby breaking up the energy-destructive standing waves). In addition, the use of both nonparallel and diffusion wall construction can reduce extreme, recurring reflections and smooth out the reverberation characteristics of a room by building more complex acoustical pathways.

Flutter echo (also called *slap echo*) is a condition that occurs when parallel boundaries are spaced far enough apart that the listener is able to discern a number of discrete echoes. Flutter echo often produces a "boingy," hollow sound that greatly affects a room's sound character as well as its frequency response. A larger room (which might contain delayed echo paths of 50 msec or more) can have its echoes spaced far enough apart in time that the discrete reflections produce echoes that actually interfere with the intelligibility of the direct sound, often resulting in a jumble of noise. In these cases, the proper application of absorption and acoustic dispersion becomes critical.

When speaking of reflections within a studio control room, one long-held design concept relates to the concept of designing the room such that the rear of the room is largely reflective and diffuse in nature (acoustically "live"), while the front of the room is largely or partially absorptive (acoustically "dead"). This philosophy (Figure 3.28) argues for the fact that the rear of the room should be largely reflective providing for a balanced environment that can help reinforce

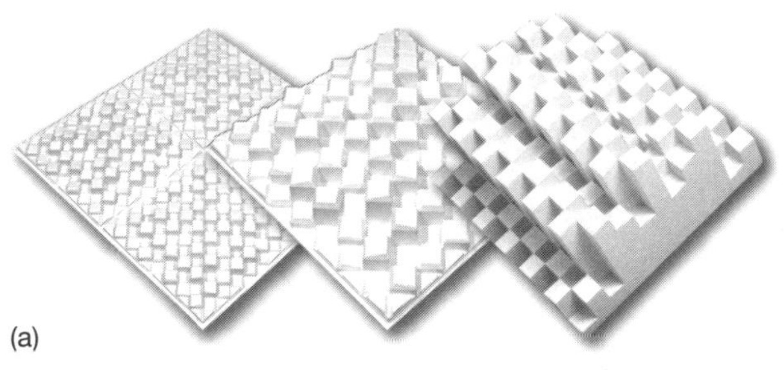

(a)

(b)

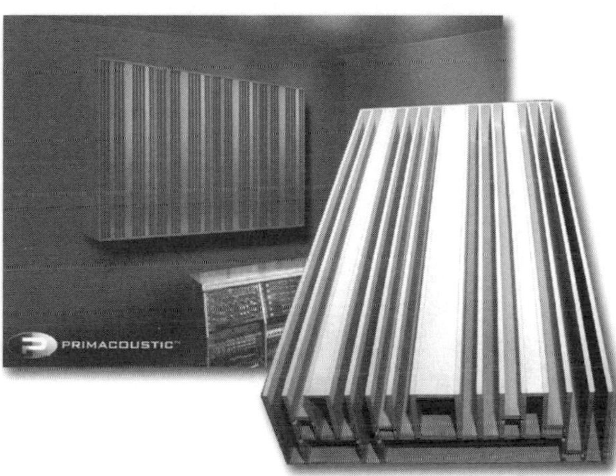

(c)

FIGURE 3.27
Commercial diffuser examples: (a) Art Diffusor sound diffusers Models F, C & E (courtesy of acousticsfirst, www.acousticsfirst.com); (b) SpaceArray sound diffusers (courtesy of pArtScience, www.partscience.com); (c) open-ended view of a Primacoustic™ Razorblade quadratic diffuser (courtesy of Primacoustic Studio Acoustics, www.primacoustic.com).

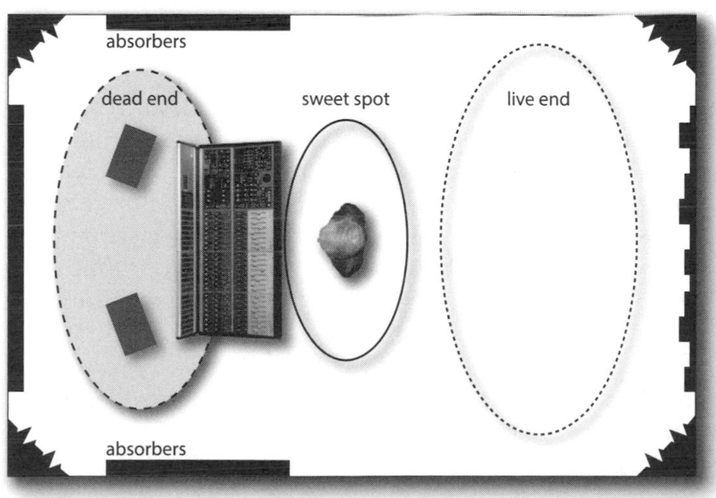

FIGURE 3.28
Control-room layout showing the live end toward the back of the room and the dead end toward the front of the room.

FIGURE 3.29
Placing bookshelves along the rear wall can provide both diffusion and a place for lots of storage.

positive reflections that can add acoustic "life" to the mix experience (Figure 3.29). The front of the room would tend more toward the absorptive side in a way that reduces standing-wave and flutter reflections that would interfere with the overall response of the room.

It's important to realize that no two rooms will be acoustically the same or will necessarily offer the same design challenges. The one constant is that careful planning, solid design and ingenuity are the foundation of any good-sounding room. You should also keep in mind that numerous studio design and commercial acoustical product firms are available that offer assistance for both large and small projects. Getting professional advice is a good thing.

ABSORPTION

Another factor that often has a marked effect on an acoustic space involves the use of surface materials and designs that can absorb unwanted sounds (either across the entire audible band or at specific frequencies). The *absorption* of acoustic energy is, effectively, the inverse of reflection (Figure 3.30). Whenever sound strikes a material, the amount of acoustic energy that's absorbed relative to the amount that's reflected can be expressed as a simple ratio known as the material's *absorption coefficient*. For a given material, this can be represented as:

$$A = I_a/I_r$$

where I_a is the sound level (in dB) that is absorbed by the surface (often dissipated in the form of physical heat), and I_r is the sound level (in dB) that is reflected back from the surface.

The factor $(1 - a)$ is a value that represents the amount of reflected sound. This makes the coefficient a decimal percentage value between 0 and 1. If we say that a surface material has an absorption coefficient of 0.25, we're actually saying that the material absorbs 25% of the original acoustic energy and reflects 75% of the total sound energy at that frequency. A sample listing of these coefficients is provided in Table 3.2.

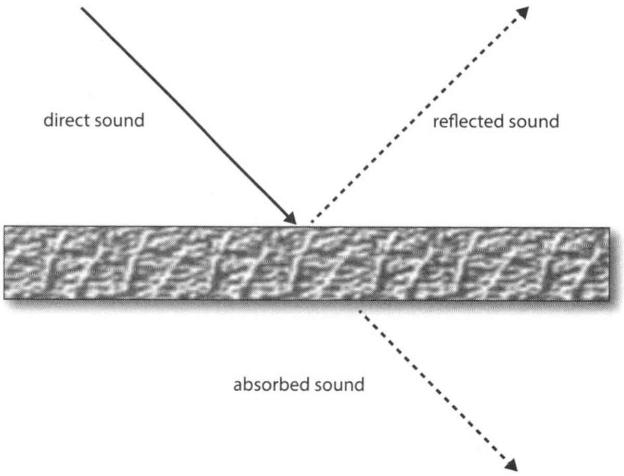

FIGURE 3.30
Absorption occurs when only a portion of the incident acoustic energy is reflected back from a material's surface.

To determine the total amount of absorption that's obtained by the sum of all the absorbers within a total volume area, it's necessary to calculate the average absorption coefficient for all of the surfaces together. The *average absorption coefficient* (A_{ave}) of a room or area can be expressed as:

$$A_{ave} = s_1a_1 + s_2a_2 + \dots s_na_n/S$$

where $s_1, s_2, \dots, s_n$ are the individual surface areas; $a_1, a_2, \dots, a_n$ are the individual absorption coefficients of the individual surface areas; and S is the total square surface area.

On the subject of absorption, one common misconception is that the use of large amounts of sound-deadening materials will reduce room reflections and therefore make a room sound "good." In fact, the overuse of absorption will often have the effect of reducing high frequencies, creating a skewed room response that is dull and bass-heavy, as well as reducing constructive room reflections that are important to a properly designed room. In fact, with regard to the balance between reflection, diffusion and absorption, many designers agree that a balance of 25% absorption and 25% diffuse reflections is a good ratio that can help preserve the "life" of a room, while reducing unwanted buildups.

HIGH-FREQUENCY ABSORPTION

The absorption of high frequencies is accomplished through the use of dense porous materials, such as fiberglass, dense fabric and carpeting. These materials generally exhibit high absorption values at higher frequencies, which can be used to control room reflections in a frequency-dependent manner. Specially designed foam and acoustical treatments are also commercially available that can be attached easily to recording studio, production room or control-room walls as a means of taming multiple room reflections and/or dampening high-frequency reflections (Figure 3.31).

In addition to buying commercial absorbers, it's very possible to put your handy shop tools to work by building your own absorber panels (of any shape, depth

Table 3.2 Absorption coefficients for various materials.

	Coefficients (Hz)					
Material	125	250	500	1000	2000	4000
Brick, unglazed	0.03	0.03	0.03	0.04	0.05	0.07
Carpet (heavy, on concrete)	0.02	0.06	0.14	0.37	0.60	0.65
Carpet (with latex backing, on 40-oz hair-felt or foam rubber)	0.03	0.04	0.11	0.17	0.24	0.35
Concrete or terrazzo	0.01	0.01	0.015	0.02	0.02	0.02
Wood	0.15	0.11	0.10	0.07	0.06	0.07
Glass, large heavy plate	0.18	0.06	0.04	0.03	0.02	0.02
Glass, ordinary window	0.35	0.25	0.18	0.12	0.07	0.04
Gypsum board nailed to 2 × 4 studs on 16-inch centers	0.013	0.015	0.02	0.03	0.04	0.05
Plywood (3/8 inch)	0.28	0.22	0.17	0.09	0.10	0.11
Air (sabins/1000 ft^3)	—	—	—	—	2.3	7.2
Audience seated in upholstered seats	0.08	0.27	0.39	0.34	0.48	0.63
Concrete block, coarse	0.36	0.44	0.31	0.29	0.39	0.25
Light velour (10 oz/yd^2 in contact with wall)	0.29	0.10	0.05	0.04	0.07	0.09
Plaster, gypsum, or lime (smooth finish on tile or brick)	0.44	0.54	0.60	0.62	0.58	0.50
Wooden pews	0.57	0.61	0.75	0.86	0.91	0.86
Chairs, metal or wooden, seats unoccupied	0.15	0.19	0.22	0.39	0.38	0.30

Note: These coefficients were obtained by measurements in the laboratories of the Acoustical Materials Association. Coefficients for other materials may be obtained from Bulletin XXII of the association.

and style). One straightforward way to do this is to build a wooden box side-frame and then stretch a light fabric (of your favorite color or pattern) around the frame, then place commercially-available rockwool bats (or resized sections) inside. When done right, these absorbers (Figure 3.32) can look very pro and fit your specific needs at a fraction of their commercial equivalents … sometimes with better results.

LOW-FREQUENCY ABSORPTION

As shown in Table 3.2, materials that are absorptive in the high frequencies often provide little resistance to the low-frequency end of the spectrum (and vice

FIGURE 3.31
Various commercial absorption and diffusion wall treatments (courtesy of Auralex Acoustics, www.auralex.com)

FIGURE 3.32
Homemade absorber panel showing fabric that has been stretched over a wooden frame. Once the rockwool is placed inside, the frame (which can be of any size or form) can be hung on the wall, lowered from the ceiling or placed in a corner.

versa). This occurs because low frequencies are best damped by pliable materials, meaning that low-frequency energy is absorbed by the material's ability to bend and flex with the incident waveform (Figure 3.33). Rooms that haven't been built to the shape and dimensions to properly handle the low end will need to be controlled by using bass traps that are tuned to reduce the room's resonance frequencies.

Another absorber type can be used to reduce low-frequency buildup at specific frequencies (and their multiples) within a room. This type of attenuation device (known as a *bass trap*) is available in a number of design types:

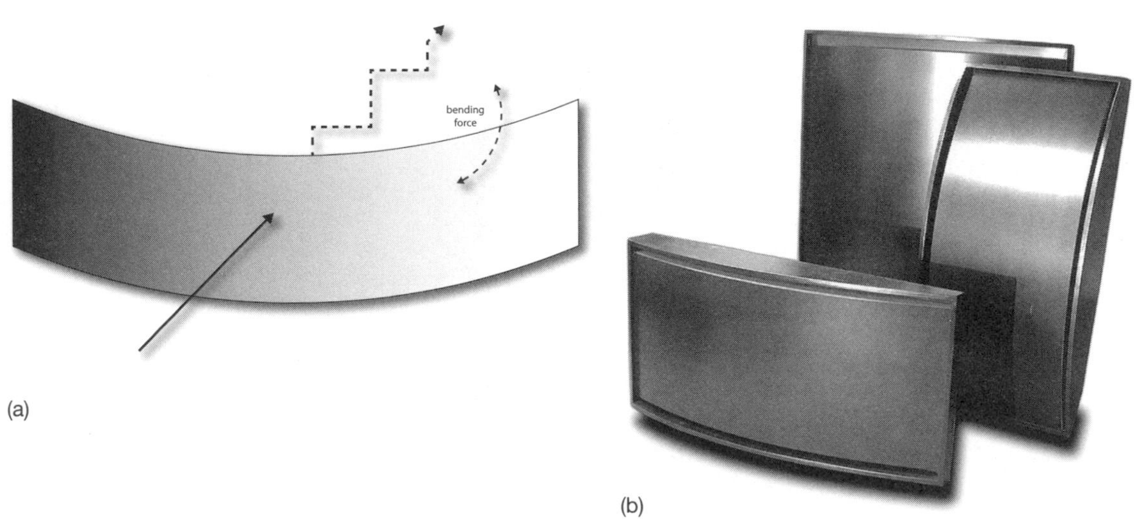

FIGURE 3.33
Low-frequency absorption. (a) A carefully designed surface that can be "bent" by oncoming soundwaves can be used to absorb low frequencies. (b) Primacoustic™ Polyfuser, a combination diffuser and bass trap (courtesy of Primacoustic Studio Acoustics, www.primacoustic.com).

- Quarter-wavelength trap
- Pressure-zone trap
- Functional trap.

The quarter-wavelength trap The quarter-wavelength bass trap (Figure 3.34) is an enclosure with a depth that's one-fourth the wavelength of the offending frequency's fundamental frequency and is often built into the rear facing wall, ceiling or floor structure and covered by a metal grating to allow foot traffic. The physics behind the absorption of a calculated frequency (and many of the harmonics that fall above it) rests in the fact that the pressure component of a sound wave will be at its maximum at the rear boundary of the trap … when the wave's velocity component is at a minimum. At the mouth of the bass trap (which is at a one-fourth wavelength distance from this rear boundary), the overall acoustic pressure will be at its lowest, while the velocity component (molecular movement) will be at its highest potential. Because the wave's motion (force) is greatest at the trap's opening, much of the signal can be absorbed by placing an absorptive material at that opening point. A low-density fiberglass lining can also be placed inside the trap to increase absorption (especially at harmonic intervals of the calculated fundamental).

Pressure-zone trap The pressure-zone bass trap absorber (Figure 3.35) works on the principle that sound pressure is doubled at large boundary points that are at 90° angles (such as walls and ceilings). By placing highly absorptive material at a boundary point (or points, in the case of a corner/ceiling intersection), the built-up pressure can be partially absorbed.

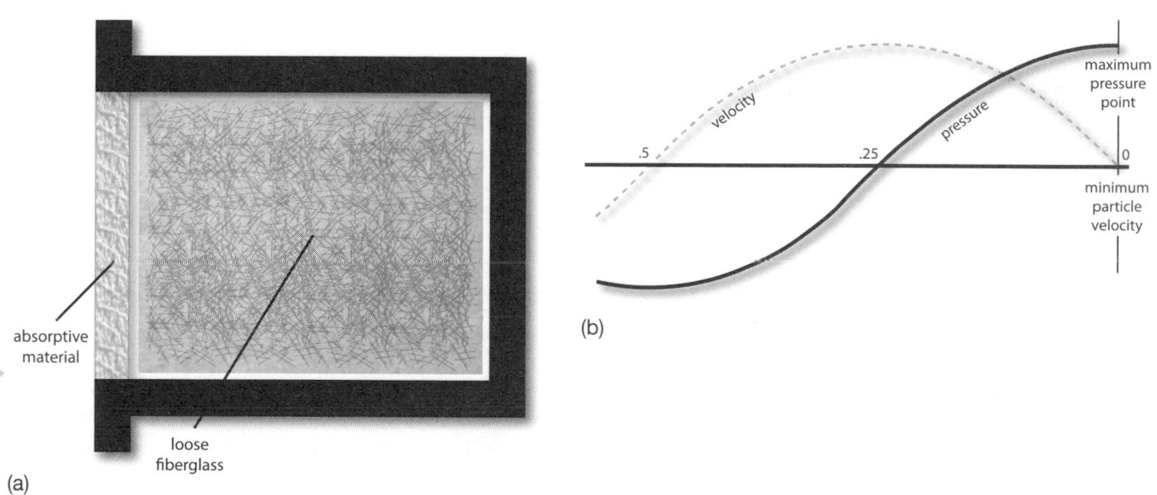

FIGURE 3.34
A quarter-wavelength bass trap: (a) physical concept design; (b) sound is largely absorbed as heat, since the particle velocity (motion) is greatest at the trap's quarter-wavelength opening.

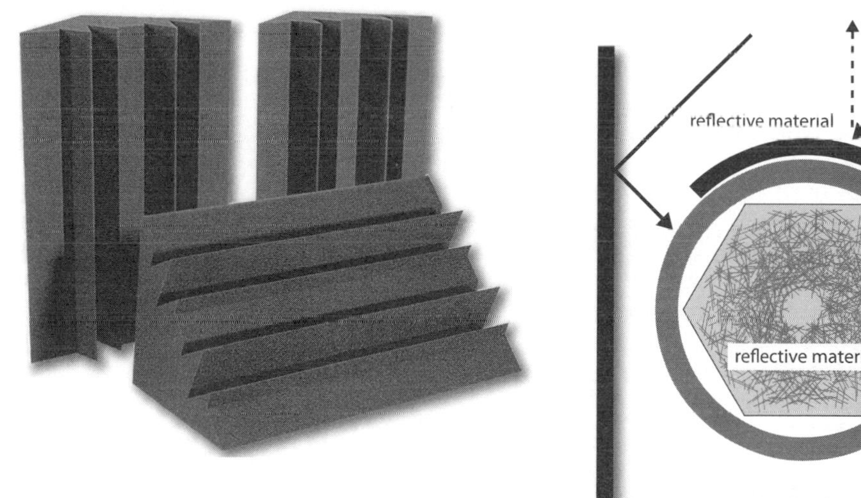

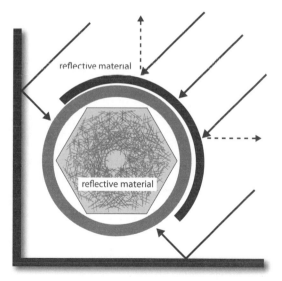

FIGURE 3.35
LENRD™ bass traps. (Courtesy of Auralex Acoustics, www.auralex.com).

FIGURE 3.36
A functional bass trap that has been placed in a corner to prevent bass buildup.

Functional trap Originally created in the 1950s by Harry F. Olson (former director of RCA Labs), the functional bass trap (Figure 3.36) uses a material generally formed into a tube or half-tube structure that is rigidly supported so as to reduce structural vibrations. By placing these devices into corners, room boundaries or in a freestanding spot, a large portion of the undesired

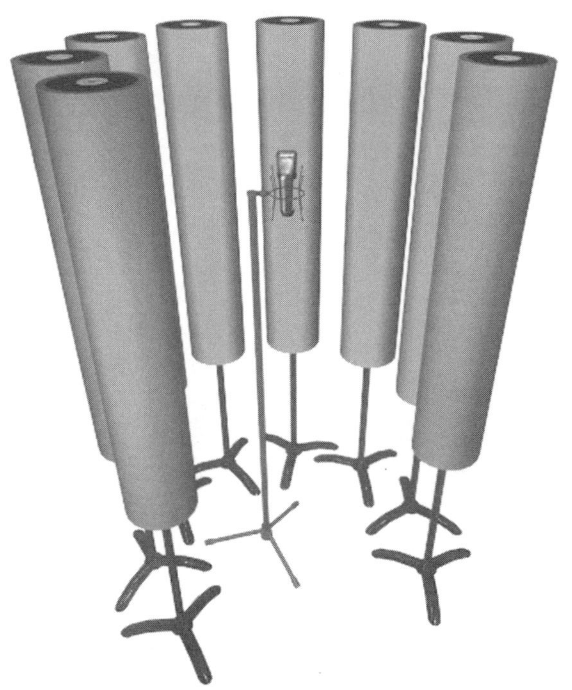

FIGURE 3.37
Quick Sound Field.
(Courtesy of Acoustic Sciences Corporation, www.tubetrap.com)

bass buildup frequencies can be absorbed. By placing a reflective surface over the portion of the trap that faces into the room, frequencies above 400 Hz can be dispersed back into the room or focal point. Figure 3.37 shows how these traps can be used in the studio to break up reflections and reduce bass buildup.

ROOM REFLECTIONS AND ACOUSTIC REVERBERATION

Another criterion for studio design is the need for a desirable room ambience and intelligibility, which is often contradictory to the need for good acoustic separation between instruments and their pickup. Each of these factors is governed by the careful control and tuning of the reverberation constants within the studio over the frequency spectrum.

Reverberation (*reverb*) is the persistence of a signal (in the form of reflected waves within an acoustic space) that continues after the original sound has ceased. The effect of these closely spaced and random multiple echoes gives us perceptible cues as to the size, density and nature of an acoustic space. Reverb also adds to the perceived warmth and spatial depth of recorded sound and plays an extremely important role in the perceived enhancement of music.

As was stated in the latter part of Chapter 2, the reverberated signal itself can be broken down into three components:

- Direct sound
- Early reflection
- Reverb

The direct signal is made up of the original, incident sound that travels from the source to the listener. Early reflections consist of the first few reflections that are projected to the listener off of major boundaries within an acoustic space; these reflections generally give the listener subconscious cues as to the size of the room. (It should be noted that strong reflections off of large, nearby surfaces can potentially have detrimental cancellation effects that can degrade a room's sound and frequency response at the listening position.) The last set of signal reflections makes up the actual reverberation characteristic. These signals are composed of random reflections that travel from boundary to boundary in a room and are so closely spaced that the brain can't discern the individual reflections. When combined, they are perceived as a single decaying signal.

Technically, reverb is considered to be the time that's required for a sound to die away to a millionth of its original intensity (resulting in a decrease over time of 60 dB), as shown by the following formula:

$$RT_{60} = V \text{ X } 0.049/AS$$

where RT is the reverberation time (in sec), *V* is the volume of the enclosure (in ft^3), *A* is the average absorption coefficient of the enclosure, and *S* is the total surface area (in ft^2). As you can see from this equation, reverberation time is directly proportional to two major factors: the volume of the room and the absorption coefficients of the studio surfaces. A large environment with a relatively low absorption coefficient (such as a large cathedral) will have a relatively long RT_{60} decay time, whereas a small studio (which might incorporate a heavy amount of absorption) will have a very short RT_{60}.

The style of music and the room application will often determine the optimum RT_{60} for an acoustical environment. Reverb times can range from 0.25 sec in a smaller absorptive recording studio environment to 1.6 sec or more in a larger music or scoring studio. In certain designs, the RT_{60} of a room can be altered to fit the desired application by using movable panels or louvers or by placing carpets in a room. Other designs might separate a studio into sections that exhibit different reverb constants. One side of the studio (or separate iso-room) might be relatively non-reflective or dead, whereas another section or room could be much more acoustically live. The more reflective, live section is often used to bring certain instruments that rely heavily on room reflections and reverb, such as strings or an acoustic guitar, to "life." The recording of any number of instruments (including drums and percussion) can also greatly benefit from a well-designed acoustically live environment.

Isolation between different instruments and their pickups is extremely important in the studio environment. If leakage isn't controlled, the room's effectiveness becomes severely limited over a range of applications. The studio designs of the 1960s and 1970s brought about the rise of the "sound sucker" era in studio design. During this time, the absorption coefficient of many rooms was raised almost to an anechoic (no reverb) condition. With the advent of the music styles of the 1980s and a return to the respectability of live studio acoustics, modern studio and control-room designs have begun to increase in size and "liveness" (with a corresponding increase in the studio's RT_{60}). This has reintroduced the buying public to the thick, live-sounding music production of earlier decades, when studios were larger structures that were more attuned to capturing the acoustics of a recorded instrument or ensemble.

Acoustic echo chambers

Another physical studio design that was used extensively in the past (before the invention of artificial effects devices) for re-creating room reverberation is the *acoustic echo chamber*. A traditional echo chamber is an isolated room that has highly reflective surfaces into which speakers and microphones are placed.

The speakers are fed from an effects send, while the mic's reverberant pickup is fed back into the mix via an input strip of effects return. By using one or more directional mics that have been pointed away from the room speakers, the direct sound pickup can be minimized. Movable partitions also can be used to vary the room's decay time. When properly designed, acoustic echo chambers have a very natural sound quality to them. The disadvantage is that they take up space and require isolation from external sounds; thus, size and cost often make it unfeasible to build a new echo chamber, especially those that can match the caliber and quality of high-end digital reverb devices.

An echo chamber doesn't have to be an expensive, built-from-the-ground-up design. Actually, a temporary chamber can be made from a wide range of acoustic spaces to pepper your next project with a bit of "acoustic spice." For example:

- An ambient-sounding chamber can be built by placing a Blumlein (crossed figure-8) pair or spaced stereo pair of mics in the main studio space and feeding a send to the studio playback monitors.
- A speaker/mic setup could be placed in an empty garage (as could a guitar amp/mic, for that matter).
- An empty stairwell often makes an excellent chamber.
- Any vocalist could tell you what'll happen if you place a singer or guitar speaker/mic setup in the shower.

From the above, it's easy to see that ingenuity and experimentation are often the name of the makeshift chamber game. In fact, there's nothing that says that the chamber has to be a real-time effect … for example, you could play back a song's effects track from a laptop DAW into a church's acoustic space and record the effect back to stereo tracks on the DAW. The options and limitless experimentations are totally up to you!

CHAPTER 4

Microphones: Design and Application

THE MICROPHONE: AN INTRODUCTION

A *microphone* (often called a *mic*) is usually the first device in a recording chain. Essentially, a mic is a transducer that changes one form of energy (sound waves) into another corresponding form of energy (electrical signals). The quality of its pickup will often depend on external variables (such as placement, distance, instrument and the acoustic environment), as well as on design variables (such as the microphone's operating type, design characteristics and quality). These interrelated elements tend to work together to affect the overall sound quality.

In order to deal with the wide range of musical, acoustic and situational circumstances that might come your way (not to mention your own personal taste), a large number of mic types, styles and designs can be pulled out of our "sonic toolbox." Because the particular characteristics of a mic might be best suited to a specific range of applications, engineers and producers use their artistic talents to get the best possible sound from an acoustic source by carefully choosing the mic or mics that fit the specific pickup application at hand.

The road to considering microphone choice and placement is best traveled by considering a few simple rules:

1 Rule 1: *There are no rules, only guidelines*. And although guidelines can help you achieve a good pickup, don't hesitate to experiment in order to get a sound that best suits your needs or personal taste.

2 Rule 2: *The overall sound of an audio signal is no better than the weakest link in the signal path.* If a mic or its placement doesn't sound as good as it could, make the changes to improve it *BEFORE* you commit it to disk, tape or whatever. More often than not, the concept of "fixing it later in the mix" will often put you in the unfortunate position of having to correct a situation after the fact, rather than recording the best sound and/or performance during the initial session.

3 Rule 3: *Whenever possible, use the "Good Rule": Good musician + good instrument + good performance + good acoustics + good mike + good placement =*

good sound. This rule refers to the fact that a music track will only be as good as the performer, instrument, mic placement and the mic itself. If any of these elements falls short of its potential, the track will suffer accordingly. However, if all of these links are the best that they can be, the recording will almost always be something that you'll be proud of!

THE "GOOD RULE"
Good musician + good instrument + good performance + good acoustics + good mic + good placement = good sound.

The miking of vocals and instruments (both in the studio and onstage) is definitely an art form. It's often a balancing act to get the most out of the Good Rule. Sometimes you'll have the best of all of the elements; at others, you'll have to work hard to make lemonade out of a situational lemon. The best rule of all is to use common sense and to trust your instincts.

Before delving into placement techniques and facts that deal with the finer points of microphone technology, I'd like to take a basic look at how microphones (and their operational characteristics) work. Why do I put this in the book? Well, from a personal standpoint, having a basic understanding of what happens "under the hood" has helped me to get a better "mental" image of how a particular mic or mic technique will sound in a given situation. Basically it helps to make technical and sonic judgments that can be combined with my own intuition to make the best artistic judgment at the time … I hope it helps you, as well.

MICROPHONE DESIGN

A microphone is a device that converts acoustic energy into corresponding electrical voltages that can be amplified and recorded. In audio production, three transducer mic types are used:

- Dynamic mic
- Ribbon mic
- Condenser mic

The dynamic microphone

In principle, the *dynamic mic* (Figure 4.1) operates by using electromagnetic induction to generate an output signal. The simple *theory of electromagnetic induction* states that whenever an electrically conductive metal cuts across the flux lines of a magnetic field, a current of a specific magnitude and direction will be generated within that metal.

Dynamic mic designs (Figure 4.2) generally consist of a stiff Mylar diaphragm of roughly 0.35-mil thickness. Attached to the diaphragm is a finely wrapped core of wire (called a *voice coil*) that's precisely suspended within a high-level magnetic field. Whenever an acoustic pressure wave hits the diaphragm's face, the attached voice coil is displaced in proportion to the amplitude and frequency

FIGURE 4.1
The Shure 58 dynamic mic. (Courtesy of Shure Incorporated, www.shure.com, Images © 2012, Shure Incorporated – used with permission).

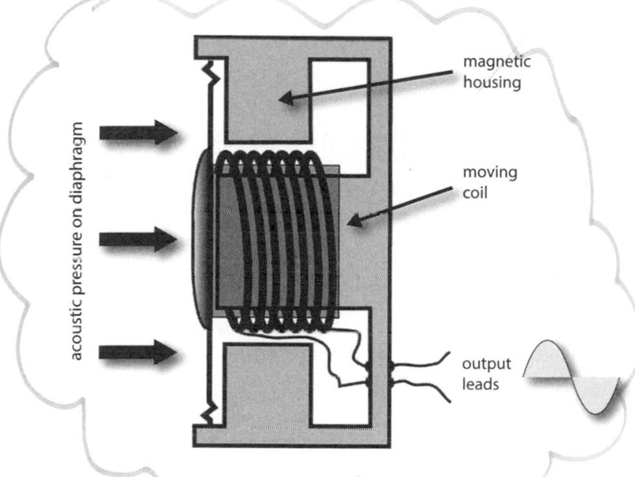

FIGURE 4.2
Inner workings of a dynamic microphone.

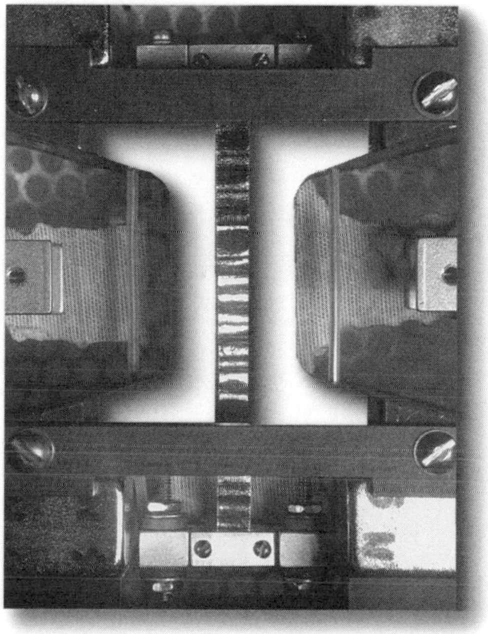

FIGURE 4.3
Cutaway detail of a ribbon microphone. (Courtesy of Audio Engineering Associates, www.ribbonmics.com)

of the wave, causing the coil to cut across the lines of magnetic flux that's supplied by a permanent magnet. In doing so, an analogous electrical signal (of a specific magnitude and direction) is induced into the coil and across the output leads, thus producing an analog audio output signal.

The ribbon microphone

Like the dynamic microphone, the *ribbon mic* also works on the principle of electromagnetic induction. Older ribbon design types, however, use a diaphragm of extremely thin aluminum ribbon (2 microns). Often, this diaphragm is corrugated along its width and is suspended within a strong field of magnetic flux (Figure 4.3). Sound-pressure variations between the front and the back of the diaphragm cause it to move and cut across these flux lines, thereby inducing a current into the ribbon that's proportional to the amplitude and frequency of the acoustic waveform. Because the ribbon generates a small output signal (when compared to the larger output that's generated by the multiple wire turns of a moving coil), its output signal is too low to drive a microphone input stage directly; thus, a step-up transformer must be used to boost the output signal and impedance to an acceptable range.

FIGURE 4.4
The AEA A440 ribbon mic. (Courtesy of Audio Engineering Associates, www.ribbonmics.com)

FIGURE 4.5
Sontronics Delta phantom powered ribbon microphone. (Courtesy of Sontronics and Omnisonics International Ltd, www.sontronics.com)

FIGURE 4.6
The Beyerdynamic M160 ribbon mic. (Courtesy of Beyerdynamic, www.beyerdynamic.com)

Until recently, traditional ribbon technology could be only found on the original, vintage mics (such as the older RCA and Cole ribbon mics); however, with the skyrocketing price of vintage mics and a resurgence in the popularity of the smooth, transient quality of the "ribbon sound," newer mics that follow the traditional design philosophies have begun to spring up on the market (Figures 4.4 and 4.5).

DEVELOPMENTS IN RIBBON TECHNOLOGY

During the past several decades, certain microphone manufacturers have made changes to original ribbon technologies by striving to miniaturize and improve their basic operating characteristics. The popular M160 (Figure 4.6) and M260 ribbon mics from Beyerdynamic use a rare-earth magnet to produce a capsule that's small enough to fit into a 2-inch grill ball (much smaller than a traditional ribbon-style mic). The ribbon (which is corrugated along its length to give it added strength and at each end to give it flexibility) is 3 microns thick, about 0.08 inch wide, 0.85 inch long and weighs only 0.000011 ounce. A plastic throat is fitted above the ribbon, which houses a pop-blast filter. Two additional filters and the grill greatly reduce the ribbon's potential for blast and wind damage, a feature that has made these designs suitable for outdoor and handheld use.

Another relatively recent advance in ribbon technology has been the development of the printed ribbon mic. In principle, the printed ribbon operates in precisely the same manner as the conventional ribbon pickup; however, the rugged diaphragm is made from a polyester film that has a spiral aluminum ribbon printed onto it. Ring magnets are then placed at the diaphragm's front and back, thereby creating a wash of magnetic flux that makes the electromagnetic induction process possible.

Other alterations to traditional ribbon technology make use of phantom power to supply power to an active, internal amplifier, so as to boost the mic's output to that of a dynamic or condenser mic, without the need for a passive transformer (an explanation of phantom power can be found in the next section on condenser mics).

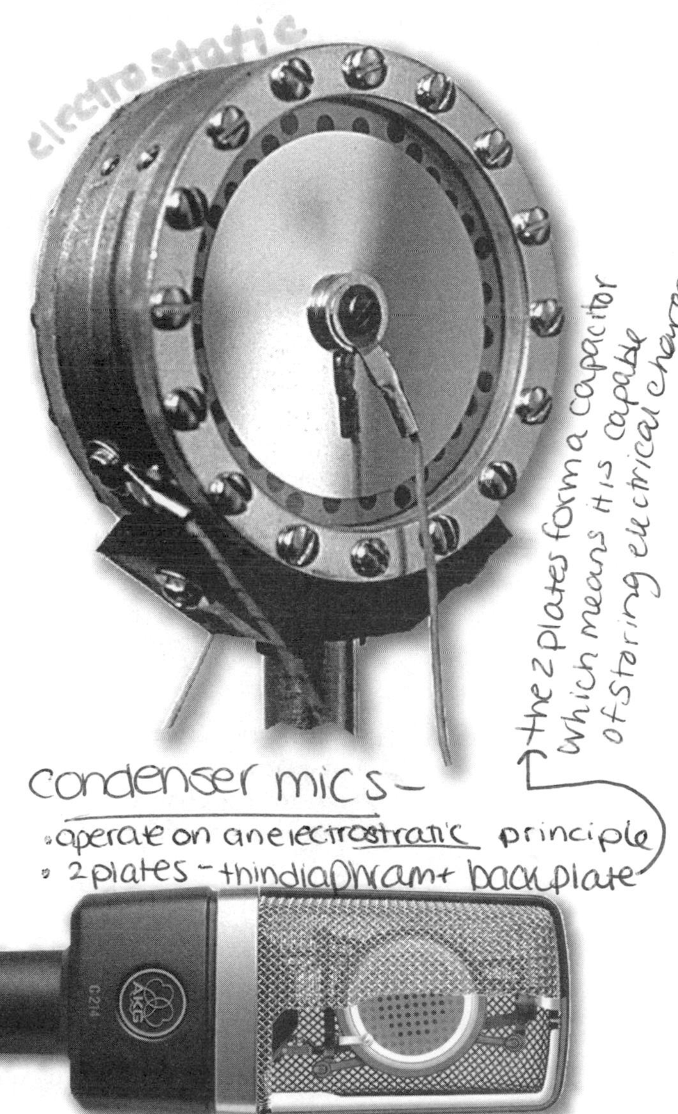

FIGURE 4.7
Exposed example of a condenser diaphragm. (Courtesy of ADK, www.adkmic.com; photograph by K. Bujack)

FIGURE 4.8
Inner detail of an AKG C214 condenser mic. (Courtesy of AKG Acoustics GmbH., www.akg.com)

The condenser microphone

Condenser mics (like those having capsules which are shown in Figures 4.7 and 4.8) operate on an *electrostatic principle* rather than the electromagnetic principle used by a dynamic or ribbon mic. The capsule of a basic condenser mic consists of two plates: one very thin movable diaphragm and one fixed backplate. These two plates form a capacitor (or condenser, as it is still called in the UK and in many parts of the world). A capacitor is an electrical device that's capable of storing an electrical charge. The amount of charge that a capacitor can store is determined by its capacitance value and the voltage that's applied to it, according to the formula:

$$Q = CV$$

where Q is the charge (in coulombs), C is the capacitance (in farads), and V is the voltage (in volts).

At its most basic level, a condenser mic operates when a regulated DC power supply is applied between its diaphragm plates to create a capacitive charge. When sound acts upon the movable diaphragm, the varying distance between the plates will likewise create a change in the device's capacitance (Figure 4.9). According to the above equation, if Q (the power supply charge) is constant and C (the diaphragm's capacitance) changes, then V (voltage across the diaphragm) will change in a proportional and inverse fashion.

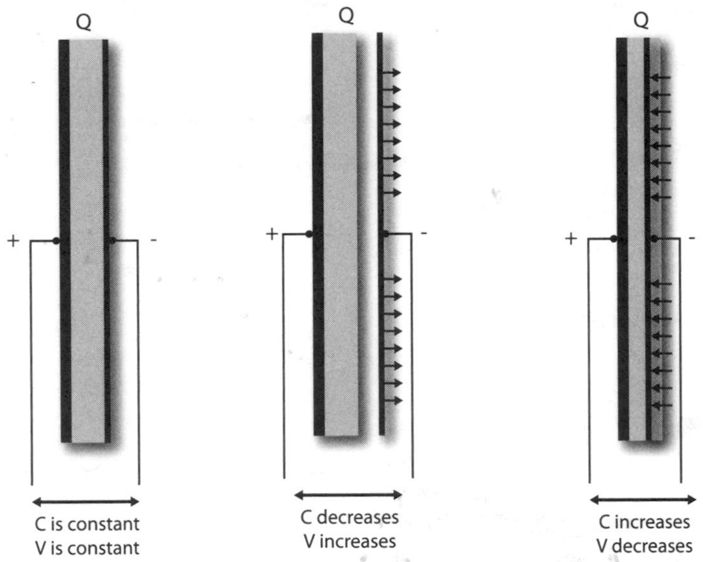

FIGURE 4.9
Output and potential relationships as a result of changing capacitance.

Since the charge (*Q*) is known to be constant and the diaphragm's capacitance (*C*) changes with differences in sound pressure, the voltage (*V*) must change in inverse proportion. Given that the capsule's voltage now changes in proportion to the sound waves that act upon it, *voilà* ... we have a condenser mic that has an audio output signal!

The next trick is to tap into the circuit to capture the changes in output voltage, by placing a high-value resistor across the circuit. Since the voltage across the resistor will change in inverse proportion to the capacitance across the capsule plates, this signal will then become the feed for the mic's output (Figure 4.10).

Since the resulting signal has an extremely high impedance, it must be fed through a preamplifier in order to preserve the mic's frequency response characteristics. As this amp must be placed at a point just following the resistor (often at a distance of 2 inches or less), it is almost always placed within the mic body in order to prevent hum, noise pickup and signal-level losses. In addition to the need for a polarizing voltage, the preamp is another reason why conventional condenser microphones require a supply voltage in order to operate.

PHANTOM POWER

Most modern professional condenser (and some ribbon) mics don't require internal batteries, external battery packs or individual AC power supplies in order to operate. Instead, they are designed to be powered directly from the console through the use of a *phantom power* supply. Phantom power works by supplying a positive DC supply voltage of +48 V through both audio conductors (pins 2 and 3) of a balanced mic line to the condenser capsule and preamp. This voltage is equally distributed through identical value resistors, so that no differential exists between the two leads. The ground side of the circuit is supplied to the capsule and preamp through the balanced cable grounding wire (pin 1).

Since the audio is only affected by potential differences between pins 2 and 3 (and not the ground signal on pin 1), the carefully matched +48V potential at these leads is therefore not electrically "visible" to the input stage of a balanced mic preamp. Instead, only the balanced, alternating audio signal that's being simultaneously carried along the two audio leads will be detected (Figure 4.11).

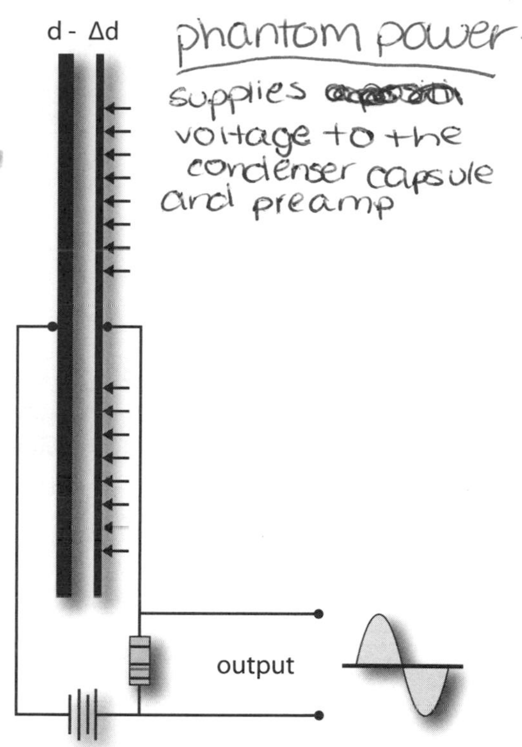

FIGURE 4.10
As a sound wave decreases the condenser spacing by *d* the capacitance will increase, causing the voltage to proportionately fall (and vice versa).

The resistors (R) used for distributing power to the signal leads should be 1/4W resistors with a ±1% tolerance and have the following values for the following supply voltages (because some mics can also be designed to work at voltages lower than 48V): 6.8 kΩ for 48V, 1.2 kΩ for 2 V, and 680Ω for a 12V supply. In addition to precisely matching the supply voltages, these resistors also provide a degree of power isolation between other mic inputs on a console. If a signal lead were accidentally shorted to ground (which could happen if defective cables or unbalanced XLR cables were used), the power supply should still be able to deliver power to other mics in the system. If two or more inputs were accidentally shorted, however, the phantom voltage could drop to levels that would be too low to be usable.

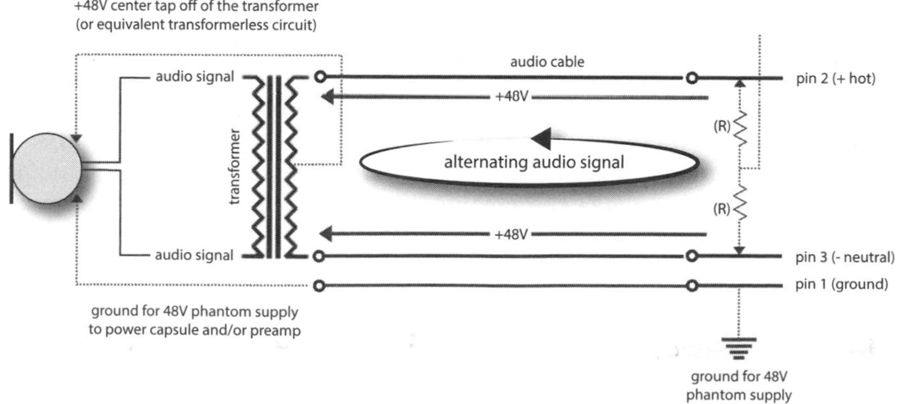

FIGURE 4.11
Schematic drawing of a phantom power system.

Although most modern condensers use some form of a field effect transistor (FET) to reduce the capsule impedance, an increasing number of original era and "revival" models use an internally housed vacuum tube to amplify and change the impedance of the condenser capsule. These mics are generally valued by studios and collectors alike for their "tube" sound, which results from even-harmonic distortion and other sonic characteristics that occur whenever tubes are used.

THE ELECTRET-CONDENSER MICROPHONE

Electret-condenser mics work on the same operating principles as their externally polarized counterparts, with the exception that a static polarizing charge has been permanently set up between the mic's diaphragm and its backplate. Since the charge (*Q*) is built into the capsule, no external source is required to power the diaphragm. However, as with a powered condenser mic, the capsule's output impedance is so high that a preamp will still be required to reduce it to a standard value. As a result, a battery, external powering source or standard phantom supply must be used to power the low-current amp.

Tutorial: Mic Types

1. Go to the Tutorial section of www.modrec.com, click on Mic Types and download the soundfiles (which include examples of each mic operating type).
2. Listen to the tracks. If you have access to an editor or digital audio workstation (DAW), import the files and look at the waveform amplitudes for each example. If you'd like to DIY, then …
3. Pull out several mics from each operating type and plug them in (if you don't have several types, maybe a studio, your school or a friend has a few you can take out for a spin). Try each one on an instrument and/or vocal. Are the differences between operating types more noticeable than between models in the same family?

MICROPHONE CHARACTERISTICS

In order to handle the wide range of applications that are encountered in studio, project and on-location recording, microphones will often differ in their overall sonic, electrical and physical characteristics. The following section highlights many of these characteristics in order to help you choose the best mic for a given application.

Directional response

The *directional response* of a mic refers to its sensitivity (output level) at various angles of incidence with respect to the front (on-axis) of the microphone (Figure

4.12). This angular response can be graphically charted in a way that shows a microphone's sensitivity with respect to direction and frequency over 360°. Such a chart is commonly referred to as the mic's *polar pattern*. Microphone directionality can be classified into two categories:

- Omnidirectional polar response
- Directional polar response.

An *omnidirectional mic* (Figure 4.13) is a pressure-operated device that's responsive to sounds which emanate from all directions. In other words, the diaphragm will react equally to all sound-pressure fluctuations at its surface, regardless of the source's location. Pickups that display *directional* properties are pressure-gradient devices, meaning that the pickup is responsive to relative differences in pressure between the front, back and sides of a diaphragm. For example, a purely pressure-gradient mic will exhibit a *bidirectional* polar pattern (commonly called a *figure-8 pattern*), as shown in Figure 4.14. Many of the older ribbon mics exhibit a bidirectional pattern. Since the ribbon's diaphragm is often exposed to sound waves from both the front and rear axis, it's equally sensitive to sounds that emanate from either direction (Figures 4.15a and 4.15b). Sounds from the rear will produce a signal that's 180° out of phase with an equivalent on-axis signal. Sound waves arriving 90° off-axis produce equal but opposite pressures at both the front and rear of the ribbon (Figure 4.15c), resulting in a cancellation at the diaphragm and no output signal.

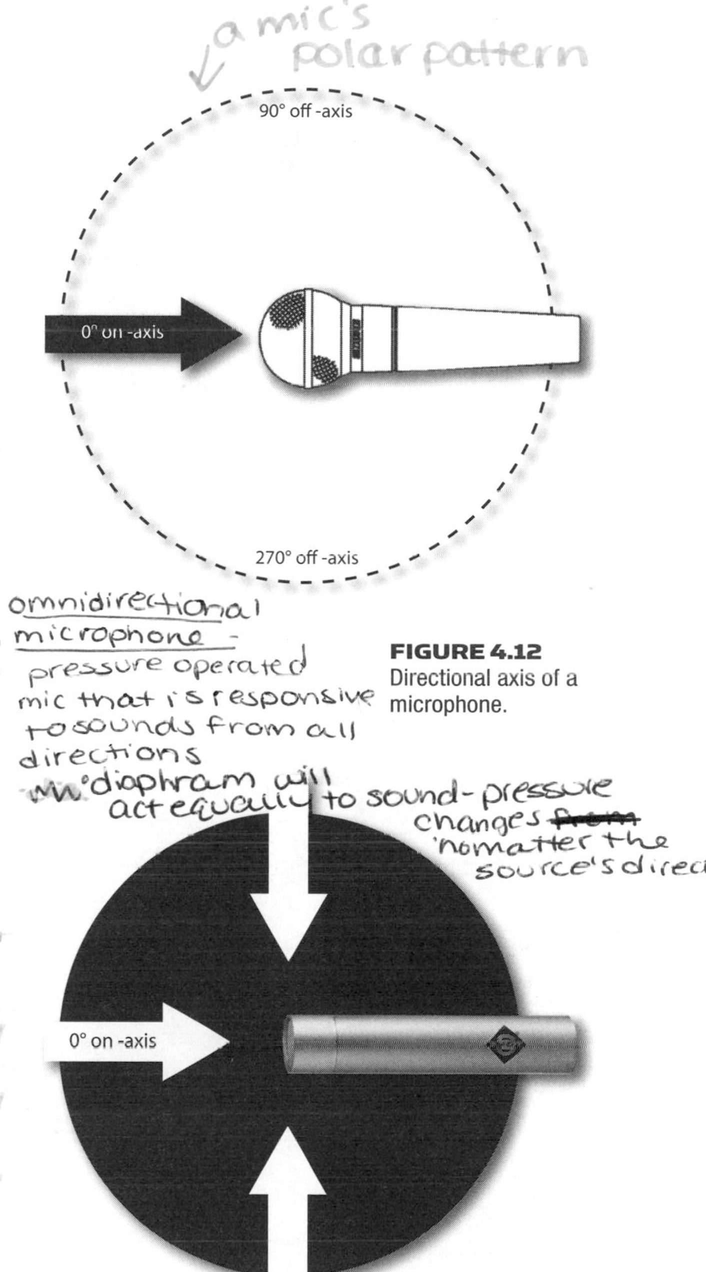

FIGURE 4.12
Directional axis of a microphone.

FIGURE 4.13
Graphic representation of a typical omnidirectional pickup pattern.

Figure 4.16 graphically illustrates how the acoustical combination (as well as electrical and mathematical combination, for that matter) of a bidirectional (pressure-gradient) and omnidirectional (pressure) pickup can be combined to obtain other directional pattern types. Actually, an infinite number of directional patterns can be obtained from this mixture, with the most widely known patterns being the *cardioid*, *supercardioid* and *hypercardioid* polar patterns (Figure 4.17).

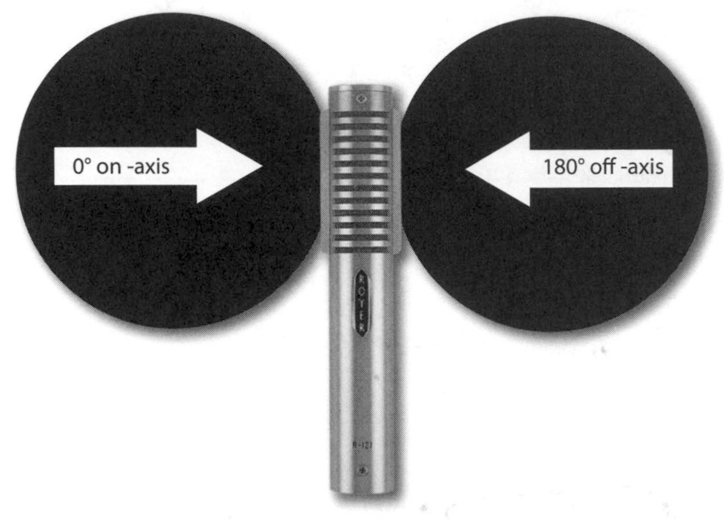

FIGURE 4.14
Graphic representation of a typical bidirectional pickup pattern.

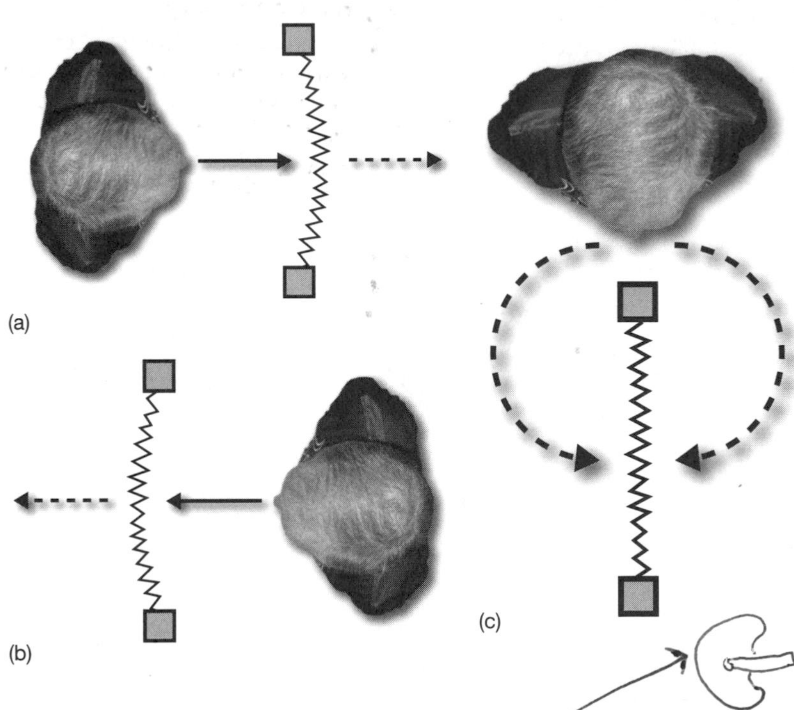

FIGURE 4.15
Sound sources on-axis and 90° off-axis at the ribbon's diaphragm. (a) The ribbon is sensitive to sounds at the front (b) … and at the rear. (c) While sound waves from the sides (90° and 270°) off-axis are canceled.

Often, dynamic mics achieve a cardioid response (named after its heart-shaped polar chart, as shown in Figure 4.18) by incorporating a rear port into their design. This port serves as an acoustic labyrinth that creates an acoustic resistance (delay). In Figure 4.19a, a dynamic pickup having a cardioid polar response is shown receiving an on-axis (0°) sound signal. In effect, the diaphragm receives

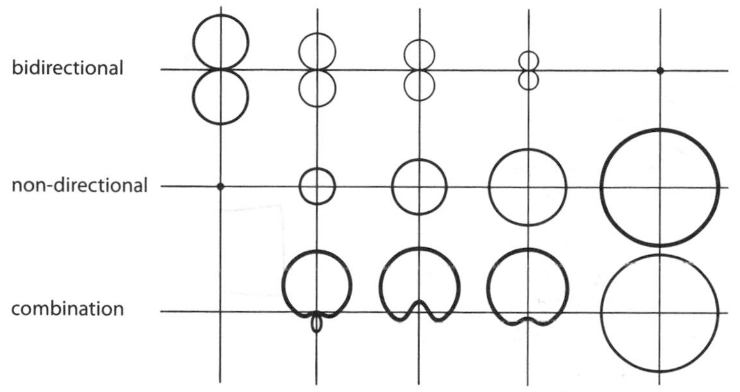

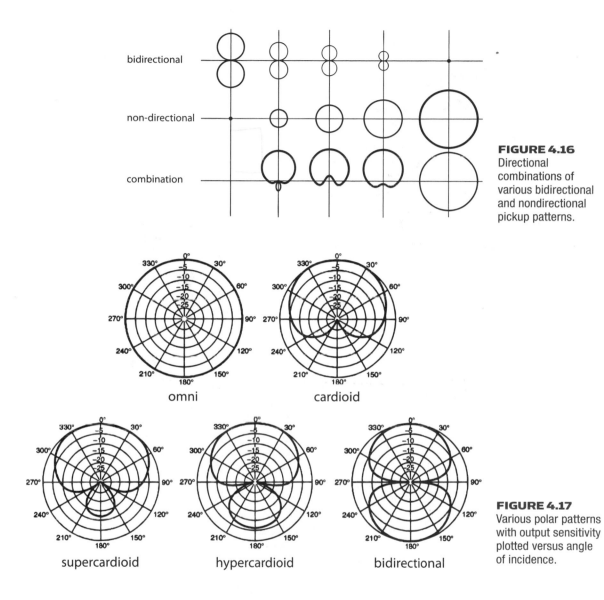

FIGURE 4.16 Directional combinations of various bidirectional and nondirectional pickup patterns.

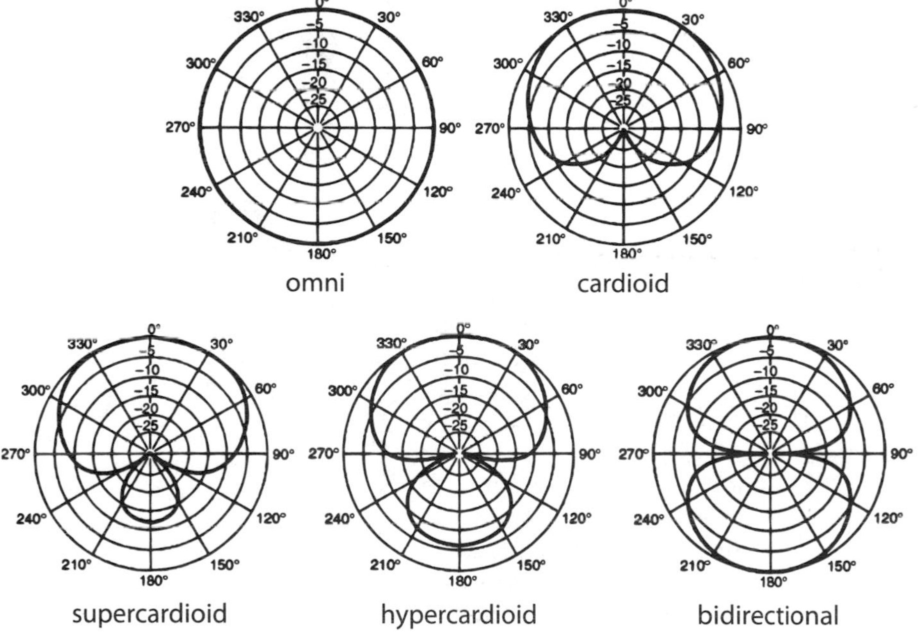

FIGURE 4.17 Various polar patterns with output sensitivity plotted versus angle of incidence.

two signals: the incident signal, which arrives from the front, and an acoustically delayed rear signal. In this instance, the on-axis signal exerts a positive pressure on the diaphragm and begins its travels 90° to a port located on the side of the pickup. At this point, the signal is delayed by another 90° (using an internal, acoustically resistive material or labyrinth). In the time it takes for the delayed signal to reach the rear of the diaphragm (180°), the on-axis signal moves on to the negative portion of its acoustic cycle and then begins to exert a negative pressure on the diaphragm (pulling it outward). Since the delayed rear signal is

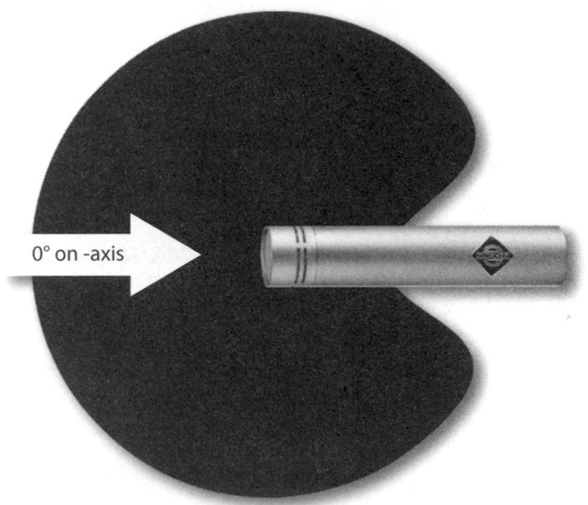

FIGURE 4.18
Graphic representation of a typical cardioid pickup pattern.

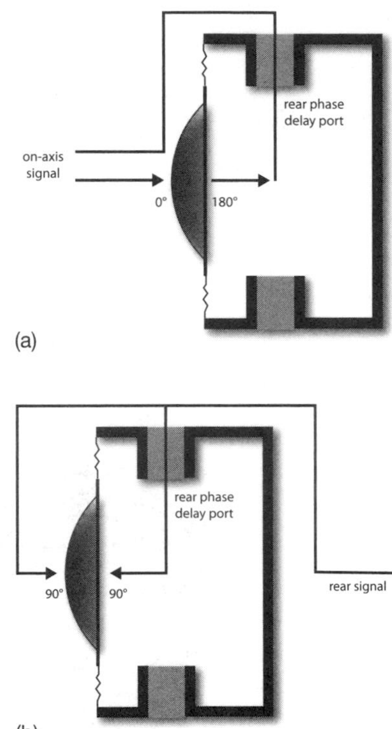

FIGURE 4.19
The directional properties of a cardioid microphone. (a) Signals arriving at the front (on-axis) of the diaphragm will produce a full output level. (b) Signals arriving at the rear of the diaphragm (180°) will cancel each other out, resulting in a greatly reduced output.

180° out of phase at this point in time, it will also begin to push the diaphragm outward, resulting in an output signal.

Conversely, when a sound arrives at the rear of the mic, it begins its trek around to the mic's front. As the sound travels 90° to the side of the pickup, it is again delayed by another 90° before reaching the rear of the diaphragm. During this delay period, the sound continues its journey around to the front of the mic, a delay shift that's also equal to 90°. Since the acoustic pressures at the diaphragm's front and rear sides are equal and opposite, the sound is being simultaneously pushed inward and outward with equal force, resulting in little or no movement … and therefore will have little or no output signal (Figure 4.19b). The attenuation of such an off-axis signal, with respect to an equal on-axis signal, is known as its *front-to-back discrimination* and is rated in decibels.

Certain condenser mics can be electrically switched from one pattern to another by using a second capsule that's mounted on both sides of a central backplate. Configuring these dual-capsule systems electrically in phase will create an omnidirectional pattern, while configuring them out of phase results in a bidirectional pattern. A number of intermediate patterns (such as cardioid and hypercardioid) can be created by electrically varying between these two polar states (in either continuous or stepped degrees), as was seen earlier in Figure 4.16.

Frequency response

The on-axis *frequency–response curve* of a microphone is the measurement of its output over the audible frequency range when driven by a constant, on-axis input signal. This response curve (which is generally plotted in output level [dB] over the 20 to 20,000Hz frequency range) will often yield valuable information and can give clues as to how a microphone will react at specific frequencies. It should be noted that a number of other variables also determine how a mic will sound, some of which have no measurement standards the final determination should always be your own ears.

A mic that's designed to respond equally to all frequencies is said to exhibit a flat frequency response (shown as the top curve in Figure 4.20a). Others can be made to emphasize or de-emphasize the high-, mid- or low-end response of the audio spectrum (shown as the boost in the high-end curve in Figure 4.20b) so as to give it a particular sonic character. The solid frequency–response curves (as shown in both parts a and b) were measured on-axis and exhibit an acceptable response. However, the same mics might exhibit a "peaky" or erratic curve when measured off-axis. These signal colorations could affect their sound when

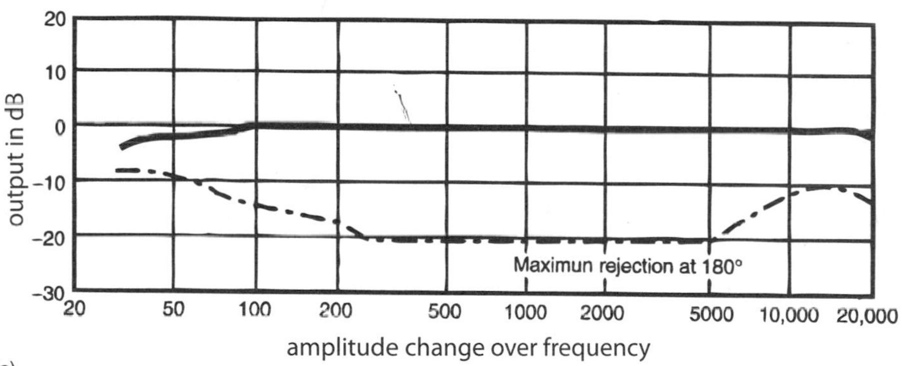

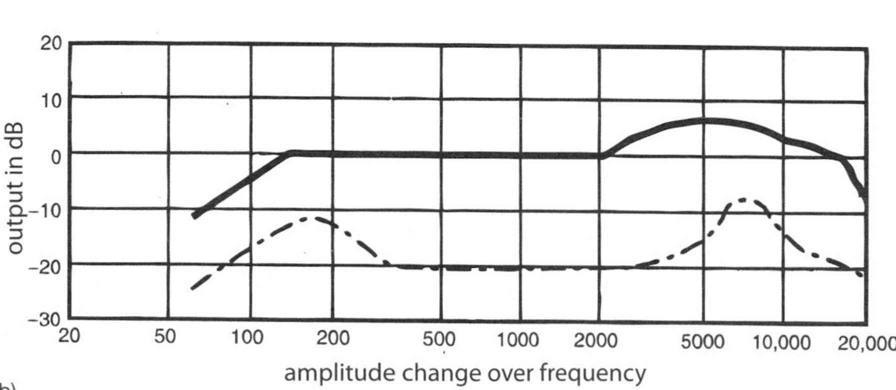

FIGURE 4.20
Frequency response curves: (a) response curve of the AKG C460B/CK61 ULS; (b) response curve of the AKG D321. (Courtesy of AKG Acoustics GmbH., www.akg-acoustics.com)

operating in an area where off-axis sound (in the form of leakage) arrives at the pickup (shown as the dotted curves in Figure 4.20a and b), and will often result in a tone quality change, when the leaked signal is mixed in with other properly miked signals.

At low frequencies, *rumble* (high-level vibrations that occur in the 3 to 25Hz region) can be easily introduced into the surface of a large unsupported floor space, studio or hall from any number of sources (such as passing trucks, air conditioners, subways or fans). They can be reduced or eliminated in a number of ways, such as:

- Using a shock mount to isolate the mic from the vibrating surface and floor stand.
- Choosing a mic that displays a restricted low-frequency response.
- Restricting the response of a wide-range mic by using a low-frequency roll-off filter.

Another low-frequency phenomenon that occurs in most directional mics is known as *proximity effect*. This effect causes an increase in bass response whenever a directional mic is brought within 1 foot of the sound source. This bass boost (which is often most noticeable on vocals) proportionately increases as the distance decreases. To compensate for this effect (which is somewhat greater for bidirectional mics than for cardioids), a low-frequency roll-off filter switch (which is often located on the microphone body) can be used. If none exists, an external roll-off or equalizer can be used to reduce the low end. Any of these tools can be used to help restore the bass response to a flat and natural-sounding balance. Another way to reduce or eliminate proximity effect and its associated "popping" of the letters "p" and "b" is to replace the directional microphone with an omnidirectional mic when working at close distances. On a more positive note, this increase in bass response has long been appreciated by vocalists and DJs for their ability to give a full, "larger-than-life" quality to voices that are otherwise thin. In many cases, the use of a directional mic has become an important part of the engineer, producer and vocalist's toolbox.

Tutorial: Proximity Effect

1. Pull out omnidirectional, cardioid and bidirectional mics (or one that can be switched between these patterns).
2. Move in on each mic pattern type from distances of 3 feet to 6 inches (being careful of volume levels and problems that can occur from popping).
3. Does the bass response increase as the distance is decreased with the cardioid? ... the bidirectional? ... the omni?

Transient response

A significant piece of data (which currently has no accepted standard of measure) is the *transient response* of a microphone (Figure 4.21). Transient response is the measure of how quickly a mic's diaphragm will react when it is hit by an acoustic wavefront. This figure varies wildly among microphones and is a major reason for the difference in sound quality among the three pickup types. For example, the diaphragm of a dynamic mic can be quite large (up to 2.5 inches). With the additional weight of the coil of wire and its core, this combination can be a very large mass when compared to the power of the sound wave that drives it. Because of this, a dynamic mic can be very slow in reacting to a waveform, often giving it a rugged, gutsy, and less accurate sound.

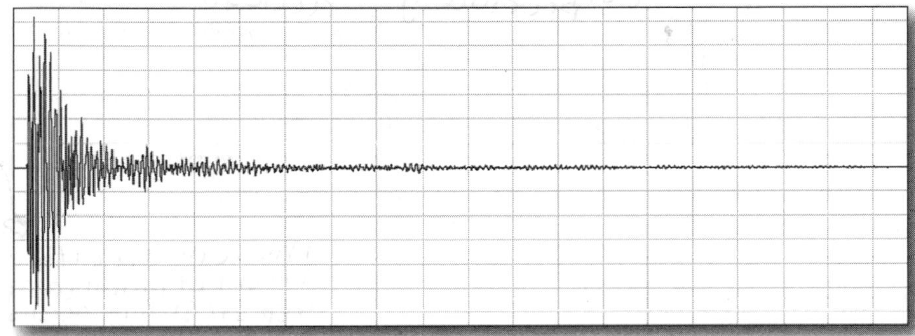

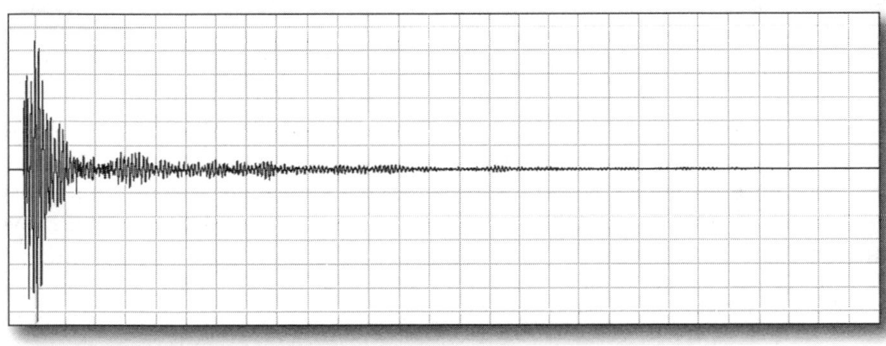

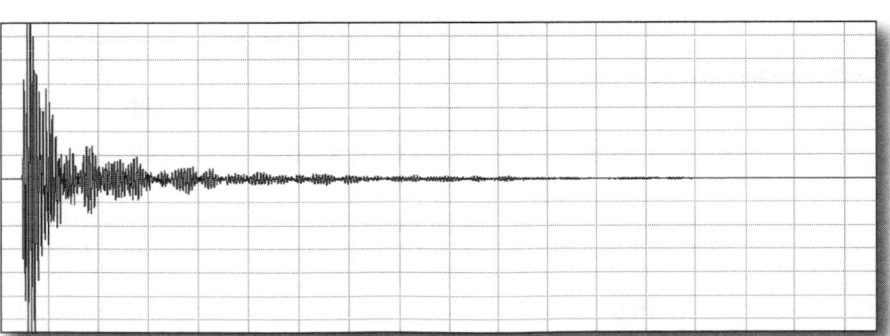

FIGURE 4.21
Transient response characteristics of a percussive woodblock using various microphone types: (a) Shure SM58 dynamic; (b) RCA 44BX ribbon; (c) AKG C3000 condenser.

By comparison, the diaphragm of a ribbon mic is much lighter, so its diaphragm can react more quickly to a sound waveform, resulting in a clearer sound. The condenser pickup has an extremely light diaphragm, which varies in diameter from 2.5 inches to less than 0.25 inch and has a thickness of about 0.0015 inch. This means that the diaphragm offers very little mechanical resistance to a sound-pressure wave, allowing it to accurately track the wave over the entire frequency range.

Output characteristics

A microphone's *output characteristics* refer to its measured sensitivity, equivalent noise, overload characteristics, impedance and other output responses.

SENSITIVITY RATING

A mic's *sensitivity rating* is the output level (in volts) that a microphone will produce, given a specific and standardized acoustic signal at its input (rated in dB SPL). This figure will specify the amount of amplification that's required to raise the mic's signal to line level (often referenced to −10 dBv or +4 dBm) and allows us to judge the relative output levels between any two mics. A microphone with a higher sensitivity rating will produce a stronger output signal voltage than one with a lower sensitivity.

EQUIVALENT NOISE RATING

The *equivalent noise rating* of a microphone can be viewed as the device's electrical self-noise. It is expressed in dB SPL or dBA (a weighted curve) as a signal that would be equivalent to the mic's self-noise voltage. As a general rule, the mic itself doesn't contribute much noise to a system when compared to the mixer's amplification stages, the recording system or media (whether analog or digital). However, with recent advances in mic preamp/mixer technologies and overall reductions in noise levels produced by digital systems, these noise ratings have become increasingly important. Interestingly enough, the internal noise of a dynamic or ribbon pickup is actually generated by the electrons that move within the coil or ribbon itself. Most of the noise that's produced by a condenser mic is generated by the built-in preamp. It almost goes without saying that certain microphone designs will have a higher degree of self-noise than will others; thus, care should be taken in your microphone choices for critical applications (such as with distant and/or classical recording techniques).

OVERLOAD CHARACTERISTICS

Just as a microphone is limited at low levels by its inherent self-noise, it's also limited at high sound-pressure levels (SPLs) by *overload distortion*. In terms of distortion, the dynamic microphone is an extremely rugged pickup, often capable of an overall dynamic range of 140 dB. Typically, a condenser microphone won't distort, except under the most severe sound-pressure levels; however, the condenser system differs from the dynamic in that at high acoustic levels the capsule's output might be high enough to overload the mic's preamplifier. To prevent this, most condenser mics offer a switchable attenuation pad that immediately follows the capsule output and serves to reduce the signal level at the preamp's input, thereby reducing or eliminating overload distortion. When inserting such an attenuation pad into a circuit, keep in mind that the mic's signal-to-noise ratio will be degraded by the amount of attenuation; therefore, it's always wise to remove the inserted pad when using the microphone under normal conditions.

MICROPHONE IMPEDANCE

Microphones are designed to exhibit different *output impedances*. Output impedance is a rating that's used to help you match the output resistance of one device to the rated input resistance requirements of another device (so as to provide the best-possible level and frequency response matching).

Impedance is measured in ohms (with its symbol being Ω or Z). The most commonly used microphone output impedances are 50, 150 and 250 Ω (low) and 20 to 50 k Ω (high). Each impedance range has its advantages. In the past, high-impedance mics were used because the input impedances of most tube-type amplifiers were high. A major disadvantage to using high-impedance mics is the likelihood that their cables will pick up electrostatic noise (like those caused by motors and fluorescent lights). To reduce such interference, a shielded cable is necessary, although this begins to act as a capacitor at lengths greater than 20 to 25 feet, which serves to short out much of the high-frequency information that's picked up by the mic. For these reasons, high-impedance microphones are rarely used in the professional recording process.

Most modern-day systems, on the other hand, are commonly designed to accept a low-impedance microphone source. The lines of very-low-impedance mics (50 Ω) have the advantage of being fairly insensitive to electrostatic pickup. They are, however, sensitive to induced hum pickup from electromagnetic fields (such as those generated by AC power lines). This extraneous noise can be greatly reduced through the use of a twisted-pair cable, because the interference that's magnetically induced into the cable will flow in opposite directions along the cable's length and will cancel out at the console or mixer's balanced microphone input stage. Mic lines of 150 to 250 Ω are less susceptible to signal losses and can be used with cable lengths of up to several thousand feet. They're also less susceptible to electromagnetic pickup than the 50 Ω lines but are more susceptible to electrostatic pickup. As a result, most professional mics operate with an impedance of 200 Ω, use a shielded twisted-pair cable and have reduced noise through the use of a balanced signal line.

BALANCED/UNBALANCED LINES

In short, a *balanced line* uses three wires to properly carry the audio signal. Two of the wires are used to carry the signal voltage, while a third lead is used as a neutral ground wire. Since neither of the two signal conductors of a balanced line is directly connected to the signal ground, the alternating current of an audio signal will travel along the two independent wires. From a noise standpoint, whenever

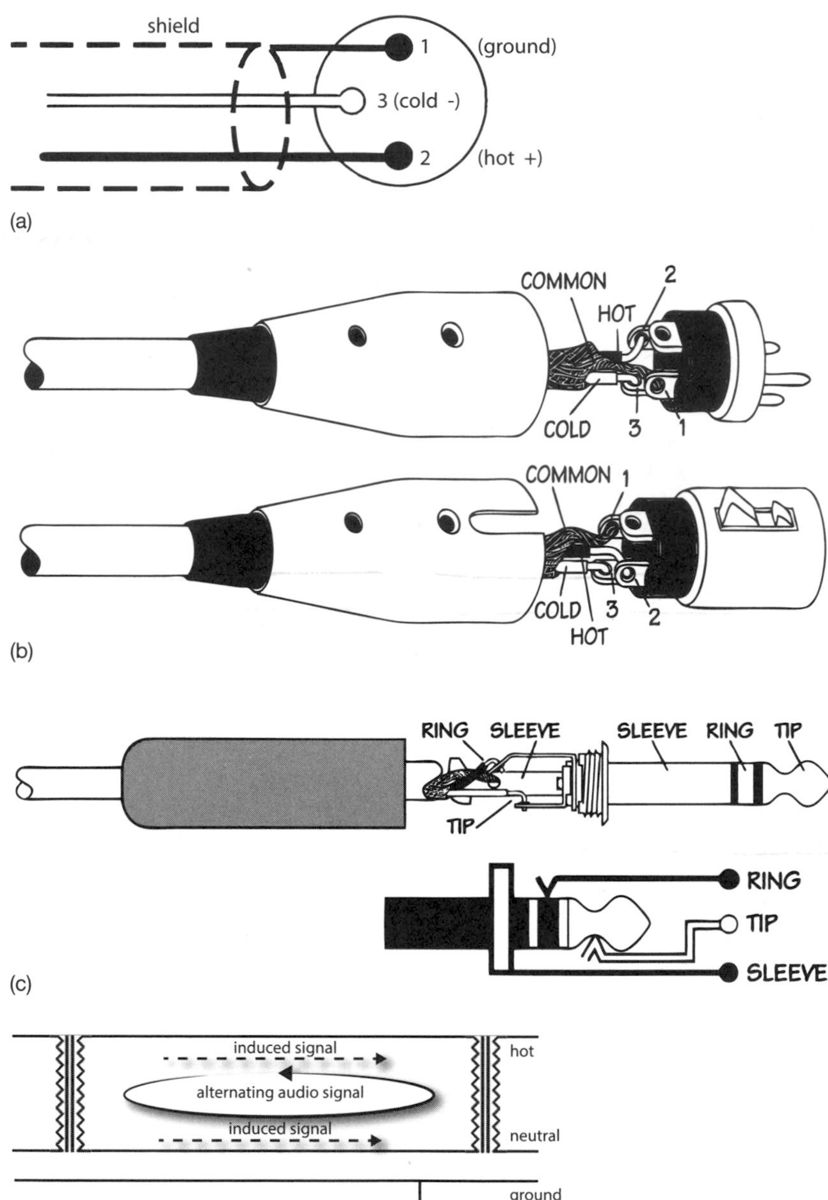

FIGURE 4.22
Wiring detail of a balanced microphone cable (courtesy of Loud Technologies Inc., www.mackie.com): (a) diagram for wiring a balanced microphone (or line source) to a balanced XLR connector; (b) physical drawings; (c) diagram for wiring a balanced 1/4-inch phone connector; (d) equivalent circuit, where the induced signals travel down the wires in equal polarities that cancel at the transformer, whereby the AC audio signals are of opposing polarities that generate an output signal.

an electrostatic or electromagnetic signal is induced across the audio leads, it will be induced into both of the audio leads at an equal level (Figure 4.22). Since the input of a balance device will only respond to the alternating voltage potentials between the two leads, the unwanted noise (which is equal and opposite in polarity) will be canceled.

The standard that has been widely adopted for the proper polarity of two-conductor, balanced, XLR connector cables specifies pin 2 as being positive (+ or hot) and pin 3 as being negative (– or neutral), with the cable ground being connected to pin 1.

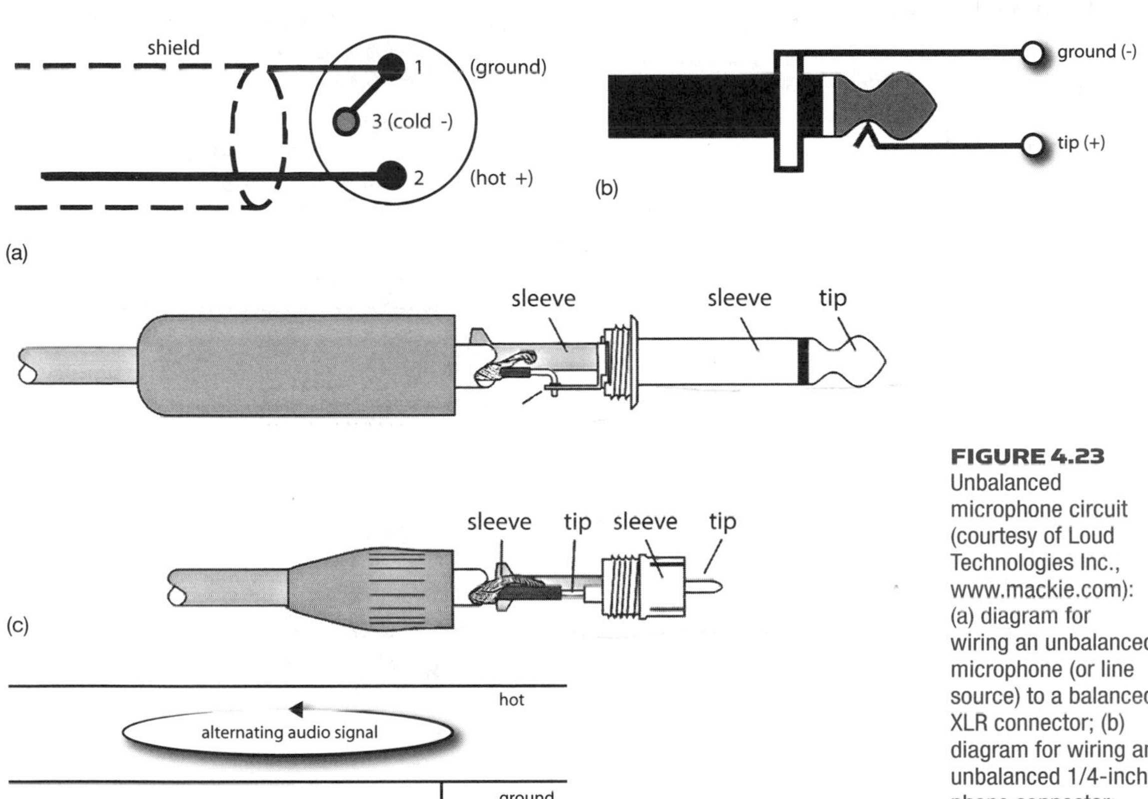

FIGURE 4.23 Unbalanced microphone circuit (courtesy of Loud Technologies Inc., www.mackie.com): (a) diagram for wiring an unbalanced microphone (or line source) to a balanced XLR connector; (b) diagram for wiring an unbalanced 1/4-inch phone connector; (c) physical drawings; (d) equivalent circuit.

If the hot and neutral pins of balanced mic cables are haphazardly pinned in a music or production studio, it's possible that any number of mics (and other equipment, for that matter) could be wired in opposite, out-of-phase polarities. For example, if a single instrument were picked up by two mics using two improperly phased cables, the instrument might totally or partially cancel when mixed to mono. For this reason, it's always wise to use a phase tester or volt–ohm meter to check the cable wiring throughout a pro or project studio complex.

High-impedance mics and most line-level instrument lines use *unbalanced lines* (Figure 4.23) to transmit signals from one device to another. In an unbalanced circuit, a single signal lead carries a positive current potential to a device, while a second, grounded shield (which is tied to the chassis ground) is used to complete the circuit's return path. When working at low signal levels (especially at mic levels), any noises, hums, buzzes or other types of interference that are induced into the signal path will be amplified along with the input signal.

MICROPHONE PREAMPS

Since the output signals of most microphones are at levels far too low to drive the line-level input stage of most recording systems, a mic preamplifier must be used to boost its signal to acceptable levels (often by 30 to 70 dB). With the advent of improved technologies in analog and digital console design, hard-disk recorders, DAWs, signal processors and the like, low noise and distortion figures have become more important than ever. To many professionals, the stock mic pres (pronounced "preeze") that are designed into many console types don't have that special "sound," aren't high enough in quality to be used in critical applications or don't have enough of a special, boutique cache for that special application. As a result, outboard mic preamps are chosen instead (Figures 4.24 through 4.27) for their low-noise, low-distortion specs and/or their unique sound. These devices might make use of tube, FET and/or integrated circuit technology, and offer advanced features in addition to the basic variable input gain,

FIGURE 4.24 Radial Power Pre 500 series-style mic preamplifier (side view). (Courtesy of Radial Engineering ltd, www.radialeng.com).

FIGURE 4.25 Grace M103 Channel Strip mic preamplifier/3 band EQ/optical compressor. (Courtesy of Grace Design, www.gracedesign.com).

FIGURE 4.26 PreSonus DigiMAX D8 8-channel microphone preamp with lightpipe. (Courtesy of Presonus Audio Electronics, Inc., www.presonus.com)

FIGURE 4.27
Rupert Neve Designs Portico 5024 4-channel mic preamp. (Courtesy of Rupert Neve Designs llc, www.rupertneve.com).

phantom power and high-pass filter controls. As with most recording tools, the sound, color scheme, retro style, tube or transistor type and budget level are up to the individual, the producer and the artist … it's totally a matter of personal style and taste. Note that mic pres have also tapped into the growing market of those systems that are based around a DAW, which doesn't need a console or mixer but does require a quality pre (or set of pres) for plugging mic signals directly into the interface.

MICROPHONE TECHNIQUES

Most microphones have a distinctive sound character that's based on its specific type and design. A large number of types and models can be used for a variety of applications, and it's up to the engineer to choose the right one for the job. Over the years, I've come to the realization that there are two particular paths that one can take when choosing the types and models of microphones for a studio's production toolbox. These can basically be placed into the categories of:

- Selecting a limited range of mics that are well suited for a wide range of applications
- Acquiring a larger collection of mics that are commonly perceived as being individually suited for a particular instrument or situation.

The first approach is ideal for the project studio and those who are just starting out and are on a limited budget. It is also common practice among seasoned professionals who swear by a limited collection of their favorite mics that are chosen to cover a wide range of applications. These dynamic and/or condenser mics can be used both in the project studio and in the professional studio to achieve the best possible sound on a budget.

The second approach (I often refer to it as the "Alan Sides" approach) is better suited to the professional studio (and to personal collectors) who actually have a need or desire to amass their own "dream collection" and offer it to their clients. In the end, both approaches have their merits. … Indeed, it's usually wise to keep an open mind and choose a range of mic types that best fit your needs, budget and personal style.

Choosing the appropriate mic, however, is only half the story. The placement of a microphone will often play just as important a role, and is one of the engineer's most valued tools. Because mic placement is an art form, there is no right or wrong. Placement techniques that are currently considered "bad" might easily be the accepted as being standard practice five years from now … and as new musical styles develop, new recording techniques will also tend to evolve, helping to breathe new life into music and production. The craft of recording should always be open to change and experimentation—two of the strongest factors that keep the music and the biz of music alive and fresh.

Here are several pieces of practical advice.

- If you're recording a musician (especially an experienced one) you might consider asking him or her how they've been recorded in the past. Do they have a favorite mic or technique that's often worked best for them. This tactic can help to put the artist at ease and give you insights into new studio miking techniques that can be a helpful production and educational tool.
- Think carefully before "printing" a recorded signal directly to a track. The recording of an instrument with effects or dynamics can't be undone at a later time … so, unless you're absolutely sure, its' often wise to add effects to the track later during mixdown … or … as an option, you could print the signal to two tracks, one with and one without effects.
- Make use of the various tools of a room's acoustics when recording an instrument. This includes the use of distance as a tool for changing the size and character of an instrument or group (which could be recorded to separate tracks for later blending within the mix).
- Although there are no rules for this type of creativity, you might keep in mind the instrument and traditional pickup style, so as to not go too far afield from the expected norm (unless you want to break the rules to lay a new path).

Pickup characteristics as a function of working distance

In studio and sound-stage recording, four fundamental styles of microphone placement are directly related to the working distance of a microphone from its sound source. These extremely important placement styles are as important as any tool in the toy box:

- Distant miking
- Close miking
- Accent miking
- Ambient miking.

DISTANT MICROPHONE PLACEMENT

With distant microphone placement (Figure 4.28), one or more mics can be positioned at a distance of 3 feet or considerably more from the intended signal source. This technique (whose distance will vary with room and instrument size) will often yield the following results:

- It can pick up a large portion of a musical instrument or ensemble, thereby preserving the overall tonal balance of that source. Often, a natural tone balance can be achieved by placing the mic at a distance that's roughly equal to the size of the instrument or sound source.
- It allows the room's acoustic environment to be picked up (and naturally mixed in) with the direct sound signal.

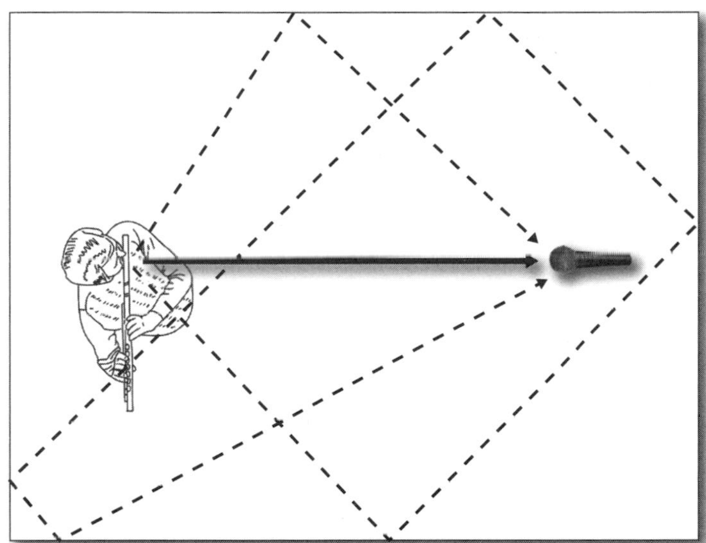

FIGURE 4.28
Example of an overall distant pickup.

Distant miking is often used to pick up large instrumental ensembles (such as a symphony orchestra or choral ensemble). In this application, the pickup will largely rely on the acoustic environment to help achieve a natural, ambient sound. The mic should be placed at a distance so as to strike an overall balance between the ensemble's direct sound and the room's acoustics, giving a balance that's determined by a number of factors, including the size of the sound source, its overall volume level, mic distance and placement and the reverberant characteristics of the room.

Although this technique is often used in classical and jazz techniques, never underestimate its power and effect with modern productions. Placing a distant mic or mic pair within a large room can bring instruments such as drums, guitars, pianos, and other percussion to life. Again, it's often a good idea to place these pickups onto separate tracks, thereby allowing decisions to be better made within mixdown.

This technique tends to add a live, open feeling to a recorded sound; however, it could put you at a disadvantage if the acoustics of a hall, church or studio aren't particularly good. Improper or bad room reflections can create a muddy or poorly defined recording. To avoid this, the engineer might take one of the following actions:

- Temporarily correct for bad or excessive room reflections by using absorptive and/or offset reflective panels (to break up the problematic reflections).
- Place the mic closer to its source and add a degree of artificial ambience.

If a distant mic is used to pick up a portion of the room sound, placing it at a random height can result in a hollow sound due to phase cancellations that occur between the direct sound and delayed sounds that are reflected off the floor and other nearby surfaces (Figure 4.29). If these delayed reflections arrive at the mic at

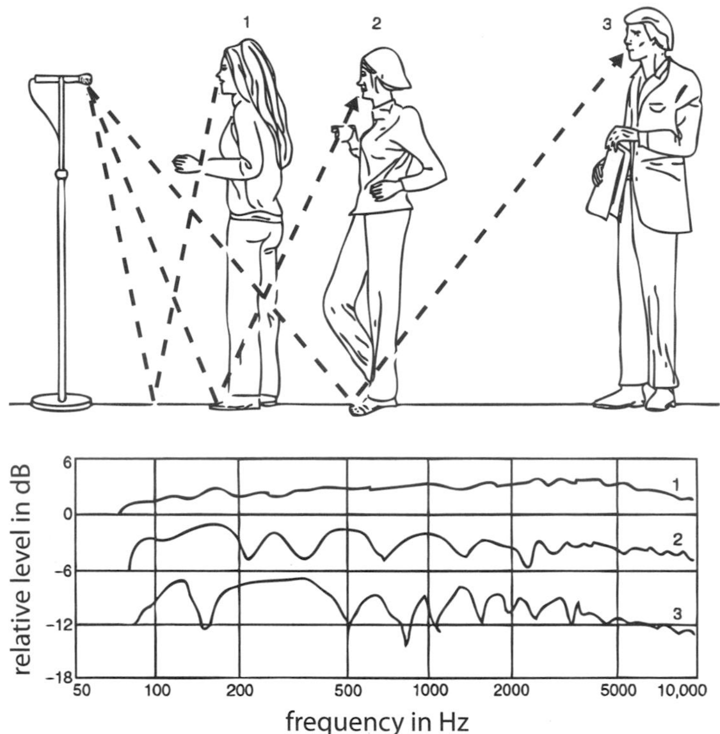

FIGURE 4.29
Resulting frequency response from a microphone that receives a direct and delayed sound from a single source.

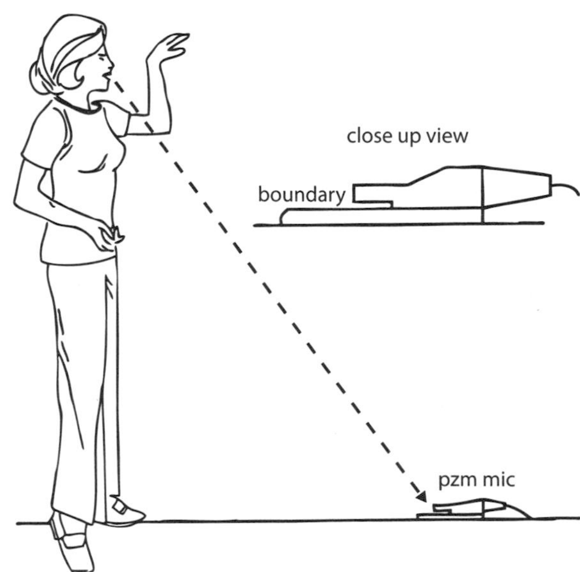

FIGURE 4.30
The boundary microphone system.

a time that's equal to one-half a wavelength (or at odd multiples thereof), the reflected signal will be 180° out of phase with the direct sound. This could produce dips in the signal's pickup response that could adversely color the signal. Since the reflected sound is at a lower level than the direct sound (as a result of traveling farther and losing energy as it bounces off a surface), the cancellation will only be partially complete. Raising the mic will have the effect of reducing reflections (due to the increased distances that the reflected sound must travel), while moving the mic close to the floor will conversely reduce the path length and raise the range in which the frequency cancellation occurs. In practice, a height of 1/8 to 1/16 inch will raise the cancellation above 10 kHz. One such microphone design type, known as a *boundary microphone* (Figures 4.30 and 4.31), places an electret-condenser or condenser diaphragm well within these low height restrictions. For this reason, this mic type might be a good choice for use as an overall distant pickup, when the mics need to be out of sight (i.e., when placed on a floor, wall or large boundary).

FIGURE 4.31
The PZM-6D boundary microphone. (Courtesy of Crown International, Inc., www.crownaudio.com.)

CLOSE MICROPHONE PLACEMENT

When a *close microphone* placement is used, the mic is often positioned about 1 inch to 3 feet from a sound source. This commonly used technique generally yields two results:

- It creates a tight, present sound quality.
- It effectively excludes the acoustic environment from being picked up.

Because sound diminishes with the square of its distance from the sound source, a sound that originates 3 inches from the pickup will be much higher in level than one that originates 6 feet from the mic (Figure 4.32). Therefore, whenever *close miking* is used, only the desired on-axis sound will be recorded; extraneous, distant sounds (for all practical purposes) won't be picked up. In effect, the distant pickup will be masked by the closer sounds and/or will be reduced to a relative level that's well below the main pickup.

Whenever an instrument's mic also picks up the sound of a nearby instrument, a condition known as *leakage* occurs (Figure 4.33). Whenever a signal is picked up by both its intended mic and a nearby mic (or mics), it's easy to see how the signals could be combined together within the mixdown process. When this occurs, level and phase cancellations often make it more difficult to have control over the volume and tonal character of the involved instruments within a mix.

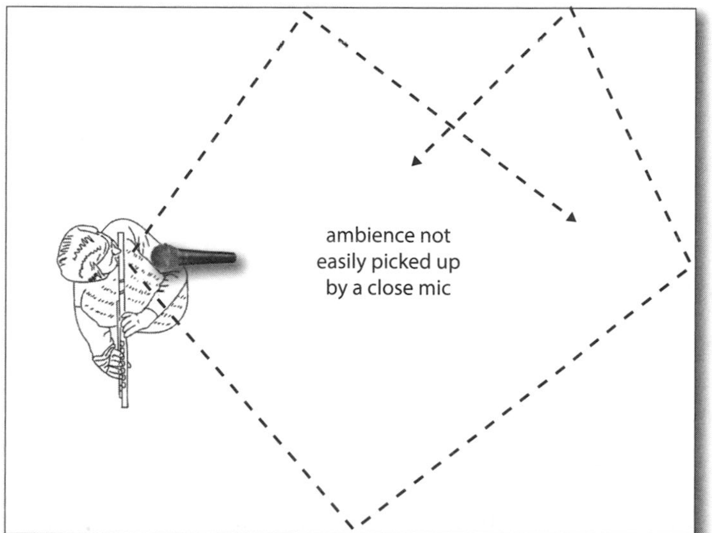

FIGURE 4.32
Close miking reduces the effects of the acoustic environment.

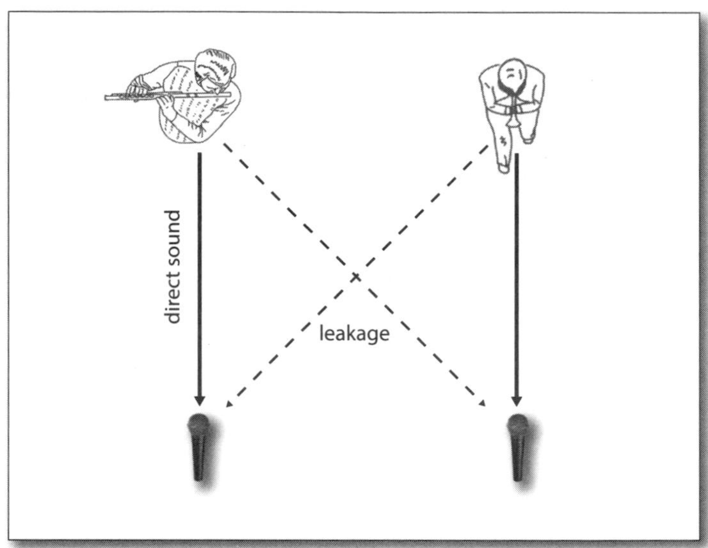

FIGURE 4.33
Leakage due to indirect signal pickup.

To avoid the problems that can be associated with leakage, try the following:

- Place the mics closer to their respective instruments (Figure 4.34a).
- Use directional mics.
- Place an acoustic barrier (known as a flat, gobo, or divider) between the instruments (Figure 4.34b). Alternatively, mic/instruments can be surrounded on several sides by sound baffles and (if needed) a top can be draped over them.
- Spread the instruments farther apart.
- An especially loud (or quieter) instrument can be isolated by putting it in an unused iso-room or vocal or instrument booth. Electronic amps that are played at high volumes can also be recorded in such a room. An amp and the mic can be covered with a blanket or other flexible sound-absorbing material, so that there's a clear path between the amplifier and the mic.
- Separation can be achieved by plugging otherwise loud electronic instruments directly into the console via a direction injection (DI) box, thereby bypassing the miked amp.

Obviously, these examples can only suggest the number of possibilities that can occur during a session. For example, you might choose not to isolate the instruments and instead, place them in an acoustically "live" room. This approach will require that you carefully place the mics in order to control leakage; however, the result will often yield a live and present sound. As an engineer, producer and/or artist, the choices belong to you. Remember, the idea is to work out the kinks beforehand and to simplify technology as much as possible in the studio because Murphy's law is always alive and well in any production facility.

Whenever individual instruments are being miked close (or semi-close), it's generally wise to follow the *3:1 distance rule.*

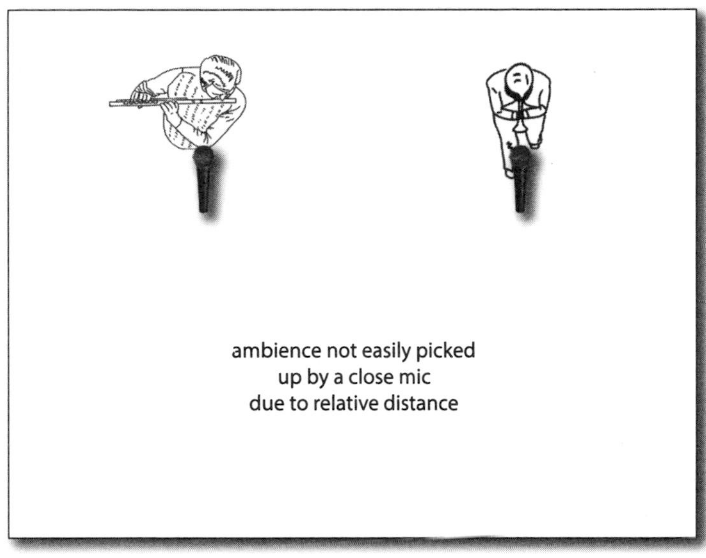

(a)

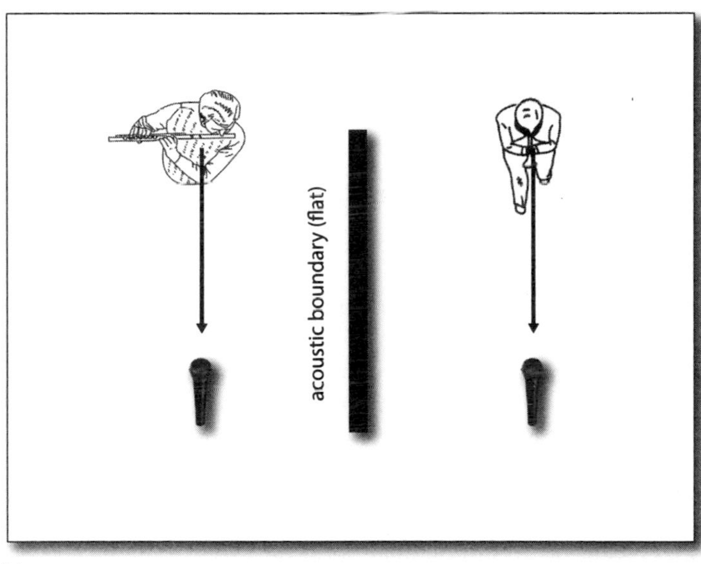

(b)

FIGURE 3.34
Two methods for reducing leakage: (a) Place the microphones closer to their sources. (b) Use an acoustic barrier to reduce leakage.

3:1 Distance Rule

To reduce leakage and maintain phase integrity, this rule states that for every unit of distance between a mic and its source, a nearby mic (or mics) should be separated by at least three times that distance (Figure 4.35).

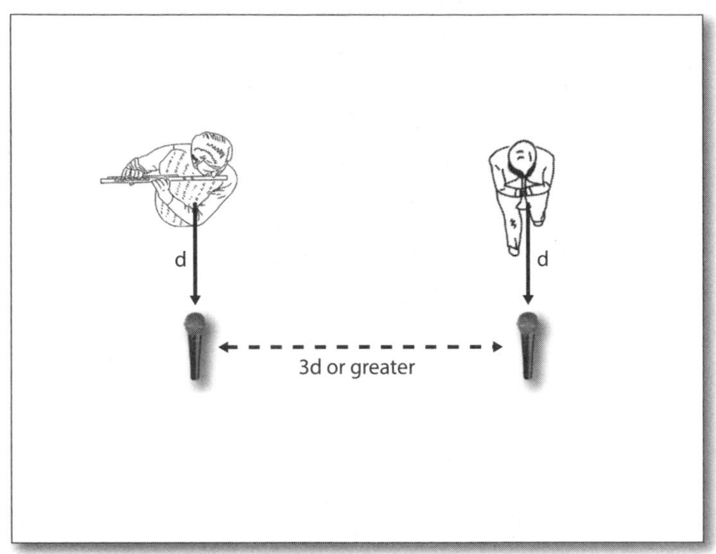

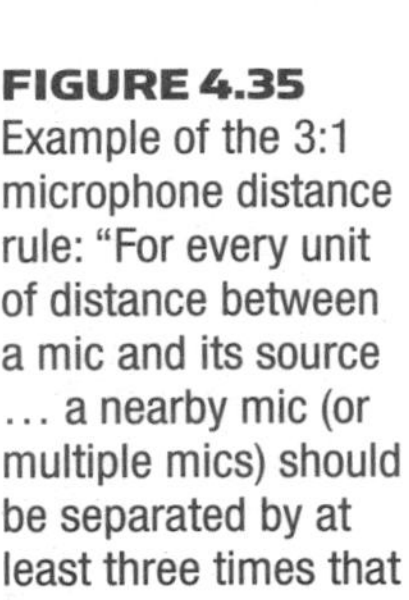

FIGURE 4.35
Example of the 3:1 microphone distance rule: "For every unit of distance between a mic and its source … a nearby mic (or multiple mics) should be separated by at least three times that distance."

Some err on the side of caution and avoid leakage even further by following a 5:1 distance rule. As always, experience will be your best teacher. Although the close miking of a sound source offers several advantages, a mic should be placed only as close to the source as is necessary, not as close as possible. Miking too close can color the recorded tone quality of a source, unless care is taken and careful experimentation is done.

It should be noted, however, that a bit of "bleed" (a slang word for leakage) between mics just might be a good thing. With semi-distant and even multiple mics that are closely spaced, the pickup of a source by several pickups can add a sense of increased depth and sonic space. Having an overall distant set of mics in the studio can add a dose of natural ambience that can actually help to "glue" a mix together. The concept of minute phase cancellations and leakage in a mix isn't always something to be feared; it's simply important that you be aware of the effects that it can have on a mix … and use that knowledge to your advantage.

Because close mic techniques commonly involve distances of 1 to 6 inches, the tonal balance (timbre) of an entire sound source often can't be picked up; rather, the mic might be so close to the source that only a small portion of the surface is actually picked up, giving it a tonal balance that's very area specific (much like hearing the focused parts of an instrument through an acoustic microscope). At these close distances, moving a mic by only a few inches can easily change the pickup tonal balance. If this occurs, try using one or more of the following remedies:

- Move the microphone along the surface of the sound source until the desired balance is achieved.
- Place the mic farther back from the sound source to allow for a wider angle (thereby picking up more of the instrument's overall sound).

- Change the mic.
- Equalize the signal until the desired balance is achieved.

Tutorial: Close Mic Experimentation

1. Mic an acoustic instrument (such as a guitar or piano) at a distance of 1 to 3 inches.
2. Move (or have someone move) the mic over the instrument's body as it's being played, while listening to variations in the sound. Does the sound change? What are your favorite and least favorite positions?

In addition to all of the above considerations, the placement of musicians and instruments will often vary from one studio and/or session to the next because of the room, people involved, number of instruments, isolation (or lack thereof) among instruments, and the degree of visual contact that's needed for creative communication. If additional isolation (beyond careful microphone placement) is needed, flats and baffles can be placed between instruments in order to prevent loud sound sources from spilling over into other open mikes. Alternatively, the instrument or instruments could be placed into separate isolation (iso) rooms and/or booths, or they could be overdubbed at a later time.

During a session that involves several musicians, the setup should allow them to see and interact with each other as much as possible. It's extremely important that they be able to give and receive visual cues and otherwise "feel the vibe." The instrument/mic placement, baffle arrangement, and possibly room acoustics (which can often be modified by placing absorbers in the room) will depend on the engineer's and artists' personal preferences, as well as on the type of sound the producer wants.

ACCENT MICROPHONE PLACEMENT

Often, the tonal and ambient qualities will sound very different between a distant- and close-miked pickup. Under certain circumstances, it's difficult to obtain a naturally recorded balance when mixing the two together. For example, if a solo instrument within an orchestra needs an extra mic for added volume and presence, placing the mic too close would result in a pickup that sounds overly present, unnatural and out of context with the distant, overall orchestral pickup. To avoid this pitfall, a compromise in distance should be struck. A microphone that has been placed within a reasonably close range to an instrument or section within a larger ensemble (but not so close as to have an unnatural sound) is known as an *accent pickup* (Figure 4.36). Whenever accent miking is used, care should be exercised in placement and pickup choices. The amount of accent signal that's introduced into the mix should sound natural relative to the overall pickup, and

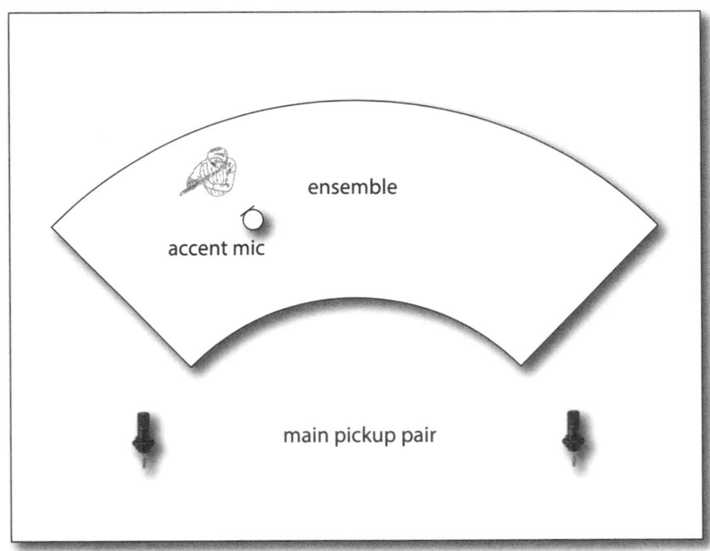

FIGURE 4.36
Accent microphone placed at proper compromise distance.

a good accent mic should only add presence to a solo passage and not stick out as separate, identifiable pickup.

AMBIENT MICROPHONE PLACEMENT

Ambient miking places the pickup at such a distance that the reverberant or room sound is equally or more prominent than the direct signal. The ambient pickup is often a cardioid stereo pair or crossed figure-8 (Blumlein) pair that can be mixed into a stereo or surround-sound production to provide a natural reverb and/or ambience. To enhance the recording, you can use ambient mic pickups in the following ways:

- In a live concert recording, ambient mics can be placed in a hall to restore the natural reverberation that is often lost with close miking techniques.
- In a live concert recording, ambient microphones can be placed over the audience to pick up their reaction and applause.
- In a studio recording, ambient microphones can be used in the studio to add a sense of space or natural acoustics back into the sound.

DIY Tutorial: Ambient Miking

1. Mic an instrument or its amp (such as an acoustic or electric guitar) at a traditional distance of 6 inches to 1 foot.
2. Place a stereo mic pair (in an X/Y and/or spaced configuration) in the room, away from the instrument.
3. Mix the two pickup types together. Does it "open" the sound up and give it more space? Does it muddy the sound up or breathe new life into it?
4. If you're lucky enough to be surround-capable, place the ambient tracks to the rear. Does it add an extra dimension of space? Does it alter the recording's definition?

Stereo miking techniques

For the purpose of this discussion, the term *stereo miking technique* refers to the use of two microphones in order to obtain a coherent stereo image. These techniques can be used in either close or distant miking of single instruments, vocals, large or small ensembles, within on-location or studio applications … in fact, the only limitation is your imagination. The four fundamental stereo miking techniques are:

- Spaced pair
- X/Y
- M/S
- Decca tree

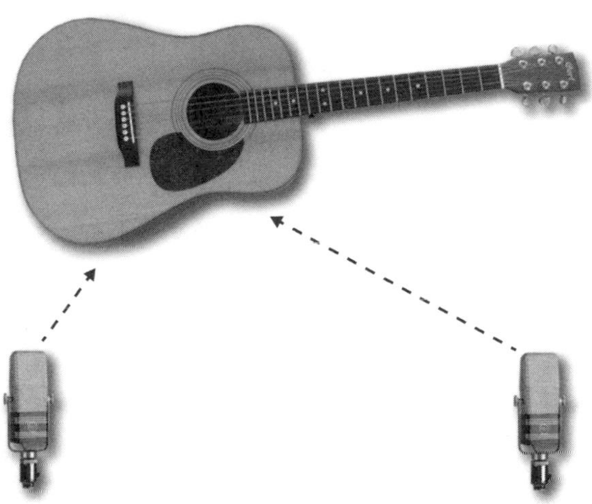

FIGURE 4.37
Spaced stereo miking technique.

SPACED PAIR

Spaced microphones (Figure 4.37) can be placed in front of an instrument or ensemble (in a left/right fashion) to obtain an overall stereo image. This technique places the two mics (of the same type, manufacturer and model) anywhere from only a few feet to more than 30 feet apart (depending on the size of the instrument or ensemble) and uses time and amplitude cues in order to create a stereo image. The primary drawback to this technique is the strong potential for phase discrepancies between the two channels due to differences in a sound's arrival time at one mic relative to the other. When mixed to mono, these phase discrepancies could result in variations in frequency response and even the partial cancellation of instruments and/or sound components in the pickup field.

X/Y

X/Y stereo miking is an intensity-dependent system that uses only the cue of amplitude to discriminate direction. With the X/Y coincident-pair technique (Figure 4.38), two directional microphones of the same type, manufacture and model are placed with their grills as close together as possible (without touching) and facing at angles to each other (generally between 90° and 135°). The midpoint between the two mics is pointed toward the source, and the mic outputs are equally panned left and right. Even though the two mics are placed together, the stereo imaging is excellent—often better than that of a spaced pair. In addition, due to their proximity, no appreciable phase problems arise. Most commonly, X/Y pickups use mics that have a cardioid polar pattern, although the Blumlein technique is being increasingly used. This technique (which is named after the unheralded inventor, Alan Dower Blumlein) uses two crossed bidirectional mics that are offset by 90° to each other. This simple technique often yields excellent ambient results for the pickup of the overall ambience within a studio or concert hall, while also being a good choice for picking up sources that are placed "in the round."

FIGURE 4.38
X/Y stereo miking technique using an X/Y crossed cardioid pair.

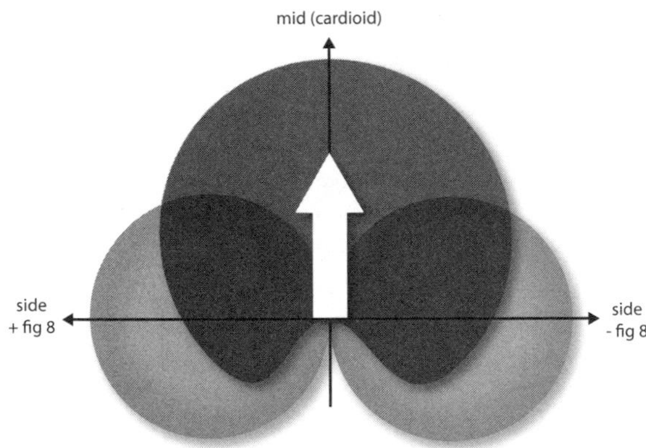

FIGURE 4.39
M/S stereo microphone technique.

Stereo microphones that contain two diaphragms in the same case housing are also available on the new and used market. These mics are either fixed (generally in a 90° or switchable X/Y pattern) or are designed so that the top diaphragm can be rotated by 180° (allowing for the adjustment of various coincident X/Y angles).

M/S

Another coincident-pair system, known as the M/S (or mid-side) technique (Figure 4.39), is similar to X/Y in that it uses two closely spaced, matched pickups. The M/S method differs from the X/Y method, however, in that it requires the use of an external transformer, active matrix, or software plug-in in order to work. In the classic M/S stereo miking configuration, one of the microphone capsules is designated the *M* (mid) position pickup and is generally a cardioid pickup pattern that faces forward, toward the sound source. The *S* (side) capsule is generally chosen as a figure-8 pattern that's oriented sideways (90° and 270°) to the on-axis pickup (i.e., with the null facing to the side, away from the cardioid's main axis). In this way, the mid capsule picks up the direct sound, while the side figure-8 capsule picks up ambient and reverberant sound. These outputs are then combined through a sum-and-difference decoder matrix either electrically (through a transformer matrix) or mathematically (through a digital M/S plug-in), which then resolves them into a conventional X/Y stereo signal: ($M + S$ = left) and ($M - S$ = right).

(a)

(b)

FIGURE 4.40
M/S decoder matrix: (a) AEA MS-38 Mark II dual-mode stereo width controller and Matrix MS processor (courtesy of Audio Engineering Associates, www.ribbonmics.com); (b) Waves S1 Stereo Imager plug-in includes True Blumlein shuffling and MS/LR processing (courtesy of Waves, www.waves.com).

One advantage of this technique is its absolute monaural compatibility. When the left and right signals are combined, the sum of the output will be $(M + S) + (M - S) = 2M$. That's to say, the side (ambient) signal will be canceled, but the mid (direct) signal will be accentuated. Since it is widely accepted that a mono signal loses its intelligibility with added reverb, this tends to work to our advantage. Another amazing side benefit of using M/S is the fact that it lets us continuously vary the mix of mid (direct) to side (ambient) sound that's being picked up either during the recording (from the console location) … or even at a later time during mixdown, after it's been recorded! These are both possible by simply mixing the ratio of mid to side that's being sent to the decoder matrix (Figure 4.40). In a mixdown scenario, all that's needed is to record the mid on one track and the side on another. (It's often best to use a digital recorder, because phase delays associated with the analog recording process can interfere with decoding.) During mixdown, routing the M/S tracks to the decoder matrix allows you to make important decisions regarding stereo width and depth at a later, more controlled date.

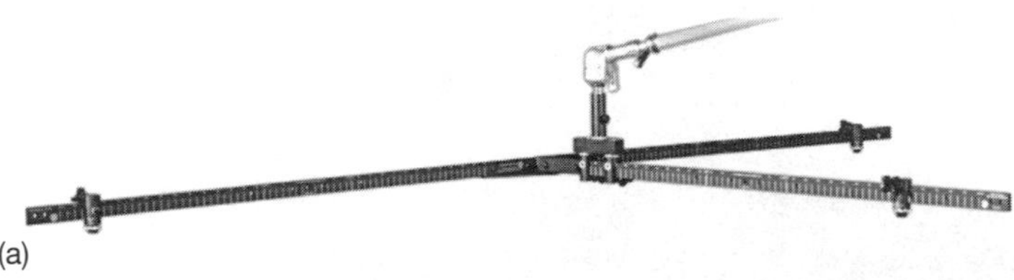
(a)

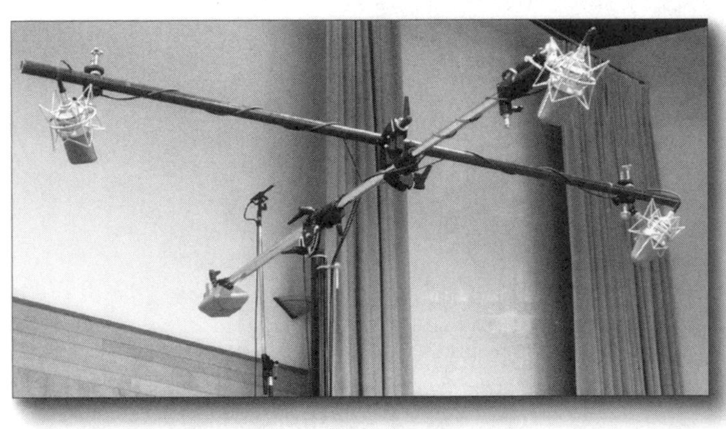
(b)

FIGURE 4.41
Decca tree microphone array. (a) hardware mount. (Courtesy of Audio Engineering Associates, www.ribbonmics.com) (b) in a studio setting. (Courtesy of Galaxy Studios, http://www.galaxy.be)

DECCA TREE

Although not as commonly used as the preceding stereo techniques, the *Decca tree* is a time-tested, classical miking technique that uses both time and amplitude cues in order to create a coherent stereo image. Attributed originally to Decca engineers Roy Wallace and Arthur Haddy in 1954, the Decca tree (Figure 4.41) consists of three omnidirectional mics (originally, Neumann M50 mics were used). In this arrangement, a left and right mic pair is placed 3 feet apart, and a third mic is placed 1.5 feet out in front and panned in the center of the stereo field. Still favored by many in orchestral situations as a main pickup pair, the Decca tree is most commonly placed on a tall boom, above and behind the conductor. According to lore, when Haddy first saw the array, he remarked, "It looks like a bloody Christmas tree!" The name stuck.

Surround miking techniques

With the advent of 5.1 surround-sound production, it's certainly possible to make use of a surround console or DAW to place sources that have been recorded in either mono or stereo into a surround image field. Under certain situations, it's also possible to consider using multiple-pickup surround miking techniques in order to capture the actual acoustic environment and then translate that into a surround mix. Just as the number of techniques and personal styles increases when miking in stereo compared to mono, the number of placement and technique choices will likewise increase when miking a source in surround. Although guidelines have been and will continue to be set, both placement and mixing styles are definitely an art and not a science. For more info on surround, see Chapter 18—Surround Sound.

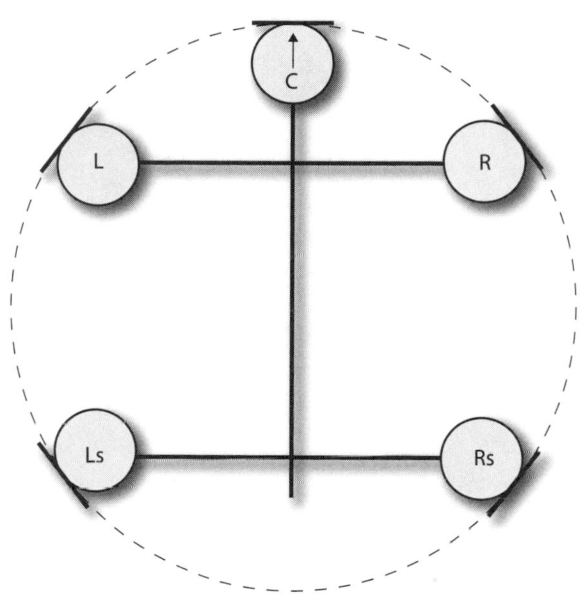

FIGURE 4.42
Five cardioid microphones can be arranged in a circular pattern (with the center microphone facing toward the source) to create a modified, mini-surround Decca tree. A four-microphone (quad) approach to surround miking can be easily made by simply eliminating the center pickup.

AMBIENT SURROUND MICS

A relatively simple, yet effective way to capture the surround ambience of a live or studio session is to simply place a spaced or coincident mic pair out in the studio at a distance from the sound source. These can be facing toward or away from the sound source, and placement is totally up to experimentation. During a surround mixdown, placing distant mics into the studio or hall can work wonders to add a sense of space to an ensemble group, drum set or instrument overdub.

SURROUND DECCA TREE

One of the most logical techniques for capturing an ensemble or instrument in a surround setting places five mics onto a modified Decca tree. This ingenious and simple system adds two rear-facing mics to the existing three-mic Decca tree system. Another simpler approach is to place five cardioid mics in a circle, such that the center channel faces toward the source, thereby creating a simple setup that can be routed L–C–R–Ls–Rs (Figure 4.42).

One last approach (which doesn't actually fall under the Decca tree category) involves the use of four cardioid mics that are spaced at 90° angles, representing L–R–Ls–Rs, with the on-axis point being placed 45° between the L and R mics. This "quad" configuration can be easily made by mounting the mics on two stereo bars that are offset by 90°.

RECORDING DIRECT

As an alternative, the signal of an electric or electronic instrument (guitar, keyboard, etc.) can be directly "injected" into a console, recorder or DAW without the use of a microphone. This option often produces a cleaner, more present sound by bypassing the distorted components of a head/amp combination. It also reduces leakage into other mics by eliminating room sounds. In the project or recording studio, the *direct injection* (DI) box (Figure 4.43) serves to interface an instrument with an analog output signal to a console or recorder in the following ways:

- It reduces an instrument's line-level output to mic level for direct insertion into the console's mic input jack.
- It changes an instrument's unbalanced, high-source impedance line to a balanced, low-source impedance signal that's needed by the console's input stage.

FIGURE 4.43
Radial JDI passive direct box. (Courtesy of Radial Engineering, www.radialeng.com)

- It often can electrically isolate the audio signal paths between the instrument and mic/line preamp stages (thereby reducing the potential for ground-loop hum and buzzes).

Most commonly, the instrument's output is plugged directly into the DI box (where it's stepped down in level and impedance), and the box's output is then fed into the mic pre of a console or DAW. If a "dirtier" sound is desired, certain boxes will allow high-level input signals to be taken directly from the amp's speaker output jack. It's also not uncommon for an engineer, producer and/or artist to combine the punchy, full sound of a mic with the present crispness of a direct sound. These signals can then be combined onto a single tape track or recorded to separate tracks (thereby giving more flexibility in the mixdown stage). The ambient image can be "opened up" even further by mixing a semi-distant or distant mic (or stereo pair) with the direct (and even with the close miked amp) signal. This ambient pickup can be either mixed into a stereo field or at the rear of a surround field to fill out the sound.

When recording a guitar, the best tone and lowest hum pickup for a direct connection occurs when the instrument volume control is fully turned up. Because guitar tone controls often use a variable treble roll-off, leaving the tone controls at the treble setting and using a combination of console EQ and different guitar pickups to vary the tone will often yield the maximum amount of control over the sound. Note that if the treble is rolled off at the guitar, boosting the highs with EQ will often increase pickup noise.

REAMPING IT IN THE MIX

Another way to alter the sound of a recorded track or to inject a new sense of acoustic space into an existing take is to reamp a track. The "reamp" process (originally conceived in 1993 by recording engineer John Cuniberti and now owned by Radial Engineering; www.radialeng.com) lets us record a guitar's signal directly to a track using a DI during the recording session and then play this cleanly recorded track back through a miked guitar amp/speaker, allowing it to be re-recorded to new tracks at another time (Figure 4.44).

The re-recording of an instrument that has been recorded directly gives us total flexibility for changing the final, recorded amp and mic sound at a later time. For example, it's well known that it's far easier to add an effect to a "dry" track that doesn't have effects during mixdown than to attempt to remove an effect after it's been printed to track. Whenever reamping is used at a later time, it's possible

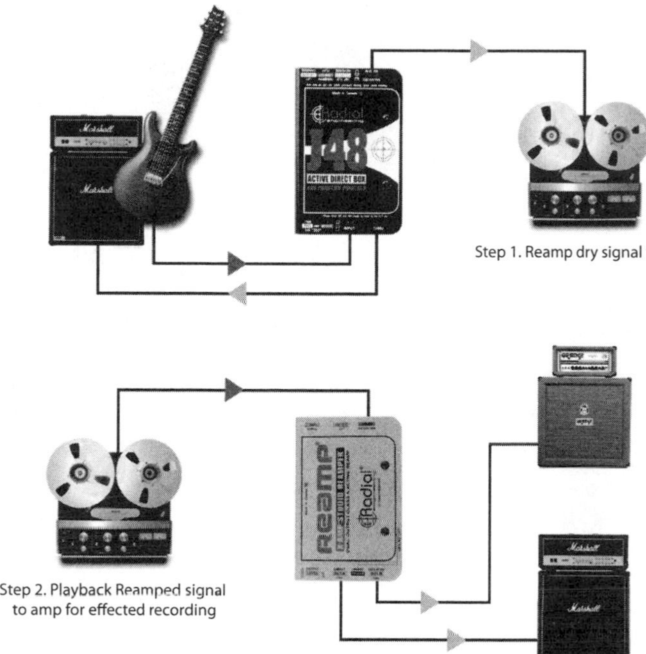
Step 1. Reamp dry signal to track
Step 2. Playback Reamped signal to amp for effected recording

FIGURE 4.44
Figure showing how a direct recording can be "reamped" in a studio, allowing for complete tonal, mic placement, and acoustical control, after the fact! (Courtesy of Radial Engineering, www.radialeng.com.)

to audition any number of amps, using any number of effects and/or mic settings, until the desired sound has been found. This process allows the musician to concentrate solely on getting the best recorded performance, without having to spend extra time getting the perfect guitar, amp, mic and room sound. Leakage problems in the studio are also reduced, because no mikes are used in the process.

Although the concept of recording an instrument directly and playing the track back through a miked amp at a later time is relatively new, the idea of using a room's sound to fill out the sound of a track or mix isn't. The reamp concept takes this idea a bit further by letting you go as wild as you like. For example, you could use the process to re-record a single, close-miked guitar amp and then go back and layer a larger stack at a distance. An electronic guitarist could take the process even further by recording his or her MIDI guitar both directly and to a sequenced MIDI track. In this way, the reamp and patch combinations would be virtually unlimited.

MICROPHONE PLACEMENT TECHNIQUES

The following sections are meant to be used as a general guide to mic placement for various acoustic and popular instruments. It's important to keep in mind that these are only guidelines. Several general application and characteristic notes are detailed in Table 4.1, and descriptions of several popular mics are

Table 4.1 Microphone selection guidelines

Needed Application	Required Microphone Choice and/or Characteristic
Natural, smooth tone quality	Flat frequency response
Bright, present tone quality	Rising frequency response
Extended lows	Dynamic or condenser with extended low-frequency response
Extended highs (detailed sound)	Condenser
Increased "edge" or midrange detail	Dynamic
Extra ruggedness	Dynamic or modern ribbon/condenser
Boosted bass at close working distances	Directional microphone
Flat bass response up close	Omnidirectional microphone
Reduced leakage, feedback, and room acoustics	Directional microphone, or omnidirectional microphone at close working distances
Enhanced pickup of room acoustics	Place microphone or stereo pair at greater working distances
Reduced handling noise	Omnidirectional, vocal microphone, or directional microphone with shock mount
Reduced breath popping	Omnidirectional or directional microphone with pop filter
Distortion-free pickup of very loud sounds	Dynamic or condenser with high maximum SPL rating
Noise-free pickup of quiet sounds	Condenser with low self-noise and high sensitivity

outlined toward the end in the "Microphone Selection" section to help give insights into placement and techniques that might work best in a particular application.

As a general rule, choosing the best mic for an instrument or vocal will ultimately depend on the sound you're searching for. For example, a dynamic mic will often yield a "rugged" or "punchy" character (which is often further accentuated by the proximity of bass boost that's generally associated with a directional mic). A ribbon mic will often yield a mellow sound that ranges from being open and clear to slightly "croony" … depending on the type and distances involved. Condenser mics are often characterized as having a clear, present and full-range sound that varies with mic design, grill options and capsule size. Before jumping into this section, I'd like to again take time to point out the "Good Rule" to anyone who wants to be a better engineer, producer and/or musician:

As a rule, starting with an experienced, rehearsed and ready musician who has a quality instrument that's well tuned is the best insurance toward getting the best possible sound. Let's think about this for a moment. Say that we have a live rhythm session that involves drums, piano, bass guitar and scratch vocals. All of the players are the best around, except for the drummer, who is new to the studio process. Unfortunately, you've now signed on to teach the drummer the ropes of proper drum tuning, studio interaction and playing under pressure. It goes without saying that the session might go far less smoothly than it otherwise would, as you'll have to take the extra time to work with the player to get the best possible sound. Once you're rolling, it'll also be up to you or the producer to pull a professional performance out of someone who's new to the field.

Don't get me wrong, musicians have to start somewhere … but an experienced studio musician who comes into the studio with a great instrument that's tuned and ready to go (and who might even clue you in on some sure-fire mic and placement techniques for the instrument) is simply a joy from a sound, performance, time and budget-saving standpoint. Simply put, if you and/or the project's producer have prepared enough to get all your "goods" lined up, the track will have a much better chance of being something that everyone can be proud of. Just as with the art of playing an instrument, careful mic choice, placement and "style" in the studio are also subjective … and are a few of the fundamental calling cards of a good engineer. Experience simply comes with time and the willingness to experiment. Be patient, learn, listen and have fun … and you too will eventually rise to the professional occasion.

Brass instruments

The following sections describe many of the sound characteristics and miking techniques that are encountered in the brass family of instruments.

TRUMPET

The fundamental frequency of a trumpet ranges from E3 to D6 (165 to 1175 Hz) and contains overtones that stretch upward to 15 kHz. Below 500 Hz, the sounds emanating from the trumpet project uniformly in all directions; above 1500 Hz, the projected sounds become much more directional; and above 5 kHz, the dispersion emanates at a tight 30° angle from in front of the bell. The formants of a trumpet (the relative harmonic and resonance frequencies that give an instrument its specific character) lie at around 1 to 1.5 kHz and at 2 to 3 kHz. Its tone can be radically changed by using a mute (a cup-shaped dome that fits directly over the bell), which serves to dampen frequencies above 2.5 kHz. A conical mute (a metal mute that fits inside the bell) tends to cut back on frequencies below 1.5 kHz while encouraging frequencies above 4 kHz. Because of the high sound-pressure levels that can be produced by a trumpet (up to 130 dB SPL), it's best to place a mic slightly off the bell's center at a distance of 1 foot or more (Figure 4.45). When closer placements are needed, a –10- to –20-dB pad can help prevent input overload at the mic or console preamp input. Under such close working conditions, a windscreen can help protect the diaphragm from windblasts.

FIGURE 4.45
Typical microphone placement for a single trumpet.

TROMBONE

Trombones come in a number of sizes; however, the most commonly used "bone" is the tenor, which has a fundamental note range spanning from E2 to C5 (82 to 523 Hz) and produces a series of complex overtones that range from 5 kHz (when played medium loud) to 10 kHz (when overblown). The trombone's polar pattern is nearly as tight as the trumpet's: Frequencies below 400 Hz are distributed evenly, whereas its dispersion angle increases to 45° from the bell at 2 kHz and above. The trombone most often appears in jazz and classical music. The *Mass in C Minor* by Mozart, for example, has parts for soprano, alto, tenor and bass trombones. This style obviously lends itself to the spacious blending that can be achieved by distant pickups within a large hall or studio. On the other hand, jazz music often calls for closer miking distances. At 2 to 12 inches, for example, the trombonist should play slightly to the side of the mic to reduce the chance of overload and wind blasts. In the miking of a trombone section, a single mic might be placed between two players, acoustically combining them onto a single channel and/or track.

TUBA

The bass and double-bass tubas are the lowest pitched of the brass/wind instruments. Although the bass tuba's range is actually a fifth higher than the double bass, it's still possible to obtain a low fundamental of B (31 Hz). A tuba's overtone structure is limited; it's top response ranges from 1.5 to 2 kHz. The lower frequencies (around 75 Hz) are evenly dispersed; however, as frequencies rise, their distribution angles reduce. Under normal conditions, this class of instruments isn't miked at close distances. A working range of 2 feet or more, slightly off-axis to the bell, will generally yield the best results.

FRENCH HORN

The fundamental tones of the French horn range from B1 to B5 (62 to 700 Hz). Its "oo" formant gives it a round, broad quality that can be found at about 340 Hz, with other frequencies falling between 750 Hz and 3.5 kHz. French horn players often place their hands inside the bell to mute the sound and promote a formant at about 3 kHz. A French horn player or section is traditionally placed at the rear of an ensemble, just in front of a rear, reflective stage wall. This wall serves to reflect the sound back toward the listener's position (which tends to create a fuller, more defined sound). An effective pickup of this instrument can be achieved by placing an omni- or bidirectional pickup between the rear, reflecting wall and the instrument bells, thereby receiving both the direct and reflected sound. Alternatively, the pickups can be placed in front of the players, thereby receiving only the sound that's being reflected from the rear wall.

Guitar

The following sections describe the various sound characteristics and techniques that are encountered when miking the guitar.

ACOUSTIC GUITAR

The popular steel-strung, acoustic guitar has a bright, rich set of overtones (especially when played with a pick). Mic placement and distance will often vary from instrument to instrument and may require experimentation to pick up the best tonal balance. A balanced pickup can often be achieved by placing the mic (or an X/Y stereo pair) at a point slightly off-axis and above or below the sound hole at a distance of between 6 inches and 1 foot (Figure 4.46). Condenser mics are often preferred for their smooth, extended frequency response and excellent transient response. The smaller-bodied classical guitar is normally strung with nylon or gut and is played with the fingertips, giving it a warmer, mellower sound than its steel-strung counterpart. To make sure that the instrument's full range is picked up, place the mic closer to the center of the bridge, at a distance of between 6 inches and 1 foot.

Miking near the sound hole

The sound hole (located at the front face of a guitar) serves as a bass port, which resonates at the lower frequencies (around 80 to 100 Hz). Placing a

FIGURE 4.46
Typical microphone placement for the guitar.

mic too close to the front of this port might result in a boomy and unnatural sound; however, miking close to the sound hole is often popular on stage or around high acoustic levels because the guitar's output is highest at this position. To achieve a more natural pickup under these conditions, the microphone's output can be rolled off at the lower frequencies (5 to 10 dB at 100 Hz).

Surround guitar miking

An effective way to translate an acoustic guitar to the wide stage of surround (if a big, full sound is what you're after) is to record the guitar using X/Y or spaced techniques stereo (panned front L/R) and pan the guitar's electric pickup (or added contact pickup) to the rear center of the surround field. Extra ambient surround mics could also be used in an all-acoustic session.

THE ELECTRIC GUITAR

The fundamentals of the average 22-fret guitar extend from E2 to D6 (82 to 1174 Hz), with overtones that extend much higher. All of these frequencies might not be amplified, because the guitar chord tends to attenuate frequencies above 5 kHz (unless the guitar has a built-in low impedance converter or low-impedance pickups). The frequency limitations of the average guitar loudspeaker often add to this effect, because their upper limit is generally restricted to below 5 or 6 kHz.

Miking the guitar amp

The most popular guitar amplifier used for recording is a small practice-type amp/speaker system. These high-quality amps often help the guitar's suffering high end by incorporating a sharp rise in the response range at 4 to 5 kHz, thus helping to give it a clean, open sound. High-volume, wall-of-sound speaker stacks are less commonly used in a session, because they're harder to control in the studio and in a mix. By far the most popular mic type for picking up an electric guitar amp is the cardioid dynamic. A dynamic tends to give the sound a full-bodied character without picking up extraneous amplifier noises. Often guitar mics will have a pronounced presence peak in the upper frequency range, giving the pickup an added clarity. For increased separation, a microphone can be placed at a working distance of 2 inches to 1 foot. When miking at a distance of less than 4 inches, mic/speaker placement becomes slightly more critical (Figure 4.47). For a brighter sound, the mic should face directly into the center of the speaker's cone. Placing it off the cone's center tends to produce a more mellow sound while reducing amplifier noise.

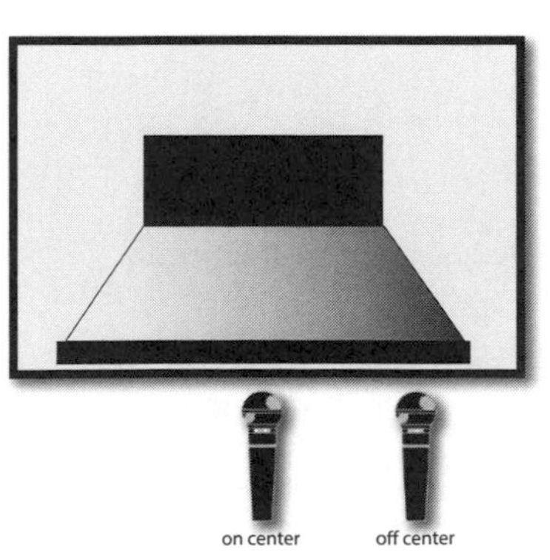

FIGURE 4.47
Miking an electric guitar cabinet directly in front of and off-center to the cone.

Isolation cabinets have also come onto the market that are literally sealed boxes that house a speaker or guitar amp/cabinet system, as well as an internal mic mount. These systems are used to reduce leakage and to provide greater control over instrument levels within a recording studio or control room during a session.

Recording direct

A DI box is often used to feed the output signal of an electric guitar directly into the mic input stage of a recording console or mixer. By routing the direct output signal to a track, a cleaner, more present sound can be recorded (Figure 4.48a). This technique also reduces the leakage that results from having a guitar amp in the studio and even makes it possible for the guitar to be played in the control room or project studio. A combination of direct and miked signals often results in a sound that adds the characteristic fullness of a miked amp to the extra "bite" that a DI tends to give. These may be combined onto a single track or, whenever possible, can be assigned to separate tracks, allowing for greater control during mixdown (Figure 4.48b). During an overdub, the ambient image can be "opened up" even further by mixing a semidistant or distant mic (or stereo pair) with the direct mic (and even with the close miked amp signal). This ambient pickup can be either mixed into a stereo field or at the rear of a surround field to fill out the sound.

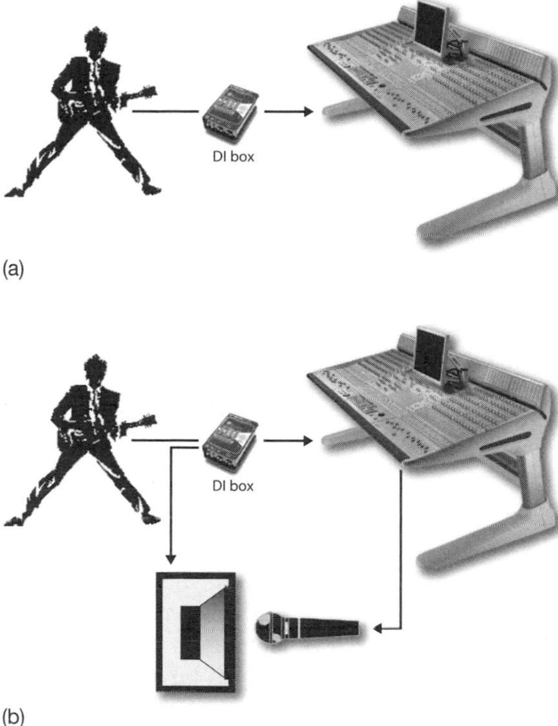

FIGURE 4.48
Direct recording of an electric guitar: (a) direct recording; (b) combined direct and miked signal.

THE ELECTRIC BASS GUITAR

The fundamentals of an electric bass guitar range from about E1 to F4 (41.2 to 343.2 Hz). If it's played loudly or with a pick, the added harmonics can range upward to 4 kHz. Playing in the "slap" style or with a pick gives a brighter, harder attack, while a "fingered" style will produce a mellower tone. In modern music production, the bass guitar is often recorded direct for the cleanest possible sound. As with the electric guitar, the electric bass can be either miked at the amplifier or picked up through a DI box. If the amp is miked, dynamic mics usually are chosen for their deep, rugged tones. The large-diaphragm dynamic designs tend to subdue the high-frequency transients. When combined with a boosted response at around 100 Hz, these large diaphragm dynamics give a warm, mellow tone that adds power to the lower register. Equalizing a bass can sometimes increase its clarity, with the fundamental being affected from 125 to 400 Hz and the harmonic punch being from 1.5 to 2 kHz. A compressor is commonly used on electric and acoustic basses. It's a basic fact that the signal output from the instrument's notes often varies in level, causing some notes to stand out while others dip in volume. A compressor having a smooth input/output ratio of roughly 4:1, a fast attack (8 to 20 milliseconds), and a slower release time (1/4 to 1/2 second) can often smooth out these levels, giving the instrument a strong, present and smooth bass line.

Keyboard instruments

The following sections describe the various sound characteristics and techniques that are encountered when miking keyboard instruments.

GRAND PIANO

The grand piano is an acoustically complex instrument that can be miked in a variety of ways, depending on the style and preferences of the artist, producer and/or engineer. The overall sound emanates from the instrument's strings, soundboard and mechanical hammer system. Because of its large surface area, a minimum miking distance of 4 to 6 feet is needed for the tonal balance to fully develop and be picked up; however, leakage from other instruments often means that these distances aren't practical or possible. As a result, pianos are often miked at distances that favor such instrument parts as:

- *Strings and soundboard*, often yielding a bright and relatively natural tone
- *Hammers*, generally yielding a sharp, percussive tone
- *Soundboard holes alone*, often yielding a sharp, full-bodied sound.

In modern music production, two basic grand piano styles can be found in the recording studio: the concert grand, which traditionally has a rich and full-bodied tone (often used for classical music and ranging in size up to 9 feet in length), and the studio grand, which is more suited for modern music production and has a sharper, more percussive edge to its tone (often being about 7 feet in length).

Figure 4.49 shows a number of miking positions that can be used in recording a grand piano. Although several mic positions are illustrated, it's important to

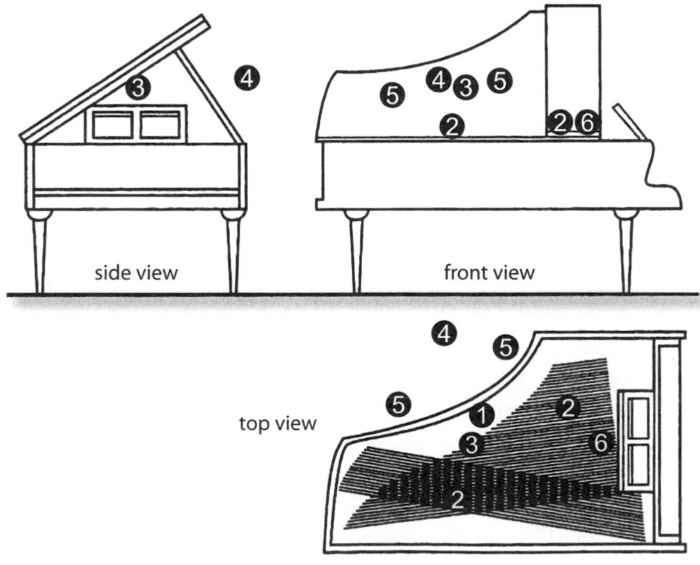

FIGURE 4.49
Possible miking combinations for the grand piano.

keep in mind that these are only guidelines from which to begin. Your own personal sound can be achieved through mic choice and experimentation with mic placement.

Position 1: The mic is attached to the partially or entirely open lid of the piano. The most appropriate choice for this pickup is the boundary mic, which can be permanently attached or temporarily taped to the lid. This method uses the lid as a collective reflector and provides excellent pickup under restrictive conditions (such as on stage and during a live video shoot).

Position 2: Two mics are placed in a spaced stereo configuration at a working distance of 6 inches to 1 inch. One mic is positioned over the low strings and one is placed over the high strings.

Position 3: A single mic or coincident stereo pair is placed just inside the piano between the soundboard and its fully or partially open lid.

Position 4: A single mic or stereo coincident pair is placed outside the piano, facing into the open lid (this is most appropriate for solo or accent miking).

Position 5: A spaced stereo pair is placed outside the lid, facing into the instrument.

Position 6: A single mic or stereo coincident pair is placed just over the piano hammers at a working distance of 4 to 8 inches to give a driving pop or rock sound.

A condenser or extended-range dynamic mic is most often the preferred choice when miking an acoustic grand piano, as those types of mics tend to accurately represent the transient and complex nature of the instrument. Should excessive leakage be a problem, a close-miked cardioid (or cardioid variation) can be used; however, if leakage isn't a problem, backing away to a compromise distance (3 to 6 feet) can help capture the instrument's overall tonal balance.

Separation

Separation is often a problem associated with the grand piano whenever it is placed next to noisy neighbors. Separation, when miking a piano, can be achieved in the following ways:

- Place the piano inside a separate isolation room.
- Place a flat (acoustic separator) between the piano and its louder neighbor.
- Place the mics inside the piano and lower the lid onto its short stick. A heavy moving or other type of blanket can be placed over the lid to further reduce leakage.
- Overdub the instrument at a later time. In this situation, the lid can be removed or propped up by the long stick, allowing the mics to be placed at a more natural-sounding distance.

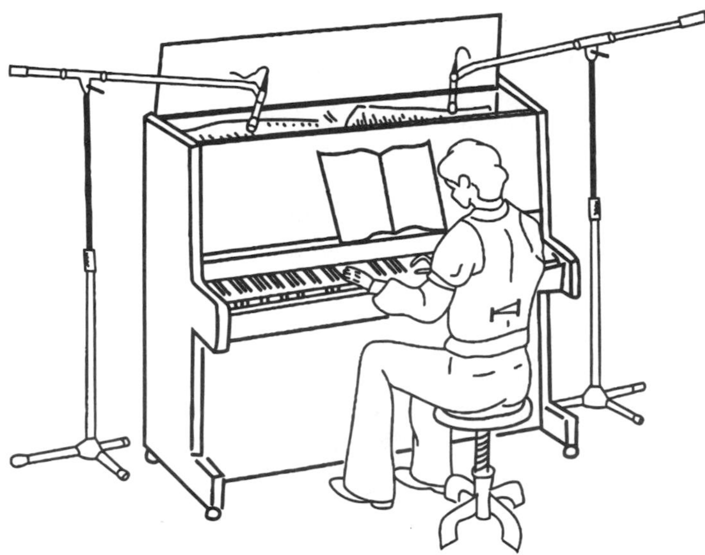

FIGURE 4.50
One possible pickup combination places the mics over the top of an upright piano.

UPRIGHT PIANO

You would expect the techniques for this seemingly harmless piano type to be similar to those for its bigger brother. This is partially true. However, because this instrument was designed for home enjoyment and not performance, the mic techniques are often very different. Since it's often more difficult to achieve a respectable tone quality when using an upright, you might want to try the following methods:

- *Miking over the top:* Place two mics in a spaced fashion just over and in front of the piano's open top, with one over the bass strings and one over the high strings. If isolation isn't a factor, remove or open the front face that covers the strings in order to reduce reflections and, therefore, the instrument's characteristic "boxy" quality. Also, to reduce resonances you might want to angle the piano out and away from any walls.
- *Miking the kickboard area:* For a more natural sound, remove the kickboard at the lower front part of the piano to expose the strings. Place a stereo spaced pair over the strings (one each at a working distance of about 8 inches over the bass and high strings). If only one mic is used, place it over the high-end strings. Be aware, though, that this placement can pick up excessive foot-pedal noise.
- *Miking the upper soundboard area:* To reduce excessive hammer attack, place a microphone pair at about 8 inches from the soundboard, above both the bass and high strings. In order to reduce muddiness, the soundboard should be facing into the room or be moved away from nearby walls.

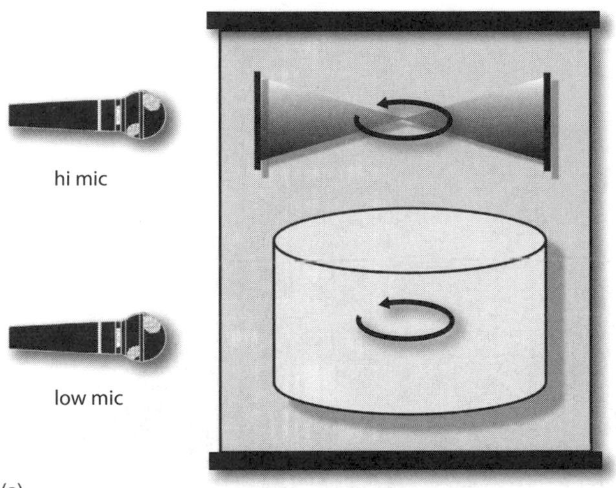

FIGURE 4.51
A Leslie speaker cabinet creates a unique vibrato effect by using a set of rotating speaker baffles that spin on a horizontal axis.
(a) Miking the rotating speakers of a Leslie cabinet;
(b) modern portable rotary amp with built-in microphones and three XLR outputs. (Courtesy of Motion Sound, www.motion-sound.com)

ELECTRONIC KEYBOARD INSTRUMENTS

Signals from most electronic instruments (such as synthesizers, samplers and drum machines) are often taken directly from the device's line-level output(s) and inserted into a console, either through a DI box or directly into a channel's line-level input. Alternatively, the keyboard's output can be plugged directly into the recorder or interface line-level inputs. The approach to miking an electronic organ can be quite different from the techniques just mentioned. A good Hammond or other older organ can sound wonderfully "dirty" through miked loudspeakers. Such organs are often played through a Leslie cabinet (Figure 4.51), which adds a unique, Doppler-based vibrato. Inside the cabinet is a set of rotating speaker baffles that spin on a horizontal axis and, in turn, produce a pitch-based vibrato as the speakers accelerate toward and away from the mics. The upper high-frequency speakers can be picked up by either one or two mics (each panned left and right), with the low-frequency driver being picked up by one mic. Motor and baffle noises can produce quite a bit of wind, possibly creating the need for a windscreen and/or experimentation with placement.

Percussion

The following sections describe the various sound characteristics and techniques that are encountered when miking drums and other percussion instruments.

DRUM SET

The standard drum kit (Figure 4.52) is often at the foundation of modern music, because it provides the "heartbeat" of a basic rhythm track; consequently, a proper drum sound is extremely important to the outcome of most music projects. Generally, the drum kit is composed of the kick drum, snare drum, high-toms, low-tom (one or more), hi-hat and a variety of cymbals. Since a full kit

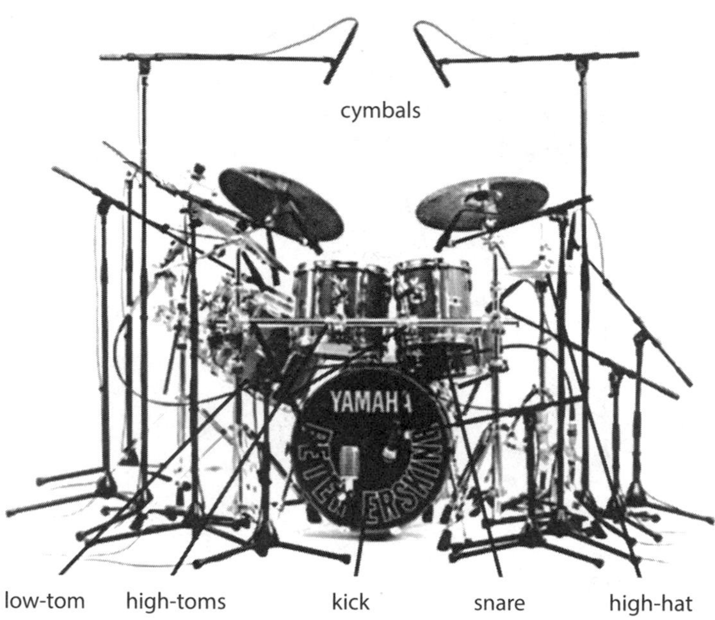

FIGURE 4.52
Peter Erskine's studio drum kit. (Courtesy of Beyerdynamic, www.beyerdynamic.com)

is a series of interrelated and closely spaced percussion instruments, it often takes real skill to translate the proper spatial and tonal balance into a project. The larger-than-life driving sound of the acoustic rock drum set that we've all become familiar with is the result of an expert balance among playing techniques, proper tuning and mic placement.

During the past several decades, drums have undergone a substantial change with regard to playing technique, miking technique and choice of acoustic recording environment. In the 1960s and 1970s, the drum set was placed in a small isolation room called a drum booth. This booth acoustically isolated the instrument from the rest of the studio and had the effect of tightening the drum sound because of the limited space (and often dead acoustics). The drum booth also physically isolated the musician from the studio, which often caused the musician to feel removed and less involved in the action. Today, many engineers and producers have moved the drum set out of smaller iso-rooms and back into larger open studio areas where the sound can fully develop and combine with the studio's own acoustics. In many cases, this effect can be exaggerated by placing a distant mic pair in the room (a technique that often produces a fuller, larger-than-life sound, especially in surround).

Before a session begins, the drummer should tune each drum while the mics and baffles for the other instruments are being set up. Each drumhead should be adjusted for the desired pitch and for constant tension around the rim by hitting the head at various points around its edge and adjusting the lugs for the same pitch all around the head. Once the drums are tuned, the engineer should listen to each drum individually to make sure that there are no buzzes, rattles, or resonant after-rings. Drums that sound great in live performance may not sound nearly as good when being close miked. In a live performance, the rattles and rings are covered up by the other instruments and are lost before the sound reaches the listener. Close miking, on the other hand, picks up the noises as well as the desired sound.

If tuning the drums doesn't bring the extraneous noises or rings under control, duct or masking tape can be used to dampen them. Pieces of cloth, dampening rings, paper towels, or a wallet can also be taped to a head in various locations (which is determined by experimentation) to eliminate rings and buzzes. Although head damping has been used extensively in the past, present methods use this damping technique more discreetly and will often combine dampening with proper design and tuning styles (all of which are the artist's personal call).

During a session, it's best to remove the damping mechanisms that are built into most drum sets, because they apply tension to only one spot on the head and unbalance its tension. These built-in dampeners often vibrate when the head is hit and are a chief source of rattles. Removing the front head and placing a blanket or other damping material inside the drum (so that it's pressing against the head) can often dampen the kick drum. Adjusting the amount of material can vary the sound from being a resonant boom to a thick, dull thud. Kick drums are usually (but not always) recorded with their front heads removed, while other drums are recorded with their bottom heads either on or off. Tuning the drums is more difficult if two heads are used because the head tensions often interact; however, they will often produce a more resonant tone. After the drums have been tuned, the mikes can be put into position. It's important to keep the mics out of the drummer's way, or they might be hit by a stick or moved out of position during the performance.

Miking the drum set

After the drum set has been optimized for the best sound, the mics can be placed into their pickup positions (Figure 4.53). Because each part of the drum set is so different in sound and function, it's often best to treat each grouping as an individual instrument. In its most basic form, the best place to start when miking a drum set is to start with the fundamental "groups." These include placing a mic on the kick (1) and on the snare drum (2). At an absolute minimum, the entire drum set can be adequately picked up using only four mics by adding two overhead pickups, either spaced (3) or coincident (4). In fact, this "bare bones" placement was (and continues to be) commonly used on many classic jazz recordings. If more tracks are available (or required), additional mics can be placed on the various toms, hi-hat and even individual cymbals.

A mic's frequency response, polar response, proximity effect and transient response should be taken into account when matching it to the various drum groups. Dynamic range is another important consideration when miking drums. Since a drum set is capable of generating extremes of volume and power (as

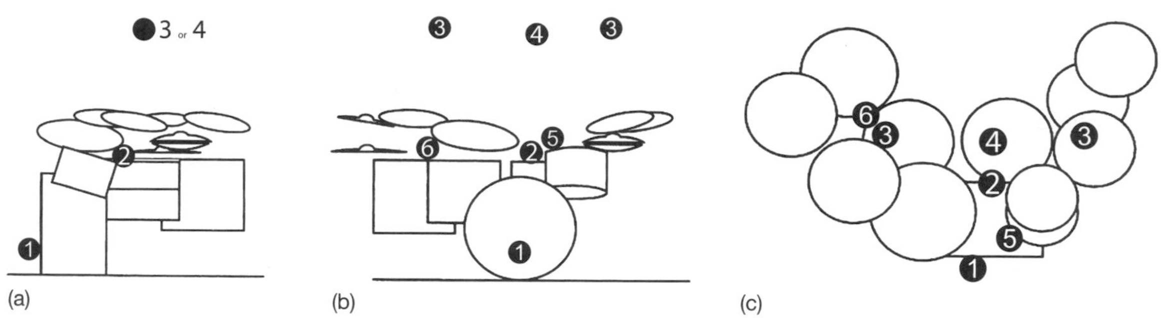

FIGURE 4.53
Typical microphone placements for a drum set: (a) side view; (b) front view; (c) top view.

well as softer, more subtle sounds), the chosen mics must be able to withstand strong peaks without distorting, and yet still be able to capture the more delicate nuances of a sound.

Since the drum set usually is one of the loudest sound sources in a studio setting, it's often wise to place it on a solidly supported riser. This reduces the amount of low-end "thud" that can otherwise leak through the floor into other parts of the studio. Depending on the studio layout, the following drum scenarios may occur:

- The drums could be placed in their own room, isolated from other instruments.
- To achieve a bigger sound, the drums could be placed in the large studio room while the other instruments are placed in smaller iso-rooms or are recorded direct.
- To reduce leakage, the drums could be placed in the studio, while being enclosed by 4-foot (or higher) divider flats.

Kick drum

The kick drum adds a low-energy drive or "punch" to a rhythm groove. This drum has the capability to produce low frequencies at high sound-pressure levels, so it's necessary to use a mic that can both handle and faithfully reproduce these signals. Often the best choice for the job is a large-diaphragm dynamic mic. Since proximity effect (bass boost) occurs when using a directional mic at close working distances and because the drum's harmonics vary over its large surface area, even a minor change in placement can have a profound effect on the pickup's overall sound. Moving the mic closer to the head (Figure 4.54) can add a degree of warmth and fullness, while moving it farther back often emphasizes the high-frequency "click." Placing the mic closer to the beater emphasizes the hard "thud" sound, whereas an off-center pickup captures more of the drum's characteristic skin tone. A dull and loose kick sound can be tightened to produce a sharper, more defined transient sound by placing a blanket or other damping material inside the drum shell firmly against the beater head. Cutting back on the

FIGURE 4.54
Placing the microphone at a distance just outside the kick drumhead to bring out the low end and natural fullness.

kick's equalization at 300 to 600 Hz can help reduce the dull "cardboard" sound, whereas boosting from 2.5 to 5 kHz adds a sharper attack, "click" or "snap." It's also often a good idea to have a can of WD-40® or other light oil handy in case squeaks from some of the moving parts (most often the kick pedal) gets picked up by the mics.

Snare drum

Commonly, a snare mic is aimed just inside the top rim of the snare drum at a distance of about 1 inch (Figure 4.55). The mic should be angled for the best possible separation from other drums and cymbals. Its rejection angle should be aimed at either the hi-hat or rack-toms (depending on leakage difficulties). Usually, the mic's polar response is cardioid, although bidirectional and super-cardioid responses might offer a tighter pickup angle. With certain musical styles (such as jazz), you might want a crisp or "bright" snare sound. This can be achieved by placing an additional mic on the snare drum's bottom head and then combining the two mics onto a single track. Because the bottom snare head is 180° out of phase with the top, it's generally a wise idea to reverse the bottom mic's phase polarity. When playing in styles where the snare springs are turned off, it's also wise to keep your ears open for snare rattles and buzzes that can easily leak into the snare mic (as well as other mics). The continued ringing of an "open" snare note (or any other drum type, for that matter) can be dampened in several ways. Dampening rings, which can be purchased at music stores, are used to reduce the ring and to deepen the instrument's tone. If there are no dampening rings around, the tone can be dampened by taping a billfold or similar-sized folded paper towel to the top of a drumhead, a few inches off its edge.

FIGURE 4.55
Typical microphone positioning for the snare drum.

Overheads

Overhead mics are generally used to pick up the high-frequency transients of cymbals with crisp, accurate detail while also providing an overall blend of the entire drum kit. Because of the transient nature of cymbals, a condenser mic is often chosen for its accurate high-end response. Overhead mic placement can be very subjective and personal. One type of placement is the spaced pair, whereby two mics are suspended above the left and right sides of the kit. These mics are equally distributed about the L/R cymbal clusters so as to pick up their respective instrument components in a balanced fashion (Figure 4.56a). Another placement

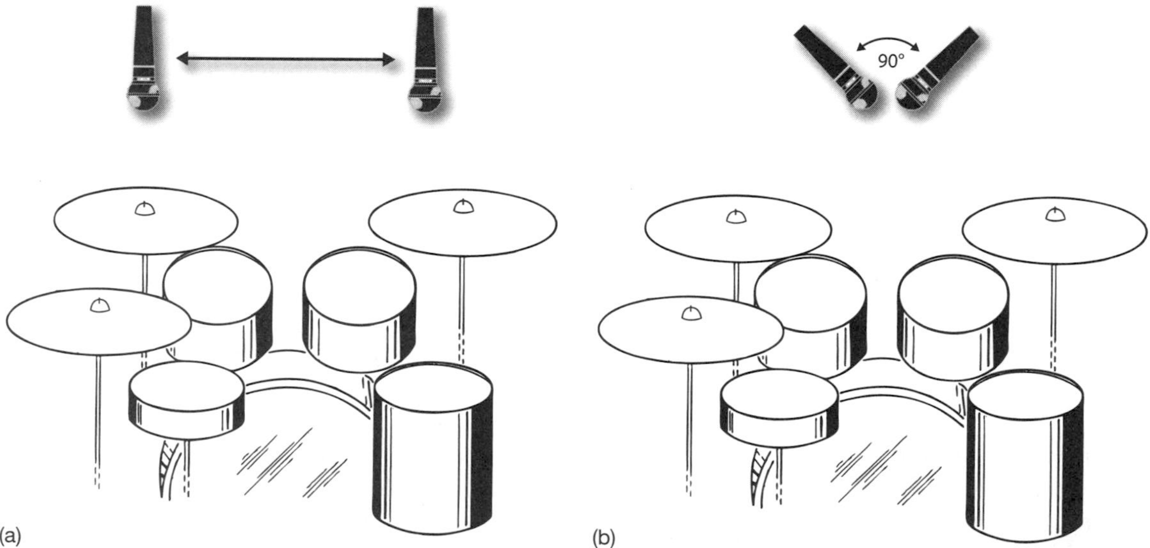

FIGURE 4.56
Typical stereo overhead pickup positions: (a) spaced pair technique; (b) X/Y coincident technique.

method is to suspend the mics closely together in a coincident fashion (Figure 4.56b). This often yields an excellent stereo overhead image with a minimum of the phase cancellations that might otherwise result when using spaced mics. Again, it's important to remember that there are no rules for getting a good sound. If only one overhead mic is available, place it at a central point over the drums. If you're using a number of pickups to close mic individual components of a kit, there might be times when you won't need overheads at all (the leakage spillover just might be enough to do the trick).

Rack-toms

The upper rack-toms can be miked either individually (Figure 4.57) or by placing a single mic between the two at a short distance (Figure 4.58). When miked

FIGURE 4.57
Individual miking of a rack-tom.

FIGURE 4.58
Single microphone placement for picking up two toms.

FIGURE 4.59
Typical microphone placement for the floor-tom.

individually, a "dead" sound can be achieved by placing the mic close to the drum's top head (about 1 inch above and 1 to 2 inches in from the outer rim). A sound that's more "live" can be achieved by increasing the height above the head to about 3 to 6 inches. If isolation or feedback is a consideration, a hypercardioid pickup pattern can be chosen. Another way to reduce leakage and to get a deep, driving tone (with less attack) is to remove the tom's bottom head and place the mic inside, 1 to 6 inches away from the top head.

Floor-tom

Floor-toms can be miked similarly to the rack-toms (Figure 4.59). The mic can be placed 2 to 3 inches above the top and to the side of the head, or it can be placed inside 1 to 6 inches from the head. Again, a single mic can be placed above and between the two floor-toms, or each can have its own mic pickup (which often yields a greater degree of control over panning and tonal color).

Hi-hat

The "hat" usually produces a strong, sibilant energy in the high-frequency range, whereas the snare's frequencies often are more concentrated in the midrange. Although moving the hat's mic won't change the overall sound as much as it would on a snare, you should still keep the following three points in mind:

- Placing the mic above the top cymbal will help pick up the nuances of sharp stick attacks.
- The open and closing motion of the hi-hat will often produce rushes of air; consequently, when miking the hat's edge, angle the mic slightly above or below the point where the cymbals meet.

- If only one mic is available (or desired), both the snare and hi-hat can be simultaneously picked up by carefully placing the mic between the two, facing away from the rack-toms as much as possible. Alternatively, a figure-8 mic can be placed between the two with the null axis facing toward the cymbals and kick.

TUNED PERCUSSION INSTRUMENTS

The following sections describe the various sound characteristics and techniques that are encountered when miking tuned percussion instruments.

Congas and hand drums

Congas, tumbas and bongos are single-headed, low-pitched drums that can be individually miked at very close distances of 1 to 3 inches above the head and 2 inches in from the rim, or the mics can be pulled back to a distance of 1 foot for a fuller, "live" tone. Alternatively, a single mic or X/Y stereo pair can be placed at a point about 1 foot above and between the drums (which are often played in pairs). Another class of single-headed, low-pitched drums (known as hand drums) isn't necessarily played in pairs but is often held in the lap or strapped across the player's front. Although these drums can be as percussive as congas, they're often deeper in tone and often require that the mic(s) be backed off in order to allow the sound to develop and/or fully interact with the room. In general, a good pickup can be achieved by placing a mic at a distance of 1 to 3 feet in front of the hand drum's head. Since a large part of the drum's sound (especially its low-end power) comes from its back hole, another mic can be placed at the lower port at a distance of 6 inches to 2 feet. Since the rear sound will be 180° out of phase from the front pickup, the mic's phase should be reversed whenever the two signals are combined.

Xylophone, vibraphone and marimba

The most common way to mic a tuned percussion instrument is to place two high-quality condenser or extended-range dynamic pickups above the playing bars at a spaced distance that's appropriate to the instrument size (following the 3:1 general rule). A coincident stereo pair can help eliminate possible phase errors; however, a spaced pair will often yield a wider stereo image.

Stringed instruments

Of all the instrumental families, stringed instruments are perhaps the most diverse. Ethnic music often uses instruments that range from being single stringed to those that use highly complex and developed systems to produce rich and subtle tones. Western listeners have grown accustomed to hearing the violin, viola, cello and double bass (both as solo instruments and in an ensemble setting). Whatever the type, stringed instruments vary in their design type and in construction to enhance or cut back on certain harmonic frequencies. These variations are what give a particular stringed instrument its own characteristic sound.

FIGURE 4.60
Example of a typical microphone placement for the violin.

VIOLIN AND VIOLA

The frequency range of the violin runs from 196 Hz to above 10 kHz. For this reason, a good mic that displays a relatively flat frequency response should be used. The violin's fundamental range is from G3 to E6 (196 to 1300 Hz), and it is particularly important to use a mic that's flat around the formant frequencies of 300 Hz, 1 kHz, and 1200 Hz. The fundamental range of the viola is tuned a fifth lower and contains fewer harmonic overtones. In most situations, the violin or viola's mic should be placed within 45° of the instrument's front face. The distance will depend on the particular style of music and the room's acoustic condition. Miking at a greater distance will generally yield a mellow, well-rounded tone, whereas a closer position might yield a scratchy, more nasal quality … the choice will depend the instrument's tone quality. The recommended miking distance for a solo instrument is between 3 and 8 feet, over and slightly in front of the player (Figure 4.60). Under studio conditions, a closer mic distance of between 2 and 3 feet is recommended. For a fiddle or jazz/rock playing style, the mic can be placed at a close working distance of 6 inches or less, as the increased overtones help the instrument to cut through an ensemble. Under PA (public address) applications, distant working conditions are likely to produce feedback (since less amplification is needed). In this situation, an electric pickup, contact, or clip-type microphone can be attached to the instrument's body or tailpiece.

CELLO

The fundamental range of the cello is from C2 to C5 (56 to 520 Hz), with overtones up to 8 kHz. If the player's line of sight is taken to be 0°, then the main direction of sound radiation lies between 10° and 45° to the right. A quality mic can be placed level with the instrument and directed toward the sound holes. The chosen microphone should have a flat response and be placed at a working distance of between 6 inches and 3 feet.

DOUBLE BASS

The double bass is one of the orchestra's lowest-pitched instruments. The fundamentals of the four-string type reach down to E1 (41 Hz) and up to around middle C (260 Hz). The overtone spectrum generally reaches upward to 7 kHz, with an overall angle of high-frequency dispersion being ±15° from the player's line of sight. Once again, a mic can be aimed at the *f* holes at a distance of between 6 inches and 1.5 feet.

Voice

From a shout to a whisper, the human voice is a talented and versatile sound source that displays a dynamic and timbrel range that's matched by few other instruments. The male bass voice can ideally extend from E2 to D4 (82 to 294 Hz) with sibilant harmonics extending to 12 kHz. The upper soprano voice can range upward to 1050 Hz with harmonics that also climb to 12 kHz.

When choosing a mic and its proper placement, it's important to step back for a moment and remember that the most important "device" in the signal chain is the vocalist. Let's assume that the engineer/producer hasn't made the classic mistake of waiting until the last minute (when the project goes over budget and/or into overtime) to record the vocals. … Good, now the vocalist can relax and concentrate on a memorable performance. Next step is to concentrate on the vocalist's "creature comforts": How are the lighting and temperature settings? Is the vocalist thirsty? Once done, you can go about the task of choosing your mic and its placement to best capture the performance.

The engineer/producer should be aware of the following traps that are often encountered when recording the human voice:

- *Excessive dynamic range:* This can be solved either by mic technique (physically moving away from the mic during louder passages) or by inserting a compressor into the signal path. Some vocalists have dynamics that range from whispers to normal volumes to practically screaming . . . all in a single passage. If you optimize your recording levels during a moderate-volume passage and the singer begins to belt out the lines, then the levels will become too "hot" and will distort. Conversely, if you set your recording levels for the loudest passage, the moderate volumes will be buried in the music. The solution to this dilemma is to place a compressor in the mic's signal path. The compressor automatically "rides" the signal's gain and reduces excessively loud passages to a level that the system can effectively handle. (See Chapter 12 for more information about compression and devices that alter dynamic range.)
- *Sibilance:* This occurs when sounds such as *f*, *s* and *sh* are overly accentuated. This often is a result of tape saturation and distortion at high levels or slow tape speeds. Sibilance can be reduced by inserting a frequency-selective compressor (known as a de-esser) into the chain or through the use of moderate equalization.
- *Excessive bass boost due to proximity effect:* This bass buildup often occurs when a directional mic is used at close working ranges. It can be reduced or compensated for by increasing the working distance between the source and the mic, by using an omnidirectional mic (which doesn't display a proximity bass buildup), or through the use of equalization.

MIC TOOLS FOR THE VOICE

Some of the most common tools in miking are used for fixing problems that relate to picking up the human voice and to room isolation.

Explosive popping *p* and *b* sounds often result when turbulent air blasts from the mouth strike the mic diaphragm. This problem can be avoided or reduced by:

- Placing a pop filter over the mic
- Placing a mesh windscreen between the mic and the vocalist
- Taping a pencil in front of the mic capsule, so as to break up the "plosive" air blasts
- Using an omnidirectional mic (which is less sensitive to popping, but might cause leakage issues)

Reducing problems due to leakage and inadequate isolation can be handled in any number of situational ways, including:

- Choice of directional pattern (i.e., choosing a tighter cardioid or hyper-cardioid pattern can help reduce unwanted leakage)
- Isolating the singer with a flat or portable isolation device
- Isolating the singer in a separate iso-booth
- Overdubbing the vocals at a later time, keeping in mind that carefully isolated "scratch" vocals can help glue the band together and give the vocalist a better feel for the song

Woodwind instruments

The flute, clarinet, oboe, saxophone and bassoon combine to make up the woodwind class of instruments. Not all modern woodwinds are made of wood nor do they produce sound in the same way. For example, a flute's sound is generated by blowing across a hole in a tube, whereas other woodwinds produce sound by causing a reed to vibrate the air within a tube.

Opening or covering finger holes along the sides of the instrument controls the pitch of a woodwind by changing the length of the tube and, therefore, the length of the vibrating air column. It's a common misunderstanding that the natural sound of a woodwind instrument radiates entirely from its bell or mouthpiece. In reality, a large part of its sound often emanates from the finger holes that span the instrument's entire length.

CLARINET

The clarinet commonly comes in two pitches: the B flat clarinet, with a lower limit of D3 (147 Hz), and the A clarinet, with a lower limit of C3 (139 Hz). The highest fundamental is around G6 (1570 Hz), whereas notes an octave above middle C contain frequencies of up to 150 Hz when played softly. This spectrum can range upward to 12 kHz when played loudly. The sound of this reeded woodwind radiates almost exclusively from the finger holes at frequencies between 800 Hz and 3 kHz; however, as the pitch rises, more of the sound emanates from the bell. Often, the best mic placement occurs when the pickup is aimed toward the lower finger holes at a distance of 6 inches to 1 foot (Figure 4.61).

FLUTE

The flute's fundamental range extends from B3 to about C7 (247 to 2093 Hz). For medium loud tones, the upper overtone limit ranges between 3 and 6 kHz. Commonly, the instrument's sound radiates along the player's line of sight for frequencies up to 3 kHz. Above this frequency, however, the radiated direction often moves outward 90° to the player's right. When miking a flute, placement depends on the type of music being played and the room's overall acoustics. When recording classical flute, the mic can be placed on-axis and slightly above the player at a distance of between 3 and 8 feet. When dealing with modern musical styles, the distance often ranges from 6 inches to 2 feet. In both circumstances, the microphone should be positioned at a point 1/3 to 1/2 the distance from the instrument's mouthpiece to its footpiece. In this way, the instrument's overall sound and tone quality can be picked up with equal intensity (Figure 4.62). Placing the mic directly in front of the mouthpiece will increase the level (thereby reducing feedback and leakage); however, the full overall body sound won't be picked up and breath noise will be accentuated. If mobility is important, an integrated contact pickup can be used or a clip mic can be secured near the instrument's mouthpiece.

FIGURE 4.61
Typical microphone position for the clarinet.

SAXOPHONE

Saxophones vary greatly in size and shape. The most popular sax for rock and jazz is the S-curved B-flat tenor sax, whose fundamentals span from B2 to F5 (117 to 725 Hz), and the E-flat alto, which spans from C3 to G5 (140 to 784 Hz). Also within this family are the straight-tubed soprano and sopranino, as well as the S-shaped baritone and bass saxophones. The harmonic content of these instruments ranges up to 8 kHz and can be extended by breath noises up to 13 kHz. As with other woodwinds, the mic should be placed roughly in the middle of the instrument at the desired distance and pointed slightly toward the bell (Figure 4.63). Keypad noises are considered to be a part of the instrument's sound; however, even these can be reduced or eliminated by aiming the microphone closer to the bell's outer rim.

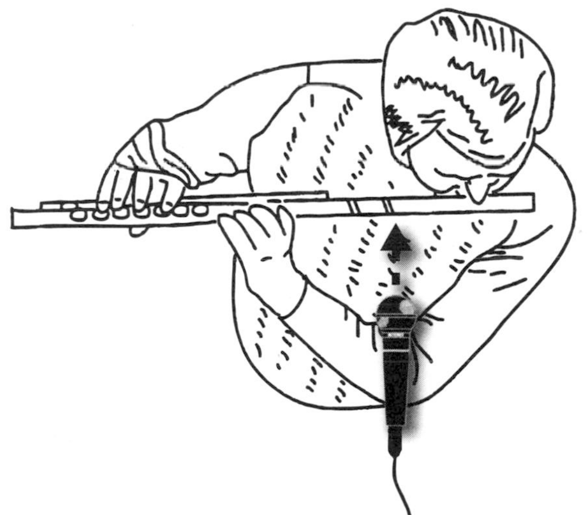

FIGURE 4.62
Typical microphone position for the flute.

HARMONICA

Harmonicas come in all shapes, sizes and keys ... and are divided into two basic types: the diatonic and the chromatic. Their pitch is determined purely by the

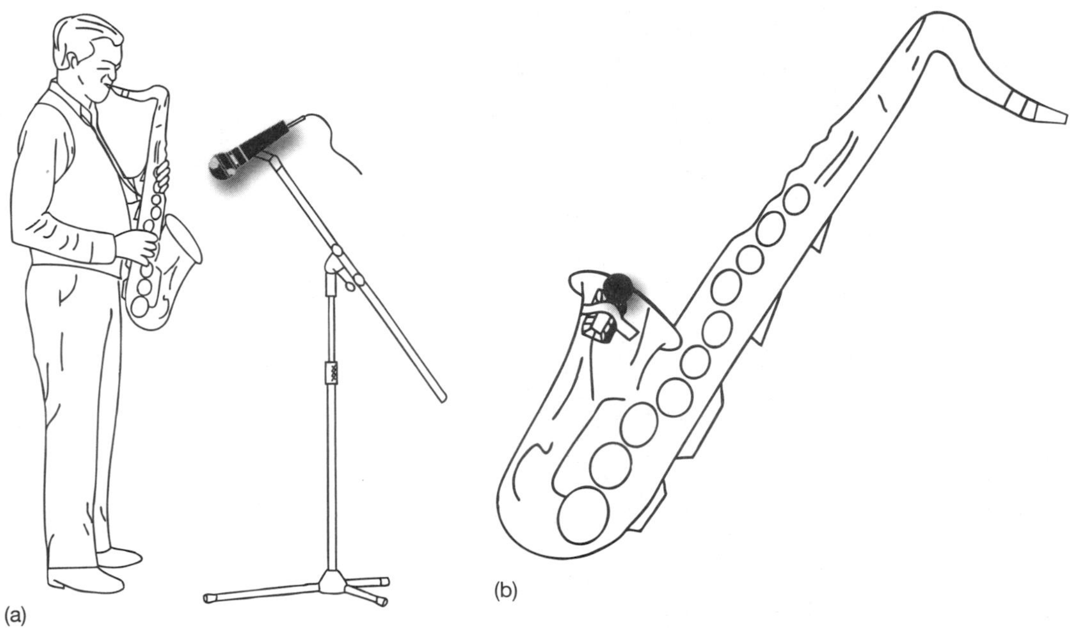

FIGURE 4.63
Typical microphone positions for the saxophone: (a) standard placement; (b) typical "clip-on" placement.

length, width and thickness of the various vibrating metal reeds. The "harp" player's habit of forming his or her hands around the instrument is a way to mold the tone by forming a resonant cavity. The tone can be deepened and a special "wahing" effect can be produced by opening and closing a cavity that's formed by the palms; consequently, many harmonica players carry their preferred microphones with them rather than being stuck in front of an unfamiliar mic and stand.

MICROPHONE SELECTION

The following information is meant to provide insights into a limited number of professional mics that are used for music recording and professional sound applications. This list is by no means complete, as literally hundreds of mics are available, each with its own particular design, sonic character and application.

Shure SM57

The SM57 (Figure 4.64) is widely used by engineers, artists, touring sound companies, etc., for instrumental and remote recording applications. The SM57's midrange presence peak and good low-frequency response make it useful for use with vocals, snare drums, toms, kick drums, electric guitars and keyboards.

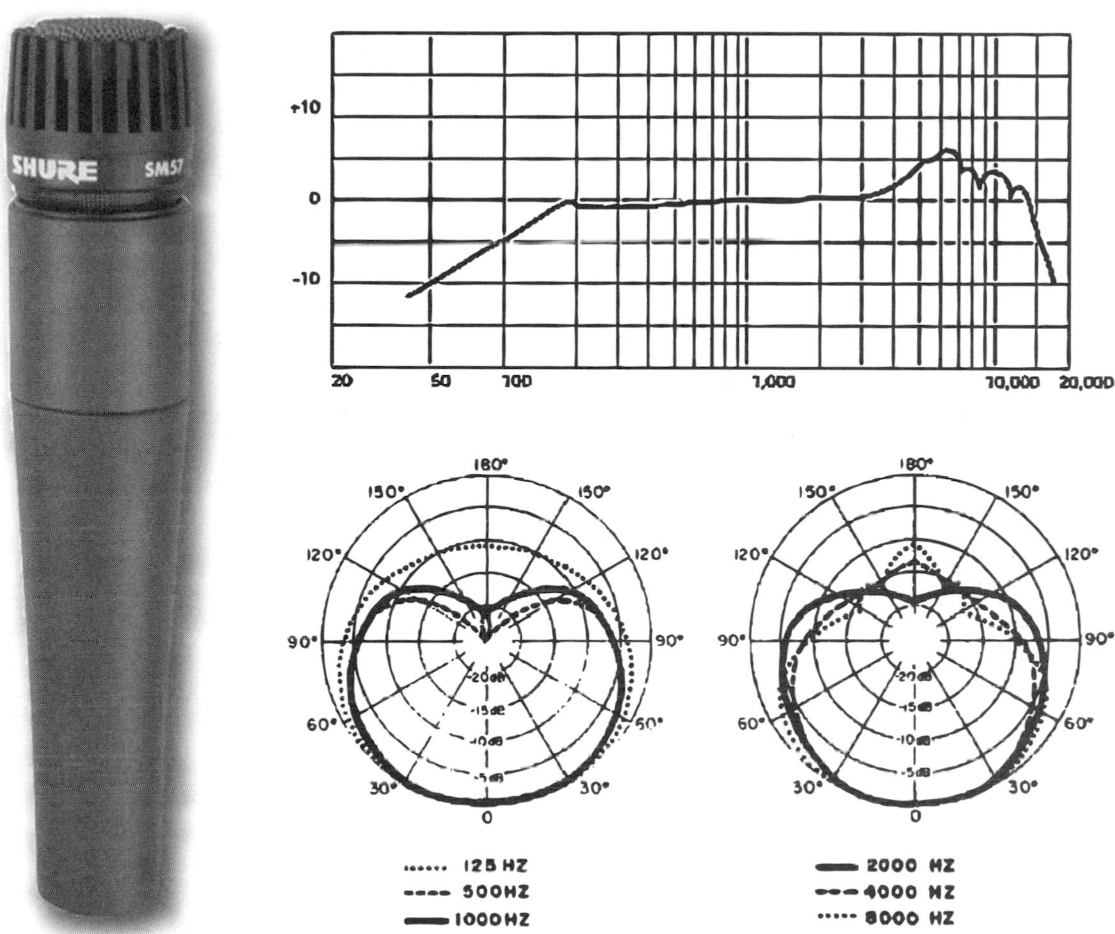

FIGURE 4.64
Shure SM57 dynamic microphone. (Courtesy of Shure Brothers, Inc., www.shure.com, Images © 2012, Shure Incorporated – used with permission)

Specifications:

- *Transducer type:* moving-coil dynamic
- *Polar response:* cardioid
- *Frequency response:* 40 to 15,000 Hz
- *Equivalent noise rating:* –7.75 dB (0 dB = 1 V/microbar).

AKG D112

Large-diaphragm cardioid dynamic mics, such as the AKG D112 (Figure 4.65), are often used for picking up kick drums, bass guitar cabinets, and other low-frequency, high-output sources.

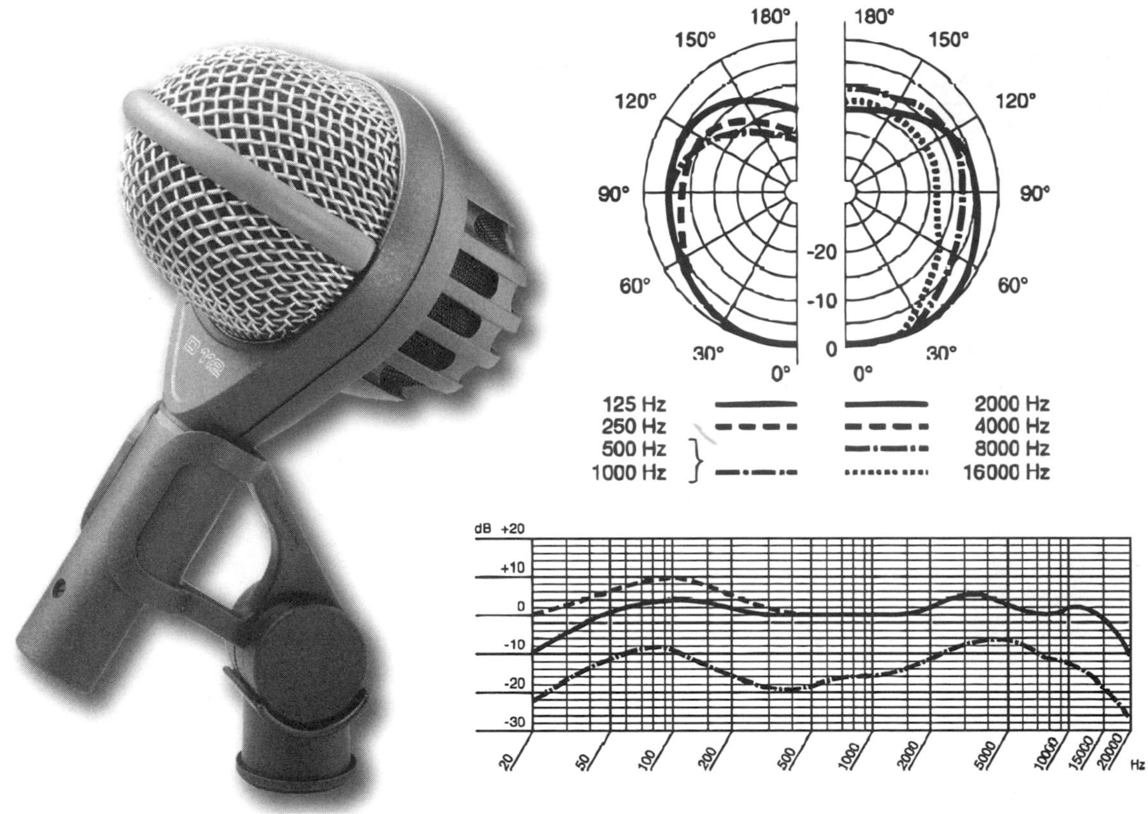

FIGURE 4.65
AKG D112 dynamic microphone. (Courtesy of AKG Acoustics GmbH., www.akg.com)

Specifications:

- *Transducer type*: moving-coil dynamic
- *Polar response:* cardioid
- *Frequency response:* 30 to 17,000 Hz
- *Sensitivity:* –54 dB ± 3 dB re. 1 V/microbar.

Beyerdynamic M-160

The Beyer M-160 ribbon microphone (Figure 4.66) is capable of handling high sound pressure levels without sustaining damage while providing the transparency that often is inherent in ribbon mics. Its hypercardioid response yields a wide-frequency response/low-feedback characteristic for both studio and stage.

Specifications

- *Transducer type:* Ribbon dynamic
- *Polar response:* Hypercardioid

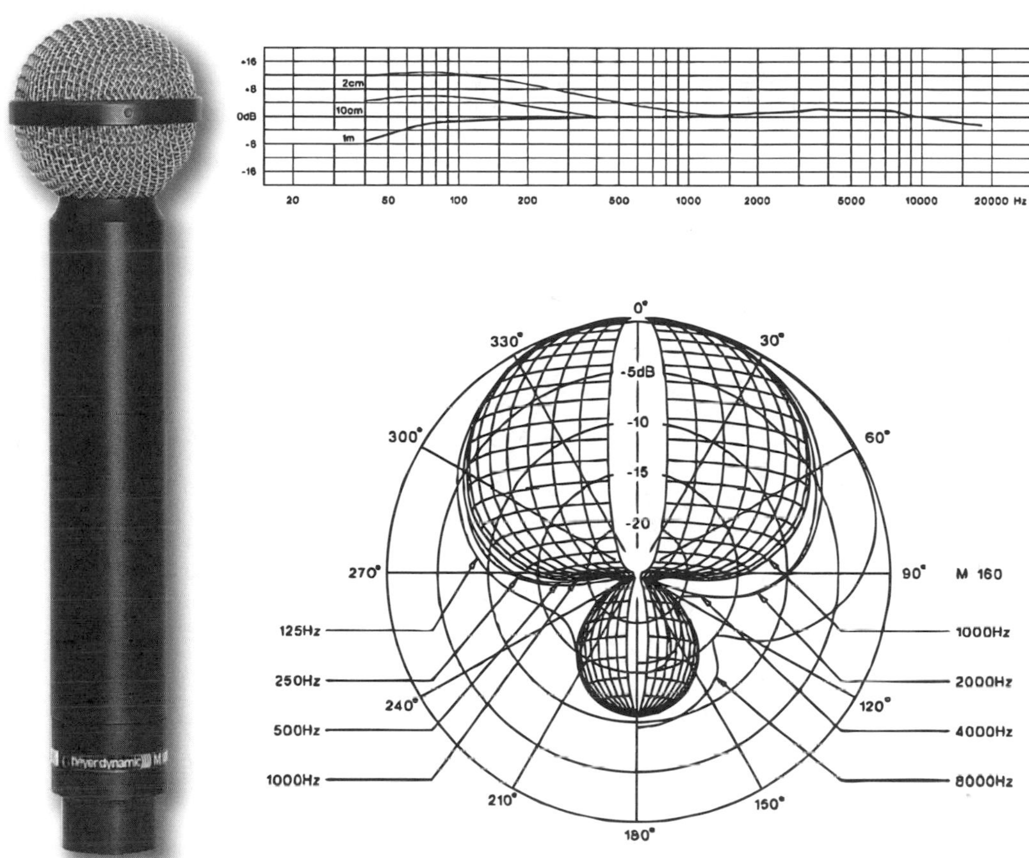

FIGURE 4.66
Beyerdynamic M-160 ribbon microphone. (Courtesy of Beyerdynamic, www.beyerdynamic.com).

- *Frequency response:* 40 – 18,000Hz
- *Sensitivity:* 52 dB (0 dB = 1 mW/Pa)
- *Equivalent noise rating:* –145dB
- *Output impedance:* 200Ω

AEA A440

The AEA model R44 and A440 ribbon microphones (Figure 4.4) carry forward the classic tradition of the venerable 1930s RCA-44 into the 21st century. The R44 series even uses "new old-stock" RCA ribbon material and features a re-engineered output transformer that recaptures the sonic signature of the original RCA-44 and surpasses it in frequency response and dynamics.

Specifications:

- *Transducer type:* ribbon dynamic
- *Polar response:* bidirectional

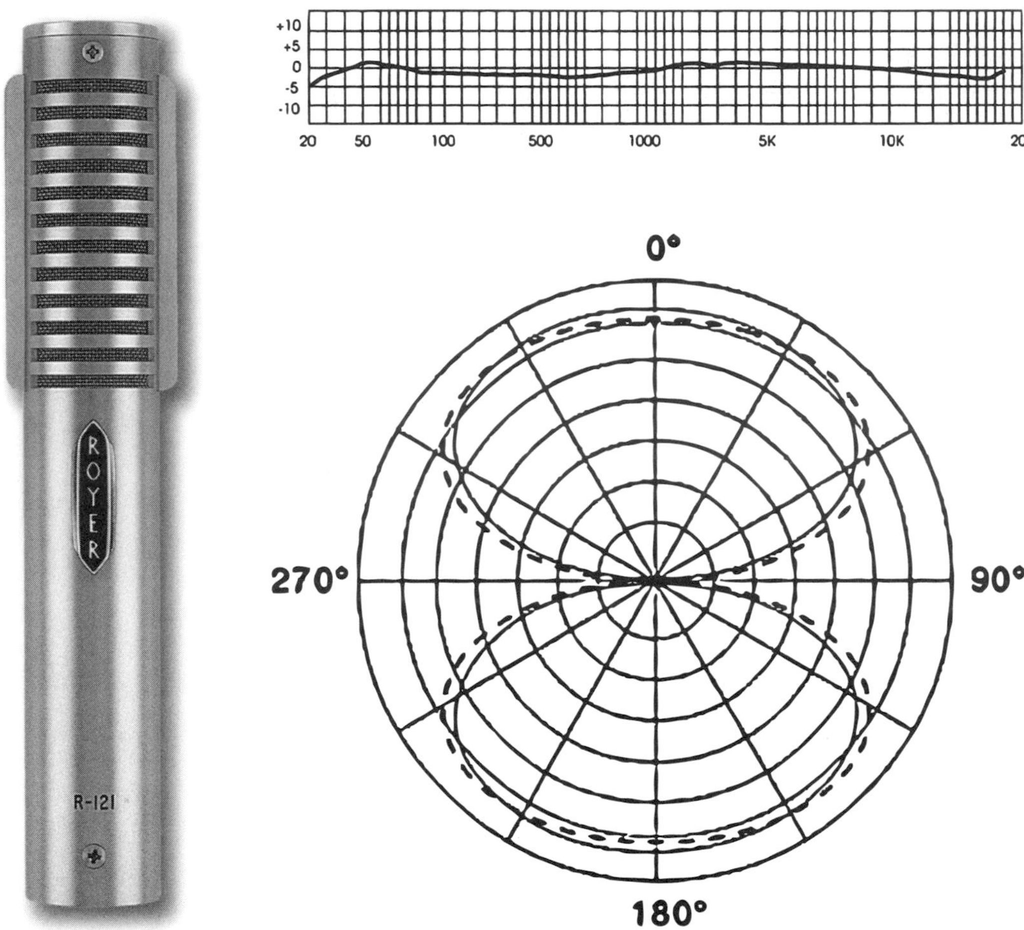

FIGURE 4.67
Royer Labs R-121 ribbon microphone. (Courtesy of Royer Labs, www.royerlabs.com)

- *Frequency response:* 20 to 20,000 Hz
- *Sensitivity:* 30 mV (–33.5 dBV)/Pa 1 Pa = 94 dB SPL
- *Equivalent noise rating:* Signal to Noise Ratio: 88 dB (A) (94 dB SPL minus equivalent noise)
- *Output impedance:* 92 Ω

Royer Labs R-121

The R-121 is a ribbon mic with a figure-8 pattern (Figure 4.67). Its sensitivity is roughly equal to that of a good dynamic mic, and it exhibits a warm, realistic tone and flat frequency response. Made using advanced materials and cutting-edge construction techniques, its response is flat and well balanced; the low end is deep and full without getting boomy, mids are well defined and realistic, and the high-end response is sweet and natural sounding.

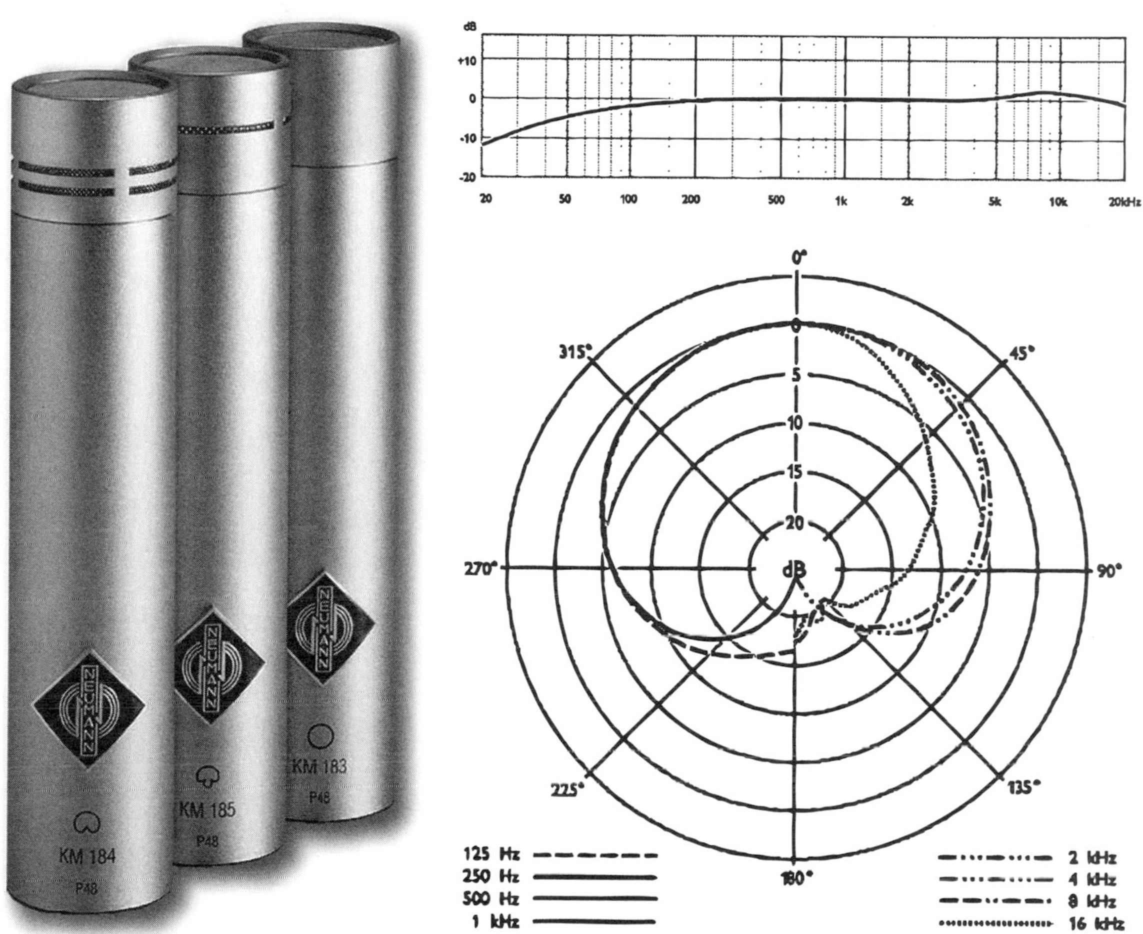

FIGURE 4.68
Neumann KM 180 Series condenser microphones. (Courtesy of Georg Neumann GMBH, www.neumann.com)

Specifications

- *Acoustic operating principle:* electrodynamic pressure gradient
- *Polar pattern:* figure 8
- *Generating element:* 2.5-micron aluminum ribbon
- *Frequency response:* 30 to 15,000 Hz ± 3 dB
- *Sensitivity:* –54 dBV re. 1 V/Pa ± 1 dB
- *Output impedance:* 300 Ω at 1 K (nominal); 200 Ω optional
- *Maximum SPL:* >135 dB

Neumann KM 180 Series

The 180 Series consists of three compact miniature microphones (Figure 4.68): the KM 183 omnidirectional and KM 185 hypercardioid microphones and the

FIGURE 4.69
AKG C 214 condenser microphone. (Courtesy of AKG Acoustics GMBH, www.akg.com)

successful KM 184 cardioid microphone. All 180 Series microphones are available with either a matte black or nickel finish and come in a folding box with a windshield and two stand mounts that permit connection to the microphone body or the XLR-connector.

Specifications

- *Transducer type:* condenser
- *Polar response:* cardioid (183), cardioid (184) and hypercardioid (185)
- *Frequency response:* 20 to 20 kHz
- *Sensitivity:* 12/15/10 mV/Pa
- *Output impedance:* 50 Ω
- *Equivalent noise level:* 16/16/18 dB –A

AKG C 214

The AKG C214 (Figure 4.69) is the younger brother of the legendary C414 condenser. Engineered for highest linearity and neutral sound, this mic uses the same one-inch capsule as the C414 in a single-diaphragm, cardioid-only design.

Specifications

- *Transducer type:* condenser
- *Polar response:* omnidirectional, wide cardioid, cardioid, hypercardioid, figure eight and four intermediate settings
- *Bass cut filter slope*: 12 dB/octave at 40 Hz and 80 Hz; 6 dB/octave at 160 Hz

FIGURE 4.70
Neumann TLM 102 condenser microphone. (Courtesy of Georg Neumann GMBH, www.neumann.com)

- *Frequency response:* 20 to 20,000 Hz
- *Sensitivity:* 20 mV/Pa.

Neumann TLM 102

The TLM 102 (Figure 4.70) has a newly developed large-diaphragm capsule (cardioid) with a maximum sound pressure level of 144 dB, which permits the recording of percussion, drums, amps and other very loud sound sources. For vocals and speech; a slight boost above 6 kHz provides for excellent presence of the voice in the overall mix. Up to 6 kHz the frequency response is extremely linear, ensuring minimal coloration and a clearly defined bass range. The capsule has an elastic suspension for the suppression of structure-borne noise.

Specifications

- *Transducer type:* condenser
- *Polar response:* cardioid
- *Frequency response:* 20 to 20,000 Hz
- *Sensitivity:* 11 mV/Pa.

Telefunken U47, C12 and ELA M251E

Telefunken (Figure 4.71) offers up historic recreations of classic microphones (in addition to their own line of microphones) that are based around the

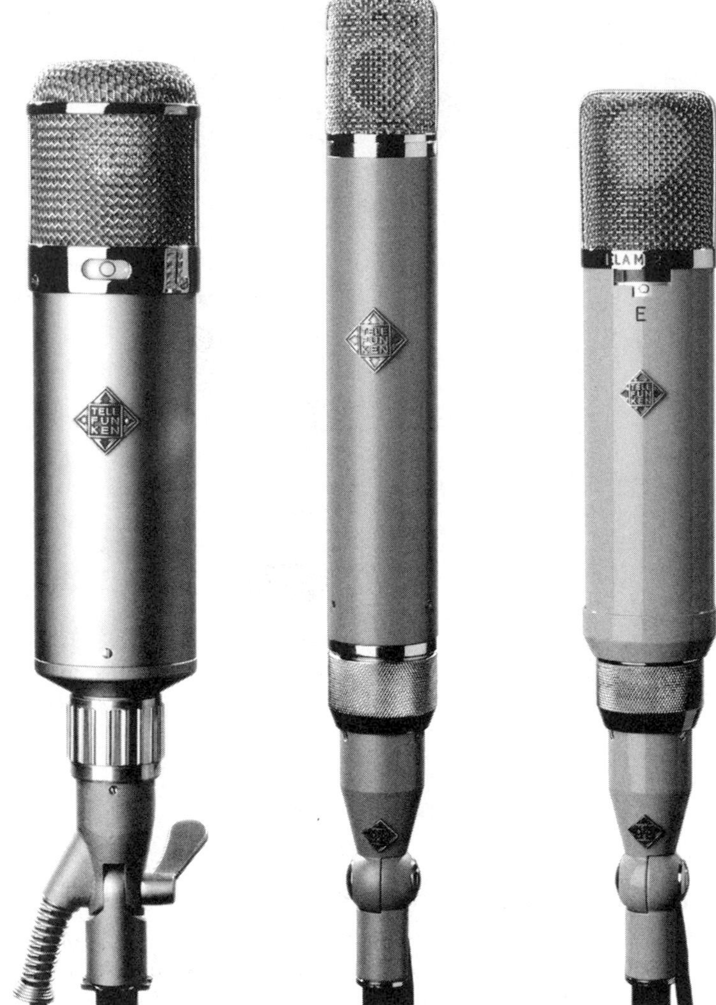

FIGURE 4.71
Telefunken U47, C12 and ELA M251E classic condenser microphone recreations. (Courtesy of Telefunken Elektroakoustik, www.telefunken-elektroakustik.com)

distinctive tube mic sound, blending vintage style and sound with the reliability of a modern-day microphone.

Sontronics Apollo stereo ribbon microphone

The Sontronics Apollo (Figure 4.72) offers up two phantom-powered ribbon microphones in a single housing that are configured as a Blumlein X/Y stereo pair. Specifications

- *Transducer type:* stereo ribbon
- *Polar response:* Blumlein stereo/figure eight
- *Frequency response:* 20 to 18,000 Hz
- *Sensitivity:* 18 mV/Pa.

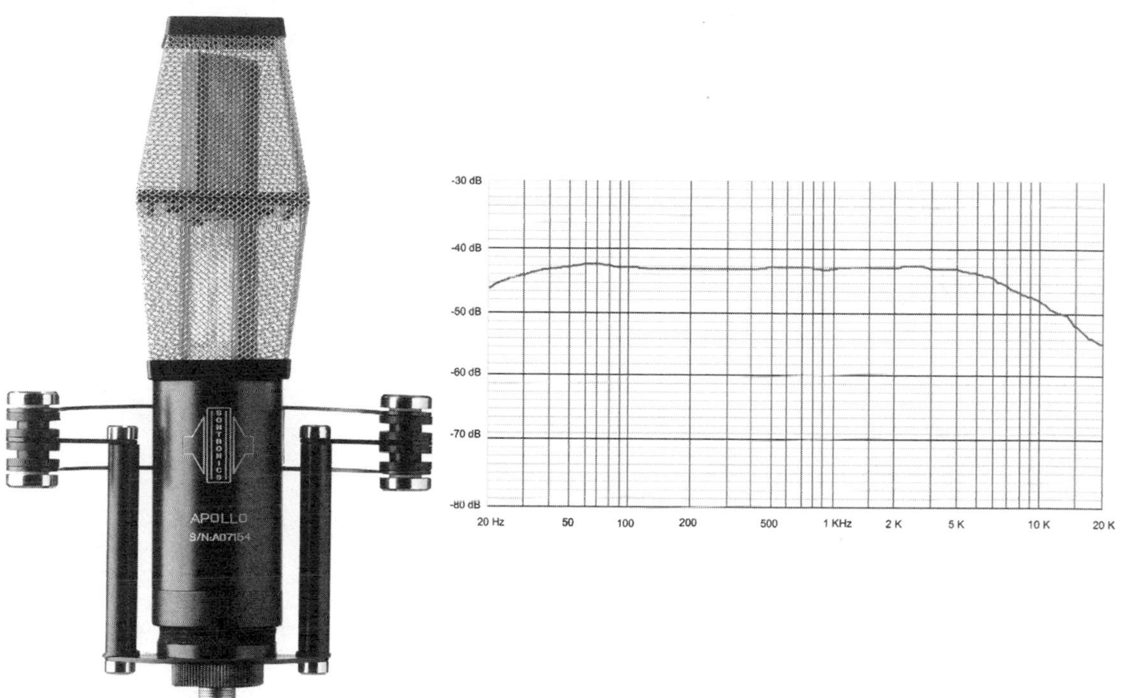

FIGURE 4.72
Sontronics Apollo Stereo ribbon microphone. (Courtesy of Sontronics and Omnisonics International Ltd, www.sontronics.com)

CHAPTER 5
The Analog Tape Recorder

From its inception in Germany in the late 1920s and its American introduction by Jack Mullin in 1945 (Figure 5.1), the *analog tape recorder* (or *ATR*) had steadily increased in quality and universal acceptance to the point that professional and personal studios had totally relied upon magnetic media for the storage of analog sound onto reels of tape. With the dawning of the project studio and computer-based DAWs, the use of two-channel and multitrack ATRs has steadily dwindled to the point where no new analog tape machine models are currently being manufactured. In short, recording to analog tape has steadily become a high-cost, future-retro, "specialty" process for getting a certain sound. This being said, the analog recording process is still highly regarded and even sought after by many studios for its special sound … and by others as a raised fist against the onslaught of the "evil digital empire." Without delving into the ongoing debate of the merits of analog versus digital, I think it's fair to say that each has its own distinct type of sound and application in audio and music production. Although professional analog recorders are usually much more expensive than their digital counterparts, as a general rule, a properly aligned, professional analog deck will have a particular sound that's often described as being full, punchy, gutsy and "raw" (when used on drums, vocals, entire mixes or anything that you want to throw at it). In fact, the limitations of tape are often used as a form of "artistic expression." From this, it's easy to see and hear why the analog tape recorder isn't dead yet … and probably won't be for some time.

TO 2-INCH OR NOT TO 2-INCH?

Before we delve into the inner workings of the analog tape recorder, let's take a moment to discuss ways in which the analog tape sound can be taken advantage of in the digital and project studio environment. Before you go out and buy your own deck, however, there are other cost-effective ways to get "that sound" on your own projects. For example:

- In recent times, a number of plug-ins have become available that can emulate (or approximate) the harmonic and overdriven sound of an analog tape track. Further info on these tape emulation plug-ins can be found later in this chapter and in Chapter 15.

FIGURE 5.1
John T. (Jack) Mullin (on the left) proudly displaying his two WWII vintage German Magnetophones, which were the first two tape-based recorders in the United States. (Courtesy of the late John T. Mullin)

- Rent a studio that has an analog multitrack for a few hours or days. You could record specific tracks to tape, transfer existing digital tracks to tape or dump an entire final mixdown to tape. For the cost of studio time and a reel of tape, you could inject your project with an entirely new type of sound (you might consider buying a single reel of multitrack tape that can be erased and reused once the takes have been transferred to disk).
- Rent an analog machine from a local studio equipment service. For a rental fee and basic cartage charges, you could reap the benefits of having an analog ATR for the duration of a project, without any undue financial and maintenance overhead.

A few guidelines should also be kept in mind when recording and/or transferring tracks to or from a multitrack recorder:

- Obviously, high recording levels add to that sought-after "overdriven" analog sound; however, driving a track too hard (hot) can actually kill a track's definition or "air." The trick is often to find a center balance between the right amount of saturation and downright distortion.
- Noise reduction can be a good thing, but it can also diminish what is thought of as that "classic analog sound." Newer, wide tape width recorders (such as ATR Services' ATR-102 1-inch, two-track and the 108C 2-inch, eight-track recorder), as well as older 2-inch, 16-track recorders, can provide improved definition without the need for noise reduction.

MAGNETIC RECORDING AND ITS MEDIA

At a basic level, an analog audio tape recorder can be thought of as a sound recording device that has the capacity to store audio information onto a magnetizable

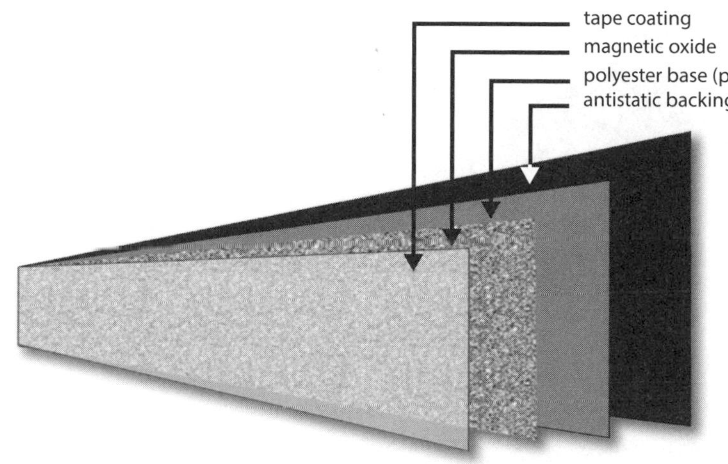

FIGURE 5.2
Structural layers of magnetic tape.

tape-based medium and then play this information back at a later time. By definition, analog refers to something that's "analogous," similar to or comparable to something else. An ATR is able to transform an electrical input signal directly into a corresponding magnetic energy that can be stored onto tape in the form of magnetic remnants. Upon playback, this magnetic energy is then reconverted back into a corresponding electrical signal that can be amplified, mixed, processed and heard.

The recording media itself is composed of several layers of material, each serving a specific function (Figure 5.2). The base material that makes up most of a tape's thickness is often composed of polyester or polyvinyl chloride (PVC), which is a durable polymer that's physically strong and can withstand a great deal of abuse before being damaged. Bonded to the PVC base is the all-important layer of magnetic oxide. The molecules of this oxide combine to create some of the smallest known permanent magnets, which are called *domains* (Figure 5.3a). On an unmagnetized tape, the polarities of these domains are randomly oriented over the entire surface of the tape. The resulting energy force of this random magnetization at the reproduce head is a general cancellation of the combined domain energies, resulting in no signal at the recorder's output (except for the tape noise that occurs due to the residual domain energy output).

When a signal is recorded, the magnetization from the record head polarizes the individual domains (at varying degrees in positive and negative angular directions) in such a way that their average magnetism produces a much larger combined magnetic flux (Figure 5.3b). When the tape is pulled across the playback head at the same, constant speed at which it was recorded, this alternating magnetic output is then converted back into an alternating signal that can then be amplified and further processed for reproduction.

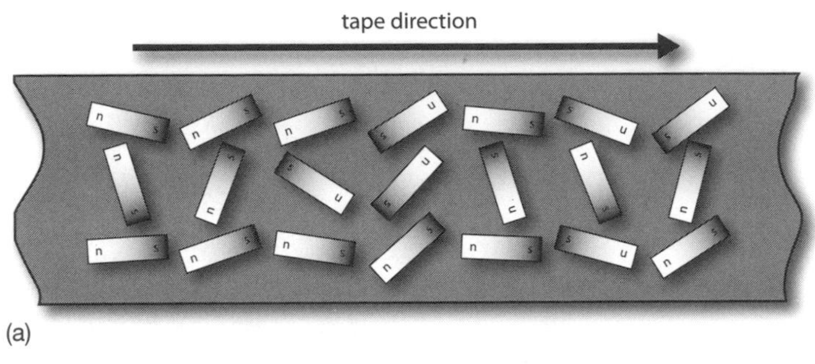

(a)

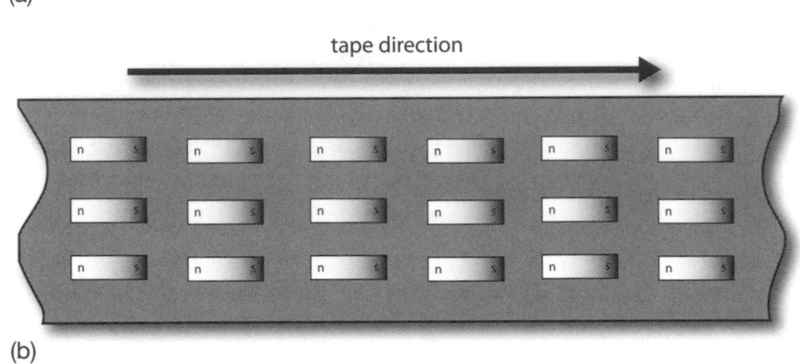

(b)

FIGURE 5.3
Orientation of magnetic domains on unmagnetized and magnetized recording tape: (a) The random orientation of an unmagnetized tape results in no output. (b) Magnetized domains result in an average flux output at the magnetic head.

FIGURE 5.4
Otari MX-5050 B3 two-channel recorder. (Courtesy of Otari Corporation, www.otari.com.).

THE PROFESSIONAL ANALOG ATR

Professional analog ATRs can be found in 2-, 4-, 8-, 16- and 24-track formats. Each configuration is generally best suited to a specific production and postproduction task. For example, a 2-track ATR is generally used to record the final stereo mix of a project (Figures 5.4 and 5.5), whereas 8-, 16- and 24-track machines are obviously used for multitrack recording (Figures 5.6 and 5.7). Although no professional analog machines are currently being manufactured, quite a few decks can be found on the used market in varying degrees of working condition. Certain recorders (such as the ATR-108C 2-inch, multitrack/mastering recorder) can be switched between tape width and track formats, allowing the machine to be converted to handle a range of multitrack, mixdown and mastering tasks.

THE TAPE TRANSPORT

The process of recording audio onto magnetic tape depends on the transport's capability to pass the tape across a head path at a constant speed and with a uniform tension. In simpler words, a recorder must uniformly pass a precise length of tape over the record head within a specific

FIGURE 5.5
ATR-102 1-inch stereo mastering recorder. (Courtesy of ATR Service Company, www.atrservice.com.)

time period (Figure 5.8). During playback, this relationship is maintained by again moving the tape across the heads at the same speed, thereby preserving the program's original pitch, rhythm and duration.

FIGURE 5.6
ATR-108C 2-inch multitrack/mastering recorder. (Courtesy of ATR Service Company, www.atrservice.com.)

This constant speed and tension movement of the tape across a head's path is initiated by simply pressing the Play button. The drive can be disengaged at any time by pressing the Stop button, which applies a simultaneous breaking force to both the left and right reels. The Fast Forward and Rewind buttons cause the tape to rapidly shuttle in the respective directions in order to locate a specific point. Initiating either of these modes engages the tape lifters, which raise the tape away from the heads (definitely an ear-saving feature). Once the play mode has been engaged, pressing the Record button allows audio to be recorded onto any selected track or tracks.

Beyond these basic controls, you might expect to run into several differences between transports (often depending on the machine's age). For example, older recorders might require that both the Record and Play buttons be simultaneously

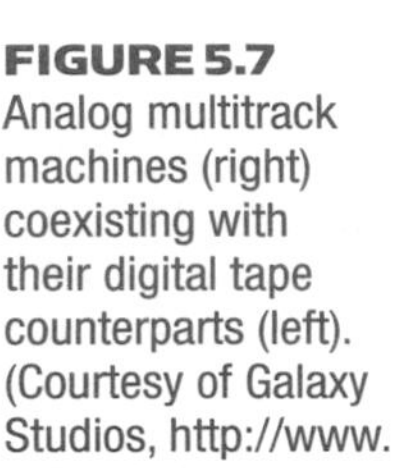

FIGURE 5.7
Analog multitrack machines (right) coexisting with their digital tape counterparts (left). (Courtesy of Galaxy Studios, http://www.galaxy.be)

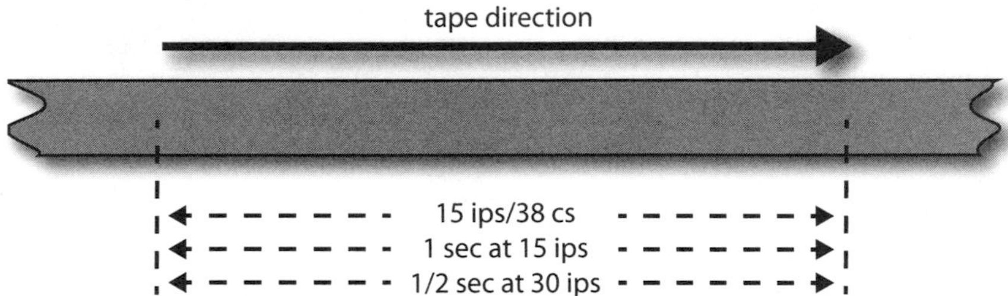

FIGURE 5.8
Relationship of time to the physical length of recording tape.

pressed in order to go into record mode; while others may begin recording when the Record button is pressed while already in the Play mode.

On certain older professional transports (particularly those wonderful Ampex decks from the 1950s and 1960s), stopping a fast-moving tape by simply pressing the Stop button can stretch or destroy a master tape, because the inertia is simply too much for the mechanical brake to deal with. In such a situation, a procedure known as "rocking" the tape is used to prevent tape damage. The deck can be rocked to its stop position by engaging the fast-wind mode in the direction opposite the current travel direction until the tape slows down to a reasonable speed … at which point it's safe to press the Stop button.

In recent decades, tape transport designs have incorporated total transport logic (TTL), which places transport and monitor functions under microprocessor control. This has a number of distinct advantages in that you can push the Play or Stop buttons while the tape is in fast-wind mode without fear of tape damage. With TTL, the recorder can sense the tape speed and direction and then automatically rock the transport until the tape can safely be stopped or can slow the tape to a point where the deck can seamlessly slip into play or record mode.

Most modern ATRs are equipped with a shuttle control that enables the tape to be shuttled at various wind speeds in either direction. This allows a specific cue point to be located by listening to the tape at varying play speeds, or the control can be used to gently and evenly wind the tape onto its reel at a slower speed for long-term storage. The Edit button (which can be found on certain professional machines) often has two operating modes: stop-edit and dump-edit. If the Edit button is pressed while the transport is in the stop mode, the left and right tape reel brakes are released and the tape sensor is bypassed. This makes it possible for the tape to be manually rocked back and forth until the edit point is found. Often, if the Edit button is pressed while in the play mode, the take-up turntable is disengaged and the tape sensor is bypassed. This allows unwanted sections of tape to be spooled off the machine (and into the trash can) while listening to the material as it's being dumped during playback.

A safety switch, which is incorporated into all professional transports, initiates the stop mode when it senses the absence of tape along its guide path; thus, the recorder stops automatically at the end of a reel or should the tape accidentally break. This switch can be built into the tape-tension sensor, or it might exist in the form of a light beam that's interrupted when tape is present.

Most professional ATRs are equipped with automatic tape counters that accurately read out time in hours, minutes, seconds and sometimes frames (00:00:00:00). Many of these recorders have digital readout displays that double as tape-speed indicators when in the "vari-speed" mode. This function incorporates a control that lets you vary the tape speed from fixed industry standards. On many tape transports, this control can be continuously varied over a ±20% range from the 7 1/2, 15 or 30 ips (inches per second) standard.

THE MAGNETIC TAPE HEAD

Most professional analog recorders use three magnetic tape heads, each of which performs a specialized task:

- Record
- Reproduce
- Erase

The function of a *record head* (Figure 5.9) is to electromagnetically transform analog electrical signals into corresponding magnetic fields that can be permanently stored onto magnetic tape. In short, the input current flows through coils of wire that are wrapped around the head's magnetic pole pieces. Since the theory

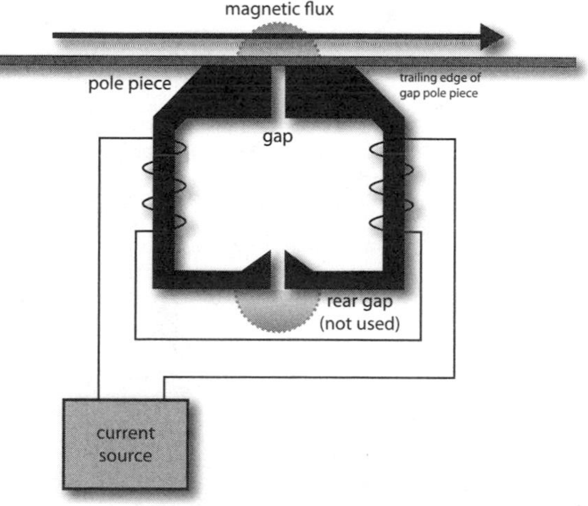

FIGURE 5.9
The record head.

of magnetic induction states that "whenever a current is injected into metal, a magnetic field is created within that metal," a magnetic force is caused to flow through the coil, into the pole pieces and across the head gap. Like electricity, magnetism flows more easily through some media than through others. The head gap between poles creates a break in the magnetic field, thereby creating a physical resistance to the magnetic "circuit." Since the gap is in physical contact with the moving magnetic tape, the tape's magnetic oxide offers a lower resistance path to the field than does the nonmagnetic gap. Thus, the flux path travels from one pole piece, into the tape and to the other pole. Since the magnetic domains retain their polarity and magnetic intensity as the tape passes across the gap, the tape now has an analogous magnetic "memory" of the recorded event.

The reproduce or *playback head* (Figure 5.10) operates in a way that's opposite to the record head. When a recorded tape track passes across the reproduce head gap, a magnetic flux is induced into the pole pieces. Since the theory of magnetic induction also states that "whenever a magnetic field cuts across metal, a current will be set up within that metal," an alternating current is caused to flow through the pickup coil windings, which can then be amplified and processed into a larger output signal.

Note that the reproduce head's output is nonlinear because this signal is proportional to both the tape's average flux magnitude and the rate of change of this magnetic field. This means that the rate of change increases as a direct function of the recorded signal's frequency. Thus, the output level of a playback head

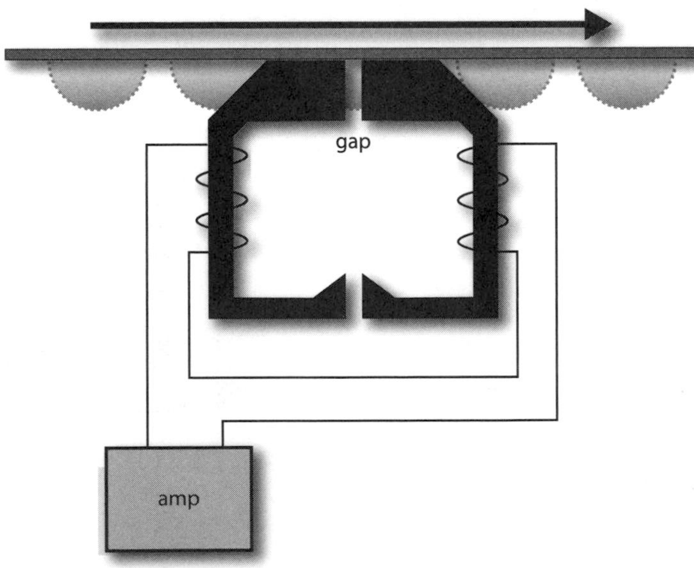

FIGURE 5.10
The playback head.

effectively doubles for each doubling in frequency, resulting in a 6 dB increase in output voltage for each increased octave. The tape speed and head gap width work together to determine the reproduce head's upper-frequency limit, which in turn determines the system's overall bandwidth. The wavelength of a signal that's recorded onto tape is equal to the speed at which tape travels past the reproduce head, divided by the frequency of the signal; therefore, the faster the tape speed, the higher the upper-frequency limit. Similarly, the smaller the head gap, the higher the upper-frequency limit.

The function of the *erase head* is to effectively reduce the average magnetization level of a recorded tape track to zero, thereby allowing the tape track to be re-recorded. After a track is placed into the record mode, a high-frequency and high-intensity sine-wave signal is fed into the erase head (resulting in a tape that's being saturated in both the positive- and negative-polarity directions). This alternating saturation occurs at such a high speed that it serves to confuse any magnetic pattern that existed on the tape. As the tape moves away from the erase head, the intensity of the magnetic field decreases, leaving the domains in a random orientation, with a resulting average magnetization or output level that's as close to zero as tape noise will allow.

EQUALIZATION

Equalization (EQ) is the term used to denote an intentional change in relative amplitudes at different frequencies. Because the analog recording process isn't linear, equalization is needed to achieve a flat frequency-response curve when using magnetic tape. The 6 dB-per-octave boost that's inherent in the playback head's response curve requires that a complementary equalization cut of 6 dB per octave be applied within the playback circuit (see Figure 5.11).

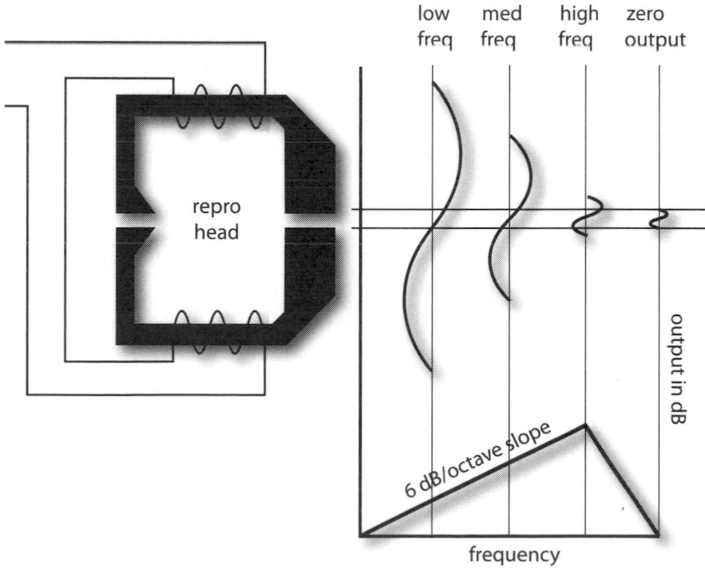

FIGURE 5.11
A flat frequency playback curve results due to complementary equalization in the playback circuit.

BIAS CURRENT

In addition to the nonlinear changes that occur in playback level relative to frequency, another discrepancy in the recording process exists between the amount of magnetic energy that's applied to the record head and the amount of magnetism that's retained by the tape after the initial recording field has been removed. As Figure 5.12a shows, the magnetization curve of tape is linear between points A and B, as well as between points C and D. Signals greater than A and D have reached the saturation level and are subject to clipping distortion. Signals falling within the B to C range are too low in flux level to adequately magnetize the domains during the recording process. For this reason, it's important that low-level signals be boosted so that they're pushed into the linear range. This boost is applied by mixing an AC *bias current* (Figure 5.12b) with the audio signal. This bias current is applied by mixing the incoming audio signal with an ultrasonic sine-wave signal (often between 75 and 150 kHz). The combined signals are amplitude modulated in such a way that the overall magnetic flux levels are given an extra "oomph," which effectively boosts the signal above the nonlinear zero-crossover range and into the linear portion of the curve. In fact, if this bias signal weren't added, distortion levels would be so high as to render the recording process useless.

MONITORING MODES

The output signal of a professional ATR channel can be switched between three working modes:

- Input
- Reproduce
- Sync

In the *input* (source) *mode*, the signal at the selected channel output is derived from its input signal. Thus, with the ATR transport in any mode (including stop), it's possible to meter and monitor a signal that's present at a channel's selected input. In the reproduce mode, the output and metering signal is derived from the playback head. This mode can be useful in two ways: It allows previously recorded tapes to be played back, and it enables the monitoring of material off of the tape while in the record mode. The latter provides an immediate quality check of the ATR's entire record and reproduce process. The *sync mode* (originally known as selective synchronization, or sel-sync) is a required feature in analog multitrack ATRs because of the need to record new material on one or more tracks while simultaneously monitoring tracks that have been previously recorded (during a process called *overdubbing*). Here's the deal … using the record head to lay down one or more tracks while listening to previously recorded tracks through the reproduce head would actually cause the newly recorded track(s) to be out of sync with the others on final playback (due to the physical distance between the two, as shown in Figure 5.13a). To prevent such a

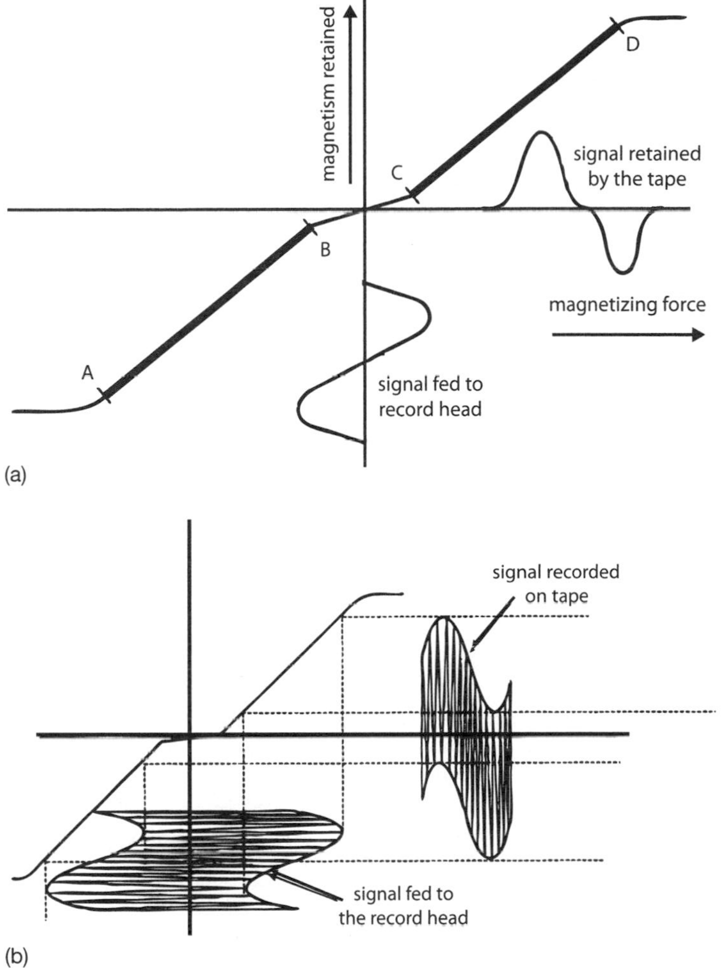

FIGURE 5.12
The effects of bias current on recorded linearity: (a) magnetization curve showing distortion at lower levels; (b) after bias, when the signal is boosted back into the curve's linear regions.

time lag, all of the reproduced tracks must be monitored off of the record head at the same time that record tracks are being laid down onto the same head. Since the record head is used for both recording and playback, there is no physical time lag and, thus, no signal delay (Figure 5.13b).

TO PUNCH OR NOT TO PUNCH

You've all heard the age-old adage … "$%& happens." Well, it happens in the studio—a lot! Whenever a mistake or bad line occurs during a multitrack session, it's often (but not always) possible to *punch-in* on a specific track or set of tracks. Instead of going back and re-recording an entire song or overdub,

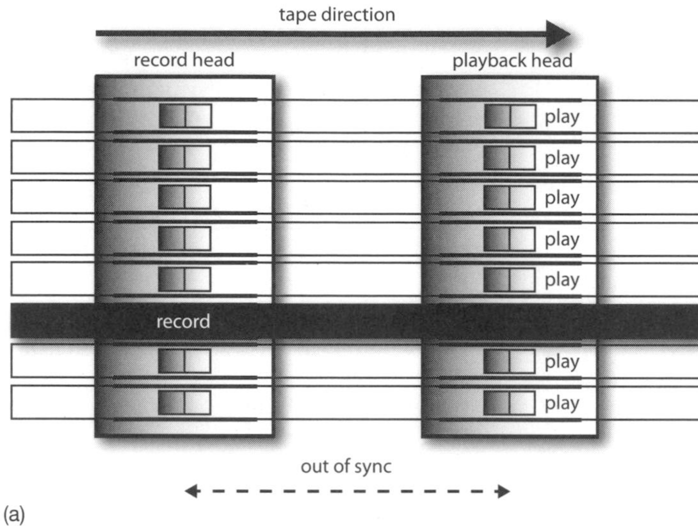

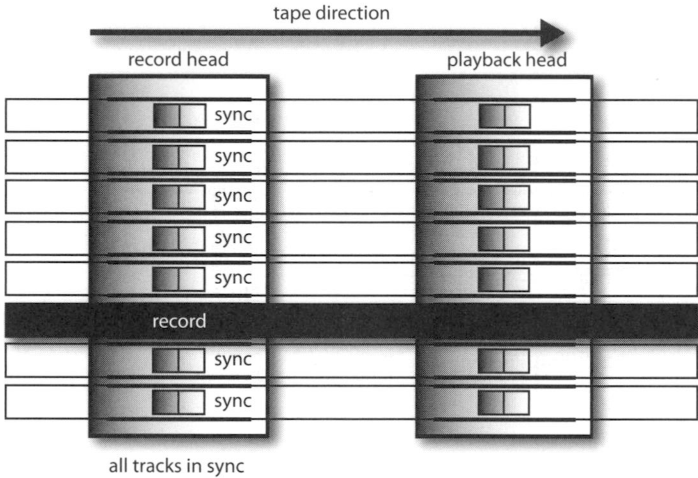

FIGURE 5.13
The sync mode's function. (a) In the monitored playback mode, the recorded signal lags behind the recorded signal, thereby creating an out-of-sync condition. (b) In the sync mode, the record head acts as both record and playback head, bringing the signals into sync.

performing a punch involves going back and re-recording over a specific section in order to fix a bad note, musical line—you name it. This process is done by cueing the tape at a logical point before the bad section and then pressing Play. Just before the section to be fixed, pressing the Record button (or entering record under automation) will place the track into record mode. At the section's end, pressing the Play button again causes the track to fall back out of record, thereby preserving the section following the punch. From a monitor standpoint, the recorder begins playback in the sync mode; once placed into record, the track switches to monitor the input source. This lets the performers hear themselves during the punch while listening to playback both before and after the take.

When performing a punch, it's often far better to "fix" the track immediately after the take has been recorded, while the levels, mic positions and performance vibe are still the same. This also makes it easier to go back and re-record the entire song or a larger section should the punch not work. If the punch can't be performed at that time, however, it's generally a good idea to take detailed notes about mic selection, placement, preamps and so on to recreate the session's setup without having to guess the details from memory.

As any experienced engineer/producer knows, performing a punch can be tricky. In certain situations, it's a complete no-brainer—for example, when a stretch of silence the size of a Mack truck exists both before and after the bad section, you'll have plenty of room to punch in and out. At other times, a punch can be very tight or problematic (e.g., if there's very little time to punch in or out, when trying to keep vocal lines fluid and in-context, when it's hard to feel the beat of a song or if it has a fast rhythm). In short, punching-in shouldn't be taken too lightly … nor so seriously that you're afraid of the process. Talk it over with the producer and/or musicians. Is this an easy punch? Does the section really need fixing? Do we have the time right now? Or, is it better just to redo the song? In short, the process is totally situational and requires attention, skill, experience and sometimes a great deal of luck.

- Before committing the punch to tape, it's often a wise idea to rehearse the punch, without actually committing the fix to tape. This has the advantage of giving both you and the performer a chance to practice beforehand.
- Some analog decks (and most DAWs) will let you enter the punch-in and punch-out times under automation, thereby allowing the punch to be surgically performed automatically.
- If you're recording onto the same track, a fudged punch may leave you with few options other than to re-record the entire song or section of a song. An alternative to this dilemma would be to record the fix into a separate track and then switch between tracks in mixdown.
- The track(s) could be transferred to a DAW, where the edits could be performed in the digital domain.

TAPE, TAPE SPEED AND HEAD CONFIGURATIONS

Professional analog ATRs are currently available in a wide range of track- and tape-width configurations. The most common analog configurations are 2-track mastering machines that use tape widths of 1/4 inch, 1/2 inch, and even 1 inch, as well as 16- and 24-track machines that use 2-inch tape. Figure 5.14 details many of the tape formats that can be currently found. Optimal tape-to-head performance characteristics for an analog ATR are determined by several parameters: track width, head-gap width and tape speed. In general, track widths are on the order of 0.080 inch for a 1/4-inch 2-track ATR; 0.070 inch for 1/2-inch

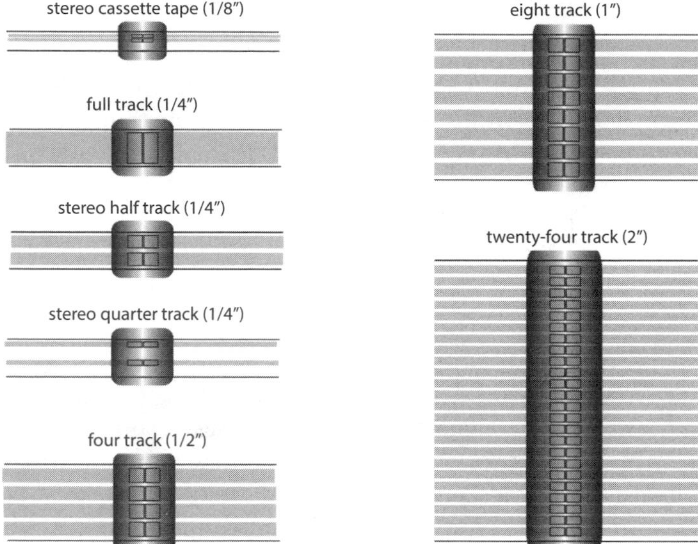

FIGURE 5.14
Analog track configurations for various tape widths.

4-track, 1-inch 8-track, and 2-inch 16-track formats or 0.037 inch for the 2-inch 24-track format. As you might expect, the greater the recorded track width, the greater the amount of magnetism that can be retained by the magnetic tape, resulting in a higher output signal and an improved signal-to-noise ratio. The use of wider track widths also makes the recorded track less susceptible to signal-level dropouts.

The most common tape speeds used in audio production are 15 ips (38 cm/sec) and 30 ips (76 cm/sec). Although 15 ips will eat up less tape, 30 ips has gained wide acceptance in recent years for having its own characteristic sound (often having a tighter bottom end), as well as a higher output and lower noise figures (which in certain cases eliminate the need for noise reduction). On the other hand, 15 ips has a reputation for having a "gutsy," rugged sound.

PRINT-THROUGH

A form of deterioration in a recording's quality, known as *print-through*, begins to occur after a recording has been made. This effect is the result of the transfer of a recorded signal from one layer of tape to an adjacent track layer by means of magnetic induction, which gives rise to an audible false signal or pre-echo on playback. The effects of print-through are greatest when recording levels are very high, and the effect decreases by about 2 dB for every 1-dB reduction in signal level. The extent of this condition also depends on such factors as length of storage, storage temperature and tape thickness (tapes with a thicker base material are less likely to have severe print-through problems).

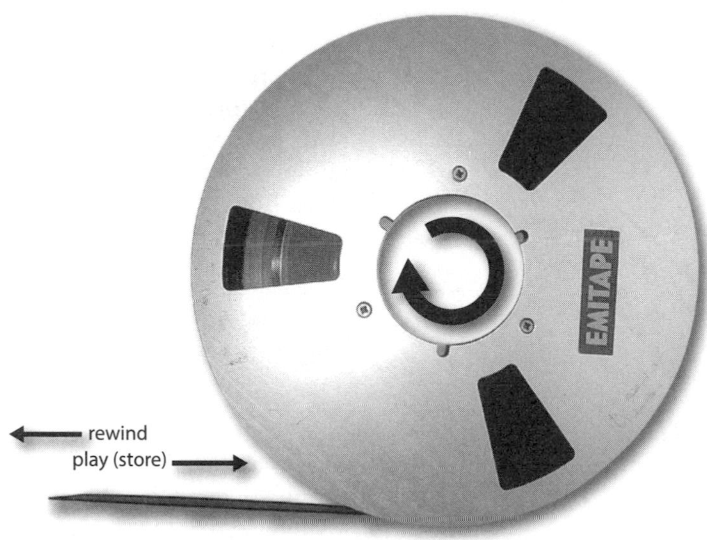

FIGURE 5.15
Recorded analog tapes should always be stored in the tails-out position.

Because of the effects of print-through, the standard method of professionally storing a recorded analog tape is in the *tails-out* position. Remember:

- Professional analog tape should *always* be stored tails-out (on the right-hand take-up reel).
- Upon playback, the tape should be wound onto the left-most "supply reel."
- During playback, feed the tape back onto the right-hand take-up reel, after which time it can again be removed for storage.
- If the tape has been continuously wound and rewound during the session, it's often wise to rewind the tape and then smoothly play or slow-wind the tape onto the take-up reel, after which time it can be removed for storage.

So why do we go through all this trouble? When a tape is stored tails-out (Figure 5.15), the print-through will bleed to the outer layers, a condition that causes the echo to follow the original signal in a way that's similar to the sound's natural decay and is subconsciously perceived by the listener as reverb instead of as an easily-audible pre-echo.

ANALOG TAPE NOISE

The roughly 60-dB signal-to-noise (S/N) limitation that's imposed on a conventional analog ATR audio track is dictated by saturation (at the high signal level end) and tape hiss at the low-end (which is heard when the overall recorded level of the program is too low). Should an optimum level produce an unacceptable amount of noise, the engineer is faced with several options: record at a higher level (with the possibility of increased distortion) or change the signal's overall dynamic range by raising low-level signals above the noise (compression or limiting).

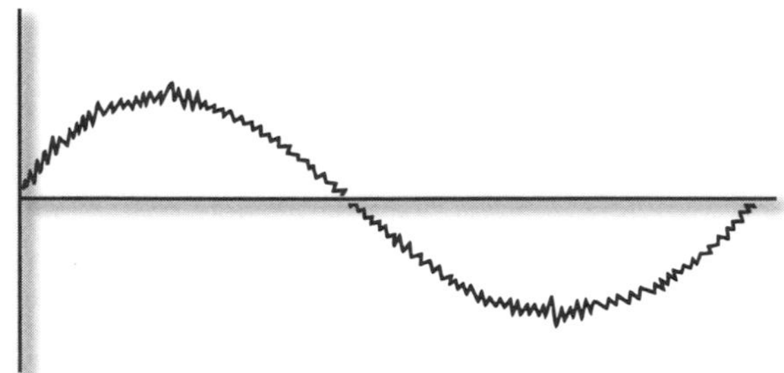

FIGURE 5.16
Modulation noise of a recorded analog sine wave.

Analog tape noise might not be much of a problem when dealing with one or two tracks in an audio production, but the combined noise and other distortions that can occur when 8, 16, 24 or 48 tracks are combined can range from being bothersome to downright unacceptable. The following types of noises are often major contributors to the problem:

- Tape and amplifier noise
- Crosstalk between tracks
- Print-through
- Modulation noise

Modulation noise is a high-frequency component that causes sonic "fuzziness" by introducing sideband frequencies that can distort the signal (Figure 5.16). This noise-based distortion is due to the magnetic and mechanical properties of the analog recording process itself, and actually increases as recorded levels rise. This noise is often higher in level than you might expect, and when combined with *asperity noise* (sideband frequencies that are also introduced by the analog record/playback process) can definitely play a role in what could be called the "analog sound."

Tutorial: Analog Tape Modulation and Asperity Noise

1. Feed a 0-VU, 1-kHz test tone to a track on a professional analog recorder.
2. Listen to the recorder's source (input) signal through the monitors at a moderate level.
3. Switch the recorder to monitor the tone from the track's playback (tape) head. Does it sound different?

These analog-based noises can be reduced to acceptable levels by using different combinations of the following actions:

- Improving the dynamic range of professional recording tape and/or increasing tape speed in order to record at higher flux levels
- Using an ATR with wider recorded tracks (i.e., allowing for higher record/playback levels and reduced crosstalk specs
- Using a thicker tape base to reduce print-through

By using tape formulations that combine low noise and high output (which have generally increased the S/N ratio by 3 dB or more), noise levels can be reduced even further. However, when all's said and done, making most or all of the above improvements is simply too costly and impractical. In the end, it's important to realize that noise is simply an inherent part of the analog recording process … and the goal is to minimize its effects.

CLEANLINESS

It's very important for the magnetic recording heads and moving parts of an ATR transport deck to be kept free from dirt and oxide shed. Oxide shed occurs when friction causes small particles of magnetic oxide to flake off and accumulate on surface contacts. This accumulation is most critical at the surface of the magnetic recording heads, since even a minute separation between the magnetic tape and heads can cause high-frequency *separation loss*. For example, a signal that's recorded at 15 ips and has an oxide shed buildup of 1 mil (0.001 inch) on the playback head will be 55 dB below its standard level at 15 kHz. Denatured (isopropyl) alcohol or an appropriate cleaning solution should be used to clean transport tape heads and guides (with the exception of the machine's pinch roller and other rubber-like surfaces) at regular intervals.

DEGAUSSING

Magnetic tape heads are made from a magnetically soft metal, which means that the alloy is easily magnetized, but once the coil's current is removed, the core won't retain any of its magnetism. Small amounts of residual magnetism, however, will build up over time, which can actually partially erase high-frequency signals from a master tape. For this reason, all of the tape heads should be demagnetized after 10 hours of operation with a head demagnetizer. This handheld device works much like an erase head in that it saturates the magnetic head with a high-level alternating signal that randomizes residual magnetic flux. Once a head has been demagnetized (after 5 to 10 seconds), it's important to move the tool to a safe distance from the tape heads at a speed of less than 2 inches per second before turning it off, so as to avoid inducing a larger magnetic flux back into the head. Before an ATR is aligned, the magnetic tape heads should always be cleaned and demagnetized in order to obtain accurate readings and to protect expensive alignment tapes.

BACKUP AND ARCHIVE STRATEGIES

In this day of hard drives, CDs and digital data, we've all come to know the importance of backing up our data. With important music and media projects, it's equally important to create a tape backup copy in case of an unforeseen catastrophe or as added insurance that future generations can enjoy the fruits of your work.

Backing up your project

The one basic truth that can be said about analog magnetic tape is that this medium has withstood the test of time. With care and reconditioning, tapes that have been recorded in the 1940s have been fully restored, allowing us to preserve and enjoy some of the best music of the day. On the other hand, digital data has two points that aren't exactly its favor:

- Data that resides on hard drives isn't the most robust of media over time. Even CDRs (which are rated to last over 100 years) haven't really been proven to last.
- Even if the data remains intact, with the ever-increasing advances in computer technology, who's to say that the media, drives, programs, session formats and file formats will be around in 10 years, let alone 50!

These warnings aren't slams against digital, just precautions against the march of technology versus the need for media preservation. For the above reasons, media preservation is a top priority for such groups as the Recording Academy's Producers and Engineers Wing (P&E Wing), as well as for many major record labels—so much so that many stipulate in their contracts that multitrack sessions (no matter what the original medium) are to be transferred and archived to 2-inch multitrack analog tape.

When transferring digital tracks to an analog machine, it's always wise to make sure that the recorder has been properly calibrated and that reference tones (1 kHz, 10 kHz, 16 kHz and 100 Hz) have been recorded at the beginning of the tape. When copying from analog to analog, both machines should be properly calibrated, but the source for the newly recorded tones should be the master tape. If a SMPTE track is required, be sure to stripe the copy with a clean, jam-sync code. The backing up of analog tapes and/or digital data usually isn't a big problem … unless you've lost your original masters. In this situation, a proper safety master can be the difference between panic and peace.

Archive strategies

Just as it's important to back up your media, it's also important that both the original and backup media be treated and stored properly. Here are a few guidelines:

- As stated earlier, always store the tapes tails-out.
- Wind the tapes onto the take-up storage reel at slow-wind or play speeds.
- Store the boxes vertically. If they're stored horizontally, the outer edges could get bent and become damaged.
- Media storage facilities exist that can store your masters or backups for a fee. If this isn't an option, store them is an area that's cool and dry (e.g., no temperature extremes, in low humidity, no attics or basements).
- Store your masters and backups in separate locations. In case of a fire or other disaster … one would be lost, but not both (always a good idea with digital data, as well).

C.L.A.S.P.

One specialized, patented system has been developed that can seamlessly integrate an analog tape machine (of various type and track configurations) into

FIGURE 5.17
The C.L.A.S.P. Closed Loop Analog Signal Processor. (Courtesy of Endless Analog, www.endlessanalog.com).

a DAW, to give a disk-based digital system that "analog feel and sound." The C.L.A.S.P. (Closed Loop Analog Signal Processor) literally inserts a working analog tape and electronics chain between the recording interface and/or console and the DAW. This system (which is the only system that has latency-free analog monitoring) takes the recorded signals from your console, interface or DAW and records them to tape. The signal is then routed from the tape machine's playback head where specialized time-stamp and time-correction are used to re-align the audio (with sample accuracy) directly into the DAW session (Figure 5.17).

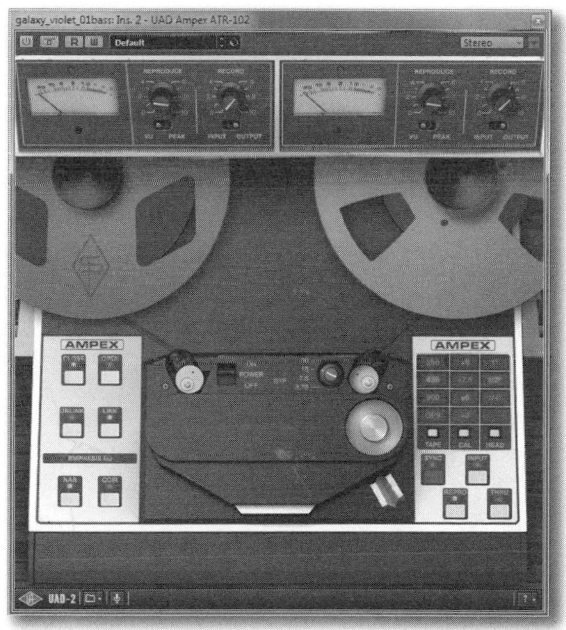

FIGURE 5.18
The Ampex ATR 102 tape emulation plugin plug-in for the Apollo and the UAD effects processing card. (Courtesy of Universal Audio, www.uaudio.com © 2013 Universal Audio, Inc. All rights reserved. Used with permission).

TAPE EMULATION PLUG-INS

As DAW technology and methods for precisely modeling hardware devices in software have advanced, new types of plug-ins that can directly emulate the sound of analog tape machines have come onto the market. These devices are capable of closely mimicing the sonic character, tape noise, distortion, changes with virtual tape formulation and bias settings of any number of tape machines and model makes … "virtually" allowing us to plug-in the sound of our favorite deck onto any workstation track or bus (Figure 5.18).

CHAPTER 6
Digital Audio Technology

There's absolutely no doubt that digital audio technology has changed the way in which all forms of media are produced, gathered and distributed. As a driving force behind human creativity, expression and connectivity, the impact of digital media production is simply HUGE, and is an integral part of both the medium and the message within modern-day communication.

As with most other media, these changes have been brought about by the integration of the personal computer and portable devices into the modern-day project studio environment. Newer generations of computers and related hardware peripherals have been integrated into both the pro and project studio to record, fold, spindle and mutilate audio with astonishing ease. This chapter is dedicated to the various digital system types and their relation to the modern-day music production studio.

THE LANGUAGE OF DIGITAL

Although digital audio is a varied and complex field of study, the basic theory behind the magic curtains isn't all that difficult to understand. At its most elementary level, it's simply a process by which numeric representations of analog signals (in the form of voltage levels) are encoded, processed, stored and reproduced over time through the use of a *binary number system*.

Just as English-speaking humans communicate by combining any of 26 letters together into groupings known as "words" and manipulate numbers using the decimal (base 10) system, the system of choice for a digital device is the binary (base 2) system. This numeric system provides a fast and efficient means for manipulating and storing digital data. By translating the alphabet, base 10 numbers or other form of information into a binary language form, a digital device (such as a computer or microprocessor) can perform calculations and tasks that would otherwise be cumbersome, less cost effective and/or downright impossible to perform in the analog domain.

To illustrate, let's take a look at how a human construct can be translated into a digital language (and back). If we type the letters C, A and T into a word processor,

C, A & T = (0100 0011)(0100 0001)(0101 0111) =

(alpha-bits) (digital words) (Conway - R.I.P!)

FIGURE 6.1 Digital and analog equivalents for a strange four-legged animal (Conway – R.I.P.).

the computer will quickly go about the task of translating these keystrokes into a series of 8-bit digital words represented as [0100 0011], [0100 0001] and [0101 0111]. On their own, these digits don't mean much, but when these groups are put together, they form a word that represents a four-legged animal that's either seldom around or always underfoot (Figure 6.1).

In a similar manner, a digital audio system works by sampling (measuring) the instantaneous voltage level of an analog signal at specific intervals over time, and then converting these samples into a series of encoded "words" that digitally represent the analogous voltage levels. By successively measuring changes in an analog signal's voltage level (over time), this stream of representative words can be stored in a form that represents the original analog signal. Once stored, the data can be processed and reproduced in ways that have changed the face of audio production forever.

It's interesting to note that binary data can be encoded as logical "1" or "0"—"on" or "off" states, using such forms of physics as:

- Voltage or no voltage (circuitry)
- Magnetic flux or no flux (hard disk or tape)
- Reflection off of a surface or no reflection (CD, DVD or other optical disc form)
- Electromagnetic waves (broadcast transmission).

From this, you'll hopefully begin to get the idea that human forms of communication (i.e., print, visual and audible media) can be translated into a digital form that can be easily understood and manipulated by a processor. Once the data has been recorded, stored and/or processed, the resulting binary data can be reconverted back into a form that can be easily understood by us humans (such as a display, readout, printed paper, lit indicator, controller interaction … you name it). If you think this process of changing one form of energy into an analogous form (and then back again) sounds like the general definition of a transducer—you're right!

DIGITAL BASICS

The following sections provide a basic overview of the various stages that are involved in the encoding of analog signals into equivalent digital data, and the subsequent converting of this data back into its original analog form.

The encoding and decoding phases of the digitization process revolve around two processes:

- Sampling (the component of time)
- Quantization (the signal-level component)

In a nutshell, sampling is a process that affects the overall bandwidth (frequency range) that can be encoded within a sound file, while quantization refers to the resolution (overall quality and distortion characteristics) of an encoded signal compared to the original analog signal at its input.

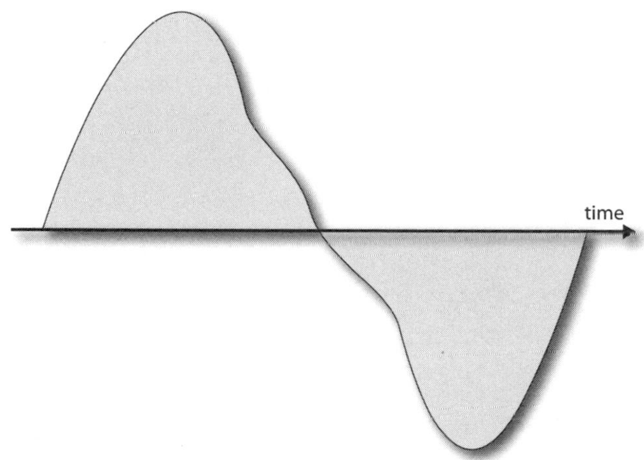

FIGURE 6.2
An analog signal is continuous in nature.

Sampling

In the world of analog audio, signals are recorded, stored and reproduced as changes in voltage levels that continuously vary over time in a continuous fashion (Figure 6.2). The digital recording process, on the other hand, doesn't operate in this manner; rather, digital recording operates by taking periodic samples of an analog audio waveform over time (Figure 6.3), and then calculating each of these snapshot samples into grouped binary words that digitally represent these voltage levels as they change over time, as accurately as possible.

During this process, an incoming analog signal is sampled at discrete and precisely timed intervals (as determined by the sample rate). At each interval, this analog signal is momentarily "held" (frozen in time), while the converter goes about the process of determining what the voltage level actually is, with a degree of accuracy that's defined by the converter's circuitry and the chosen bit rate. The converter then generates a binary-encoded word that's numerically equivalent to the analog voltage level at that point in time (Figure 6.4). Once this is done, the converter can store the representative word into a memory medium (tape, disk, disc, etc.), release its hold, and then go about the task of determining the values of the next sampled voltage. The process is then continuously repeated throughout the recording process.

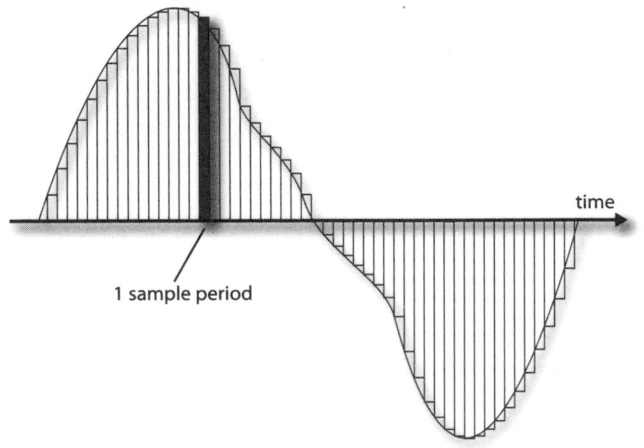

FIGURE 6.3
A digital signal makes use of periodic sampling to encode information.

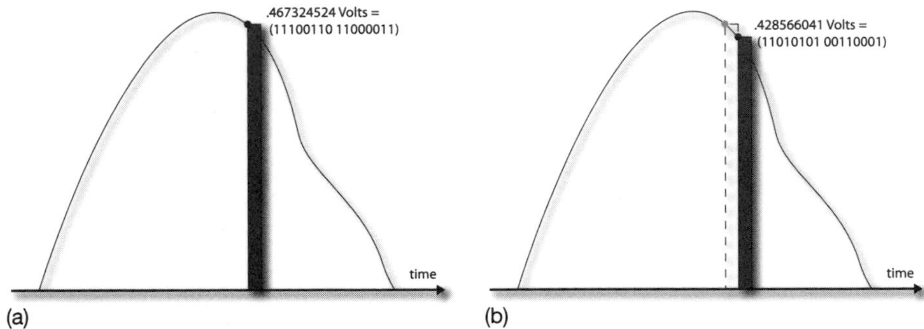

FIGURE 6.4
The sampling process. (a) The analog signal is momentarily "held" (frozen in time), while the converter goes about the process of determining the voltage level at that point in time and then converting that level into a binary-encoded word that's numerically equivalent to the original analog voltage level. (b) Once this digital information is processed and stored, the sample is released and the next sample is held, as the system again goes about the task of determining the level of the next sampled voltage … and so forth, and so forth, and so forth over the duration of the recording.

Within a digital audio system, the *sampling rate* is defined as the number of measurements (samples) that are periodically taken over the course of a second. Its reciprocal (sampling time) is the elapsed time that occurs between each sampling period. For example, a sample rate of 44.1 kHz corresponds to a sample time of 1/44,100 of a second.

This process can be likened to a photographer who takes a series of action sequence shots. As the number of pictures taken in a second increases, the accuracy of the captured event will likewise increase until the resolution is so great that you can't tell that the successive, discrete pictures have turned into a

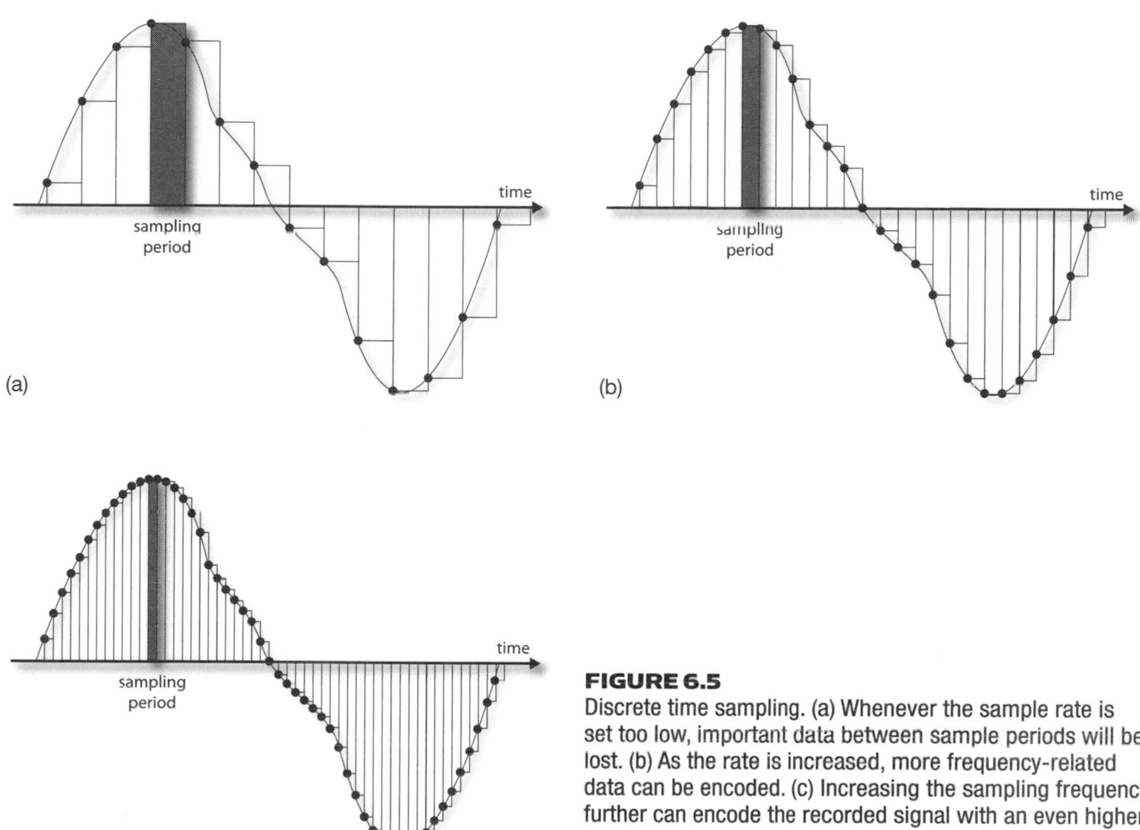

FIGURE 6.5
Discrete time sampling. (a) Whenever the sample rate is set too low, important data between sample periods will be lost. (b) As the rate is increased, more frequency-related data can be encoded. (c) Increasing the sampling frequency further can encode the recorded signal with an even higher bandwidth range.

continuous and (hopefully) compelling movie. Since the process of sampling is tied directly to the component of time, the sampling rate of a system determines its overall bandwidth (Figure 6.5), meaning that a recording made at a higher sample rate will be capable of storing a wider range of frequencies (effectively increasing the signal's bandwidth at its upper limit).

Quantization

Quantization represents the amplitude component of the digital sampling process. It is used to translate the voltage levels of a continuous analog signal (at discrete sample points over time) into binary digits (bits) for the purpose of manipulating or storing audio data in the digital domain. By sampling the amplitude of an analog signal at precise intervals over time, the converter determines the exact voltage level of the signal (during a sample interval, when the voltage level is momentarily held), and then outputs the signal level as an equivalent set of binary numbers (as a grouped word of *n*-bits length) which represent

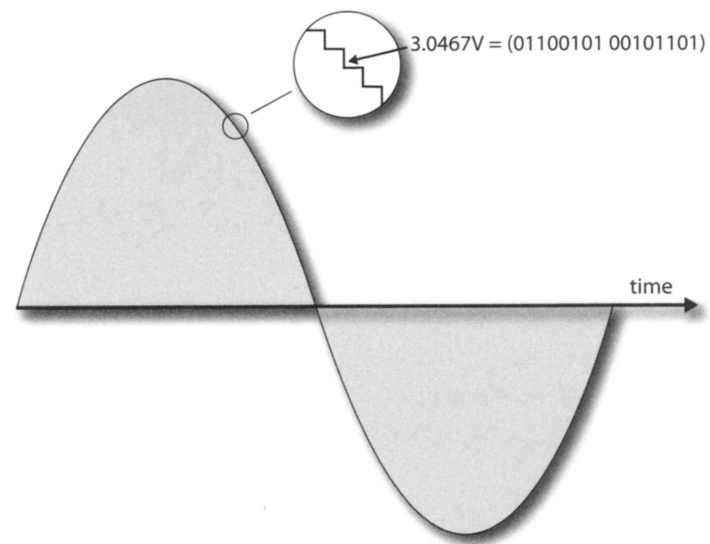

FIGURE 6.6
The instantaneous amplitude of the incoming analog signal is broken down into a series of discrete voltage steps, which are then converted into an equivalent binary-encoded word.

the originally sampled voltage level (Figure 6.6). The resulting word is used to encode the original voltage level with as high a degree of accuracy as can be permitted by the word's bit length and the system's overall design.

Currently, the most common binary word length for audio is 16-bit (i.e., [1111110101000001]); however, bit depths of 20- and 24-bit resolution are also in common use. In addition, computers and signal-processing devices are capable of performing calculations internally at the 32- and 64-bit resolution level. This added internal headroom at the bit level helps reduce errors in level and performance at low-level resolutions whenever multiple audio datastreams are mixed or processed within a digital signal processing (DSP) system. This greater internal bit resolution is used to reduce errors that might accumulate within the least significant bit (LSB; the final and smallest numeric value within a digital word). As multiple signals are mixed together and multiplied (a regular occurrence in gain change and processing functions), lower-bit-resolution numbers play a more important role in determining the signal's overall accuracy and distortion. Since the internal bit depth is higher, these resolutions can be preserved (instead of being dropped by the system's hardware or software processing functions), with a final result being an n-bit datastream that's relatively free of errors. This leads us to the conclusion that greater word lengths will often directly translate into an increased resolution (and, within reason, arguably higher quality) due to the added number of finite steps into which a signal can be digitally encoded. The following details the number of encoding steps that are encountered for the most commonly used bit lengths:

8-bit word = (*nnnnnnnn*) = 256 steps
16-bit word = (*nnnnnnnn nnnnnnnn*) = 65,536 steps
20-bit word = (*nnnnnnnn nnnnnnnn nnnn*) = 1,048,576 steps
24-bit word = (*nnnnnnnn nnnnnnnn nnnnnnnn*) = 16,777,216 steps
32-bit word = (*nnnnnnnn nnnnnnnn nnnnnnnn nnnnnnnn*) = 4,294,967,296 steps
where *n* = binary 0 or 1.

The details of the digital audio record/playback process can get quite detailed and complicated, but the essential basics are:

- Sampling (in the truest sense of the word) an analog voltage signal at precise intervals in time
- Converting these samples into a digital word value that most accurately represents these voltage levels
- Storing them within a digital memory device

Upon playback, these digital words are then converted back into discrete voltages (again, at precise intervals in time), allowing the originally recorded signal voltages to be re-created, processed and played back.

The devil's in the details

Although the basic concept behind the sample-and-hold process is relatively straightforward, delving further into the process can quickly bog you down in the language of high-level math and physics. Luckily, there are a few additional details relating to digital audio that can be discussed at a basic level.

THE NYQUIST THEOREM

This basic rule relates to the sampling process … and states that:

In order for the desired frequency bandwidth to be faithfully encoded in the digital domain, the selected sample rate must be at least *twice as high* as the highest frequency to be recorded (sample rate ≥2 × highest frequency).

In plain language, should frequencies that are greater than twice the sample rate be allowed into the sampling process, these high frequencies would have a periodic nature that's shorter than the sample rate can actually capture. When this happens, the successive samples won't be able to accurately capture these higher frequencies, and will instead actually record false or "alias" frequencies that aren't actually there, but will be heard as harmonic distortion (Figure 6.7).

From a practical point of view, this would mean that an audio signal with a bandwidth of 20 kHz would require that the sampling rate be at least 40 kHz

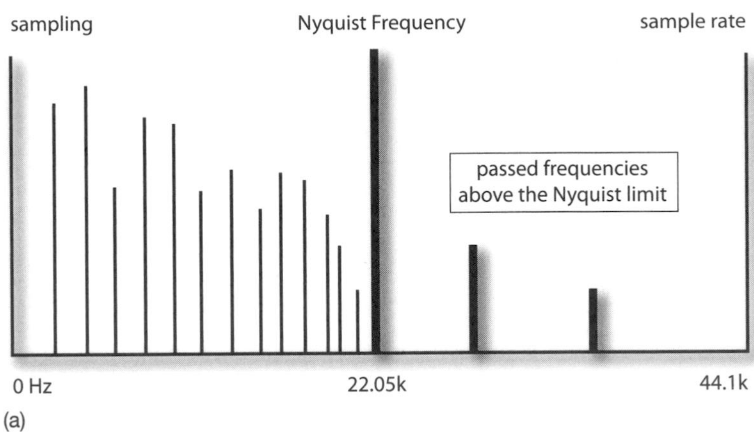

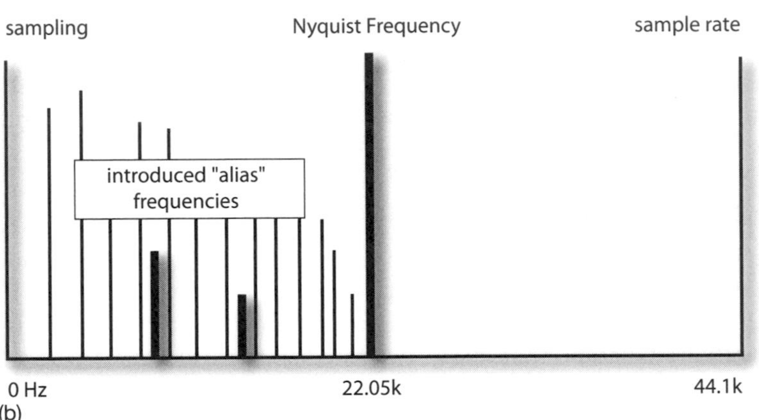

FIGURE 6.7 Frequencies that enter into the digitization process above the Nyquist half-sample frequency limit can introduce harmonic distortion: (a) frequencies greater than 2x the sampling rate limit; (b) resulting "alias" frequencies that are introduced into the audio band as distortion.

samples/sec, and a bandwidth of 22 kHz would require a sampling rate of at least 44 kHz samples/sec, etc.

To eliminate the effects of *aliasing*, a low-pass filter is placed into the circuit before the sampling process takes place. In theory, an ideal filter would pass all frequencies up to the Nyquist cutoff frequency and then prevent any frequencies above this point from passing. In the real world, however, such a "brick wall" doesn't exist. For this reason, a slightly higher sample rate must be chosen in order to account for the cutoff slope that's required for the filter to be effective (Figure 6.8). As a result, an audio signal with a bandwidth of 20 kHz will actually be sampled at a standardized rate of 44.1 samples/sec, while a bandwidth of roughly 22 kHz would require the use of a sampling rate of at least 48 kHz samples/sec, etc.

OVERSAMPLING

This sampling-related process is commonly used in professional and consumer digital audio systems to improve the Nyquist filter's anti-aliasing characteristics.

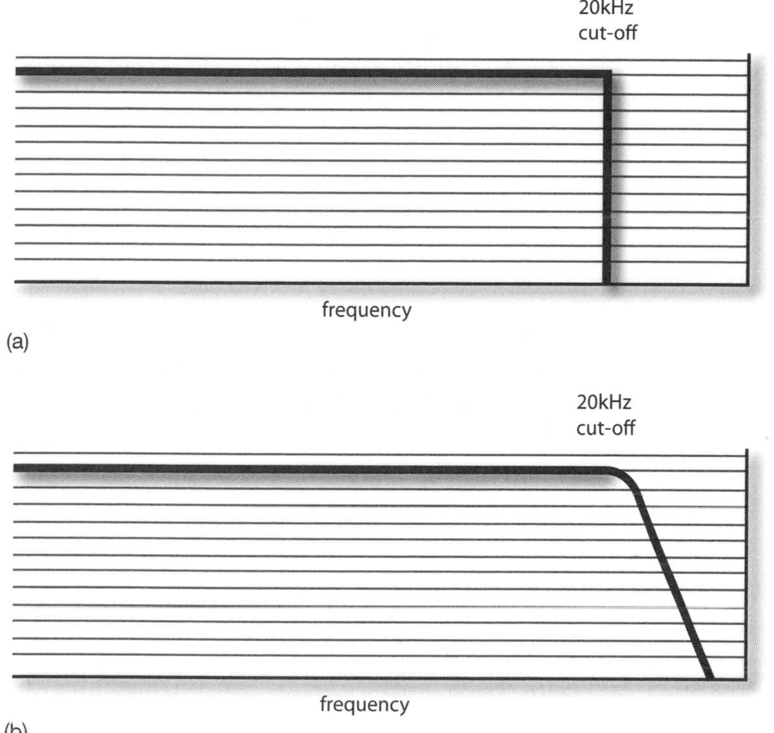

FIGURE 6.8
Anti-alias filtering. (a) An ideal filter would have an infinite attenuation at the 20 kHz Nyquist cutoff frequency. (b) Real-world filters require an additional frequency "guardband" in order to fully attenuate unwanted frequencies that fall above the half-bandwidth Nyquist limit.

Oversampling increases the effective sample rate by factors ranging between 12 and 128 times the original rate. There are three main reasons for this process:

- Nyquist filters can be expensive and difficult to properly design. By increasing the effective sample bandwidth, a simpler and less-expensive filter can be used.
- Oversampling generally results in a higher-quality analog-to-digital (A/D) and digital-to-analog (D/A) converter that sounds better.
- Since multiple samples are taken of a single sample-and-hold analog voltage, the average noise sampling will be lower.

Following the sample stage, the sampled data is digitally scaled back down to the target data rate and bandwidth for further processing and/or storage.

SIGNAL-TO-ERROR RATIO

The signal-to-error ratio is used to measure the quantization process. A digital system's *signal-to-error ratio* is closely akin (although not identical) to the analog concept of signal-to-noise (S/N) ratio. Whereas a S/N ratio is used to indicate the overall dynamic range of an analog system, the signal-to-error ratio of a digital audio device indicates the degree of accuracy that's used to capture a sampled level and its step-related effects.

Although analog signals are continuous in nature, as we've read, the process of quantizing a signal into an equivalent digital word isn't. Since the number of discrete steps that can be encoded within a digital word limits the accuracy of the quantization process, the representative digital word can only be an approximation (albeit an extremely close one) of the original analog signal level. Given a properly designed system, the signal-to-error ratio for a signal coded with *n* bits is:

$$\text{Signal-to-error ratio} = 6n + 1.8(\text{dB})$$

Therefore, the theoretical signal-to-error ratio for the most common bit rates will yield a dynamic range of:

8-bit word = 49.8 dB	20-bit word = 121.8 dB	32-bit word = 193.8 dB
16-bit word = 97.8 dB	24-bit word = 145.8 dB	64-bit word = 385.8 dB

DITHER

Also relating to quantization, dither is commonly used during the recording or conversion process to increase the overall bit resolution (and therefore low-level noise and signal clarity) of a recorded signal, when converting from a higher to a lower bit rate.

Technically, dither is the addition of very small amounts of randomly generated noise to an existing bitstream that allows the S/N and distortion figures to fall to levels that approach their theoretical limits. The process makes it possible for low-level signals to be encoded at less than the data's least significant bit level (less than a single quantization step, as shown in Figure 6.9). You heard that right … by adding a small amount of random noise into the A/D path, we can actually:

FIGURE 6.9
Values falling below the least significant bit level cannot be encoded without the use of dither.

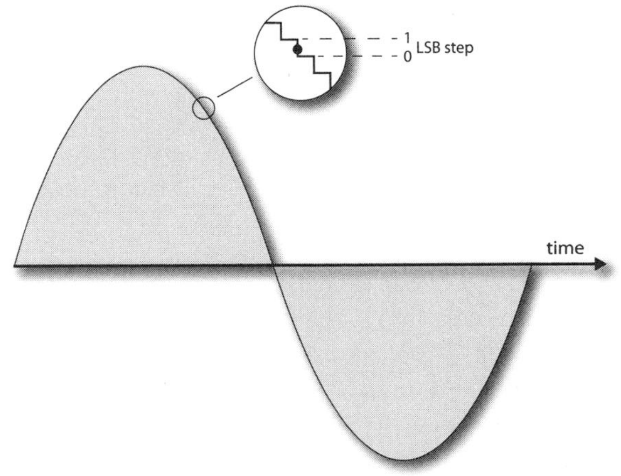

- Improve the resolution of the conversion process below the least significant bit level.
- Reduce harmonic distortion in a way that greatly improves the signal's performance.

The concept of dither relies on the fact that noise is random. By adding small amounts of randomization into the quantization process, there is an increased probability that the D/A converter will be able to guess the least significant bit of a low-level signal more accurately. This is due to the fact that the noise shapes the detected sample in such a way that the sample-and-hold (S/H) circuitry can determine the original analog value with greater precision.

Dither is often applied to an application or process to reduce quantization errors and slight increases in noise and/or fuzziness that could otherwise creep into a bitstream. For example, when multiple tracks are mixed together within a DAW, it's not uncommon for digital data to be internally processed at 32- and 64-bit depths. In situations like this, dither is often used to smooth and round the data values, so that the low-level (least significant bit) resolutions won't be lost when they are interpolated back to their original target bit depths.

Dither can also be manually applied to sound files that have been saved at 20- and 24-bit depths (or possibly greater by the time you read this). Applications and DAW plug-ins can be used to apply dither to a sound file or master mix, so as to reduce the effects of lost resolution due to the truncation of least significant bits. For example, mastering engineers might experiment with applying dither to a high-resolution file before saving or exporting it as a 16-bit final master. In this way, noise is reduced and the sound file's overall clarity can be increased.

Fixed- vs. floating-point processing

Many of the newer digital audio and DAW systems make use of floating-point arithmetic in order to process, mix and output digital audio. The advantage of floating- over fixed-point DSP calculations is that the former is able to use numeric "shorthand" in order to process a wider range of values at any point in time. In short, it's able to easily move or "float" the decimal point of a very large number in a way that can represent it as a much smaller value. By doing so, the processor is able to internally calculate much larger bit depth values (i.e., 32- or 64-bit) with relative ease and increased resolution.

THE DIGITAL RECORDING/REPRODUCTION PROCESS

The following sections provide a basic overview of the various stages within the process of encoding analog signals into equivalent digital data (Figure 6.10a) and then converting this data back into its original analog form (Figure 6.10b).

The recording process

In its most basic form, the *digital recording chain* includes a low-pass filter, a sample-and-hold circuit, an analog-to-digital converter, circuitry for signal coding and error correction. At the input of a digital sampling system, the analog signal must be band limited with a low-pass filter so as to stop frequencies that are greater than half the sample rate frequency from passing into the A/D conversion circuitry. Such a stopband (anti-aliasing) filter generally makes use of a sharp roll-off slope at its high-frequency cutoff point (oversampling might be used to simplify and improve this process).

Following the low-pass filter, a *sample-and-hold* (S/H) circuit holds and measures the analog voltage level that's present during the sample period. This period is determined by the sample rate (i.e., 1/44,100th of a second for a 44.1 rate). At

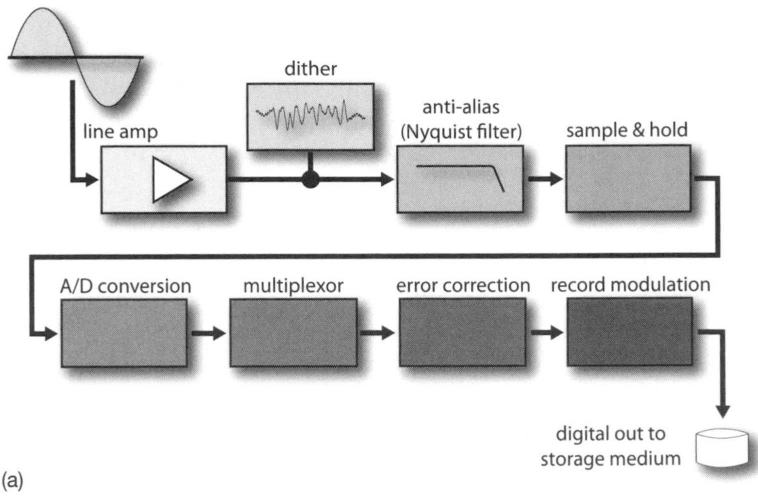

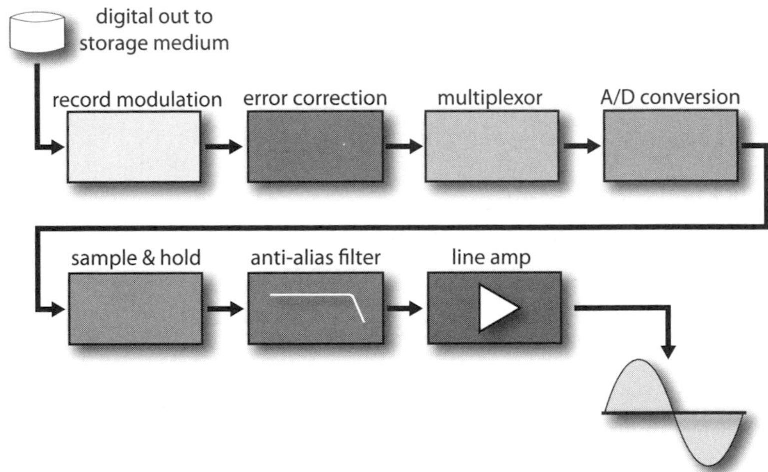

FIGURE 6.10
The digital audio chain: (a) recording; (b) reproduction.

this point, computations are performed to translate the sampled voltage into an equivalent binary word. This step in the A/D conversion is one of the most critical components of the digitization process, because the sampled DC voltage level must be quickly and accurately quantized into an equivalent digital word (to the nearest step level).

Once the sampled signal has been converted into its equivalent digital form, the data must be conditioned for further data processing and storage. This conditioning includes data coding, data modulation and error correction. In general, the binary digits of a digital bitstream aren't directly stored onto a recording medium as raw data; rather, data coding is used to translate the data (along with synchronization and address information) into a form that allows the data to be most

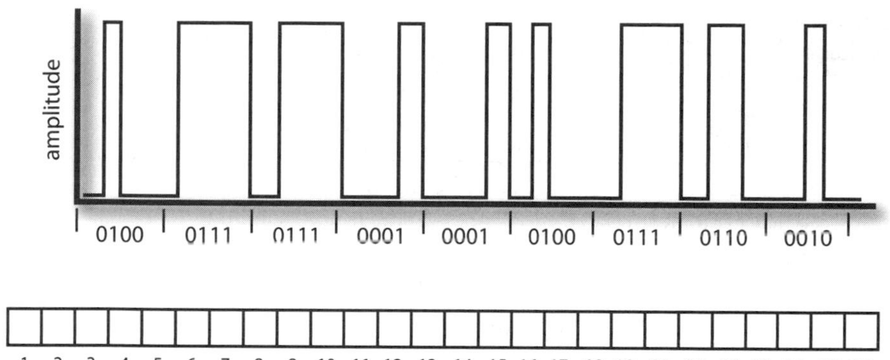

FIGURE 6.11 Pulse-code modulation.

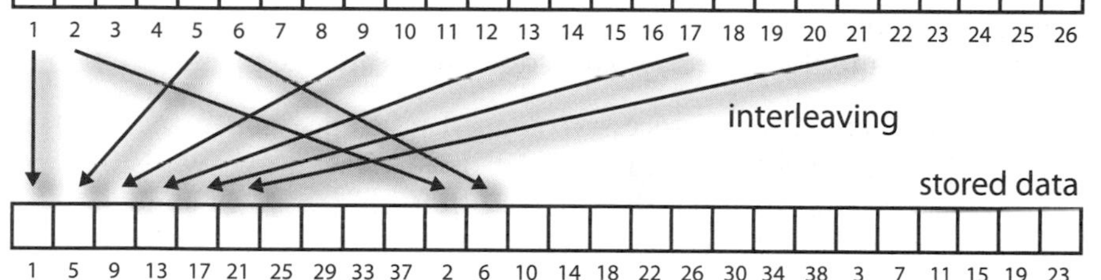

FIGURE 6.12 Example of interleaved error correction.

efficiently and accurately stored to a memory or storage media. The most common form of digital audio data coding is *pulse-code modulation*, or PCM (Figure 6.11).

The density of stored information within a PCM recording and playback system is extremely high … so much so that any imperfections (such as dust, fingerprints or scratches that might adhere to the surface of any magnetic or optical recording medium) will cause severe or irretrievable errors to be generated. To keep these errors within acceptable limits, several forms of *error correction* are used (depending on the media type). One method uses redundant data in the form of parity bits and check codes in order to retrieve and/or reconstruct lost data. A second uses error correction that involves interleaving techniques, whereby data is deliberately scattered across the digital bitstream, according to a complex mathematical pattern. The latter has the effect of spreading the data over a larger surface of the recording media, thereby making the recording media less susceptible to dropouts (Figure 6.12). In fact, it's a simple truth that without error correction, the quality of most digital audio media would be greatly reduced or (in the case of the CD and DVD) rendered almost useless.

The playback process

In many respects, the digital reproduction chain works in a manner that's complementary to the digital encoding process. Since most digital media encodes data onto media in the form of highly saturated magnetic transition states or optical reflections, the recorded data must first be reconditioned in a way that restores the digital bitstream back into its original, modulated binary state (i.e.,

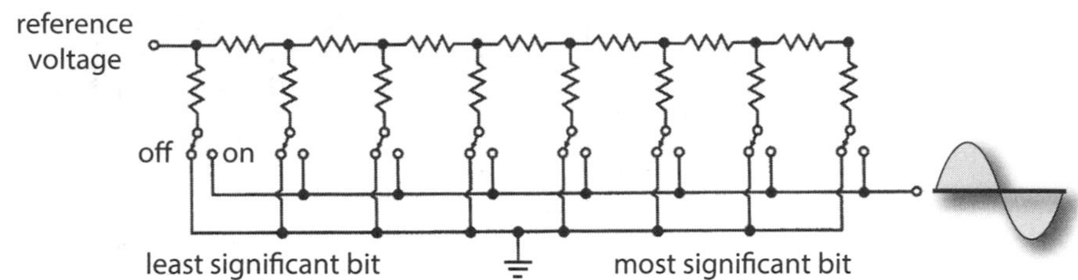

FIGURE 6.13
A stepped resistance network is a common device for accomplishing D/A conversion by assigning each word bit to a series of resistors that are scaled by factors of 2.

a transitional square wave). Once this is done, the encoded data can be de-interleaved (reassembled) back into its original form, where it can be converted back into PCM data, and the process of D/A conversion can take place.

Within the D/A conversion process, a stepped resistance network (sometimes called an R/2R network) is used to convert the representative words back into their analogous voltage levels for playback. During a complementary S/H period, each bit within the representative digital word is assigned to a resistance leg in the network (moving from the most significant to the least significant bit). Each "step" in the resistance leg is then designed to successively pass one-half the reference voltage level as it moves down the ladder toward the LSB (Figure 6.13). The presence or absence of a logical "1" in each step is then used to turn on each successive voltage leg. As you might expect, when all the resistance legs are added together, their voltages equal a precise level that matches the originally recorded voltage during the recorded sample period. As these voltages are reproduced over time in precise intervals (as determined by the sample rate) … voilà, you have a playback signal!

Following the conversion process, a final, complementary low-pass filter is inserted into the signal path. Again, following the principle of the Nyquist theorem, this filter is used to smooth out any of the step-level distortions that are introduced by the sampling process, resulting in a waveform that faithfully represents the originally recorded analog waveform.

Sound file sample rates

The *sample rate* of a recorded digital audio bitstream directly relates to the resolution at which a recorded sound will be digitally captured. Using the film analogy, if you capture more samples (frames) of a moving image as it moves through time, you'll have a more accurate representation of that recorded event. If the number of samples are too low, the resolution will be "lossy" and will distort the event. On the other hand, taking too many picture frames might result in a recorded bandwidth that's so high that the audience won't be able to discriminate any advantages that the extra information has to offer … or the storage requirements will become increasingly large as the bandwidth and file size increase.

This analogy relates perfectly to audio because the choices of sample rate will be determined by the bandwidth (number of overall frequencies that are to be captured) versus the amount of storage is needed to either save the data to a memory storage media … or the time that will be required to up/download a file through a transmission and/or online datastream. Beyond the basic adherence to certain industry sample rate standards, such are the personal decisions that must be made regarding the choice of sample rate to be used on a project. Although other sample-rate standards exist, the following are the most commonly used in the professional, project and audio production community:

- *32k:* This rate is often used by broadcasters to transmit/receive digital data via satellite. With its overall 15-kHz bandwidth and reduced data requirements, it is also used by certain devices in order to conserve on memory and is commonly used in satellite broadcast communications. Although the pro community doesn't generally use this rate, it's surprising just how good a sound can be captured at 32k (given a high-quality converter).
- *44.1k:* The long-time standard of consumer and pro audio production, 44.1 is the chosen rate of the CD-audio standard. With its overall 20-kHz bandwidth, the 44.1k rate is generally considered to be the minimum sample rate for professional audio production. Assuming that high-quality converters are used, this rate is capable of recording lossless audio, while conserving on memory storage requirements.
- *48k:* This standard was adopted early on as a standard sample rate for professional audio applications (particularly when referring to hardware digital audio devices). It's also the adopted standard rate for use within professional video and DVD production.
- *88.2k:* As a simple multiple of 44.1, this rate is often used within productions that are intended to be high-resolution products.
- *96k:* This rate has been adopted as the de facto sample rate for high-resolution recordings.
- *192k:* This high-resolution rate is uncommon within pro audio production, as the storage and media requirements are quite high.

Sound file bit rates

The *bit rate* of a digitally recorded sound file directly relates to the number of quantization steps that are encoded into the bitstream. As a result, the bit rate (or bit depth) is directly correlated to the:

- Accuracy at which a sampled level (at one point in time) is to be encoded
- Signal-to-error figure … and thus the overall dynamic range of the recorded signal.

If the bit rate is too low to accurately encode the sample, the resolution will lead to quantization errors, which will lead to distortion. On the other hand, too high of a bit depth might result in a resolution that's so high that the resulting gain in resolution is lost on the audience's ability to discriminate it … or the storage requirements might become so high that the files become inordinately large.

Although other sound file bit rate standards exist, the following are the most commonly used within the pro, project and general audio production community:

- *16 bits:* The long-time standard of consumer and professional audio production, 16 bits is the chosen bit depth of the CD-audio standard (offering a theoretical dynamic range of 97.8 dB). It is generally considered to be the minimum depth for high-quality professional audio production. Assuming that high-quality converters are used, this rate is capable of lossless audio recording, while conserving on memory storage requirements.
- *20 bits:* Before the 24-bit rate came onto the scene, 20 bits was considered to be the standard for high-bit-depth resolution. Although it's used less commonly, it can still be found in high-definition audio recordings (offering a theoretical dynamic range of 121.8 dB).
- *24 bits:* Offering a theoretical dynamic range of 145.8 dB, this standard bit rate is often used in high-definition audio applications, often in conjunction with the 96k sample rate (i.e., 96/24).

Further reading on sound file and compression codec specifics can be found in Chapter 10 (Multimedia).

Regarding digital audio levels

Over the past few decades, the trend toward making recordings that are as loud as possible has totally pervaded the industry to the point that it has be given the name of "The Loudness War." Not only is this practice used in mastering to make a song or project stand out in an on-air or in-the-pocket playlist … it has also followed in the analog tradition of recording a track as hot as possible to get the best noise figures and punch. All of this is arguably well and good, except for the fact that in digital recording a track at too "hot" a level doesn't add extra punch—it just adds really nasty distortion.

The dynamic range of a digital recording ranges from its floor signal level (00000000 00000000) for a 16-bit recording to its full-scale headroom ceiling of (11111111 11111111). Average or peak levels above full scale can easily ruin a recording. As such, since digital has a wider dynamic range than analog, it's always a good idea to reduce your levels so that they peak from −12 to −20 dB. This will accurately capture the peaks without clipping, without introducing an appreciable amount of noise into the mix. Recording at higher bit rates (i.e., 24 bits) will further reduce noise levels, allowing for increased headroom when recording at reduced levels. Of course, there are no standard guidelines or reference levels (digital meters aren't even standardized, for the most part) … so you might want to further research the subject on your own.

Digital audio transmission

In the digital age, it's become increasingly common for audio data to be distributed throughout a connected production system in the digital domain. In this way, digital audio can be transmitted in its original numeric form and (in theory) without any degradation throughout a connected path or system. When looking at the differences between the distribution of digital and analog audio, it should be kept in mind that, unlike its counterpart, the transmitted bandwidth of digital audio data occurs in the megahertz range; therefore, the transmission of digital audio actually has more in common with video signals than the lower bandwidth range that's encountered within analog audio. This means that care must be exercised to ensure that impedance is more closely matched and that quick-fix solutions don't occur (for example, using a Y-cord to split a digital signal between two devices is a major no-no). Failure to follow these precautions could seriously degrade or deform the digital signal, causing increased jitter and distortions.

Due to these tight restrictions, several digital transmission standards have been adopted that allow digital audio data to be quickly and reliably transmitted between compliant devices. These include such protocols as:

- AES/EBU
- S/PDIF
- SCMS
- MADI
- ADAT lightpipe
- TDIF

AES/EBU

The AES/EBU (Audio Engineering Society and the European Broadcast Union) protocol has been adopted for the purpose of transmitting digital audio between professional digital audio devices. This standard (which is most often referred to as simply an AES digital connection) is used to convey two channels of interleaved digital audio through a single, three-pin XLR mic or line cable in a single direction. This balanced configuration connects pin 1 to the signal ground, while pins 2 and 3 are used to carry signal data. AES/EBU transmission data is low impedance in nature (typically 110 Ω) and has digital burst amplitudes that range between 3 and 10 V. These combined factors allow for a maximum cable length of up to 328 feet (100 meters) at sample rates of less than 50 kHz without encountering undue signal degradation.

Digital audio channel data and subcode information are transmitted in blocks of 192 bits that are organized into 24 words (with each being 8 bits long). Within the confines of these data blocks, two subframes are transmitted during each sample period that convey information and digital synchronization codes for both channels in an L-R-L-R … fashion. Since the data is transmitted as a self-clocking biphase code (Figure 6.14), wire polarity can be ignored. In addition, whenever two devices are directly connected, the receiving device will usually derive its reference timing clock from the digital source device.

In the late 1990s, the AES protocol was amended to include the "stereo 96k dual AES signal" protocol. This was created to address signal degradations that can

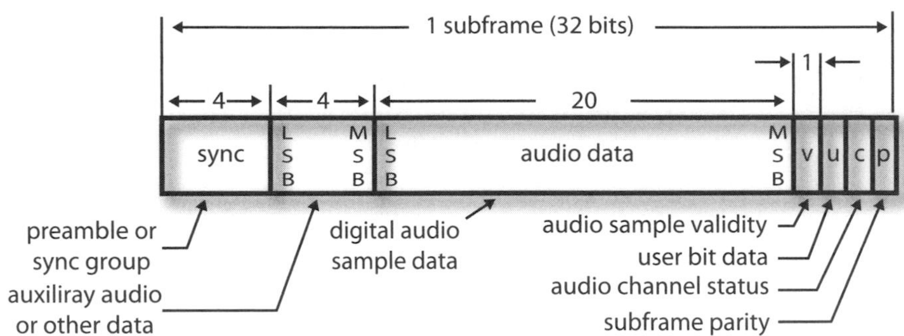

FIGURE 6.14 AES/EBU subframe format.

(a)

(b)

FIGURE 6.15 S/PDIF connectors: (a) RCA coax connection; (b) Toslink optical connection.

occur when running longer cable runs at sample rates above 50 kHz. To address the problem, the dual AES standard allows stereo sample rates above 50 kHz (such as 96/24) to be transmitted over two synchronized AES cables (with one cable carrying the L information and the other carrying the R).

S/PDIF

The S/PDIF (Sony/Phillips Digital Interface) protocol has been widely adopted for transmitting digital audio between consumer digital audio devices and their professional counterparts. Instead of using a balanced 3-pin XLR cable, the popular S/PDIF standard has adopted the single-conductor, unbalanced phono (RCA) connector (Figure 6.15a), which conducts a nominal peak-to-peak voltage level of 0.5 V between connected devices, with an impedance of 75 Ω. In addition to using standard RCA cable connections S/PDIF can also be transmitted between devices using Toslink optical connection lines (Figure 6.15b), which are commonly referred to as "lightpipe" connectors.

As with the AES/EBU protocol, S/PDIF channel data and subcode information are transmitted in blocks of 192 bits consisting of 12 words that are 16 bits long. A portion of this information is reserved as a category code that provides the necessary setup information (sample rate, copy protection status and so on) to the copying device. Another portion is set aside for transmitting audio data that's used to relay track indexing information (such as start ID and program ID numbers), allowing this relevant information to be digitally transferred from the master to the copy. It should be noted that the professional AES/EBU protocol isn't capable of digitally transmitting these codes during a copy transfer.

In addition to transmitting two channels in an interleaved L-R-L-R … fashion, S/PDIF is able to communicate multichannel data between devices. Most commonly, this shows up as a direct 5.1 surround-sound link between a DVD player and an audio receiver/amplifier playback system (via either an RCA coax or optical connection).

SCMS

Initially, certain digital recording devices (such as a DAT recorder) were intended to provide consumers with a way to make high-quality recordings for their own personal use. Soon after its inception, however, for better or for worse, the recording industry began to see this new medium as a potential source of lost royalties due to home copying and piracy practices. As a result, the RIAA (Recording Industry Association of America) and the former CBS Technology Center set out to create a "copy inhibitor." After certain failures and long industry deliberations, the result of these efforts was a process that has come to be known as the *Serial Copy Management System*, or SCMS (pronounced "scums"). This protocol was incorporated into many consumer digital devices in order to prohibit the unauthorized copying of digital audio at 44.1 kHz (SCMS doesn't apply to the making of analog copies). Note also that, due to the demise of the DAT format (both as a recording and mass-music distribution medium), SCMS has pretty much fallen out of favor in both the consumer and pro audio communities.

So, what is SCMS? Technically, it's a digital protection flag that is encoded in byte 0 (bits 6 and 7) of the S/PDIF subcode area. This flag can have only one of three possible states:

- *Status 00:* No copy protection, allowing unlimited copying and subsequent dubbing
- *Status 10:* No more digital copies allowed
- *Status 11:* Allows a single copy to be made of this product, but that copy cannot be copied.

MADI

The MADI (Multichannel Audio Digital Interface) standard was jointly proposed as an AES standard by representatives of Neve, Sony and SSL as a straightforward, clutter-free digital interface connection between multitrack devices (such as a digital tape recorder, high-end workstation or mixing console, as shown in

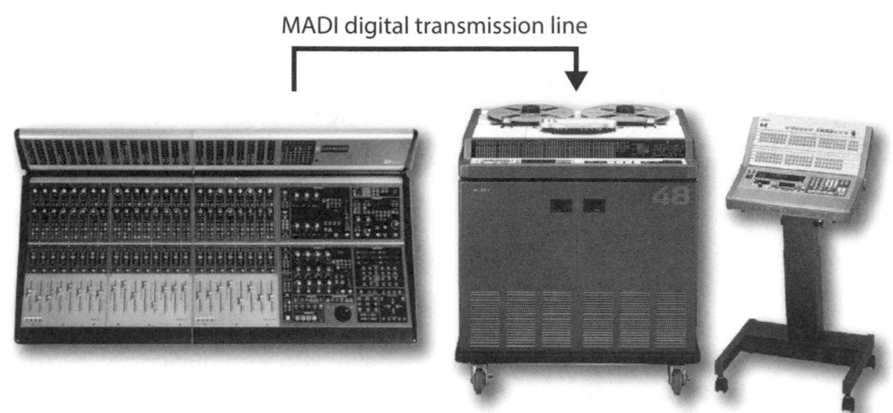

FIGURE 6.16
MADI offers a clutter-free digital interface connection between multitrack devices.

Figure 6.16). The format allows up to 56 channels of linearly encoded digital audio to be connected via a single 75 Ω, video-grade coaxial cable at distances of up to 120 feet (50 meters) or at greater distances whenever a fiber-optic relay is used.

In short, MADI makes use of a serial data transmission format that's compatible with the AES/EBU twin-channel protocol (whereby the data, Status, User and parity bit structure is preserved), and sequentially cycles through each channel (starting with Ch. 0 and ending with Ch. 55). The transmission rate of 100 Mbit/sec provides for an overall bandwidth that's capable of handling audio and numerous sync codes at various sample rate speeds (including allowances for pitch changes either up or down by 12.5% at rates between 32 and 48 kHz).

ADAT LIGHTPIPE

A wide range of modular digital multitrack recorders, audio interface, mic pre and hardware devices currently use the Alesis *lightpipe* system for transmitting multichannel audio via a standardized optical cable link. These connections make use of standard Toslink connectors and cables to transmit up to eight channels of uncompressed digital audio at resolutions up to 24 bit at sample rates up to 48k over a sequential, optical bitstream. In what is called the S/MUX IV mode, a lightpipe connection can also be used to pass up to four channels of digital audio at higher sample rates (i.e. 88.2 and 96k).

Although these connectors are identical to those that are used to optically transmit S/PDIF stereo digital audio, the datastreams are completely incompatible with each other. Lightpipe data isn't bidirectional, meaning that the connection can only travel from a single source to a destination. Thus, two cables will be needed to distribute data both to and from a device. In addition, synchronization data is imbedded within the digital datastream, meaning that no additional digital audio sync connections are necessary in order to lock device timing clocks; however, transport and timecode information is not transmitted and will require additional connections.

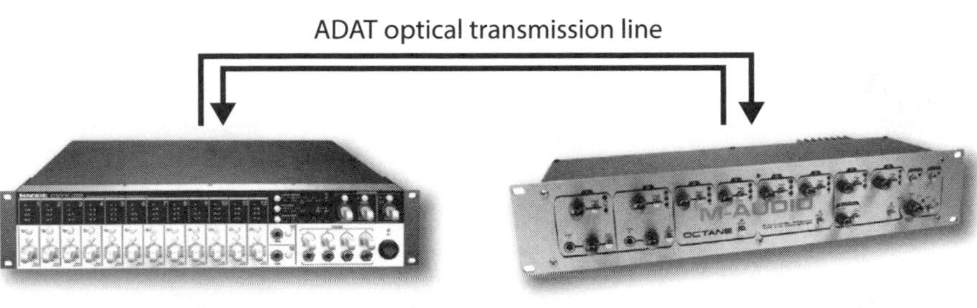

FIGURE 6.17
ADAT optical I/O Interconnection drawing.

FIGURE 6.18
Behringer UltraGain Pro-8 8-channel preamp and digital converter. (Courtesy of Behringer Int'l GmbH, www.behringer.com)

It's interesting to note that, although the Alesis ADAT tape-based format has virtually disappeared into the sunset, the lightpipe protocol lives on as a preferred way of passing multichannel digital audio to and from a digital audio interface/DAW combination (Figures 6.17 and 6.18). For example, it's common for a modern multichannel mic pre to have a lightpipe output, allowing a stream of up to eight digital outs to be easily inserted into a suitable audio interface, with a single optical cable, thus increasing the number of inputs by eight (or 16, should two ADAT I/O interconnections be available).

TDIF

The TDIF (Tascam Digital InterFace) is a proprietary format that uses a 25-pin D-sub cable to transmit and/or receive up to eight channels of digital audio between compatible devices. Unlike the lightpipe connection, TDIF is a bidirectional connection, meaning that only one cable is required to connect the eight ins and outs of one device to another. Although systems that support TDIF-1 cannot send and receive sync information (a separate wordclock connection is required for that; wordclocks are discussed below), the newer TDIF-2 protocol is capable of receiving and transmitting digital audio sync through the existing connection, without any additional cabling.

Signal distribution

If copies are to be made from a single, digital audio source, or if data is to be distributed throughout a connected network using AES/EBU, S/PDIF or MADI digital transmission cables, it's possible to daisy chain the data from one device to the next in a straightforward fashion (Figure 6.19a). This method works well only if a few devices are to be chained together. However, if several devices are

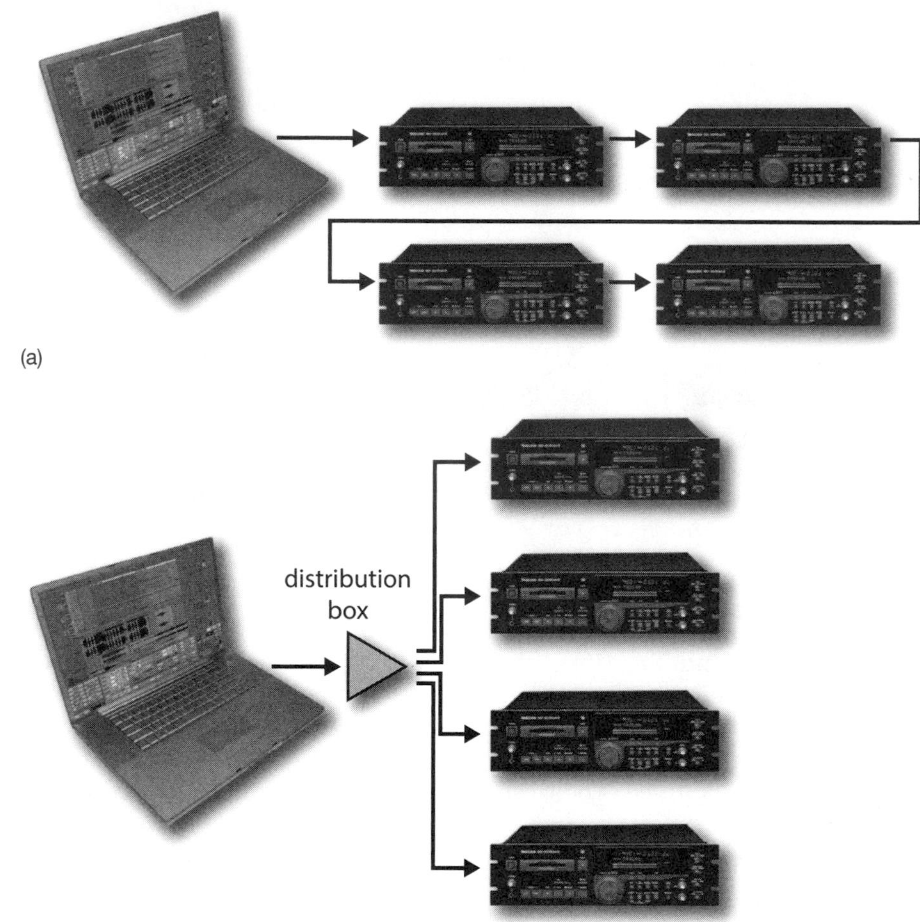

FIGURE 6.19 Digital audio distribution: (a) several devices connected in a daisy-chain fashion; (b) multiple devices connected using a distribution device.

connected together, time-base errors (known as *jitter*) might be introduced into the path, with the possible side effects being added noise, distortion and a slightly "blurred" signal image. One way to reduce the likelihood of such time-base errors is to use a digital audio distribution device that can route the data from a single digital audio source to a number of individual device destinations (Figure 6.19b).

What is jitter?

Jitter is a controversial and widely misunderstood phenomenon. To my knowledge, it's been explained best by Bob Katz of Digital Domain (www.digido.com, Orlando, FL). The following is a brief excerpt of his article "Everything You Always Wanted to Know About Jitter But Were Afraid to Ask." Further reading on digital audio and mastering techniques can be found in Bob's excellent book, *Mastering Audio: The Art and the Science*, from Focal Press (www.focalpress.com):

Jitter is time-base error. It is caused by varying time delays in the circuit paths from component to component in the signal path. The two most common causes of jitter are poorly designed Phase Locked Loops (PLLs) and waveform distortion due to mismatched impedances and/or reflections in the signal path.

Here is how waveform distortion can cause time-base distortion: The top waveform (Figure 6.20a) represents a theoretically perfect digital signal. Its value is 101010, occurring at equal slices of time, represented by the equally spaced dashed vertical lines. When the first waveform passes through long cables of incorrect impedance, or when a source impedance is incorrectly matched at the load, the square wave can become rounded, fast rise times become slow, and reflections in the cable can cause misinterpretation of the actual zero crossing point of the waveform. The second waveform (Figure 6.20b) shows some of the ways the first might change; depending on the severity of the mismatch you might see a triangle wave, a square wave with ringing or simply rounded edges. Note that the new transitions (measured at the zero line) in the second waveform occur at unequal slices of time. Even so, the numeric interpretation of the second waveform is still 101010! There would have to be very severe waveform distortion for the value of the new waveform to be misinterpreted, which usually shows up as audible errors—clicks or tics in the sound. If you hear tics, then you really have something to worry about.

If the numeric value of the waveform is unchanged, why should we be concerned? Let's rephrase the question: "When (not why) should we become concerned?" The answer is "hardly ever." The only effect of time-base distortion is in the listening; as far as it can be proved, it has no effect on the dubbing of tapes or any

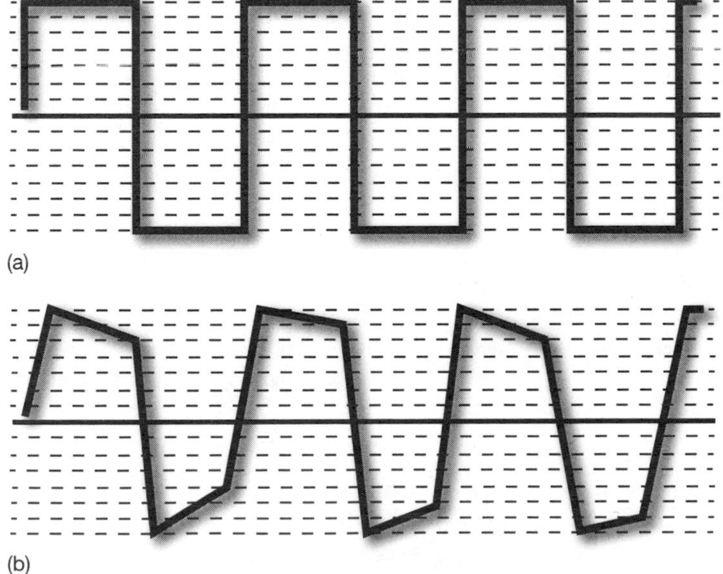

FIGURE 6.20 Example of time-base errors: (a) a theoretically perfect digital signal source; (b) the same signal with jitter errors.

digital-to-digital transfer (as long as the jitter is low enough to permit the data to be read; high jitter may result in clicks or glitches as the circuit cuts in and out). A typical D/A converter derives its system clock (the clock that controls the sample and hold circuit) from the incoming digital signal. If that clock is not stable, then the conversions from digital to analog will not occur at the correct moments in time. The audible effect of this jitter is a possible loss of low-level resolution caused by added noise, spurious (phantom) tones, or distortion added to the signal.

A properly dithered 16-bit recording can have over 120 dB of dynamic range; a D/A converter with a jittery clock can deteriorate the audible dynamic range to 100 dB or less, depending on the severity of the jitter. I have performed listening experiments on purist, audiophile-quality musical source material recorded with a 20-bit accurate A/D converter (dithered to 16 bits within the A/D). The sonic results of passing this signal through processors that truncate the signal at −110, −105 or −96 dB are increased "grain" in the image; instruments losing their sharp edges and focus; reduced sound-stage width; and apparent loss of level causing the listener to want to turn up the monitor level, even though high-level signals are reproduced at unity gain. Contrary to intuition, you can hear these effects without having to turn up the listening volume beyond normal (illustrating that low-level ambience cues are very important to the quality of reproduction). Similar degradation has been observed when jitter is present. Nevertheless, the loss due to jitter is subtle and primarily audible with the highest-grade audiophile D/A converters.

Wordclock

One aspect of digital audio recording that never seems to get enough attention is the need for locking (synchronizing) the sample clocks within a connected digital audio system to a common timing reference. Left unchecked, it's possible that such gremlins as clicks, pops and jitter (oh my!) would make their way into the audio chain. Through the use of a single, master timing reference known as *wordclock*, the overall sample-and-hold conversion states during both the record and playback process for all digital audio channels and devices within the system will occur at exactly the same point in time.

How can you tell when the wordclock between multiple devices either isn't connected or is improperly set? If you hear ticks, clicks and pops over the monitors that sound like "pensive monkeys pounding away at a typewriter," you've probably found the problem. When this happens, stop…and deal with the issue, as those monkeys will almost certainly make it into your recordings.

To further illustrate, let's assume that we're in a room that has four or five clocks and none of them read the same time! In places like this, you'll never quite know what the time really is … the clocks could be running at different speeds or at the same speed but are set to different times. Basically, trying to accurately keep track of the time would end up being a jumbled nightmare. On the other hand, if all of these clocks were locked to a single, master clock (remember those self-correcting clocks that were installed in most schools?), accurately keeping track of the time (even when moving from room to room) would be much simpler.

In effect, wordclock works in much the same fashion. If the sample clock (the timing reference that determines the sample rate and DSP traffic control) for each device were set to operate in a freewheeling, internal fashion … the timing references for each device within the connected digital audio chain wouldn't accurately match up. Even though the devices are all running at the same sample rate, the resulting mismatches in time will often result in clicks, ticks, excessive jitter and other unwanted grunge. To correct for this, the internal clocks of all the digital devices within a connected chain must be referenced to a single "master" wordclock timing element (Figure 6.21).

Similar to the distribution of timecode, there can only be one master wordclock reference within a connected digital distribution network. This reference source can be derived from a digital mixer, soundcard or any desired source that can transmit wordclock. Often, this reference pulse is chained between the involved devices through the use of BNC and/or RCA connectors, using low-capacitance cables (often 75 Ω, video-grade coax cable is used, although this cable grade isn't always necessary on shorter cable runs).

It's interesting to note that wordclock isn't generally needed when making a digital copy directly from one device to another (via such protocols as AES,

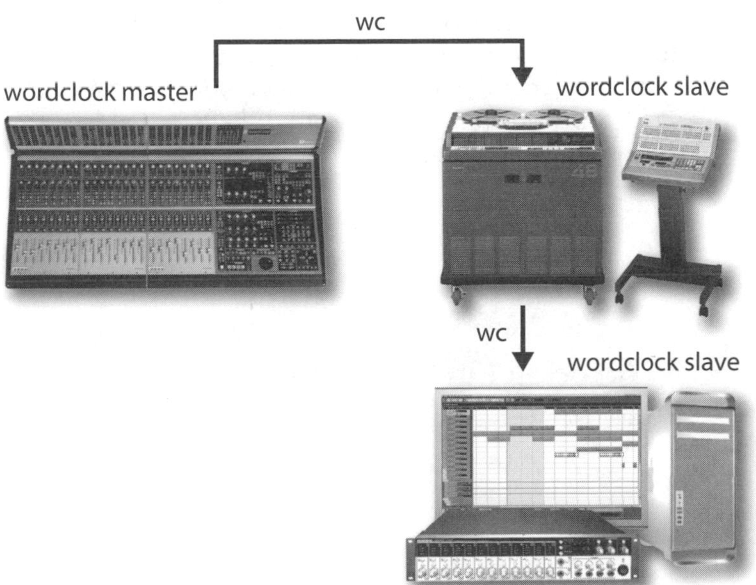

FIGURE 6.21
Example of wordclock distribution showing that there can be only one master clock within a digital production network.

S/PDIF, MADI or TDIF2), because the timing information is actually embedded within the data bitstream itself. Only when we begin to connect devices that share and communicate digital data throughout a production network do we begin to see the immediate need for wordclock. Note also that Digidesign had actually created a special version of wordclock for Pro Tools systems that ticks off 256 pulses for every sample. Superclock (also known generically as 256Fs) is transmitted over the same type and cable as wordclock and was eventually abandoned in favor of the industry standard wordclock protocol.

It almost goes without saying that there will often be differences in connections and parameter setups from one system to the next. In addition to proper cabling, impedance and termination considerations throughout the network, specific hardware and software setups may be required in order to get all the device blocks to communicate. To better understand your particular system's setup (and to keep frustration to a minimum), it's always a good idea to keep all of your device's physical or pdf manuals close at hand.

DIGITAL AUDIO RECORDING SYSTEMS

For the remainder of this chapter, we'll be looking at some of the digital audio recording device types that are currently available on the market. From my own personal viewpoint, not only do I find the here and now of recording technology to be exciting and full of cost-effective possibilities … I also love the fact that there are lots of recording media and device-type options. In other words, a digital hardware or software system that works really well for me might not be the best and easiest solution for you! As we take a look at many of these device and system choices, I hope that you'll take the time to learn about each one (and possibly even try your hand at listening to and/or working with each system type). In the earlier years of recording, there were only a few ways in which a successful recording could be made. Now, in the age of technological options … your mission (should you decide to accept it) is to research and test-drive devices and/or production systems to find the one that best suits your needs, budget and personal working style.

Tutorial: Further Research

Of course, it's important to point out that many of the explanations in this chapter are, by design, basic introductions to digital audio technology and its underlying concepts. To gain greater insight, it's always best to roll up your virtual sleeves and dive into books, magazines and the Internet. A better understanding of a technology or technique is often yours for the asking by researching and downloading appropriate archived magazine articles and user pdf manuals.

Samplers

Historically, one of the first production applications in digital audio came in the form of the drum machine, which gave us the ability to trigger prerecorded drum and percussion sounds from a single, dedicated electronic instrument. This made it possible for electronic musicians to add percussion samples to their own compositions without having to hire out a drummer/percussionist. Out of this innovation sprang a major class of sample and synthesis technology that has fueled electronic production technology over its history.

Not long after the introduction of the basic drum machine, devices were developed that were capable of recording, musically transposing, processing and reproducing segments of digitized audio directly from RAM (random access memory) … and thus, the *sampler* was born.

Over several decades, the simple hardware sample playback device has evolved into more complex hardware and (more recently) software systems (Figures 6.22 and 6.23) that are capable of streaming hundreds of carefully crafted samples directly from a computer's hard disk and RAM memory.

Assuming that sufficient memory is available, any number of audio samples can be loaded into a system in such a way that their playback can be transposed in real time (either up or down) over a number of octave ranges in a musical fashion. Since memory is often (but not always) limited in size, the digital sample segments are generally limited in length and range from only a few seconds to

FIGURE 6.22
Yamaha MOTIF XF8 keyboard workstation. (Courtesy of Yamaha Corporation of America. www.yamaha.com)

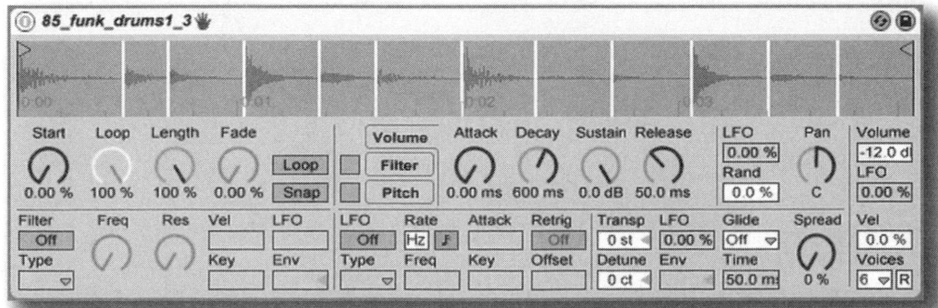

FIGURE 6.23
Ableton Live's "Simpler" software sampler. (Courtesy of Ableton, www.ableton.com)

one or more minutes. Quite simply, these musical transpositions occur by reproducing the recorded digital audio segments at sample rates that correspond to established musical intervals. These samples can then be assigned to a specific note on a MIDI controller or mapped across a keyboard range in a multiple-voice (polyphonic) musical fashion.

Hard-disk recording

As project and computer technology matured, advances in hardware and digital audio technology began to open the doors for recording audio as longer and longer sample files into RAM. It wasn't long after this that computer technology advanced to the point where specialized hardware, software and I/O interfacing could be used to record, edit and play back audio directly from a computer's hard disk. Thus, the concept of the hard-disk recorder was born.

A few of the numerous advantages to using a hard-disk recording system in an audio production environment include:

- *The ability to handle multiple audiofiles:* Hard-disk recording time is often limited only by the size of the disk, while the track count is determined by the speed and throughput of the overall system.
- *Random-access editing:* Once audio (or any type of media data) is recorded onto a disk, any segment of the program can be instantly accessed at any time, regardless of the order in which it was recorded.
- *Nondestructive editing:* A process that allows audio segments (often called *regions*) to be placed in any context and/or order within a program without changing or affecting the originally recorded sound file in any way. Once edited, these edited tracks and/or segments can be reproduced to create a single, cohesive song or project.
- *DSP:* Digital signal processing can be performed on a sound file and/or segment in either real time or non-real time (often in a nondestructive fashion).

Add to this the fact that computer-based digital audio devices can integrate various media types (such as audio, video and MIDI) into a single mean, lean production machine that's often easy to use, cost effective and time effective ... and you have a system that offers the artist and engineer an unprecedented degree of production power. Check out Chapter 7 for more information on the basics of hard-disk recording within the digital audio workstation (DAW) environment.

Hard-disk multitrack recorders

Unlike software DAWs that use the graphic user interface (GUI) of a personal computer to offer up an overall music production environment, some people opt to record using a hard-disk multitrack recording system (Figure 6.24) that mimics the basic transport, operational and remote controls of a traditional multitrack recorder. The basic allure of such a recording system is that it makes use of a simple, dedicated multitrack hardware interface in a way that combines with the speed, flexibility and nonlinear benefits of a hard-disk-based system.

Often these devices will incorporate one or more removable hard-drive bays and include some form of file compatibility with existing workstation software, allowing for direct sound file import, edit and export capabilities from within the DAW.

FIGURE 6.24
Tascam X-48 multitrack hard-disk recorder. (Courtesy of Tascam, www.tascam.com)

Portable studios

A number of all-in-one hardware recording systems exist that are capable of recording to hard disk or solid-state flash memory. These systems include all of the required hardware and the control system's interface to record, edit, mix down and play back a project virtually anywhere when using an AC adapter or batteries. However, with the growing popularity of the iPad, other pad devices, powerful apps and 3rd party supporting hardware (Figure 6.25) allow for an ever-growing number of options for recording, editing, mixing… even including options for importing/exporting full mix sessions between the pad and your favorite studio DAW. Care to create mixes of your latest hit single on a sailboat off the coast of New Zealand?

Flash memory handhelds

Recently, a new class of portable recording device (and interface options for popular iDevices) have come onto the scene. These devices, known as handhelds, are so small that they'll even fit in your pocket. Recording to solid-state flash memory, handhelds often include a set of built-in high-quality electret-condenser mics and offer many of the recording and overdub features that you might expect from a larger portable recording system, such as a USB link (allowing for easy data transfer or for directly turning the device into a USB interface with built-in mics!), built-in effects, microphone emulation (allowing the mic's character to be changed to conform to a number of mic styles), a guitar tuner, etc. Certain handhelds allow external professional mics to be plugged-in that can be phantom powered, while others are designed with four mic capsules that even allow for quad surround recording.

FIGURE 6.25
Alesis iO Dock Pro Audio Dock for the iPad. (Courtesy of Alesis, www.alesis.com)

In fact, these small, lightweight and powerful recording systems are often of surprisingly high-quality, and can be used for any number of purposes, such as:

- Recording a live concert with amazing ease
- Recording on-mic sound that can be synched up with a video camera or video project

FIGURE 6.26
Portable MiniDisc recorder.

- Recording dialog (for books-on-tape, etc.)
- Easily capturing music and nature samples
- Easily capturing samples for film (Foley and SFX)
- Crude speaker and room calibration (for matching L/R and even surround volume levels)

Older technologies

Although a number of recording technologies have come and gone over the years, there are a few recording formats that are common enough to definitely warrant mentioning.

MINIDISC

Since its introduction by Sony as a consumer device in 1992, the MiniDisc (MD) has established itself as a viable medium for recording and storing CD, MP3 tracks and original recordings (Figure 6.26). Based on established rewritable magneto-optical (MO) technology, the 64-mm disc system has a standard recording capacity of up to 74 minutes (effectively the same record/play time as the CD, at one-quarter the size). This extended record/play time is due to a compression codec known as Adaptive Transform Acoustic Coding (ATRAC), which makes use of an encoding scheme that reduces data by eliminating or reducing data that has been deemed to be imperceptible by the average listener (in much the manner of MP3 and other compression codecs). Probably the biggest selling point of the MiniDisc recording medium is its portability, although this has been challenged by the introduction of the previously mentioned flash memory handheld recorder.

THE ROTATING-HEAD DIGITAL AUDIO RECORDER

Historically, *rotating-head technology* is the backbone of a wide range of media production systems. Not only is it the basis of numerous digital audio technologies, such as DAT (digital audio tape) and MDM (modular digital multitrack) recording systems—it's the basis of early video technologies, including the videocassette and its professional counterparts. For this reason alone, a basic introduction and coverage of these technologies is warranted.

THE ROTARY HEAD

Even if it's not as relevant in this day and age of solid-state and hard-disk memory technologies, the *rotary head* is still of tremendous importance in a number of media-based devices, including the professional and consumer VCR, DAT recorders, modular digital multitracks and more.

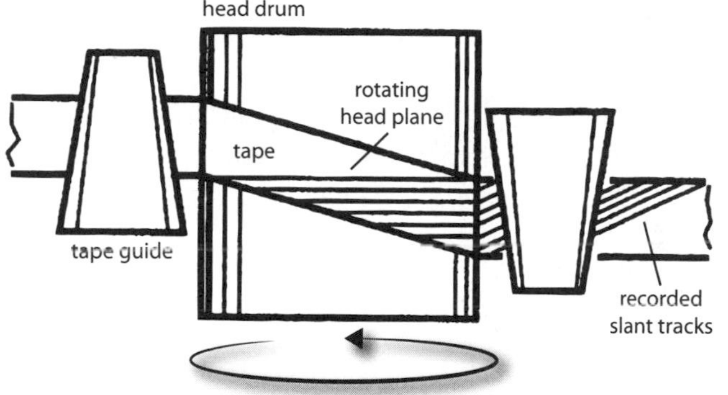

FIGURE 6.27
A helical scan path tape path is wrapped around the drum mechanism, so as to record tracks at a helical "spiraling slant" angle.

Because of the tremendous amount of data density that's required to record and reproduce video, broadband and digital audio media, the recording of data onto a single, linear tape track is impractical (for example, Vera, the first video machine ever made, used 20 ½-inch reels of wire that were running at speed of 200 inches/sec—almost 17 feet/sec—to encode video)! Of course, running a recording medium at this rate was out of the question … so, in order to get around this bandwidth limitation, a rotating-head *helical scan* path was incorporated to effectively increase the recorded data path, while making use of a slow-moving tape.

True to its name, this process uses magnetic recording heads that are mounted onto a rotating drum that's mounted at a slant with respect to the tape's moving axis. The tape is partially wrapped around the drum, so as to make contact with the rotating heads. As the head rotates, numerous closely spaced tape "scan" paths (or "swipes") are consecutively recorded along the tape path at a slanted angle. If we were able to lay these angular tracks out in a straight line, the resulting recorded track would be much longer in length than would be possible with a linear track, resulting in a system that can record wider frequency bandwidths at a slow-moving tape speed. An example of a helical transport that employs rotary head technology is shown in Figure 6.27.

DIGITAL AUDIO TAPE (DAT) SYSTEM

The digital audio tape, or DAT, recorder is a compact recording medium for recording PCM digital audio. DAT technology makes use of an enclosed compact cassette that's slightly larger than modern miniDV videotapes. DAT recorders are commonly equipped with both analog and digital input/outputs, and the specification allows for recording/playback at three standard sampling frequencies: 32, 44.1, and 48 kHz (although sample rate capabilities and system features may vary from one recorder to the next). Current DAT tapes offer running times of up to 2 hours when sampling at 44.1 and 48 kHz. As technology marches on, this tape-based medium continues to give way to newer flash memory and hard-disk-based devices.

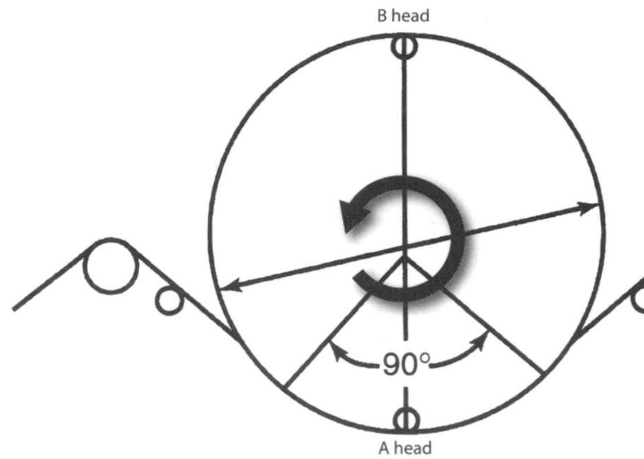

FIGURE 6.28
A half-wrap tape path, showing a 90° amount of tape-to-head contact.

DAT TAPE/TRANSPORT FORMAT

The actual track width of the DAT format's helical scan can range downward to about 1/10th the thickness of a human hair (which allows for a recording density of 114 million bits per square inch). To assist with tape tracking and to maintain the highest quality playback signal, a tracking correction system is used to center the heads directly over the tape's scan path. The head assembly of a DAT recorder uses a 90° half-wrap tape path. (Figure 6.28 shows a two-head assembly in which each head is in direct contact with the tape 25% of the time.) Such a discontinuous signal requires that the digital data be communicated to the heads in digital "bursts," which necessitates the use of a digital buffer. On playback, these bursts are again smoothed out into a continuous datastream.

THE MODULAR DIGITAL MULTITRACK

Another device that is less commonly seen in the pro and project setting is the modular digital multitrack system or MDM recording system. These historically important devices single-handedly ushered professional and project studio production out of the analog age and into the affordable age of the personal project studio. This techno explosion occurred because these compact multitrack digital audio recorders were capable of recording eight tracks of digital audio onto videotape-grade cassette cartridges that could often be bought at your favorite neighborhood drugstore. At the time of their introduction, another important aspect of the MDM revolution was their price. Amazingly, these devices offered digital audio recording quality to the pro and project studio at a cost that was often 1/10th the price of a comparable analog machine. MDMs are said to be "modular" because several of them can be linked together (in groups of eight) in a synchronous fashion that allows them to work in tandem as a large-scale multitrack recording system (often configured in the 24- and 32-track range).

ADAT MDM FORMAT

The ADAT standard, which was created by the Alesis Corporation, is no longer available on the market (although, as was previously mentioned, the 8-channel ADAT optical transmission link is still going strong). These rotary-head, eight-track modular digital multitrack recorders made use of standard S-VHS videotape to record PCM audio at 44.1k/48k at 16-/20-bit word depths.

DTRS MDM FORMAT

Created by Tascam, the rotary-head Digital Tape Recording System (DTRS; see Figure 6.29) is capable of recording up to 108 minutes of digital audio onto a

FIGURE 6.29
Tascam DA-98HR modular digital multitrack recorder. (Courtesy of Tascam, www.tascam.com)

standard 120-minute Hi-8-mm videotape, and (like the ADAT format) can be combined with other DTRS recorders to create a system that has 24, 32 or more tracks. Digital I/O is made through the proprietary TDIF connection, which uses a special 25-pin D-sub connector to link to other DRTS recorders, digital mixers, hard-disk recorders or external accessories. The TDIF digital interconnection comes in two flavors:

- TDIF-1 is capable of transmitting and receiving all eight channels of digital audio (in a bidirectional I/O fashion) over a single cable
- TDIF-2 is capable of transmitting eight channels of digital audio, as well as sync data without the need for an external connection.

CHAPTER 7

The Digital Audio Workstation

Over the history of digital audio production, the style, form and function of hard-disk recording has changed to meet the challenges of faster processors, reduced size, larger drives, improved hardware systems and the ongoing push of marketing forces to sell, sell, sell! As a result, there are numerous hard-disk system types that are designed for various purposes, budgets and production styles. As new technologies and programming techniques continue to turn out new hardware and software systems at a dizzying pace, many of the long-held production limitations have vanished as increased track counts, processing power and affordability have changed the way we see the art of production itself. In recent years, no single term signifies these changes more than that of the "DAW."

In recent years, the *digital audio workstation (DAW)* has come to signify an integrated computer-based hard-disk recording system that commonly offers a wide and ever-changing number of production features such as:

- Advanced multitrack recording, editing and mixdown capabilities
- MIDI sequencing, edit and score capabilities
- Integrated video and/or video sync capabilities
- Integration with peripheral hardware devices such as controllers, MIDI and audio interface devices
- Plug-in DSP (digital signal processing) support
- Support for plug-in virtual instruments
- Support for integrating timing and control elements with other production software (ReWire).

Truth of the matter is … by offering an astounding amount of production power for the buck, these software-based programs (Figures 7.1 through 7.3) and their associated hardware devices have revolutionized the faces of professional, project and personal studios in a way that touches almost every life within the audio and music production communities.

INTEGRATION NOW ... INTEGRATION FOREVER!

Throughout the history of music and audio production, we've grown used to the idea that certain devices were only meant to perform a single task: A

FIGURE 7.1
Pro Tools hard-disk editing workstation for the Mac® or PC. (Courtesy of Avid Technology, Inc., www.avid.com)

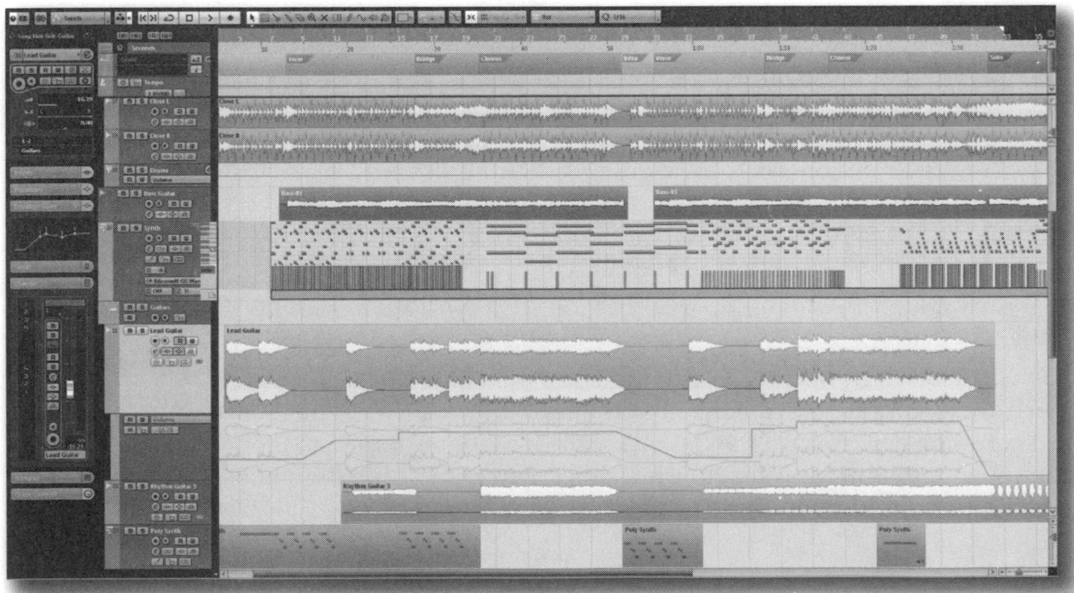

FIGURE 7.2
Cubase 6 Media Production System for the Mac® or PC. (Courtesy of Steinberg Media Technologies GmbH, www.steinberg.net)

FIGURE 7.3
Logic DAW for the Mac. (Courtesy of Apple Inc., www.apple.com)

recorder records and plays back, a limiter limits and a mixer mixes. Fortunately, the age of the microprocessor has totally broken down these traditional lines in a way that has created a breed of digital chameleons that can change their functional colors as needed to match the task at hand. Along these same lines, the digital audio workstation isn't so much a device as a systems concept that can perform a wide range of audio production tasks with relative ease and speed. Some of the characteristics that can (or should be) offered by a DAW include:

- *Integration:* One of the biggest features of a workstation is its ability to provide centralized control over the digital audio recording, editing, processing and signal routing functions within the production system. It should also provide for direct communications with production-related hardware and software systems, as well as transport and sync control to/from external media devices.
- *Communication:* A DAW should be able to communicate and distribute pertinent audio, MIDI and automation-related data throughout the connected network system. Digital timing (wordclock) and synchronization (SMPTE timecode and/or MTC) should also be supported.
- *Speed and flexibility:* These are probably a workstation's greatest assets. After you've become familiar with a particular system, most production tasks can be tackled in far less time than would be required using similar analog equipment. Many of the extensive signal processing, automation and systems communications features would be far more difficult to accomplish in the analog domain.
- *Automation:* Because all of the functions are in the digital domain, the ability to instantly save and recall a session and to instantly undo a performed action becomes a relatively simple matter.

- *Expandability:* Most DAWs are able to integrate new and important hardware and software components into the system with little or no difficulty.
- *User-friendly operation:* An important element of a digital audio workstation is its ability to communicate with its central interface unit: you! The operation of a workstation should be relatively intuitive and shouldn't obstruct the creative process by speaking too much "computerese."

I'm sure you've gathered from the points above that a software system (and its associated hardware) that's capable of integrating audio, video and MIDI under a single, multifunctional umbrella can be a major investment, both in financial terms and in terms of the time that's spent learning to master the overall program environment. When choosing a system for yourself or your facility, be sure to take the above considerations into account. Each system has its own strengths, weaknesses and particular ways of working. When in doubt, it's always a good idea to research the system as much as possible before committing to it. Feel free to contact your local dealer for a salesroom test drive. As with a new car, purchasing a DAW can be an expensive proposition that you'll probably have to live with for a while. Once you've taken the time to make the right choice, you can get down to the business of making music.

DAW HARDWARE

Keeping step with the modern-day truism "technology marches on," the hardware and software specs of a computer and the connected peripherals continue to change at an ever-increasing pace. This is usually reflected as general improvements in such areas as their:

- Need for speed (multiple processors)
- Increased computing power
- Increased disk size and speed
- Increased memory size and speed
- Operating system (OS) and peripheral integration
- General connectivity (networking and the Web).

In this day and age, it's definitely important that you keep step with the ever-changing advances in computer-related production technology. That's not to say you need to update your system every time a new hard- or soft-whizbang comes onto the market. On the contrary, it's often a wise person who knows when a system is working just fine for his or her own personal needs and who does the research to update software and fine-tune the system (to the best of his or her ability). On the other hand, there will come a time (and you'll know all too well when it arrives) that this "march" of technology will dictate a system change to keep you in step with the times. As with almost any aspect of technology, the best way to judge what will work best for you and your system is to research any piece of hardware that you're considering … quite simply, read the specs, reads the reviews, ask your friends and then make your best, most informed choice.

When buying a computer for audio production, one of the most commonly asked questions is "Which one—Mac or PC?" The answer as to which operating system (OS) will work best for you will actually depend on:

- Your preference
- Your needs
- The kind of software you currently have
- The kind of computer platform and software your working associates or friends have.

Any of these can be a determining factor—cost, the fact that you already have lots of PC or Mac software, the fact that you grew up with and are comfortable with a particular OS. Even this particular question is being sidestepped with the advent of Apple's Boot Camp, which allows a Mac to boot up under the Mac or Windows OS, giving you freedom of choice to have either or both.

Once you've decided which side of the platform tracks you'd like to live on, the more important questions that you should be asking are:

- Is my computer fast and powerful enough for the tasks at hand?
- Does it have enough hard disks that are large and fast enough for my needs?
- Is there enough random access memory (RAM)?
- Do I have enough monitor space (real estate) to see the important things at a glance?

On the "need for speed" front, it's always a good idea to buy (or build) a computer at the top of its performance range at any given time. Keeping in mind that technology marches on, the last thing that you'll want to do is buy a new computer only to soon find out that it's underpowered for the tasks ahead.

The newer dual, quad and eight-core (multiprocessor) systems allow for faster calculations. Their tasks are spread across multiple CPUs; for example, a number of DAWs allow for their odd/even track counts to be split across multiple CPUs to increase the overall processing load, for added track count and DSP capabilities.

With today's faster and higher capacity hard drives, it's a simple matter to install cost-effective drives with terabyte capacities into a system. These drives can be internal, or they can be installed in portable drive cases (Figure 7.4) that can be plugged into either a Thunderbolt®, FireWire® or USB port, making it easy to take your own personal drive with you to the studio or on-stage.

The speed at which the disc platters turn will often affect a drive's access time. Modern drives that spin at 7200 rpm or higher with 64MB or more of internal cache memory are often preferable for both audio and video production. SSD (solid state drives) are also available that don't have moving parts at all, but include solid state memory that can have access times that blaze at 6Gb a second or faster. Within a production system, it's also good to have a second drive that's strictly dedicated to your media files … this is always a good practice, because

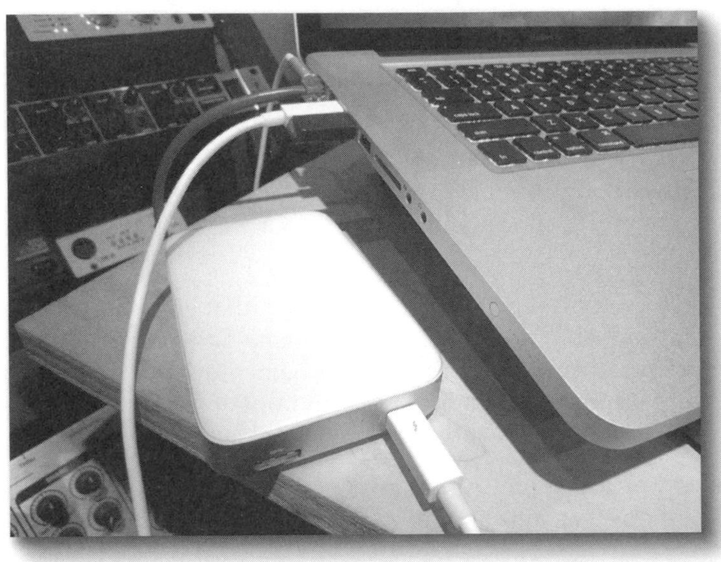

FIGURE 7.4
A Thunderbolt data connection in Action.

data could be interrupted or slowed should the main drive need to be accessed by the computer's operating system.

Regarding random access memory, it's always good to use as much RAM that is as fast as you can muster. If a system doesn't have enough RAM, data will often have to be swapped to the system's hard drive, which can slow things down and affect overall performance. When dealing with music sampling, video and digital image technologies, having a sufficient amount of RAM becomes even more of an issue.

With regard to system and application software, it's often wise to perform an update to keep your system … well … up-to-date. This holds true even if you just bought the software, because it's often hard to tell how long it's been sitting on the shelves—and even if it is brand-spanking new, chances are new revisions will still be available. Updates don't come without their potential downfalls; given the incredible number of hardware/software system combinations that are available, it's actually possible that an update might do more harm than good. In this light, it's actually not a bad idea to do a bit of research before clicking that update button. Who said all this stuff would be easy?

Just like there never seems to be enough space around the house or apartment, having a single, undersized monitor can leave you feeling cramped for visual "real estate." For starters, a sufficiently large monitor that's capable of working at higher resolutions will greatly increase the size of your visual desktop; however, if one is a good thing, two is always better! Both Windows® and Mac OS systems offer support for multiple monitors (Figure 7.5). By adding a commonly

FIGURE 7.5
You can never have enough visual "real estate"!

available "dual-head" video card, your system can easily double your working monitor space for fewer bucks than you might think. I've found that it's truly a joy to have your edit window, mixer, effects sections, and transport controls in their own places—all in plain and accessible view.

The desktop computer

Desktop computers are often (but not always) too large and cumbersome to lug around. As a result, these systems are most often found as a permanent installation in the professional, project and home studio (Figures 7.6 and 7.7). Historically, desktops have offered more processing power than their portable counterparts, but in recent times, this distinction has become less and less of a factor.

The laptop computer

One of the most amazing characteristics of the digital age is miniaturization. At the forefront of the studio-a-go-go movement is the laptop computer (Figures 7.8 and 7.9). From the creation of smaller, lighter and more powerful notebooks has come the technological Phoenix of the portable DAW and music performance machine. With the advent of USB and FireWire audio interfaces, controllers and other peripheral devices, these systems are now capable of handling most (if not all) of the edit and processing functions that can be handled in the studio. In fact, these AC/battery-powered systems are often powerful enough to

FIGURE 7.6
The Mac Pro™ with Cinema display. (Courtesy of Apple Computers, Inc., www.apple.com)

FIGURE 7.7
ION A2 Rackmount Audio Computer. (Courtesy of Rain Computers, LLC, www.raincomputers.com)

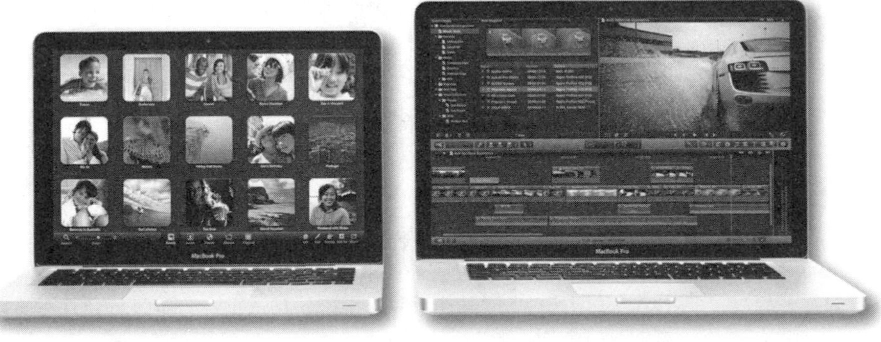

FIGURE 7.8
The MacBook Pro 13" and 15." (Courtesy of Apple Computers, Inc., www.apple.com)

handle advanced DAW edit/mixing functions, as well as being able to happily handle a wide range of plug-in effects and virtual instruments, all in the comfort of … anywhere!

FIGURE 7.9
LiveBook™ A2 laptop. (Courtesy of Rain Computers, LLC, www.raincomputers.com)

That's the good news! Now, the downside of all this portability is the fact that, since laptops are optimized to run off of a battery with as little power drain as possible, their:

- processors will often run slower, so as to conserve on battery power.
- BIOS (the important subconscious brains of a computer) might be different (again, especially with regards to battery-saving features).
- hard drives might not spin as fast (generally they're shipped with 5400-rpm speed drives).
- video display capabilities are sometimes limited when compared to a desktop (video memory is often shared with system RAM, reducing graphic quality and refresh rate).
- internal audio interface usually isn't so great (but that's why there are so many external interface options).

Although the central processing unit (CPU) will often run more slowly than a desktop (usually to reduce power consumption in the form of heat), most modern laptops are more than powerful enough to perform on the road. For this reason, it's always best to get a high-quality system with the fastest CPU(s) that you can afford.

Most often, the primary problems with a laptop lie in their basic BIOS and OS battery-saving features. When it comes to making music with a laptop, there actually is a real difference between the Mac and a PC. Basically, there's little difference between a Mac laptop and a Mac tower, because the BIOSs are virtually identical. Conversely, the BIOS of a laptop PC is often limited in power and functional capabilities when compared to a desktop. When shopping for a PC laptop, it's often good to research how a particular BIOS chipset will work for music, particularly with regard to certain audio interface devices. In this instance, it's always good to research and read up on the laptop that's caught your fancy before you buy (also, keep in mind that the store's return policy might buy you time to take it for a technological spin before committing).

Both PC and Mac laptops have an automatic power-saving feature (called "speed step" and "processor cycling," respectively) that changes the CPU's speed in much the same way that a vehicle changes gears in order to save energy. Often these gear changes can wreak havoc on many of the DSP functions of an audio workstation. Turning them off will greatly improve performance, at the expense of battery life.

Hard-drive speeds on a laptop are often limited when compared to a desktop computer, resulting in slower access times and fewer track counts on a multitrack DAW. Even though these speeds are often more than adequate for general music applications, speeds can often be improved through the use of a 7200-rpm or faster external USB, FireWire or Thunderbolt drive.

Again, when it comes to RAM, it's often good to pack as much into the laptop as you can. This will reduce data swapping to disk when using larger audio applications (especially software synths and samplers). Within certain systems, the laptop's video card capabilities will run off of the system's RAM (which can severely limit audio processing functions in a DAW). Memory-related problems might also crop up when using a motherboard to run a dual-monitor (LCD and external monitor) setup.

It almost goes without saying, it's no secret that the internal audio quality of most laptops ranges from being acceptable to abysmal. As a result, about the only true choice is to find an external audio interface that works best for you and your applications. Fortunately, there's a ton of audio interface choices for connecting via either FireWire or USB, ranging from a simple stereo I/O device to those that include multitrack audio, MIDI and controller capabilities in a small, on-the-go package.

The iOS and Pad Generation

One of the biggest changes in recent years in the project and home studio markets has been the introduction of the iPad and portable pad technology. Of course, these devices are amazingly portable, but their strength lies in their overall flexibility and in the vast number software options that are coming onto the market on a daily basis (Figure 7.10).

The most popular pad device is, of course, Apple's own iPad. Given their popularity, the number of useful software and supporting hardware systems is truly staggering.

- On-the-go recordings can easily be made using quality mics and headphone monitoring.

FIGURE 7.10
Tons of "apps" are coming onto the market each day that are able to control, generate, assist and anything else to do with the creation and production of music ... seriously fun stuff! (Courtesy of Presonus Audio Electronics, Inc., www.presonus.com)

- Multitrack recording DAWs are being written for these devices that let us record, mix and even transfer entire sessions between a pad and our main workstation with relative ease.
- Electronic instruments and entire music production environments can be taken literally anywhere, allowing us to quickly save and work on musical ideas at the drop of a hat.

In addition to these ideas (and many, many more), another huge contribution pad technology has made to music production is the pad's ability to serve as a remote controlling device. For example, in the not-too-distant past, DAW controllers would easily cost over $1000 and would need both a power and a USB connection. Now, with the introduction of pad-based DAW controllers, the same basic functionality is available for $4.99 (or free), fits in your hand and is totally wireless. In addition, such a pad can be used to trigger DJ samples, live performance triggering, lighting ... the list is virtually limitless.

Accessories and accessorize

I know it seems like an afterthought, but there's an ever-growing list of hardware and travel accessories that can help you to take your portable rig on the road. Just a small listing includes:

- Laptop backpacks for storing your computer and gear in a safe, fun case (Figure 7.11)
- Pad stands and cases
- Instrument cases and covers
- Flexible LED gooseneck lights that lets you view your keyboard in the dark or on-stage

FIGURE 7.11
Studio Pack Mobile Laptop Studio Backpack. (Courtesy of M-Audio, a division of Avid, www.m-audio.com.)

- Laptop DJ stands for raising your laptop above an equipment-packed table.

SYSTEM INTERCONNECTIVITY

In the not-too-distant past, installing a device into a computer or connecting between computer systems could be a major hassle. With the development of the USB and FireWire protocols (as well as the improved general programming of hardware drivers), hardware devices such as mice, keyboards, cameras, soundcards, modems, MIDI interfaces, CD and hard drives, MP3 players, portable fans, LED Christmas trees and cup warmers … can be plugged into an available port, installed and be up and running in no time, (usually) without a hassle.

With the development of a standardized network protocol, it's now possible to link computers together in a way that allows for the fast and easy sharing of data throughout a connected system. Using such a system, artists and businesses alike can easily share and swap files on the other side of the world, and pro or project studios can share sound files and video files using an Ethernet connection with relative ease.

USB

In recent computer history, few interconnection protocols have affected our lives like the universal serial bus (USB). In short, USB is an open specification for connecting external hardware devices to the personal computer, as well as a special set of protocols for automatically recognizing and configuring them. The first of the following two speeds are supported by USB 1.0, while all three are supported by USB 2.0:

- *USB 1.0 (1.5 Mbits/sec):* A low speed for the attachment of low-cost peripherals (such as a joystick or mouse)
- *USB 1.0 (12 Mbits/sec):* For the attachment of devices that require a higher throughput (such as data transfer, soundcards, digitally compressed video cameras and scanners)
- *USB 2.0 (480 Mbits/sec):* For high throughput and fast transfer of the above applications
- *USB 3.0 (5 Gbits/sec):* For even higher throughput and fast transfer of the above applications

The basic characteristics of USB include:

- Up to 127 external devices can be added to a system without having to open up the computer. As a result, the industry is moving toward a "sealed case" or "locked-box" approach to computer hardware design.
- Newer operating systems will often automatically recognize and configure a basic USB device that's shipped with the latest device drivers.
- Devices are "hot pluggable," meaning that they can be added (or removed) while the computer is on and running.
- The assignment of system resources and bus bandwidth is transparent to the installer and end user.
- USB connections allow data to flow bi-directionally between the computer and the peripheral.
- USB cables can be up to 5 meters in length (up to 3 meters for low-speed devices) and include two twisted pairs of wires, one for carrying signal data and the other pair for carrying a DC voltage to a "bus-powered" device. Those that use less than 500 milliamps (1/2 amp) can get their power directly from the USB cable's 5-V DC supply, while those having higher current demands will need to be externally powered.
- Standard USB cables have different connectors at each end. For example, a cable between the PC and a device would have an "A" plug at the PC (root) connection and a "B" plug for the device's receptacle.

Cable distribution and "daisy-chaining" are done via a data "hub" (Figure 7.12). These devices act as a traffic cop in that they cycle through the various USB inputs in a sequential fashion, routing the data into a single data output line. It should be noted that not all hubs are created equal. In certain situations, the chipset that's used within a hub might not be compatible with certain MIDI and audio interface systems. Should a connection problem arise, contact the manufacturer of your audio-related device for advice.

FIGURE 7.12
USB hubs in action.

FireWire

Originally created in the mid-1990s by Apple (and later standardized as IEEE-1394), the FireWire protocol is similar to the USB standard in that it uses twisted-pair wiring to communicate bidirectional, serial data within a hot-swappable, connected chain. Unlike USB (which can handle up to 127 devices per bus), up to 63 devices can be connected within a connected FireWire chain. FireWire supports two speed modes:

- FireWire 400 or IEEE-1394a (400 Mbits/sec) is capable of delivering data over cables up to 4.5 meters in length. FireWire 400 is ideal for communicating large amounts of data to such devices as hard drives, video camcorders and audio interface devices.
- FireWire 800 or IEEE-1394b (800 Mbits/sec) can communicate large amounts of data over cables up to 100 meters in length. When using fiber-optic cables, lengths in excess of 90 meters can be achieved in situations that require long-haul cabling (such as sound stages and studios).

Unlike USB, compatibility between the two modes is mildly problematic, because FireWire 800 ports are configured differently from their earlier predecessor … and therefore require adapter cables to ensure compatibility.

Thunderbolt

Originally released in 2011, Thunderbolt (Figure 7.13) combines the DisplayPort and PCIe bus into a serial data interface that runs at speeds of up to 10Gbit/sec per port. A single Thunderbolt port can support a daisy chain of up to six Thunderbolt devices (two of which can be DisplayPort display devices).

Networking

Beyond the concept of connecting external devices to a single computer, another concept hits at the heart of the connectivity age—*networking*. The ability to set up and make use of a *local area network (LAN)* can be extremely useful in the home, studio and/or office, in that it can be used to link multiple computers with various data, platforms and OS types. In short, a network can be set up in a number

FIGURE 7.13
Thunderbolt port on a Macbook Pro.

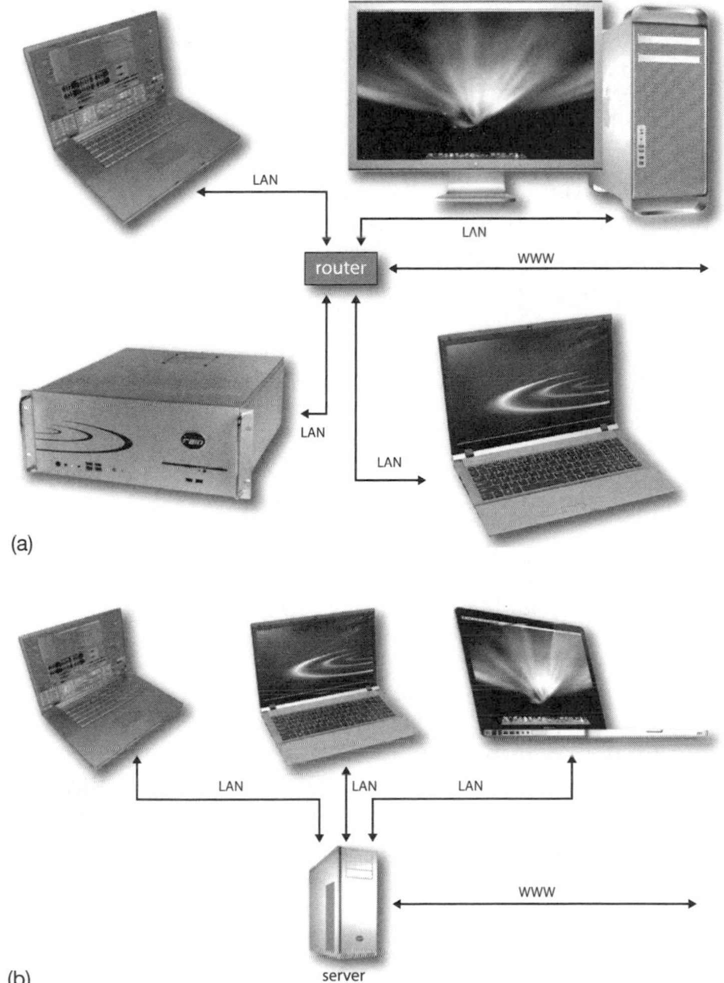

FIGURE 7.14
Local area network (LAN) connections. (a) Data can be shared between independent computers in a home or workplace LAN environment. (b) Computer terminals may be connected to a centralized server, allowing data to be stored, shared and distributed from a central location.

of different ways, with varying degrees of complexity and administrative levels. There are two common ways that data can be handled over a LAN (Figure 7.14):

- The first is a system whereby the data that's shared between linked computers resides on the respective computers and is communicated back and forth in a decentralized manner.
- The second makes use of a centralized computer (called a *server*) that uses an array of high-capacity hard drives to store "all" of the data that relates to the everyday production aspects of a facility. Often, such a system will have a redundant set of drives (RAID) that actually clones the entire system on a moment-to-moment basis as a safety backup procedure. In larger facilities where data integrity is critical, a set of backup tapes may be made on a daily basis for extra insurance and archival purposes.

No matter what level of complexity is involved, some of the more common uses for working with a network connection include:

- *Sharing files:* Within a connected household, studio or business, a LAN can be used to share files, sound files, video images—virtually anything—throughout the connected facility. This means that various productions rooms, studios and offices can simultaneously share and swap data and/or media files in a way that's often transparent to the users.
- *Shared Web connection:* One of the cooler aspects of using a LAN is the ability to share an Internet connection over the network from a single, connected computer or server. The ability to connect from any computer with ease is just another reason why you should strongly consider wiring your studio and/or house with LAN connections.
- *Archiving and backup:* In addition to the benefits of archiving and backing up data with a server system … even the simplest LAN can be a true lifesaver. For example, let's say that we need to make a backup DVD of a session but don't have the time to tie up our production DAW. In this situation, we could simply burn the disc on another computer that's connected to the system, and continue working away, without interruption.
- *Accessing sound files and sample libraries:* It goes without saying that sound and sample files can be easily accessed from any connected computer…. Actually, if you're wireless (or have a long enough cable), you can go out to the pool and soak up the sun while working on your latest project!

On a final note, those who are unfamiliar with networking are urged to learn about this powerful and easy-to-use data distribution and backup system. For a minimal investment in cables, hubs and educational reading, you might be surprised at the time, trouble and life-saving benefits that will be almost instantly realized.

THE AUDIO INTERFACE

An important device that deserves careful consideration when putting together a DAW-based production system is the digital audio interface. These devices can have a single, dedicated purpose, or they might be multifunctional in nature. In either case, their main purpose in the studio is to act as a connectivity bridge

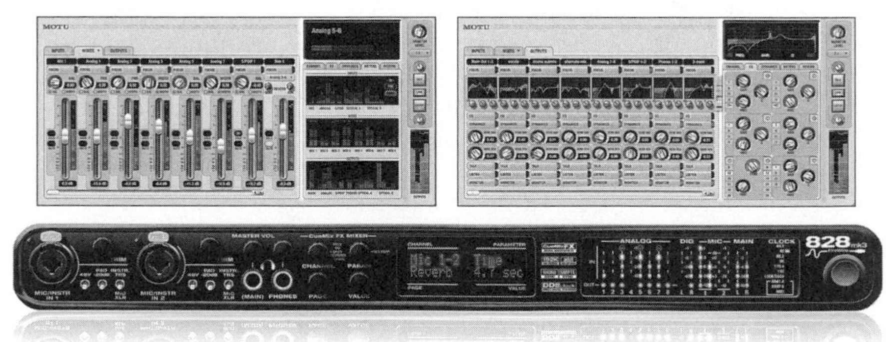

FIGURE 7.15
MOTU 828mk3 FireWire audio interface with effects. (Courtesy of MOTU, Inc., www.motu.com)

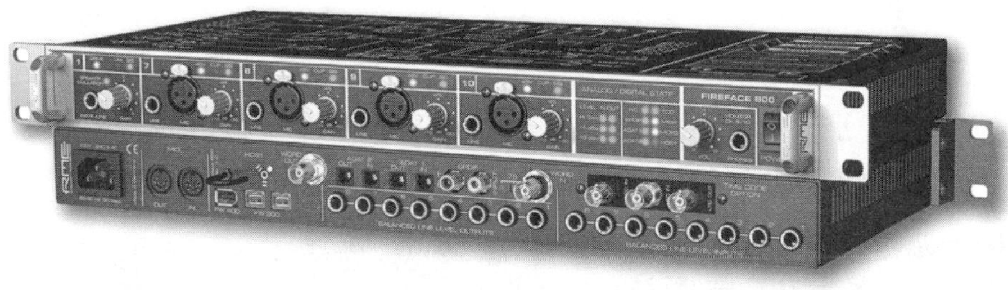

FIGURE 7.16 RME Fireface 800 192kHz/24bit 56 channel firewire interface (Courtesy of Audio AG, www.rme-audio.de)

FIGURE 7.17 Apollo FireWire and Thunderbolt audio interface with integrated UAD effects processing card. (Courtesy of Universal Audio, www.uaudio.com © 2013 Universal Audio, Inc. All rights reserved. Used with permission)

FIGURE 7.18 Pro Tools HD with 192 I/O interface. (Courtesy of Digidesign, a division of Avid, www.digidesign.com)

between the outside world of analog audio and the computer's inner world of digital audio (Figures 7.15 through 7.18). Audio interfaces come in all shapes, sizes and functionalities; for example, an audio interface can be:

- Built into a computer (although, more often than not, these devices are often limited in quality and functionality)
- A simple, two-I/O audio device

- Multichannel, offering eight analog I/Os and numerous I/O expansion options
- Fitted with one or more MIDI I/O ports
- One that offers digital I/O, wordclock and sync options
- Fitted with a controller surface (with or without motorized faders) that provides for hands-on DAW operation.

These devices may be designed as hardware cards that fit directly into the computer, or they might plug into the system via USB or FireWire. An interface might have as few as two inputs and two outputs, or it might have as many as 24. It might offer a limited number of sample rate and bit-depth options, or it might be capable of handling rates up to 96 kHz/24 bits or higher. Unless you buy a system that's been designed to operate with a specific piece of hardware, you should weigh the vast number of interface options and capabilities with patience and care—the options could easily affect your future expansion choices.

Audio driver protocols

Audio driver protocols are software programs that set standards for allowing data to be communicated between the system's software and hardware. A few of the more common protocols are:

- *WDM:* The Windows Driver Model is a robust driver implementation that's directly supported by Microsoft Windows. Software and hardware that conform to this basic standard can communicate audio to and from the computer's basic audio ports.
- *ASIO:* The Audio Stream Input/Output architecture forms the backbone of VST. It does this by supporting variable bit depths and sample rates, multichannel operation, and synchronization. This commonly used protocol offers low latency, high performance, easy setup and stable audio recording within VST.
- *MAS:* The MOTU Audio System is a system extension for the Mac that uses an existing CPU to accomplish multitrack audio recording, mixer, bussing and real-time effects processing.
- *CoreAudio:* This driver allows compatible single-client, multichannel applications to record and play back through most audio interfaces using Mac OS X. It supports full-duplex recording and playback of 16-/24-bit audio at sample rates up to 96 kHz (depending on your hardware and CoreAudio client application).

In most circumstances, it won't be necessary for you to be familiar with the protocols—you just need to be sure that your software and hardware are compatible for use with the driver protocol that works best for you. Of course, further information can always be found at the respective companies' websites.

Latency

When discussing the audio interface as a production tool, it's important that we touch on the issue of latency. Quite literally, *latency* refers to the buildup of delays (measured in milliseconds) in audio signals as they pass through the audio circuitry of the audio interface, CPU, internal mixing structure and I/O routing chains. When monitoring a signal directly through a computer's signal path, latency can be experienced as short delays between the input and monitored signal. If the delays are excessive, they can be unsettling enough to throw a performer off time. For example, when recording a synth track, you might actually hear the delayed monitor sound shortly after hitting the keys (not a happy prospect). Because we now have faster computers, improved audio drivers, and better programming, latency has been reduced to such low levels that it's not even noticeable. For example, the latency of a standard Windows audio driver can be truly pitiful (upward to 500 ms). By switching to a supported ASIO driver and by optimizing the interface/DAW buffer settings to their lowest operating size (without causing the audio to stutter), these delay values can be reduced down to an unnoticeable range.

DAW CONTROLLERS

One of the more common complaints against the digital audio editor and workstation environment (particularly when relating to the use of on-screen mixers) is the lack of a hardware controller that gives the user access to hands-on controls. In recent years, this has been addressed by major manufacturers and third-party companies in the form of a hardware DAW controller interface.

These controllers (Figure 7.19) generally mimic the design of an audio mixer in that they offer slide or rotary gain faders, pan pots, solo/mute and channel select

FIGURE 7.19
Mackie Control Universal Pro DAW controller. (Courtesy of Loud Technologies, Inc., www.mackie.com)

FIGURE 7.20
The Raven MTX large-scale Multi-touch Audio Production Console. (Courtesy of Slate Pro Audio, www.slateproaudio.com)

buttons – with the added bonus of a full transport remote. A channel select button is used to actively assign a specific channel to a section that contains a series of grouped pots and switches that relate to EQ, effects and dynamic functions. These controllers offer direct mixing control over eight input strips at a time. By switching between the banks in groups of 8 (1–8, 9–16, 17–24, etc.), any number of the grouped inputs can be accessed by the virtual hardware mixer. These devices will also often include software function keys that can be programmed to give quick and easy access to the DAW's more commonly used program keys.

In recent years the sheer number, types and price-points of digital software controllers has completely exploded. These devices can range from being an inexpensive hardware controller that can fit in your back pocket, to larger controller systems that can control any number of audio and music production software systems in the studio.

In addition to the wide range of hardware controllers that are available on the market, an ever-growing number of software-based touch-screen controllers (Figure 7.20) have begun to take over the market, due to the fact that they're computer-based devices themselves, and as such, can change its form, function and entire way of working with a single … touch.

As was mentioned, controller technology has changed with the introduction of applications for the iPad and Android pad devices. With the introduction of controller "apps", hardware controllers that used to cost hundreds or thousands of dollars can now be emulated in software and purchased from the "store" for the virtual cost of a cup of coffee (Figure 7.21).

Controller commands are most commonly transmitted between the controller and audio editor via MIDI and device-specific MIDI SysEx messages. Therefore, in order to be able to integrate a controller into your system, the DAW's current version must be specifically programmed to accept the control codes from a

FIGURE 7.21
V-Control Pro DAW controller for the iPad. (Courtesy of Neyrinck, www.vcontrolpro.com)

FIGURE 7.22
SSL Nucleus DAW controller and audio interface. (Courtesy of Solid State Logic, www.solid-state-logic.com)

particular controller, unless the DAW and controller make use of a new plug-in architecture that allows compatible devices to be freely connected. Most controller surfaces communicate these messages to the DAW host via the easy-to-use wireless, USB or FireWire protocols.

Certain controllers also offer all-in-one capabilities that can make them straightforward and cost-effective devices for first-time buyers (Figures 7.22). Often, these devices include a multichannel audio interface, MIDI interface ports, monitor capabilities and full controller functions. Others may already have an existing digital mixer that can actually be used as a fully functional controller (and in certain circumstances, as a multichannel audio interface) when connected to a DAW host program. For these and other reasons, taking the time to research your needs and current equipment capabilities can save you both time and money.

FIGURE 7.23
Digidesign ICON D-Control integrated controller/console. (Courtesy of Avid Technology, Inc., www.avid.com)

It's important to note that there are numerous controllers from which to choose—and just because others feel that the mouse is cumbersome doesn't mean that you have to feel that way; for example, I have several controllers in my own studio, but the mouse is still my favorite tool. As always, the choice is totally up to you.

SOUND FILE FORMATS

An amazingly varied number of sound file formats exist within audio and multimedia production. Here is a list of the most commonly used audio production formats that don't use data compression:

- *Wave (.wav):* The Microsoft Windows format supports both mono and stereo files at a variety of bit and sample rates. WAV files contain PCM coded audio (uncompressed pulse-code modulation formatted data) that follows the Resource Information File Format (RIFF) spec, which allows extra user information to be embedded and saved within the file itself.
- *Broadcast wave (.wav):* In terms of audio content, broadcast wave files are the same as regular wave files; however, text strings for supplying additional information can be embedded in the file according to a standardized data format.
- *Apple AIFF (.aif or .snd):* This standard sound file format from Apple supports mono or stereo, 8-bit or 16-bit audio at a wide range of sample rates. Like broadcast wave files, AIFF files can contain embedded text strings.

Format interchange and compatibility

At the sound file level, most software editors and DAWs are able to read a wide range of uncompressed and compressed formats, which can then be saved into a new format. At the session level, there are several standards that allow for the exchange of data for an entire session from one platform, OS or hardware device to another. These include the following:

- Open Media Framework Interchange (OMFI) is a platform-independent session file format intended for the transfer of digital media between different DAW applications; it is saved with an .omf file extension. OMF (as it is commonly called) can be saved in either of two ways: (1) "export all to one file," when the OMF file includes all of the sound files and session references that are included in the session (be prepared for this file to be extremely large), or (2) "export media file references," when the OMF file will not contain the sound files themselves but will contain all of the session's region, edit and mix settings; effects (relating to the receiving DAW's available plug-ins and ability to translate effects routing); and I/O settings. This second type of file will be small by comparison; however, the original sound files must be transferred into the session folders.
- One audio-export only option makes use of the broadcast wave soundfile format. By using Broadcast wave, many DAWs are able to directly read the time-stamped data that's imbedded into the soundfile and then automatically line them up within the session.
- Software options that can convert session data between DAWs also exist. Pro Convert from Solid State Logic, for example, is a stand-alone program that helps you tailor soundfile information, level and other information, and then transfer one DAW format into another format or readable XML file.

Although these systems for allowing file and session interchangeability between workstation types can be a huge time and work saver, it should be pointed out that they are more often than not far from perfect. It's not uncommon for files not to line up properly (or load at all), plug-ins can disappear and/or lose their settings … you name it, it could happen. As a result, the most highly recommended and surefire way to make sure that a session will load into any DAW platform is to make (print or export) a copy of each track, starting from the session's beginning (00:00:00:00) and going to the end of that particular track. Using this system, all you need to do is load each track into the new workstation at their respective track beginning points and get to work.

Of course, the above method won't load any of the plug-in or mixer settings (often an interchange problem anyway). Therefore, it's extremely important that you properly document the original session, making sure that:

- All tracks have been properly named (supplying additional track notes and documentation, if needed).
- All plug-in names and settings are well documented (a screenshot can go a long way toward keeping track of these settings).

- Any virtual instrument or MIDI tracks are printed to an audio track. (Make sure to include the original MIDI files in the session, and to document all instrument names and settings … again, screenshots can help.)
- Any special effects or automation moves are printed to the particular track in question (make sure this is well documented) and you should definitely consider providing an additional copy of the track without effects or automation.

Sound file sample and bit rates

While the sample rate of a recorded bitstream (samples per second) directly relates to the resolution at which a recorded sound will be digitally captured, the bit rate of a digitally recorded sound file directly relates to the number of quantization steps that are encoded into the bitstream. It's important that these rates be determined and properly set *before* starting a session. Further reading on sample and bit rate depths can be found in Chapter 6. Additional info on sound files and compression codecs can be found in Chapter 10 (see Table 10.2 for a listing of uncompressed audio bit rate and file sizes).

DAW SOFTWARE

By their very nature, digital audio workstations (Figures 7.24 through 7.27) are software programs that integrate with computer hardware and functional applications to create a powerful and flexible audio production environment. These programs commonly offer extensive record, edit and mixdown facilities through the use of such production tools as:

- Extensive sound file recording, edit and region definition and placement
- MIDI sequencing and scoring
- Real-time, on-screen mixing
- Real-time effects
- Mixdown and effects automation

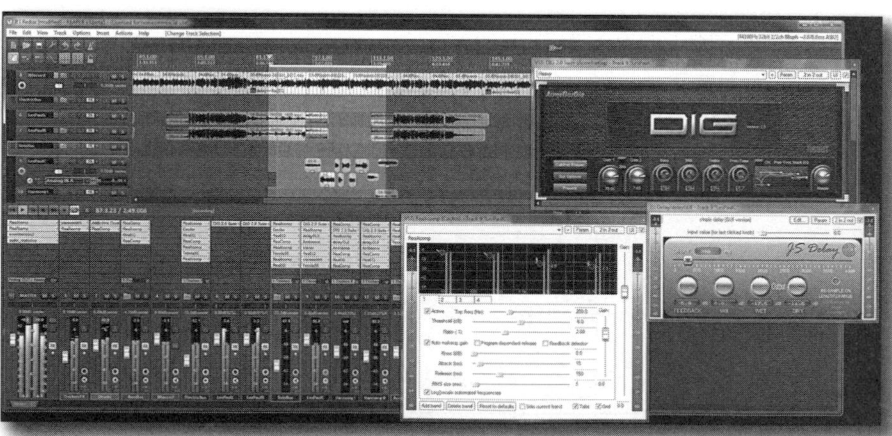

FIGURE 7.24
Reaper Digital Audio Workstation software (Courtesy of Cockos Incorporated, www.reaper.fm)

- Sound file import/export and mixdown export
- Support for video/picture synchronization
- Systems synchronization
- Audio, MIDI and sync communications with other audio programs (e.g., ReWire)
- Audio, MIDI and sync communications with other software instruments (e.g., VST technology)

FIGURE 7.25
Mark of the Unicorn Digital Performer. (Courtesy of MOTU, Inc., www.motu.com)

FIGURE 7.26
Collage of the various screens within the Cubase media production DAW. (Courtesy of Steinberg Media Technologies GmbH, www.steinberg.net)

FIGURE 7.27
Pro Tools 10 hard-disk recording software for the Mac or PC. (Courtesy of Avid Technology, Inc., www.avid.com)

This list is but a smattering of the functional capabilities that can be offered by an audio production DAW.

Suffice to say that these powerful software production tools are extremely varied in their form and function. Even with their inherent strengths, quirks, and complexities, their basic look, feel, and operational capabilities have, to some degree, become unified among the major DAW competitors. Having said this, there are enough variations in features, layout, and basic operation that individuals—from aspiring beginner to seasoned professional—will have their favorite DAW make and model. With the growth of the DAW and computer industries, people have begun to customize their computers with features, added power and peripherals that rival their love for souped-up cars and motorcycles. In the end, though (as with many things in life), it doesn't matter which type of DAW you use—it's how you use it that counts!

Sound file recording, editing, region definition and placement

Most digital audio workstations are capable of recording sound files in mono, stereo, surround or multichannel formats (either as individual files or as a single interleaved file). These production environments graphically display sound file information within a main graphic window (Figure 7.28), which contains drawn waveforms that graphically represent the amplitude of a sound file over time in a WYSIWYG (what you see is what you get) fashion. Depending on the system type, sound file length and the degree of zoom, the entire waveform may be shown on the screen. Or, only a portion will show, but the waveform will continue to scroll off one or both sides of the screen. Graphic editing differs greatly from the "razor blade" approach that's used to cut analog tape in that the waveform gives us both visual and audible cues as to precisely where an edit point should be. Using this common display technique, any position, cut/copy/paste, gain and time changes to the waveform will be instantly reflected on the

FIGURE 7.28
Main edit window within the Cubase audio production software. (Courtesy of Steinberg Media Technologies GmbH, a division of Yamaha Corporation, www.steinberg.net)

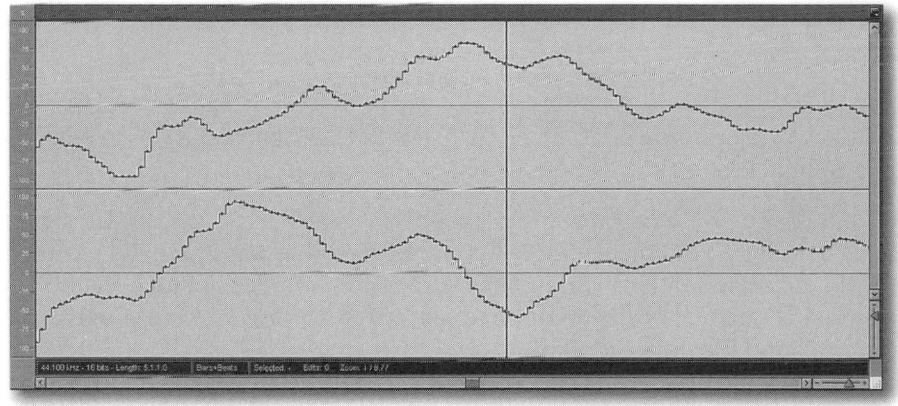

FIGURE 7.29
Zoomed-in edit window showing individual samples. (Courtesy of Steinberg Media Technologies GmbH, a division of Yamaha Corporation, www.steinberg.net)

screen. Usually, these edits are nondestructive (a process whereby the original file isn't altered—only the way in which the region in/out points are accessed or the file is processed as to gain, spectrum, etc.).

Only when a waveform is zoomed-in fully is it possible to see the individual waveshapes of a sound file (Figure 7.29). At this zoom level, it becomes simple to locate zero-crossing points (points where the level is at the 0, center-level line). In addition, when a sound file is zoomed-in to a level that shows individual sample points, the program might allow the sample points to be redrawn

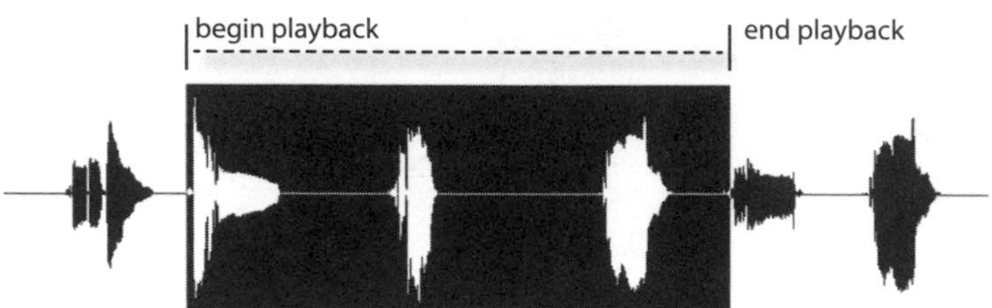

FIGURE 7.30 Nondestructive editing allows a region within a larger sound file to begin at a specific point and play until the endpoint is reached.

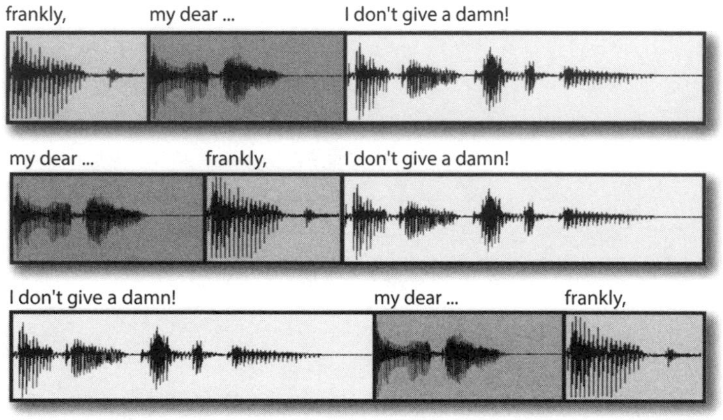

frankly, my dear ... I don't give a damn!

my dear ... frankly, I don't give a damn!

I don't give a damn! my dear ... frankly,

FIGURE 7.31 Example of how snippets from Rhett's famous *Gone with the Wind* dialogue can be easily rearranged.

in order to remove potential offenders (such as clicks and pops) or to smooth out amplitude transitions between loops or adjacent regions.

The nondestructive edit capabilities of a DAW refer to a disk-based system's ability to edit a sound file without altering the data that was originally recorded to disk. This important capability means that any number of edits, alterations or program versions can be performed and saved to disk without altering the original sound file data.

Nondestructive editing is accomplished by accessing defined segments of a recorded digital audio file (often called regions) and allowing them to be reproduced in a user-defined order or segment length in a manner other than was originally recorded. In effect, when a specific region is defined, we're telling the program to access the sound file at a point that begins at a specific memory address on the hard disk and continues until the ending address has been reached (Figure 7.30). Once defined, these regions can be inserted into the list (often called a playlist or editlist) in such a way that they can be accessed and reproduced in any order. For example, Figure 7.31 shows a snippet from *Gone With the Wind* that contains the immortal words "Frankly, my dear, I don't give a damn." By segmenting it into three regions we could use a DAW editor to output the words in several ways.

Tutorial: Recording a Sound File to Disk

1. Download a demo copy of your favorite DAW (these are generally available off the company's website for free).
2. Download the workstation's manual and familiarize yourself with its functional operating basics.
3. Consult the manual regarding the recording of a sound file.
4. Assign a track to an interface input sound source.
5. Name the track! It's almost always best to name the track (or tracks) before going into record. In this way, the file will be saved to disk within the session folder under a descriptive name instead of an automatically generated file name (e.g., killerkick.wav instead of track16-01.wav).
6. Save the session and assign the input to another track, and overdub a track along with the previously recorded track.
7. Repeat as necessary until you're having fun!
8. Save your final results for the next tutorial.

When working in a graphic editing environment, regions can usually be defined by positioning the cursor over the waveform, pressing and holding the mouse or trackball button and then dragging the cursor to the left or right, which highlights the selected region for easy identification. After the region has been defined, it can be edited, marked, named, maimed or otherwise processed.

As one might expect, the basic cut-and-paste techniques used in hard-disk recording are entirely analogous to those used in a word processor or other graphics-based programs:

- *Cut:* Places the highlighted region into clipboard memory and deletes the selected data (Figure 7.32).
- *Copy:* Places the highlighted region into memory and doesn't alter the selected waveform in any way (Figure 7.33).
- *Paste:* Copies the waveform data that's within the system's clipboard memory into the sound file beginning at the current cursor position (Figure 7.34).

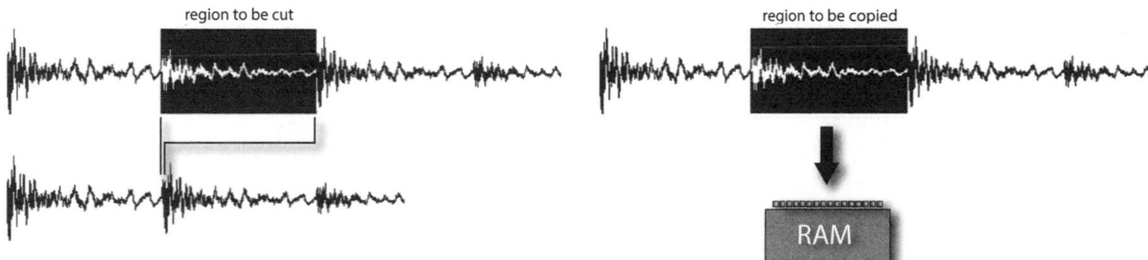

FIGURE 7.32
Cutting inserts the highlighted region into memory and deletes the selected data.

FIGURE 7.33
Copying simply places the highlighted region into memory without changing the selected waveform in any way.

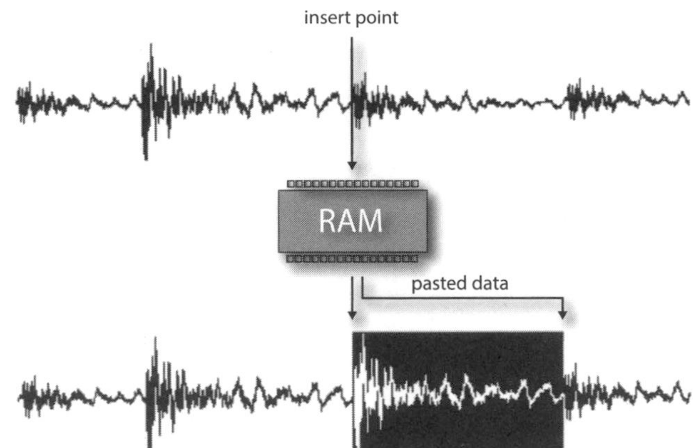

FIGURE 7.34
Pasting copies the data within the system's clipboard memory into the sound file at the current cursor position.

Tutorial: Copy and Paste

1. Open the session from the preceding tutorial, "Recording a Sound File to Disk."
2. Consult your editor's manual regarding basic cut-and-paste commands (which are almost always the standard PC and Mac commands).
3. Open a sound file and define a region that includes a musical phrase, lyric or sentence.
4. Cut the region and try to paste it into another point in the sound file in a way that makes sense (musical or otherwise).
5. Feel free to cut, copy and paste to your heart's desire to create an interesting or totally wacky sound file.

Besides basic nondestructive cut-and-paste editing techniques, the amplitude processing of a signal is one of the most common types of changes that are likely to be encountered. These include such processes as gain changing, normalization and fading.

Gain changing relates to the altering of a region or track's overall amplitude level, such that a signal can be proportionally increased or reduced to a specified level (often in dB or percentage value). To increase a sound file or region's overall level, a function known as normalization can be used. *Normalization* (Figure 7.35) refers to an overall change in a sound file or defined region's signal level, whereby the file's greatest amplitude will be set to 100% full scale (or a set percentage level of full scale), with all other levels in the sound file or region being proportionately changed in gain level.

The fading of a region (either in or out, as shown in Figure 7.36) is accomplished by increasing or reducing a signal's relative amplitude over the course of a defined duration. For example, fading in a file proportionately increases a

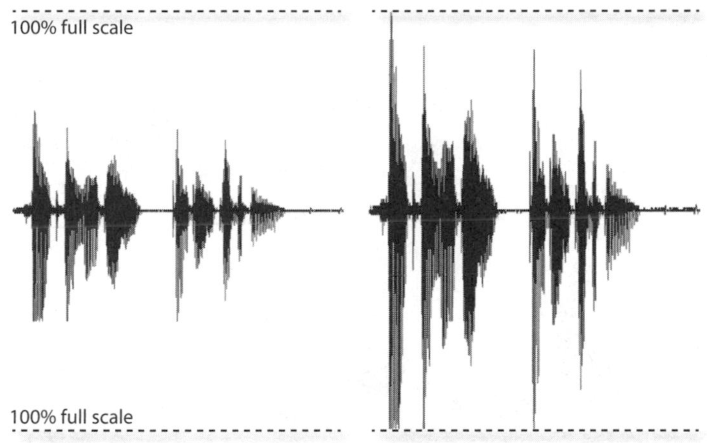

FIGURE 7.35
Original signal and normalized signal level.

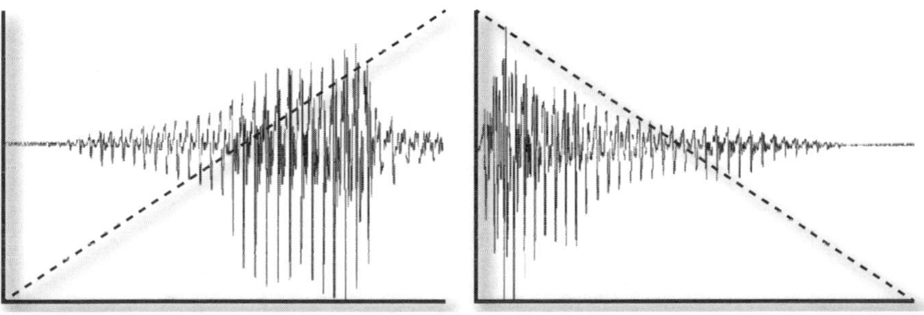

FIGURE 7.36
Examples of fade-in and fade-out curves.

region's gain from infinity (zero) to full gain. Likewise, a fade-out has the opposite effect of creating a transition from full gain to infinity. These DSP functions have the advantage of creating a much smoother transition than would otherwise be humanly possible when performing a manual fade.

A cross-fade (or X-fade) is often used to smooth the transition between two audio segments that either are sonically dissimilar or don't match in amplitude at a particular edit point (a condition that would otherwise lead to an audible "click" or "pop"). This functional tool basically overlaps a fade-in and fade-out between the two waveforms to create a smooth transition from one segment to the next (Figure 7.37). Technically, this process averages the amplitude of the signals over a user-definable length of time in order to mask the offending edit point.

FIXING THE SOUND WITH A SONIC SCALPEL

In traditional multitrack recording, should a mistake or bad take be recorded onto a new track, it's a simple matter to start over and re-record over the unwanted take. However, if only a small part of the take was bad, it's easy to go

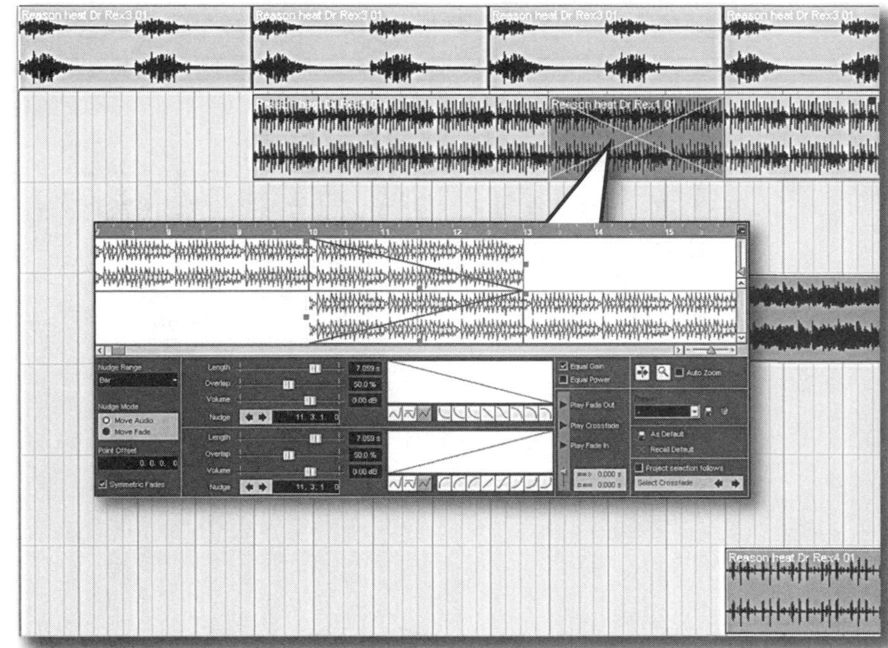

FIGURE 7.37 Example of a cross-faded sound file. (Courtesy of Steinberg Media Technologies GmbH, a division of Yamaha Corporation, www.steinberg.net)

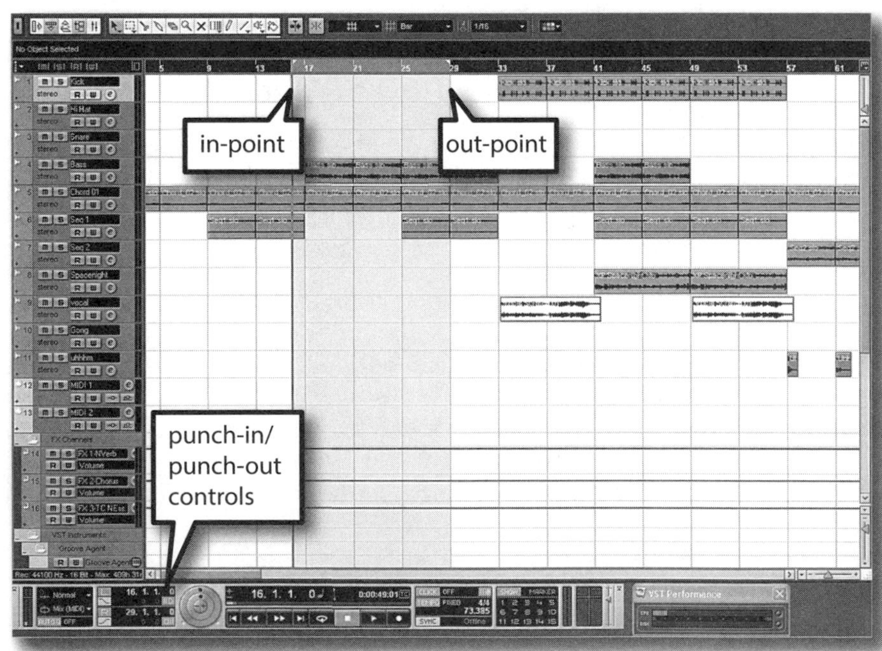

FIGURE 7.38 Punch-ins let you selectively replace material and correct mistakes. (Courtesy of Steinberg Media Technologies GmbH, a division of Yamaha Corporation, www.steinberg.net)

back and perform a punch-in (Figure 7.38). During this process, the recorder or DAW:

- Silently enters into record at a predetermined point
- Records over the unwanted portion of the take
- Silently falls back out of record at a predetermined point.

A punch can be manually performed on most recording systems; however, DAWs and newer tape machines can be programmed to automatically go into and fall out of record at a predetermined time.

When punching-in, any number of variables can come into play. If the instrument is an overdub, it's often easy to punch the track without fear of any consequences. If an offending instrument is part of a group or ensemble, leakage from the original instrument could find its way into adjacent tracks, making a punch difficult or impossible. In such a situation, it's usually best to re-record the piece, pick up at a point just before the bad section and splice or insert it into the original recording or attempt to punch the section using the entire ensemble.

From a continuity standpoint, it's often best to punch-in on a section immediately after the take has been recorded, because changes in mic choice, mic position or the general "vibe" of the session can lead to a bad punch that'll be hard to correct. If this isn't possible, make sure that you carefully document the mic choice, placement, preamp type, etc., before moving onto the next section: You'll be glad you did.

Performing a "punch" should always be done with care. Allowing the track to continue recording too long a passage could possibly cut off a section of the following, acceptable track and likewise require that the following section be redone. Stopping it short could cut off the natural reverb trail of the final note.

It needs to be pointed out that performing a punch using a DAW is often "far" easier than doing the same on an analog recorder. For example:

- If the overdub wasn't that great, you can simply click to "undo" it and start over!
- If the overdub was started early and cut into the good take (or went too long), the leading and/or trailing edge of the punch can be manually adjusted to expose or hide sections after the punch has been performed.

These beneficial luxuries can go a long way toward reducing operator error—and its associated tensions—during a session.

COMPING

When performing a musically or technically complex overdub, most DAWs will let you *comp* (short for composite) multiple overdubs together into a final, master take (Figure 7.39). Using this process, a DAW can be programmed to automatically enter into and out of record at the appropriate points. When placed

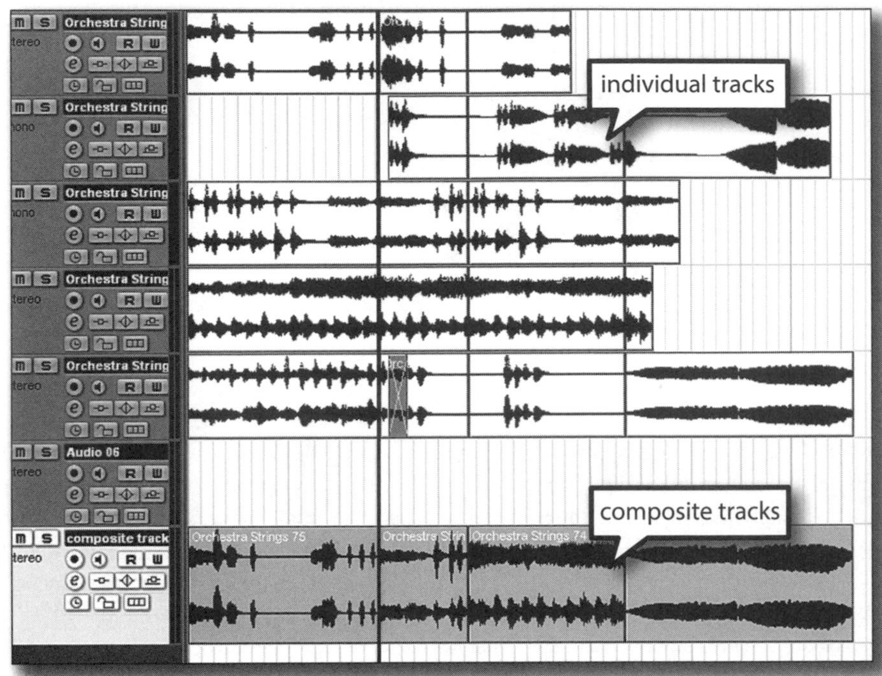

FIGURE 7.39
A single composite track can be created from several partially acceptable takes.

into record mode, the DAW will start laying down the overdub into a new and separate track. At the end of the overdub, it'll loop back to the beginning and start recording the next pass onto a separate track. This process of laying down consecutive takes will continue, until the best take is done or the artist gets tired of recording. Once done, an entire overdub might be chosen, or individual segments from the various takes might be assembled together into a final, master overdub. Such is the life of a digital microsurgeon!

MIDI sequencing and scoring

Most DAWs include extensive support for MIDI (Figures 7.40 and 7.41), allowing electronic instruments, controllers, effects devices, and electronic music software to be integrated with multitrack audio and video tracks. This important feature often includes the full implementation for:

- MIDI sequencing, processing and editing
- Score editing and printing
- Drum pattern editing
- MIDI signal processing
- Support for linking the timing and I/O elements of an external music application (ReWire)
- Support for software instruments (VSTi and RTAS).

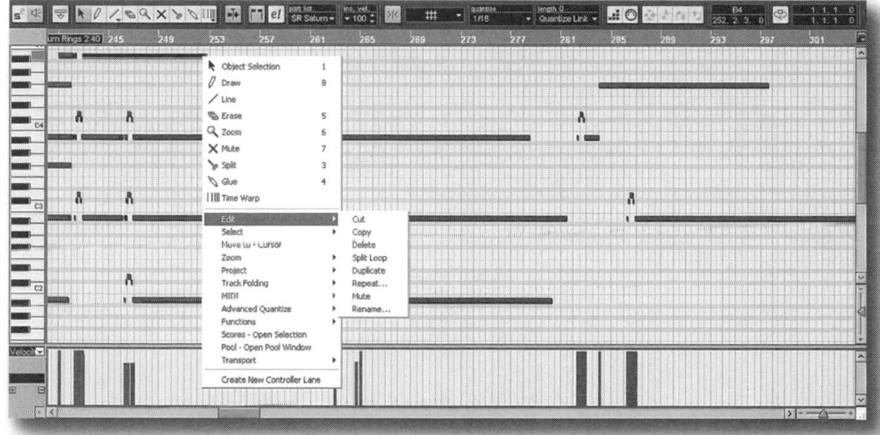

FIGURE 7.40
Steinberg Cubase/Nuendo piano roll edit window. (Courtesy of Steinberg Media Technologies GmbH, a division of Yamaha Corporation, www.steinberg.net)

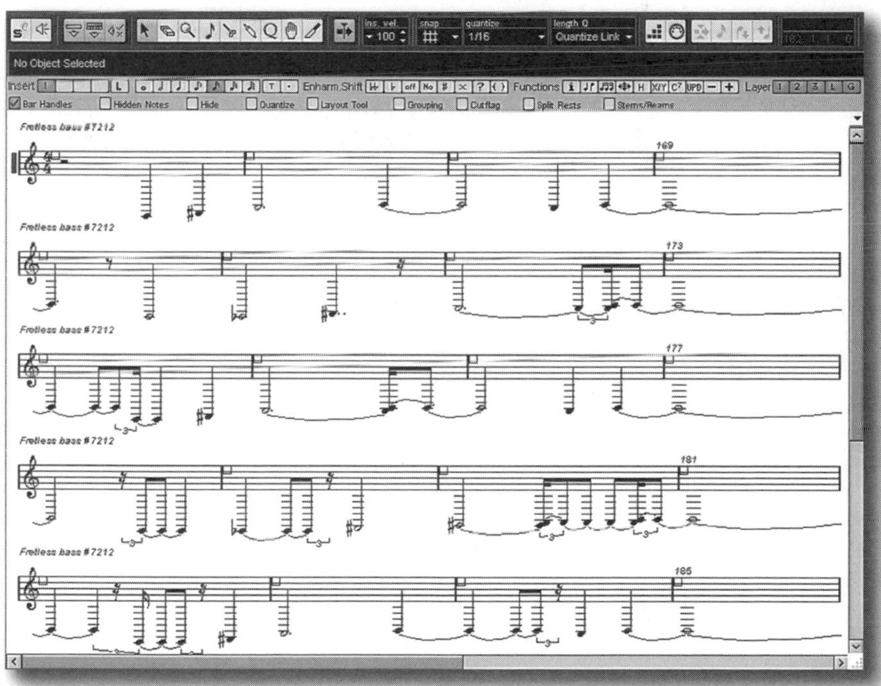

FIGURE 7.41
Steinberg Cubase/Nuendo notation edit window. (Courtesy of Steinberg Media Technologies GmbH, a division of Yamaha Corporation, www.steinberg.net)

Further reading about the wonderful world of MIDI can be found in Chapter 9.

SUPPORT FOR VIDEO AND PICTURE SYNC

Most high-end DAWs also include support for displaying a video track within a session, both as a video window that can be displayed on the desktop and

FIGURE 7.42
Most high-end DAW systems are capable of importing a video file directly into the project session window.

in the form of a video thumbnail track (which often appears as a linear guide track). Through the use of SMPTE timecode, MTC and wordclock, external video players and edit devices can be locked with the workstation's timing elements, allowing for full "mix to picture" capabilities (Figure 7.42).

Real-time, on-screen mixing

In addition to their ability to offer extensive region edit and definition, one of the most powerful cost- and time-effective features of a digital audio workstation is its ability to offer on-screen mix capabilities (Figures 7.43 and 7.44), known as "mixing inside the box." Essentially, most DAWs include an extensive digital mixer interface that offers most (if not all) of the capabilities that are offered by larger analog and/or digital consoles—without the price tag. In addition to the basic input strip fader, pan, solo/mute and select controls, most DAW software mixers offer broad support for EQ, effects plug-ins (offering a tremendous amount of DSP flexibility), spatial positioning (pan and possibly surround-sound positioning), total automation (both mixer and plug-in automation), external mix, function and transport control from a supported external hardware controller, support for exporting a mixdown to a file … the list goes on and on and on. Further reading on the mixers, consoles and the process of mixing audio can be found in Chapter 14.

FIGURE 7.43 Nuendo on-screen mixer. (Courtesy of Steinberg Media Technologies GmbH, a division of Yamaha Corporation, www.steinberg.net)

FIGURE 7.44 ProTools on-screen mixer. (Courtesy of Avid Technology, Inc., www.avid.com)

DSP EFFECTS

In addition to being able to cut, copy and paste regions within a sound file, it's also possible to alter a sound file, track or segment using digital signal processing techniques. In short, DSP works by directly altering the samples of a sound file or defined region according to a program algorithm (a set of programmed instructions) in order to achieve a desired result. These processing functions can be performed either in real time or non-real time (offline):

- *Real-time DSP:* Commonly used in most modern-day DAW systems, this process makes use of the computer's CPU or additional acceleration hardware to perform complex DSP calculations during actual playback. Because no calculations are written to disk in an offline fashion, significant savings in time and disk space can be realized when working with productions that involve complex or long processing events. In addition, the automation instructions for real-time processing are embedded within the saved session file, allowing any effect or set of parameters to be changed, undone and redone without affecting the original sound file.
- *Non-real-time DSP:* Using this method, signal processing (such as changes in level, EQ, dynamics or reverb) that is too calculation intensive to be carried out during playback will be calculated (in an offline fashion). In this way, the newly calculated file (containing the effect, submix, etc.) will be played back, without having to use up the extra resources that are now available to the CPU for other functions. DAWs will often have a specific term for tracks or processing functions that have been written to disk, such as "locking" or "freezing" a file. When DSP is performed in non-real time, it's almost always wise to save both the original and the affected sound files, just in case you need to make changes at a later time.

Most DAWs offer an extensive array of DSP options, ranging from options that are built into the basic I/O path of the input strip (e.g., basic EQ and gain-related functions) to DSP effects and plug-ins that come bundled with the DAW package, to third-party effects plug-ins that can be either inserted directly into the signal path (direct insertion) or offered as a master effect path that numerous tracks can be assigned to and/or mixed into (side chain). Although the way in which effects are implemented in a DAW will vary from one make and model to the next, the basic fundamentals will be much the same. The following notes describe but a few of the possible effects that can be plugged into the signal path of DAW; however, further reading on effects processing can be found in Chapter 14 (Signal Processing).

- *Equalization:* EQ is, of course, a feature that's often implemented at the basic level of a virtual input strip (Figures 7.45 and 7.46). Most systems give full parametric control over the entire audible range, offering overlapping control over several bands with a variable degree of bandwidth control (Q). Beyond the basic EQ options, many third-party EQ plug-ins are available on the market that vary in complexity, musicality and market appeal (Figures 7.47 and 7.48).
- *Dynamic range:* Dynamic range processors (Figures 7.49 and 7.50) can be used to change the signal level of a program. Processing algorithms are available that emulate a compressor (a device that reduces gain by a ratio that's proportionate to the input signal), limiter (reduces gain at a fixed ratio above a certain input threshold), or expander (increase the overall dynamic range of a program). These gain changers can be inserted directly into a track, used as a grouped master effect, or inserted into the final output path for use as a master gain processing block.

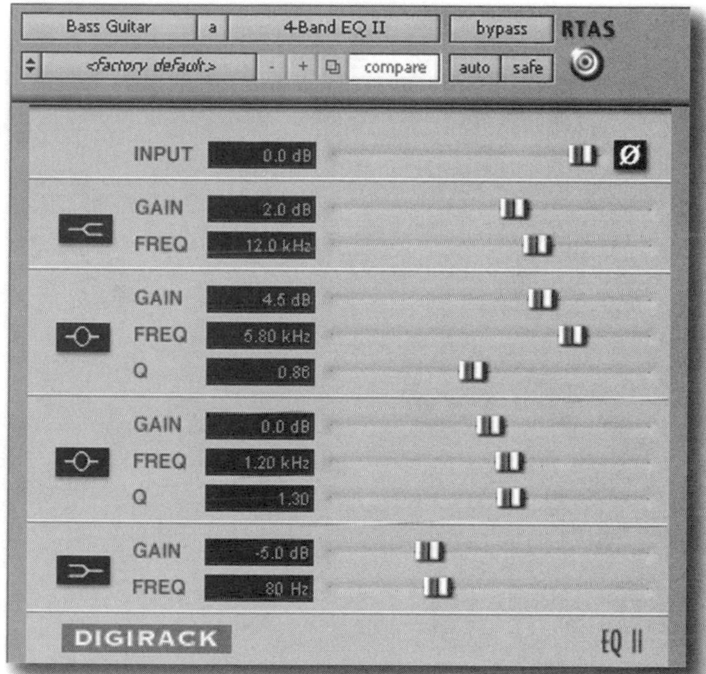

FIGURE 7.45
5-Band Digirack EQIII plug-in for Pro Tools. (Courtesy of Avid Technology, Inc., www.avid.com)

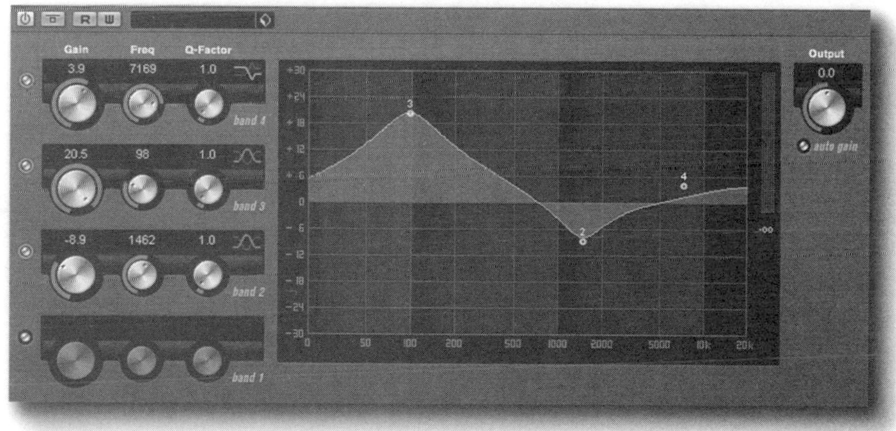

FIGURE 7.46
EQ plug-in for Cubase/Nuendo. (Courtesy of Steinberg Media Technologies GmbH, a division of Yamaha Corporation, www.steinberg.net)

In addition to the basic complement of dynamic range processors, wide assortments of multiband dynamic plug-in processors (Figure 7.51) are available for general and mastering DSP applications. These processors allow the overall frequency range to be broken down into various frequency bands. For example, a plug-in such as this could be inserted into a DAW's main output path, which

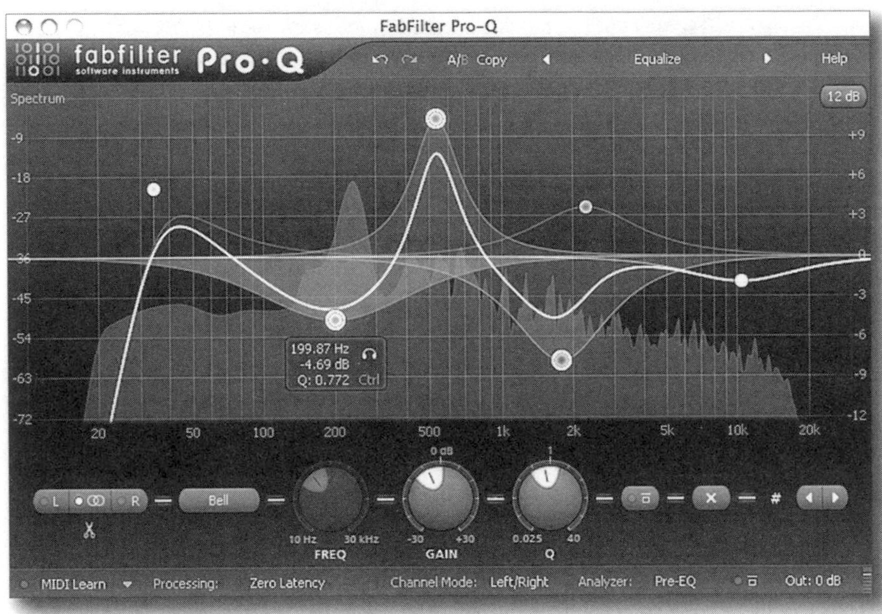

FIGURE 7.47
FabFilter Pro-Q 24-band EQ plug-in for mixing and mastering. (Courtesy of FabFilter, www.fabfilter.com)

FIGURE 7.48
Neve 31102 (top) and Harrison 32C equalizer and filters (bottom) for Apollo and the UAD effects processing card. (Courtesy of Universal Audio, www.uaudio.com © 2013 Universal Audio, Inc. All rights reserved. Used with permission)

allows the lows to be compressed while the mids are lightly limited and the highs are de-essed to reduce sibilance.

- *Delay:* Another important effects category that can be used to alter and/ or augment a signal revolves around delays and regeneration of sound over time. These time-based effects use delay (Figures 7.52 and 7.53) to add a perceived depth to a signal or change the way that we perceive the

FIGURE 7.49 Fairchild® 670 compressor Plug-In for Apollo and the UAD effects processing card. (Courtesy of Universal Audio, www.uaudio.com

FIGURE 7.50 Precision Limiter for Apollo and the UAD effects processing card. (Courtesy of Universal Audio, www.uaudio.com

FIGURE 7.51 Precision Multiband dynamics plug-in for Apollo and the UAD effects processing card. (Courtesy of Universal Audio, www.uaudio.com

dimensional space of a recorded sound. A wide range of time-based plug-in effects exist that are all based on the use of delay (and/or regenerated delay) to achieve such results as:

- Delay
- Chorus
- Flanging
- Reverb

- Further reading on the subject of delay (and signal processing in general) can be found in Chapter 15 (Signal Processing).
- *Pitch and time change:* Pitch change functions make it possible to shift the relative pitch of a defined region or track either up or down by a specific percentage ratio or musical interval. Most systems can shift the pitch of a sound file or defined region by determining a ratio between the current and the desired pitch and then adding (lower pitch) or dropping (raise pitch) samples from the existing region or sound file. In addition to raising or lowering a sound file's relative pitch, most systems can combine variable sample rate and pitch shift techniques to alter the duration of a region or track. These pitch- and time-shift combinations make it possible for such changes as:
  - *Pitch shift only:* A program's pitch can be changed while recalculating the file so that its length remains the same.
  - *Change duration only:* A program's length can be changed while shifting the pitch so that it matches that of the original program.
  - *Change in both pitch and duration:* A program's pitch can be changed while also having a corresponding change in length.

When combined with shifts in time (delay), changes in pitch make it possible for a multitude of effects to be created (such as flanging, which results from random fluctuations in delay and time shifts that are mixed with the original signal to create an ethereal "phasey" kind of sound).

DSP PLUG-INS

Workstations often offer a number of DSP effects that come bundled with the program; however, a staggering range of third-party plug-in effects can be inserted into a signal path which perform functions for any number of tasks ranging from the straightforward to the wild-'n'-zany. These effects can be programmed to seamlessly integrate into a host DAW application that conforms to such plug-in platforms as:

- *DirectX:* A DSP platform for the PC that offers plug-in support for sound, music, graphics (gaming) and network applications running under Microsoft Windows (in its various OS incarnations)
- *AU (Audio Units):* Developed by Apple for audio and MIDI technologies in OS X; allows for a more advanced GUI and audio interface

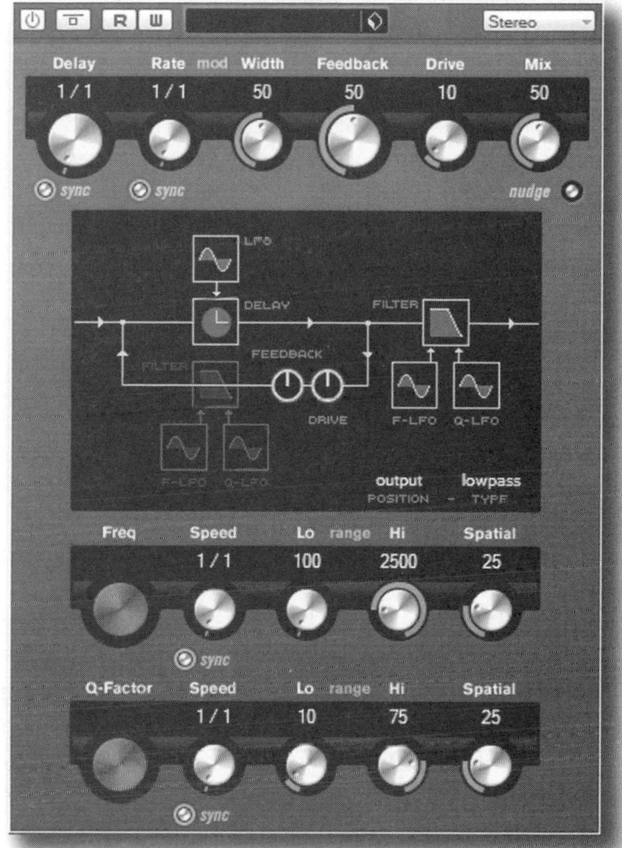

FIGURE 7.52
ModMachine VST plug-in. (Courtesy of Steinberg Media Technologies GmbH, a division of Yamaha Corporation, www.steinberg.net)

FIGURE 7.53
Emulation of the Roland RE-201 "Space Echo" for Apollo and the UAD effects processing card. (Courtesy of Universal Audio, www.uaudio.com © 2013 Universal Audio, Inc. All rights reserved. Used with permission)

- *VST (Virtual Studio Technology):* A native plug-in format created by Steinberg for use on either a PC or Mac; all functions of a VST effect processor or instrument are directly controllable and automatable from the host program
- *MAS (MOTU Audio System):* A real-time native plug-in format for the Mac that was created by Mark of the Unicorn as a proprietary plug-in format for Digital Performer; MAS plug-ins are fully automatable and do not require external DSP in order to work with the host program
- *AudioSuite:* A file-based plug-in that destructively applies an effect to a defined segment or entire sound file, meaning that a new, affected version of the file is rewritten in order to conserve on the processor's DSP overhead (when applying AudioSuite, it's often wise to apply effects to a copy of the original file so as to allow for future changes)
- *RTAS (Real-Time Audio Suite):* A fully automatable plug-in format that was designed for various flavors of Digidesign's Pro Tools and runs on the power of the host CPU (host-based processing) on either the Mac or PC
- *TDM (Time Domain Multiplex):* A plug-in format that can only be used with Digidesign Pro Tools systems (Mac or PC) that are fitted with Digidesign Farm cards; this 24-bit, 256-channel path integrates mixing and real-time digital signal processing into the system with zero latency and under full automation

These popular software applications (which are being programmed by major manufacturers and third-party startups alike) have helped to shape the face of hard-disk recording by allowing us to pick and choose the plug-ins that best fit our personal production needs. As a result, new companies, ideas and task-oriented products are constantly popping up on the market, literally on a monthly basis.

REWIRE

ReWire and ReWire2 are special protocols that were developed by Propellerhead Software and Steinberg to allow audio to be streamed between two simultaneously running computer applications. Unlike a plug-in, where a task-specific application is inserted into a compatible host program, ReWire allows the audio and timing elements of an independent client program to be seamlessly integrated into another host program. In essence, ReWire provides virtual patch chords (Figure 7.54) that link the two programs together within the computer. A few of ReWire's supporting features include:

- Real-time streaming of up to 64 separate audio channels (256 with ReWire2) at full bandwidth from one program into its host program application
- Automatic sample accurate synchronization between the audio in the two programs
- An ability to allow the two programs to share a single soundcard or interface
- Linked transport controls that can be controlled from either program (provided it has some kind of transport functionality)

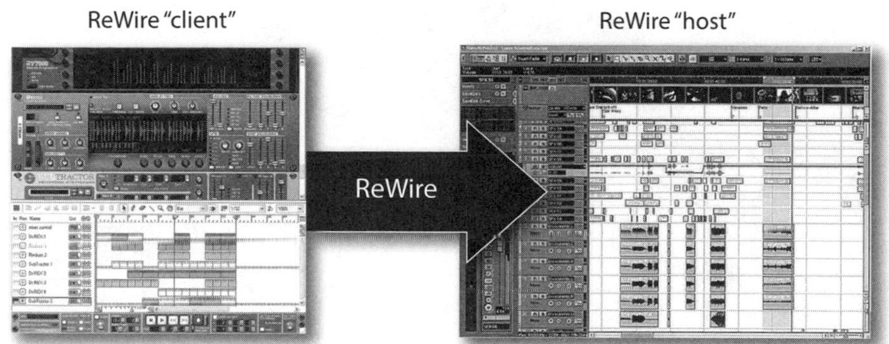

FIGURE 7.54
ReWire allows a client program to be inserted into a host program (often a DAW) so the programs can run simultaneously in tandem.

- An ability to allow numerous MIDI outs to be routed from the host program to the linked application (when using ReWire2)
- A reduction of the total number of system requirements that would be required if the programs were run independently.

This useful protocol essentially allows a compatible program to be plugged into a host program in a tandem fashion. As an example, ReWire could allow Propellerhead's Reason (client) to be "ReWired" into Steinberg's Cubase DAW (host), allowing all MIDI functions to pass through Cubase into Reason while patching the audio outs of Reason into Cubase's virtual mixer inputs. For further information on this useful protocol, consult the supporting program manuals.

ACCELERATOR PROCESING SYSTEMS

In most circumstances, the CPU of a host DAW program will have sufficient power and speed to perform all of the DSP effects and processing needs of a project. Under extreme production conditions, however, the CPU might run out of computing steam and choke during real-time playback. Under these conditions, there are a couple of ways to reduce the workload on a CPU: On one hand, the tracks could be "frozen," meaning that the processing functions would be calculated in non-real time and then written to disk as a separate file. On the other hand, an accelerator card (Figure 7.55) that's capable of adding extra CPU power can be added to the system, giving the system extra processing power to perform the necessary effects calculations. Note that in order for the plug-ins to take advantage of the acceleration they need to be specially coded for that specific DSP card. Of course, as computers have gotten faster and more powerful, native processing packages (Figure 7.56), have come onto the market, which make use of the modern computers multi-processor capabilities.

Mixdown and effects automation

One of the great strengths of the "in the box" age is how easily all of the mix and effects parameters can be automated and recalled within a mix. The ability to

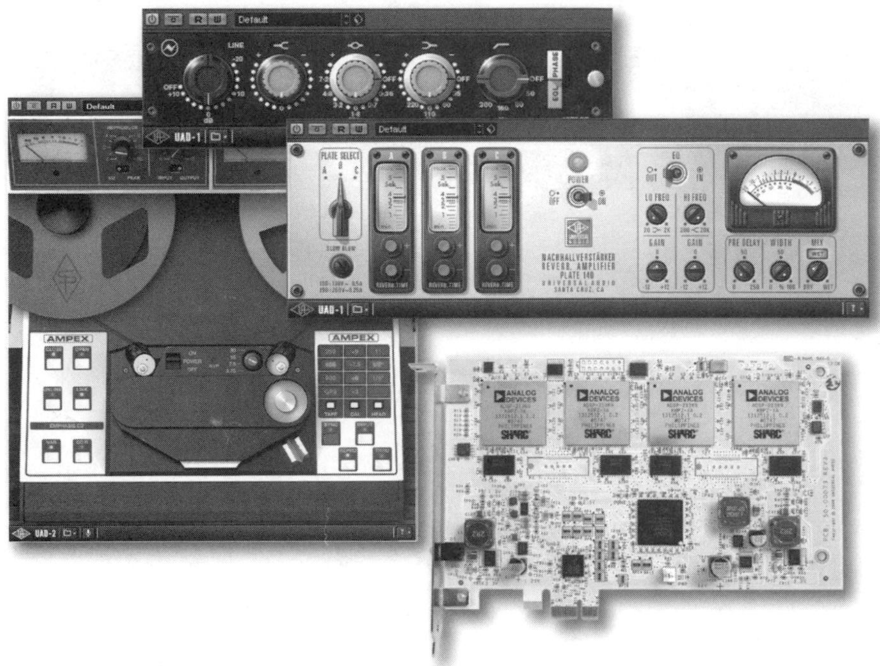

FIGURE 7.55
The UAD-2 DSP PCIe DSP processor and several plug-in examples. (Courtesy of Universal Audio, www.uaudio.com, © 2013 Universal Audio, Inc. All rights reserved. Used with permission.)

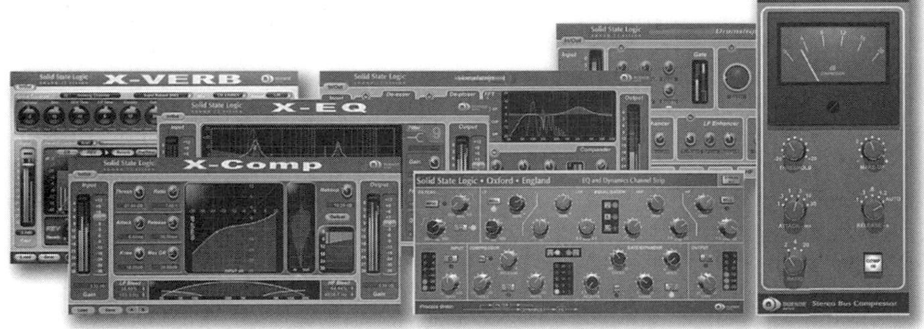

FIGURE 7.56
Solid State Logic Duende Native Audio Processing Tools. (Courtesy of Solid State Logic, www.solid-state-logic.com)

change levels, pan and virtually control any parameter within a project makes it possible for a session to be written to disk, saved and recalled at a second's notice. In addition to grabbing a control and moving it (either virtually on-screen or from a physical controller), another interface style for controlling automation parameters (known as *rubberband* controls) lets you view, draw and edit variables as a graphic line that details the various automation moves over time.

As with any automation moves, these rubberband settings can be undone, redone or recalled back to a specific point in the edit stage. Often (but not always), the moves within a mix can't be "undone" and reverted back to any specific point in the mix. In any case, one of the best ways to save (and revert to) a particular mix

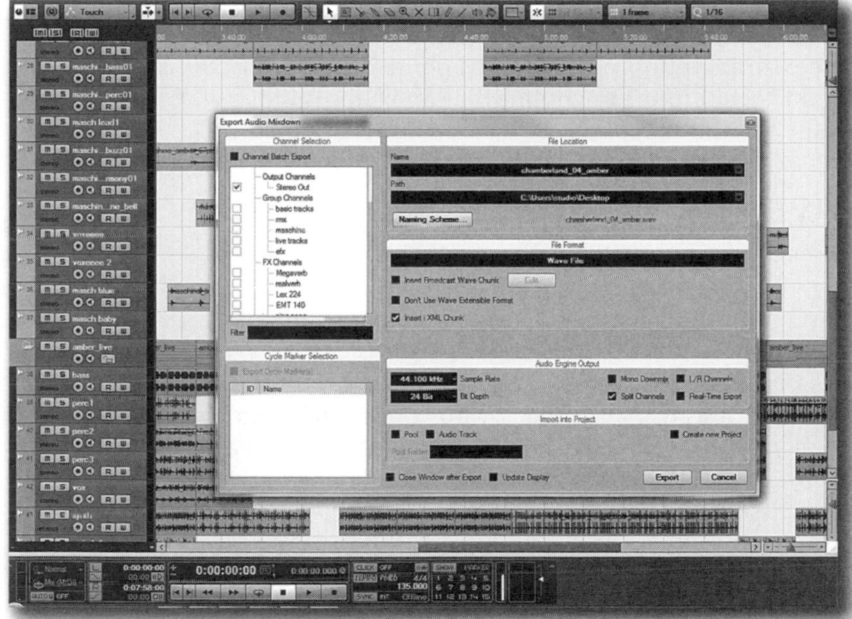

FIGURE 7.57 Most DAWs are capable of exporting session sound files, effects and automation to a final mixdown track.

version (or various alternate mix versions) is simply to save a specific mix under a unique, descriptive session file title (e.g., colabs_oceanis_radiomix01.ses) and then keep on working. By the way, it's always wise to save your mixes on a regular basis (many a great mix has been lost in a crash because it wasn't saved); also, progressively saving your mixes under various name or version numbers (song01.ses, song02.ses, etc.) can come in handy if you need to revert to a past mix version.... In short, save often and save regularly!

EXPORTING A MIXDOWN TO FILE

Once your mix is ready, most DAWs systems are able to export (print) part or all of a session to a single file or set of sound files (Figure 7.57). An entire session or defined region can be exported as a single interleaved file (containing multiple channels that are encoded into a single L-R-L-R sound file), or can be saved as separate, individual (L.wav and R.wav) sound files. Of course, a surround or multichannel mix can be likewise exported as a single interleaved file, or as separate files.

Often, the session can be exported in non-real time (a faster-than-real-time process that can include all mix, plug-in effects, automation and virtual instrument calculations) or in real time (a process that's capable of sending and receiving real-time analog signals through the audio interface so as to allow for the insertion of external effects devices, etc.). Usually, a session can be mixed down in a number of final sound file and bit/sample rate formats. Most DAWs will allow third-party plug-ins to be inserted into the final (master) output section, allowing for the export of a session into a specific output file format. (For example, a discrete surround mix can be folded down into a two-channel Dolby ProLogic surround-sound file, or the same file can be rendered as a Dolby Digital 5.1 or DTS file for insertion into a DVD video soundtrack.)

POWER TO THE PROCESSOR ... UHHH, PEOPLE!

Speaking of having enough power and speed to get the job done, there are definitely some tips and tricks that can help you get the most out of your digital audio workstation. Let's take a look at some of the more important items. It's important to keep in mind that keeping up with technology can have its triumphs and its pitfalls. No matter which platform you choose to work with, there's no substitute for reading, research, and talking with your peers about your techno needs. It's generally best to strike a balance between our needs, our desires, the current state of technology and the relentless push of marketing to grab your money. And it's usually best to take a few big breaths before making any important decisions.

1 Get a computer that's powerful enough

With the increased demand for higher bit/sample rate resolution, more tracks, more plug-ins, more of everything, you'll obviously want to make sure that your computer is fast and powerful enough to get the job done in real time without spitting and sputtering digits. This often means getting the most up-to-date and powerful computer/processor system that your budget can reasonably muster. With the advent of 32 and 64-bit OS platforms and dual, quad and eight-core processors (chips that effectively contain multiple CPUs), you'll want to make sure that your hardware will support these features before taking the upgrade plunge. The same goes for your production software and driver availability. If any part of this hardware, software and driver equation is missing, the system will not be able to make use of these advances. Therefore, one of the smartest things you can do is research the type and system requirements that would be needed to operate your production system—and then make sure that your system exceeds these figures by a comfortable margin so as to make allowances for future technological advances and the additional processing requirements that are associated with them. If you have the budget to add some of the extra bells and whistles that go with living on the cutting edge, you should take the time to research whether or not your system will actually be able to deliver the extra goods when these advances actually hit the streets.

2 Make sure you have enough memory

It almost goes without saying that your system will need to have an adequate amount of random access memory (RAM) and hard-disk storage in order for you to take full advantage of your processor's potential and your system's data storage requirements. RAM is used as a temporary storage area for data that is being processed and passed to and from the computer's central processing unit (CPU). Just as there's a "need for speed" within the computer's CPU, the general guidelines for RAM ask that we install memory with the fastest transfer speed that can be supported by the computer. It's important that you install as much memory as your computer and budget will allow. Installing too little RAM will force the OS to write this temporary data to and from the hard disk, a process that's much slower than transfer to RAM and causes the system's overall performance to slow to a crawl. For those who are making extensive use of virtual sampling technology (whereby samples are transferred to RAM), it's usually a wise idea to throw as much RAM as is possible into the system.

Hard-disk requirements for a system are certainly an important consideration. The general considerations include:

- *Need for size:* Obviously, you'll want to have drives that are large enough to meet your production storage needs. With the use of numerous tracks within a session, often at sample rates of 24 bit/96k, data storage can quickly become an important consideration.
- *Need for speed:* With the current track count and sample rate requirements that can commonly be encountered in a DAW session, it's easy to understand how lower disk access times (the time that's required for the drive heads to move from one place to another on a disk and then output that data) becomes important.

It's interesting to note that, in general, the disk drives with larger storage capacities will often have shorter access times and a higher data transfer rate. These large-capacity drives will often spin at speeds that are at least 7200 rpm, will have extremely access times and are able to sustain a transfer rate of 6Gb/sec or faster.

3 Keep your production media separate

Whenever possible, it's important to keep your program and operating system data on a separate drive from the one that holds your production data. This is due to the simple fact that a computer periodically has to check in and interact with both the currently running program and the OS. Should the production media be on the same disk, periodic interruptions in audio data will occur as the disk takes a moment out to go perform a program-related task (and vice versa), resulting in a reduction in media and program data access time, as well as throughput. For these reasons, it's also wise to keep the program and media drives off of the same data cable and onto separate data paths.

4 Update your drivers!

In this day and age of software revisions, it's always a good idea to go on the Web and search for the latest update to a piece of software or a hardware driver. Even if you've just bought a product new out of the box, it might easily have been sitting on a music store shelf for over a year. By going to the company website and downloading the latest versions, you'll be assured that it has the latest and greatest capabilities. In addition, it's always wise to save these updates to disk in your backup directories. This way, if you're outstanding in your field and there's a hardware or software problem, you'll be able to reload the software or drivers and "should" be on your way in no time.

5 Going at least dual monitor

Q: How do you fit the easy visual reference of multiple programs, documents and a digital audio workstation onto a single video monitor?
A: You often don't. Those of you who rely on your computer for recording and mixing, surfin', writing, etc., should definitely think about doubling your computer's visual real estate by adding an extra monitor to your computer system.

Folks who have never seen or thought much about adding a second monitor might be skeptical and ask, "What's the big deal?" But, all you have to do is sit down and start opening programs to see just how fast your screen can get filled up. When using a complicated production program (such as a DAW or a high-end graphics app), getting the job done with a single monitor can be an exercise in total frustration. There's just too much we need to see and not enough screen real estate to show it on. The ability to quickly make a change on a virtual mixer, edit window or software plug-in has become more of a necessity than a luxury.

Truth is, in this age of the Mac and Windows, adding an extra monitor is a fairly straightforward proposition. You could either spring for a dual-head graphics card that can support two monitors or you can visit your favorite computer store and buy a second video card. Getting hold of a second monitor could be as simple as grabbing an unused one from the attic or buying a standard large-screen LCD or LED monitor.

Once you've installed the hardware, the software side of building a dual-monitor system is relatively straightforward. Simply call up the resolution settings in the control panel or System Preferences and change the resolution settings for each monitor. Once you extend your desktop across both monitors, you should be well on your way.

Those of you who use a laptop can also enjoy many of these benefits by plugging the second monitor into the video out and following the setup steps that are recommended by your computer's operating system. You should be aware that many laptops are limited in the way they share video memory and might be restricted in the resolution levels that can be selected.

This might not seem much like a recording tip, but once you get a dual-monitor system going your whole approach to producing content (of any type) on a computer will instantly change and you'll quickly wonder how you ever got along without it!

6 Keeping your computer quiet

Noise! Noise! Noise! It's everywhere! It's in the streets, in the car, and even in our studios. It seems like we spend all those bucks getting the best sound possible, only to gunk it all up by placing this big computer box that's full of noisy fans and whirring hard drives smack in the middle of a critical listening area. Fortunately, a number of companies have begun to find ways to reduce the problem. Here are a few solutions:

- Whenever possible, use larger, low-rpm fans to reduce noise.
- Certain PC motherboards come bundled with a fan speed utility that can monitor the CPU and case heat and adjust the fan speeds accordingly.
- Route your internal case cables carefully. They could block the flow of air, which can add to heat and noise problems.
- A growing number of hard-disk drives are available as quiet drives. Check the manufacturer's noise ratings.
- You might consider placing the computer in a well-ventilated area, just outside the production room. Always pay special attention to ventilation (both inside and outside the computer box), because heat is a killer that'll reduce the life span of your computer. (*Note:* When building my own studio I designed a special alcove/glass door enclosure that houses my main computer—no muss, no fuss, and very little noise.)
- Thanks to gamers and audio-aware buyers, a number of companies exist that specialize in quiet computer cases, fans and components. These are always fun to check out on the web.

7 Backup and archive strategies

It's pretty much always true that it's not a matter of if a hard drive will fail, but when. It's not a matter of if something will happen to an irreplaceable hard disk, but when. When we least expect it, disaster will strike. It's our job to be prepared for the inevitable. This type of headache can of course be partially or completely averted by backing up your active program and media files, as well as by archiving your previously created sessions and then making sure that these files are also backed up.

As previously stated, it's generally wise to keep your computer's operating system and program data on a separate hard disk (usually the boot drive) and then store your session files on a separate media drive. Let's take this as a practical and important starting point. Beyond this premise, as most of you are quite aware, the basic rules of hard-disk management are extremely personal, and will often differ from one computer user to the next (Figure 7.58). Given these differences, I'd still like to offer up some basic guidelines:

- It's important to keep your data (of all types) well organized, using a system that's both logical and easy to follow. For example, online updates of a program or hardware driver downloads can be placed into their own directories; data relating to your studio can be placed in the "studio" directory

FIGURE 7.58
Data and hard-drive management (along with a good backup scheme) are extremely important facets of media production.

and subdirectories; documents, MP3s, and all the trappings of day-to-day studio operations can be also placed on the disk, using a system that's easy to understand.

- Session data should likewise be logical and easy to find. Each project should reside in its own directory on a separate media drive, and each song should likewise reside in its own subdirectory of the session project directory.
- Remember to save various take versions of a mix. If you just added the vocals to a song, go ahead and save the session under a new version name. This acts as an "undo" function that lets you go back to a specific point in a session. The same goes for mixdown versions. If someone likes a particular mix version or effect, go ahead and save the mix under a new name or version number (my greatest song 1 ver15.ses) or (my greatest song 1 ver15 favorite effect.ses). In fact, it's generally wise to save the various versions throughout the mix. The session files are usually small and might save your butt at a later point in time.

With regard to backup schemes, a number of options also exist. Although making backups to optical media (CD and DVD) is a common occurrence, it's been found that the most robust and long-lived backups are those that reside on a hard drive. There's also a wise saying: "Media isn't truly backed up unless it's copied in three different places." All I can do is lay out a general backup formula that has worked extremely well for me over the years:

- *CD/DVD:* I start off with the general premise that CD and DVD media is temporary media that's to be used for getting media data into the hands of clients, friends, etc.
- *Boot drive:* The boot disk is used to store the operating system data, program data, program- and driver-related downloads, documents, graphics

and non-production-related media (such as MP3 music). It can also be used as a "local" archive drive for storing past sessions, assuming that you have this data adequately backed up elsewhere.

- *Second internal or external drive:* Current media data resides on this drive. It can also be used as an archive drive for storing all of your past sessions along with your current sessions.
- *External backup drive 1:* If the disk is large enough, it can be used as a backup drive for both your boot and second media drive. If it isn't large enough, you'll want to back up to multiple drives.
- *External backup drive 2:* If the disk is large enough, it can be used as a second backup drive for both your boot and media data. You might strongly consider storing this drive offsite. In the case of a theft, fire or other unforeseen disaster, you'll still be able to recover your precious data.
- *Cloud storage:* Renting disk space from a company that specializes in data storage and backup might also be a wise decision. In this way, backup is not a big concern, as the service will automatically back your precious data up on a regular basis.

All of the above are simply suggestions. I rarely give opinions like these in a book; however, they've served me so well and for so long that I had to pass them along.

8 Session documentation

Most of us don't like to deal with housekeeping. But when it comes to recording and producing a project, documenting the creative process can save your butt after the session dust has settled—and help make your postproduction life much easier (besides, you never know when something will be reissued/remixed). So let's discuss how to document the details that crop up before, during and after the session. After all, the project you save might be your own!

DOCUMENTING WITHIN THE DAW

One of the simplest ways to document and improve a session's workflow is to name a track before you press the record button, because most DAWs will use that as a basis for the file name. For example, by naming a track "Jenny lead voc take 5," most DAWs will automatically save and place the newly recorded file into the session as "Jenny lead voc take 5.wav" (or .aif). Locating this track later would be a lot easier than rummaging through sound files only to find that the one that you want is "Audio018-05." Because some DAW track displays are limited to about 8 characters, consider putting the easily identifiable text first (i.e., "leadvoc-jenny take5," which might display as leadvoc-j).

Also, make use of your DAW's notepad (Figure 7.59). Most programs offer a scratchpad function that lets you fill in information relating to a track or project; use this to name a specific synth patch, note the mic used on a vocal, and include other info that might come in handy after the session's specifics have been long forgotten.

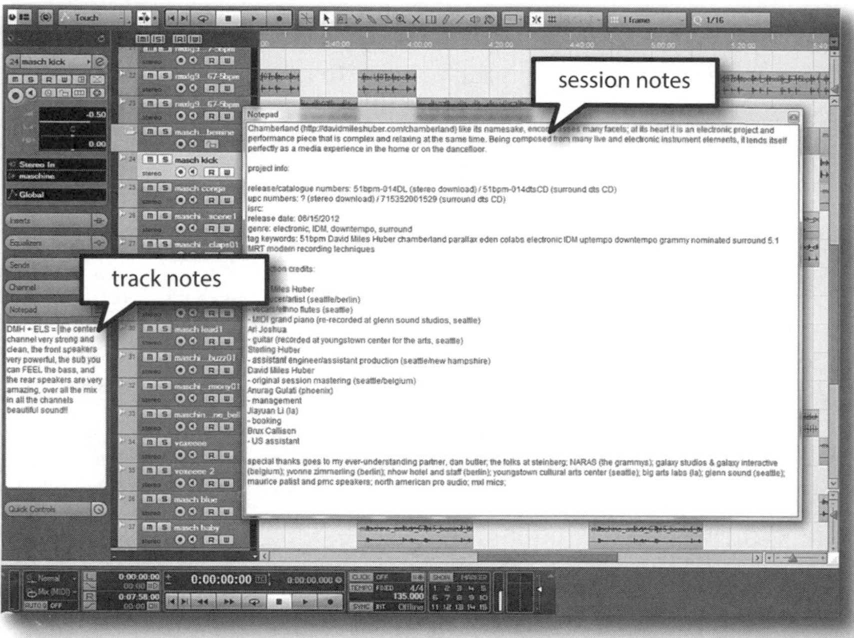

FIGURE 7.59
Cubase/Nuendo Notepad apps. (Courtesy of Steinberg Media Technologies GmbH, www.steinberg.net.)

Markers and marker tracks can also come in super-handy. These tracks can alert us to mix, tempo and other kinds of changes that might be useful in the production process. I'll often place the lyrics into a marker track, so I can sing the track myself without the need for a lead sheet, or to help indicate phrasings to another singer.

MAKE A DOCUMENTATION DIRECTORY

The next step toward keeping better track of details is to create a "MySong Documents" directory within the song's session, and fill that folder with documents and files that relate to the session such as:

- Your contact info
- Song title and basic production notes (composer, lyricist, label, business and legal contacts)
- Producer, engineer, assistant, mastering engineer, duplication facility, etc. (with contact info)
- Original and altered tempos, tempo changes, song key, timecode settings, etc.
- Original lyrics, along with any changes (changed by whom, etc.)
- Additional production notes
- Artist and supporting cast notes (including their roles, musician costs, address info, etc.)
- Lists of any software versions and plug-in types, as well as any pertinent settings (you never know if they'll be available at a future time, and a description and screenshot might help you to duplicate it within another app)

- Lists of budget notes and production dates (billing hours, studio rates and studio addresses … anything that can help you write off the $$$)
- Scans of copyright forms, session contracts, studio contracts and billings
- Anything else that's even remotely important.

In addition, I'll often take screenshots of some of my more complicated plug-in settings and place these into this "time capsule" folder. If I have to redo the track later for some reason, I refer to the JPG screenshot and start reconstruction. Photos or movie clips can also be helpful in documenting which type of mic, instrument and specific placements were used within a setup. You can even use pictures to document outboard hardware settings and patch arrangements. Composers can use the "Doc" folder to hold original scratchpad recordings that were captured on your PDA, cell phone, or message machine (I do this for copyright purposes).

Furthermore, a "MySong Graphics" directory can hold the elements, pictures and layouts that relate to the project's artwork … "MySong Business" and "MySong artwork" directory, etc. might also come in handy.

DAW GUIDELINES

The Producers and Engineers Wing of NARAS (the Grammy folks) are nailing down a wide range of guidelines that can help with aspects of documentation, session transfers, backups and other techno issues. At present, the P&E is offering general DAW guidelines for Pro Tools; although a non-platform-specific version is in the works, the information's still general enough for everyone. It's well worth downloading a copy (as well as the material on surround and mastering) from http://www.grammy.org/recording-academy/producers-and-engineers/guidelines.

IN CLOSING

At this time, I'd like to refer you to the many helpful pointers in Chapter 21. I'm doing this in the hope that you'll read this section twice (at least)—particularly the discussion on project preparation, session documentation and backup/archive strategies. I promise that the time will eventually come when you'll be glad you did.

CHAPTER 8

Groove Tools and Techniques

The expression "getting into the groove" of a piece of music often refers to a feeling that's derived from the underlying foundation of music: rhythm. With the introduction and maturation of MIDI and digital audio, new and wondrous tools have made their way into the mainstream of music production. These tools (Figure 8.1) can help us to use technology to forge, fold, mutilate and create compositions that make direct use of rhythm and other building blocks of music through the use of looping technology. Of course, the cyclic nature of loops can be repeat–repeat–repetitive in nature, but new toys and techniques for looping can inject added flexibility, control and real-time processing into a project that can be used by artists in wondrously expressive ways.

In this chapter we'll be touching on many of the approaches and software packages that have evolved (and continue to evolve) into what is one of the fastest and most accessible facets of personal music production. It's literally impossible to hit on the finer operational points of all of these systems; for that, I'll rely on your motivation and drive to:

- Download many of the software demos that are readily available
- Delve into their manuals and working tutorials
- Begin to create your own grooves and songs that can then be integrated with your own music or those of collaborators.

If you do these three things, you'll be shocked and astounded as to how much you've learned. And these experiences can directly translate into skills that'll widen your production horizons and possibly change your music.

THE BASICS

The basic idea behind groove-based tools rests with tempo matching, the idea that various rhythms, grooves, pads and any other imaginable sounds of various tempos, lengths (and often musical keys) can be artfully crafted together into a single, working song session.

FIGURE 8.1
Probably the most widely used groove tool on the planet ... GarageBand™ for the Mac and iPad.

Because groove-based tools often deal with rhythms and cyclic-based loops that are pulled from various musical sources, the factors that need to be managed are:

- Sync
- Tempo and length
- Time and pitch change techniques

The aspect of sync relates to the fact that the various loops in a groove project will need to sync up with each other (or in multiple lengths and timings of each other). It almost goes without saying that multiple loops that are successively or simultaneously triggered must have a synchronous timing relationship with one another—otherwise, it's a jumbled mess of sound.

Another relationship relates to the aspect of tempo. Just as sync is imperative, it's also necessary for the files to be adjusted in pitch (resampling) and/or length (time stretching), so that they precisely match the currently selected tempo (or are programmed to be a relative multiple of the session's tempo).

A final aspect is associated with time and pitch change techniques. This is the process of altering a sound file (often which is rhythmically repetitive and short in length) to match the current session tempo and to synchronously align them within the software by using variable sample- and pitch-shifting techniques. Using these basic digital signal processing (DSP) tools, it's possible to alter a sound file's duration (varying the length of a program by raising or lowering its playback sample rate) or to alter its relative pitch (either up or down). In this way, these loops can be matched up or musically combined by using any of three possible time and pitch change combinations:

- *Time change:* A program's length can be altered without affecting its pitch.
- *Pitch change:* A program's length can remain the same while pitch is shifted either up or down.
- *Both:* Both a program's pitch and length can be altered using resampling techniques.

By setting the loop program to a master tempo (or a specific tempo at that point in the song), an audio segment or file can be imported, examined as to sample rate or length and then recalculated to a new relative tempo that matches the current session tempo. Voilà! We now have a defined segment of audio that matches the tempo of all of the other segments, allowing it to play and interact in relative sync with the other defined segments and/or loop files. (Note: more in-depth reading on pitch and time changing can be found in Chapter 15.)

It's also important to note that these time-shifting processes are created according to a program algorithm. This means that the sound is shaped according to a basic mathematic calculation. More often than not, the music program will let you choose between a number of algorithms, according to the type of sound that's being processed. For example, a basic drum loop might sound good when a "beats" detection algorithm is used, while a long, slow pad might sound totally unnatural. In short, it's always a good idea to be aware of the various time shift options that are available to you and to take the time to listen for the processing tools that best match the music. Additionally, these algorithms vary from program to program. For example, the "pad" (a long, continuous series of chords) setting on one looping program might sound entirely different than another might sound … just something to be aware of.

Warping

In addition to basic pitch- and time-shifting, most looping tools make use of a tempo and sound processing technique, called *warping*. This process uses various time-shift tools to match the timing elements of a sound file by detecting and entering hitpoint markers into the sound file. These markers are most often automatically detected at percussive transient points, in a way that makes it easier for the time and pitch shifting process to best match the file to the session tempo (most often by adjusting the transient timings, so that they fall directly on the beat and its subdivisions).

In addition to helping the loops better match the session tempo, warp hitpoint markers can also be manually moved and manipulated to change the basic "feel" of a loop. For example, various hitpoints can be moved ahead or back in time give the loop an entirely new feel or swing. In short, take the time to experiment and learn about how you can shape and create new sounds using warp and hitpoint technology. Note: there are countless online videos on this subject that are both insightful and fun.

Beat slicing

In addition to warp pitch- and time-stretch techniques, another method, called *beat slicing*, makes use of an entirely different process to match the length and timings of a sound file segment to the session tempo. Rather than changing the speed and pitch of a sound file, the beat slicing process actually breaks an audio file into a number of small segments (not surprisingly called slices) and then changes the length and timing elements of a segment by simply adding or subtracting time between these slices. This procedure has the advantage of preserving the pitch, timbre and sound quality of a file but has the potential downside of creating noticable breaks in the sound file. From this, it's easy to understand why the beat-slicing process often works best on percussive sounds that have silence between the individual hitpoints. This process is usually carried out by detecting the transient events within the loop and then automatically placing the slices at their appropriate beat-matching points (according to automatic or user-definable sensitivity and detection controls).

As with so many other things we've discussed, the sky (and your imagination) is the limit when it comes to the tricks and techniques that can be used to match the tempos and various time-/pitch-shift elements between loops, MIDI and sound file segments. I strongly urge you to take the time to read the manuals of various loop and DAW software packages to learn more about the actual terms and procedures and then put them into practice. If you take the time, I guarantee that your production skills and your outlook on these tools will greatly expand.

For the remainder of this chapter, we'll be looking at several of the more popular groove tools and toys. This is by no means a complete listing, and I recommend that you keep reading the various trade magazines, websites, and other resources, as new and exciting technologies come onto the market regularly.

LOOPING YOUR DAW

Most digital audio workstations offer various features that can make it possible to incorporate looping into a session, along with other track-based sound, MIDI and video files. Even when features that could make a specific task much easier aren't available, it's often possible to think through the situation and find workarounds that can help tackle the problem at hand. For example, a number of workstations allow the beginning point of an edited segment to be manually placed at a specific point. And, by simply clicking on and dragging the tail end of the segment, the sound file can be time-stretched into the session's proper time relationship (while the pitch remains unchanged). Said another way, if a session has been set to a tempo of 94 bpm, an 88-bpm loop can be imported at a specific measure. Then, by turning on the DAW's snap-to-grid and automatic time-stretch functions, the segment can be time-stretched until it snuggly fits into the session's native tempo (Figure 8.2). Now that the loop fits the current tempo, it can be manually looped (i.e., copied and/or placed) to your heart's content.

As was mentioned, different DAWs and editing systems will have differing ways of tackling a time-stretch situation and with varying degrees of ease and accuracy.

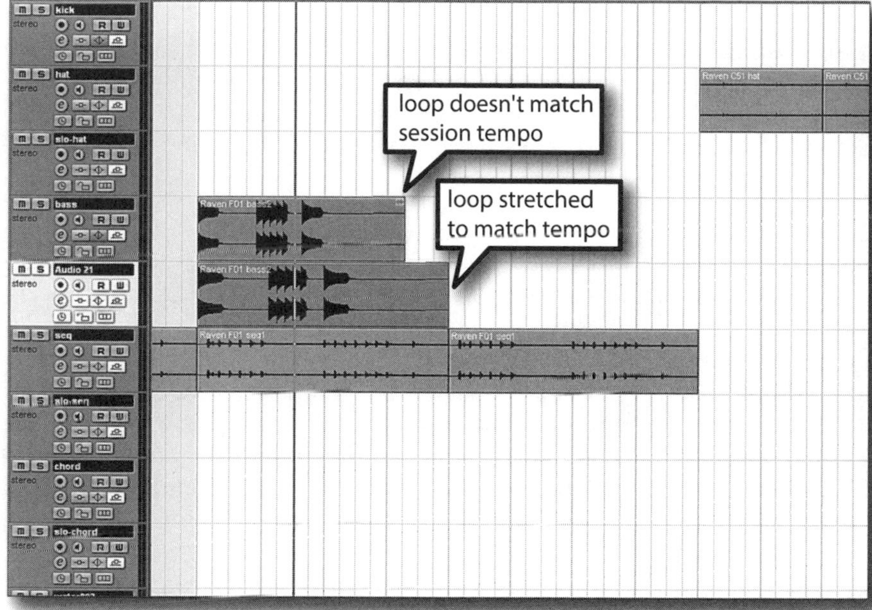

FIGURE 8.2
Most production workstations allow a loop or soundfile to be stretched to match the session's tempo.

In addition to the automated loop options that are in a number of groove and DAW programs, another (and often quite accurate) way to fit your loops into a session is to precisely set you session tempo and grid to the desired tempo (or varying tempo map) and then manually adjust your loop timings to fit that tempo grid. In this way, you can listen to the various stretching algorithms (beat, pad, vocal, etc.) and manually audition each loop for the best possible (least distorted) sound. Just remember, there are usually no hard and fast rules. With planning, ingenuity and your manual's help, you'll be surprised at the number of ways that a looping problem can be turned into an interesting opportunity.

Tutorial: Manual Looping with a DAW

1. Go to www.acidplanet.com and download an ACID 8pack set of loops.
2. Load the individual loops into your favorite DAW.
3. Set the time display to read in bars and beats (tempo).
4. Import the loops into the session.
5. Match the session tempo to a bpm setting that seems to match the 8pack's intended tempo.
6. It's possible (but not likely) that the loop lengths won't match up. If this happens, consult your DAW's manual in order to manually time-stretch the loops to fit the desired session tempo.
7. Copy and duplicate the various loop tracks, until you've made a really great song!
8. Save the session to play for your friends!

GROOVE AND LOOP SYSTEMS

Looping tools and toys don't only exist in the software world; there are also a number of hardware groove keyboards and module boxes on the market. These systems, which range widely in sounds, functionality and price, can offer up a vast range of unique sounds that can be quite useful for laying a foundation under your production.

In the past, getting a hardware grove tool to sync into a session could be time consuming, frustrating and problematic, requiring time and tons of manual reading. However, with the advent of powerful time- and pitch-shift processing within most DAWs, the sounds from these hardware devices can be pulled into a session without too much trouble. For example, a single groove loop (or multiple loops) could be recorded into a DAW (at a bpm that's close to the session's tempo), edited and then imported into the session, at which time the loop could be easily stretched into time sync, allowing it to be looped to your heart's content. Just remember, necessity is the mother of invention—patience and creativity are probably your most important tools in the looping process.

Loop-based audio software

Loop-based audio editors are groove-driven music programs (Figures 8. 3 and 8.4) that are designed to let you drag and drop prerecorded or user-created loops and audio tracks into a graphic multitrack production interface. At their basic level, these programs differ conceptually from their traditional DAW counterpart

FIGURE 8.3
Apple GarageBand™. (Courtesy of Apple Computers, Inc., www.apple.com)

FIGURE 8.4
Avid Transfuser Loop, Beat and Groove Instrument. (Courtesy of Avid Technology, Inc., www.avid.com)

in that the pitch- and time-shift architecture is so variable and dynamic that, even after the basic rhythmic, percussive and melodic grooves have been created, their tempo, track patterns, pitch, session key, etc., can be quickly and easily changed at any time. With the help of custom, royalty-free loops (available from many manufacturer and third-party companies), users can quickly and easily experiment with setting up grooves, backing tracks and creating a sonic ambience by simply dragging the loops into the program's main sound file view, where they can be arranged, edited, processed, saved and exported.

One of the most interesting aspects of the loop-based editor is its ability to match the tempo of a specially programmed loop to the tempo of the current session. Amazingly enough, this process isn't that difficult to perform, because the program extracts the length, native tempo and pitch information from the imported files and, using various digital time- and pitch-change techniques, adjusts the loop to fit the native time and pitch parameters of the current session. This means that loops of various tempos and musical keys can be automatically adjusted in length and pitch so as to fit in time with previously existing loops … just drag, drop and go!

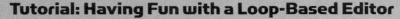

Tutorial: Having Fun with a Loop-Based Editor

1. Go to www.acidplanet.com and download their free version of ACID Xpress … OR …
2. Go to www.ableton.com and download their trial version of Ableton Suite Trial
3. Download an ACID 8pack set of loops from www.acidplanet.com (or use the loops in Abelton Live)
4. Load the individual loops.
5. Read the program's manual and begin to experiment with the loops.
6. Mess around with the tempo and musical keys.
7. Copy and duplicate the various loop tracks to your heart's content until you've made a really great song!
8. Save the session to play for your friends!

Of course, the graphic user interfaces (GUIs) between looping software editors and tools can differ greatly. Most layouts use a track-based system that lets you enter or drag a preprogrammed loop file into a track and then drag it to the right in a way that repeats the loop. Again, it's worth stressing that DAW editors will often include such looping functions that can be either basic (requiring manual editing or sound file processing) or advanced in nature (having any number of automated loop functions).

Other loop programs make use of visual objects that can be triggered and combined into a real-time mix by clicking on an icon or track-based grid. One of the most popular programs is Live from Ableton (Figure 8.5). Live is an interactive loop-based program that's capable of recalculating the time, pitch and tempo structure of a sound file or easily defined segment, and then entering that loop into the session at a defined global tempo. Basically, this means that a segment of any length or tempo that's been pulled into the session grid will be recalculated to the master tempo and can be combined, mixed and processed in perfect sync with all other loops in the project session.

In Live's Arrangement View, as in all traditional sequencing programs, everything happens along a fixed song timeline. The program's Session View breaks this limiting paradigm by allowing media files to be mapped onto a grid as buttons (called clips). Any clip can be played at any time and in any order in a random fashion that lends itself to interactive performance both in the studio and on-stage. Functionally, each vertical column, or track, can play only one clip at a time. The horizontal rows are called scenes. Any clip can be played in a loop fashion by clicking on its launch button. By clicking on a scene launch at the screen's right, every clip in a row will simultaneously play beyond the wide range of timing, edit, effects, MIDI and instrument controls.

Of course, most of the looping software packages are capable of incorporating MIDI into a project or live interactive performance (Figure 8.6). This can be

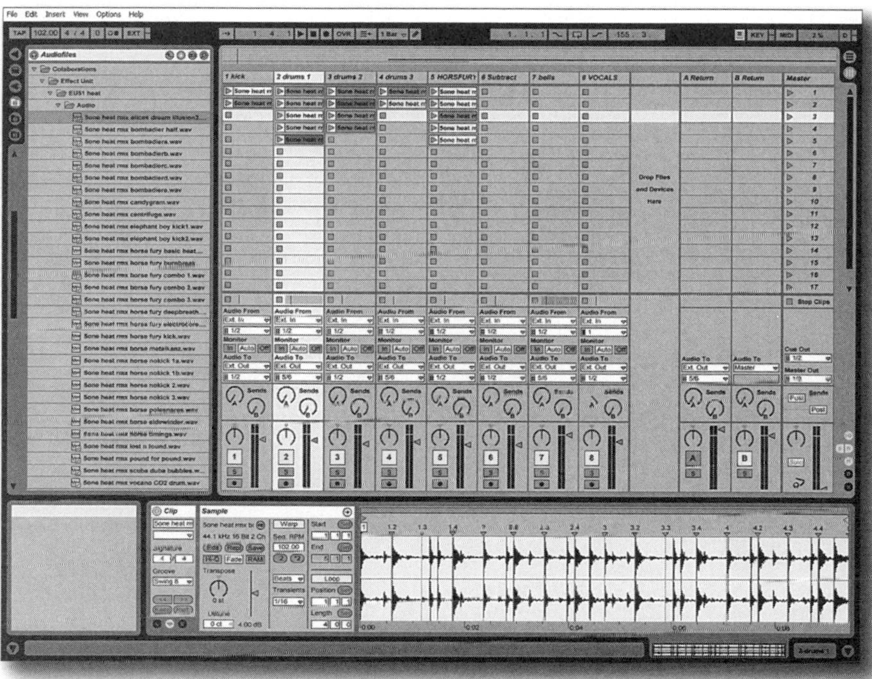

(a)

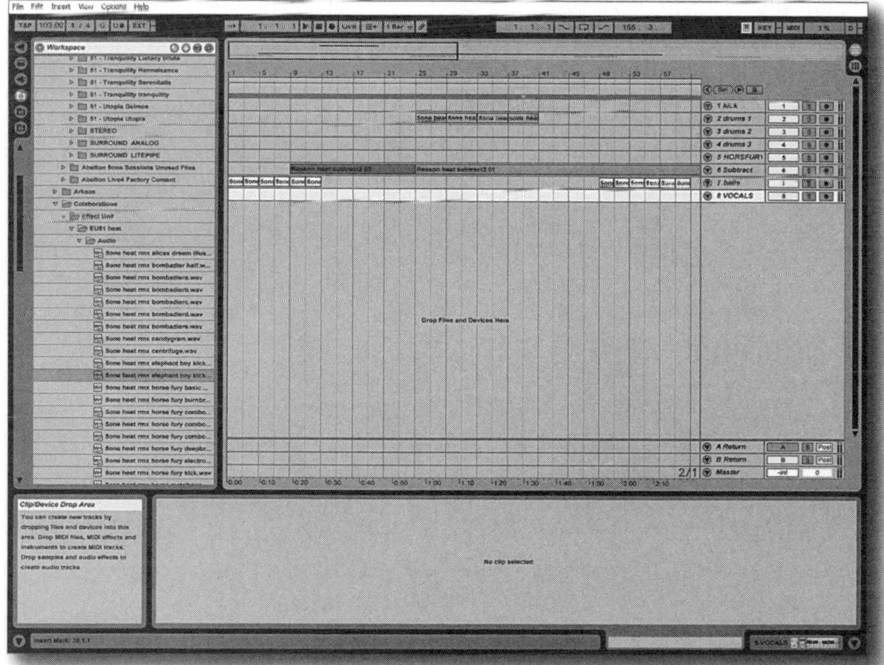

(b)

FIGURE 8.5
Ableton Live performance audio workstation: (a) Arrangement View; (b) Session View. (Courtesy of Ableton AG, www.ableton.com)

FIGURE 8.6
On-stage remote control over Ableton Live in a performance setting.

done through the creation of carefully-crafted audio loops, drum loops, and MIDI sequence/instrument data, allowing performance data to be sequenced in a traditional fashion, or played live in performance. When a studio DAW is used in conjunction with groove loop toys, MIDI hardware and various instrument plug-ins each of these tools can work together to play an important role in the never-ending creative process.

Another popular groove and production-related tool is Reason from the folks at Propellerheads. Reason differs from most soundfile-based looping programs in that it's an overall music production environment in that it includes a MIDI sequencer (Figure 8.7), as well as a wide range of software instrument modules (Figure 8.8) that can be played, mixed and combined in anto a comprehensive music production environment that can be controlled from any external keyboard or MIDI controller. Reason also includes a large number of signal processors that can be applied to any instrument or instrument group, under full and easily controlled automation.

In essence, Reason is a combination of modeled representations of vintage analog synthesis gear, mixed with the latest of digital synthesis and sampling technology (Figures 8.9 and 8.10). Combine these with a modular approach to signal and effects processing, add a generous amount of internal and remote mix and controller management (via an external MIDI controller), top this off with a quirky yet powerful sequencer, and you have an integrated software package that powerful and convenient.

Once you've finished the outline of a track, the obvious idea is to create new instrument tracks that can be combined in a traditional multitrack building-block approach, until a song begins to form. Since the loops and instruments

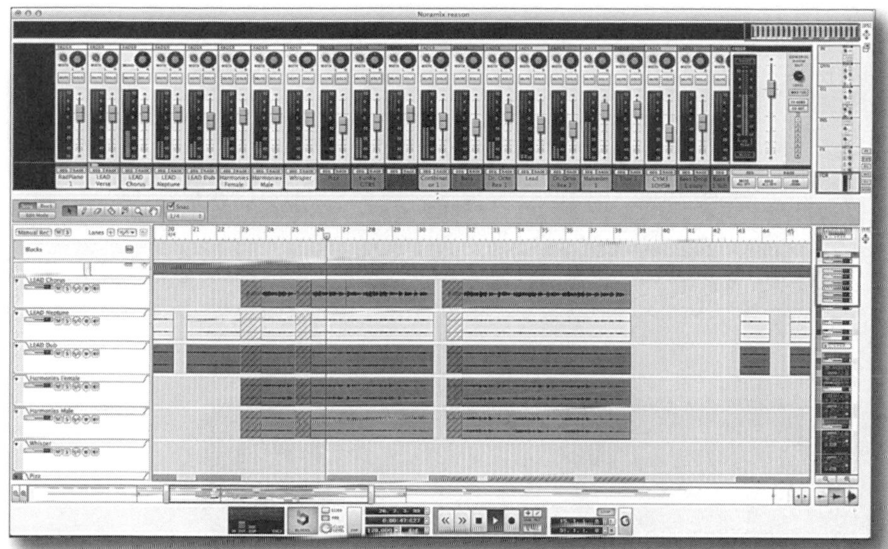

FIGURE 8.7
Reason's sequencer and mixer window. (Courtesy of Propellerhead Software, www.propellerheads.se)

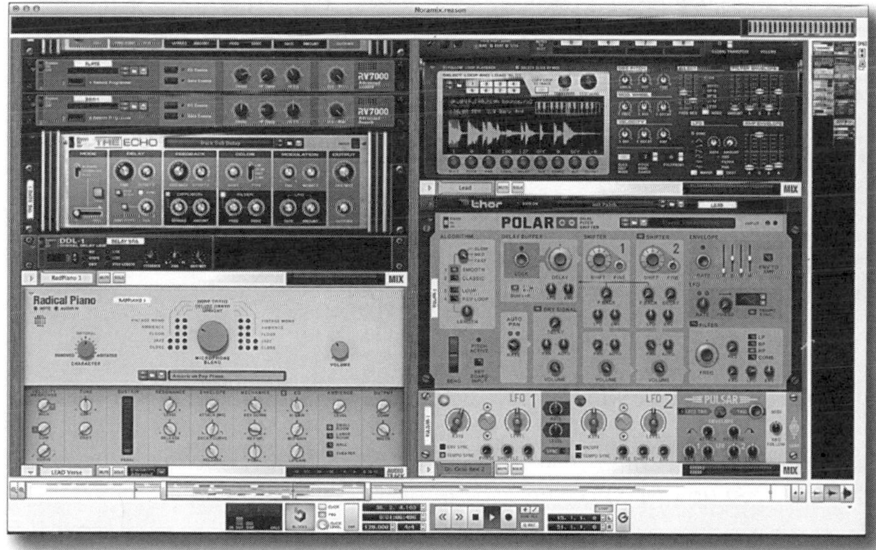

FIGURE 8.8
Reason instrument and effects music production environment. (Courtesy of Propellerhead Software, www.propellerheads.se)

are built from preprogrammed sounds and patches that reside on the hard disk, the actual Reason file (.rns) is often quite small, making this a perfect production media for publishing your songs on the Web or for collaborating with others around the world. Additional sounds, loops and patches are widely available for sale or for free as "refills" that can be added to your collection to greatly expand the software's palette.

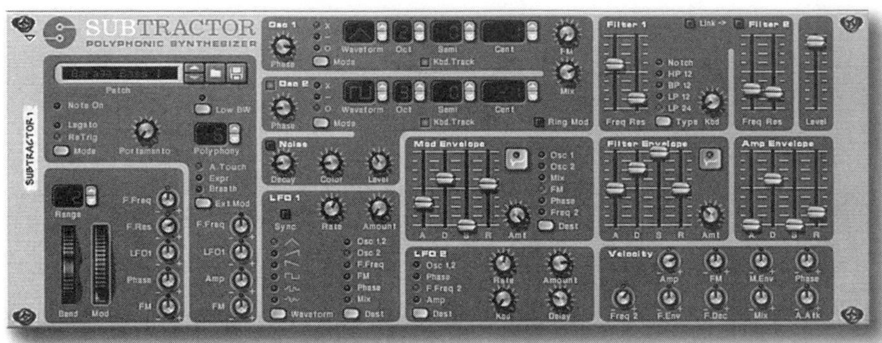

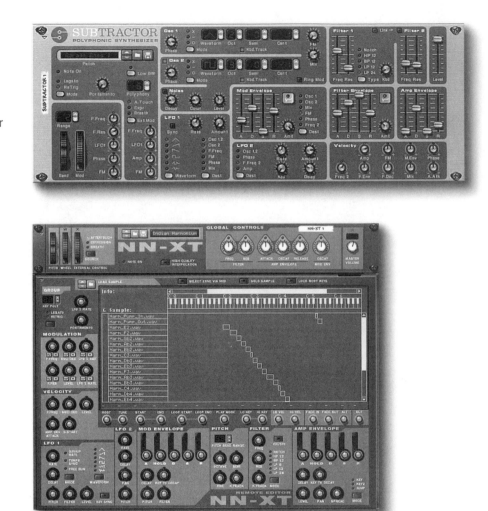

FIGURE 8.9
Reason's SubTractor polyphonic synth module. (Courtesy of Propellerhead Software, www.propellerheads.se)

FIGURE 8.10
Reason's NN-XT sampler module. (Courtesy of Propellerhead Software, www.propellerheads.se)

I definitely suggest that you download the demo and take it for a spin. You'll also want to check out the "New to Reason" video clips. Due to its open-ended nature, this program might take a while to get used to, but the journey just might open you up to a whole new world of production possibilities.

ReWire

In Chapter 7, we learned that ReWire is a special protocol that was developed by Propellerhead Software and Steinberg to allow audio to be streamed between two simultaneously running computer applications. Unlike a plug-in, where a task-specific application is inserted into a compatible host program, ReWire allows the audio and timing elements of a supporting program to be seamlessly integrated into another host program that supports ReWire. For example, DAWs such as Cubase/Nuendo and Pro Tools support ReWire, allowing production environment programs such as Ableton Live and Reason to be directly "wired" in software into the production inner-workings of a DAW session. This allows audio to be routed through the DAW's virtual mixer or I/O, making it possible for the timing elements of the client application to be synchronously controlled by the host DAW, allowing for greatly expanded instrument and production options within a studio or on-stage environment.

If you feel up to the task, download a few program demos, consult the involved program manuals, and try it out for yourself. The most important rule to remember when using ReWire is that the host program should always be opened first … and then the client. When shutting down, the client program should always be closed first.

On a final ReWire note: the virtual instruments within Reason offer up a powerful set of instruments that can be ReWired into the MIDI and performance paths of a DAW. In this way, one or more virtual instruments could be played into the DAW from your controller, allowing you to add more instruments and tonalities to your production. Once you've finished tracking and the MIDI data has been saved to their respective tracks, the ReWired instruments could be exported (bounced) to audio tracks within the DAW.

Groove and loop-based plug-ins

It's a sure bet that for every hardware looping tool, there are far more software plug-in groove tools and toys (Figures 8.11 and 8.12) that can be inserted into your DAW. These amazing software wonders often make life easier by:

- Automatically following the session tempo or tempo map
- Allowing I/O routing to plug into the DAW's mixer
- Making use of the DAW's automation and external controller capabilities
- Allowing individual or combined groove loops to be imported into a session.

FIGURE 8.11
Stylus RMX real-time groove module. (Courtesy of Spectrasonics, www.spectrasonics.net)

FIGURE 8.12
Battery 3 virtual drum and loop module. (Courtesy of Native Instruments GmbH, www.native-instruments.com)

These software instruments come in a wide range of sounds and applications that can often be edited and effected using an on-screen user interface, that can often be remotely controlled from an external MIDI controller.

Groove and loop dedicated hardware controllers

An increasing number of groove-based controllers (Figures 8.13 and 8.14) are appearing on the market that control either stand-alone and/or plug-in based groove software systems. These controllers offer both hardware control over the various software parameters and direct performance playing surfaces (often mimicking basic drum machine pad control surfaces). Although some of these controllers are dedicated to playing and manipulating drum kits and percussion sounds, others allow for samplefile data to be entered and edited in the system in an unlimited performance environment.

In addition to using the hardware surface to directly control the system's groove software, many of these controllers can be used as a controller for other 3rd-party software plug-ins. For example, a hardware controller might be able to directly control sampler plugins, synths, other drum machine plug-ins, etc. Often, all it takes is ingenuity and a willingness to wade your way through the vast number of software integration options.

Drum and drum loop plug-ins

Virtual software drum and groove machines are also part of the present-day landscape and can be used in a stand-alone, plugged-in or rewired production environment.

FIGURE 8.13
Maschine groove hardware/software production system. (Courtesy of Native Instruments GmbH, www.native-instruments.com)

FIGURE 8.14
Spark Drum Machine. (Courtesy of Arturia, www.arturia.com)

These plug-ins (Figures 8.15 and 8.16) are capable of injecting a wide range of groove and sonic spice options into a digital audio and/or MIDI project at the session's current tempo, allowing accompaniment patterns that can range from being simple and nonvarying over time, to individually complex instrument parts that can be meticulously programmed into a session or performed on the fly. In addition, most of these tools include multiple signal paths that let you route individual or groups of voices to a specific mixer input or DAW input. This makes it possible for isolated voices to be individually mixed, panned or

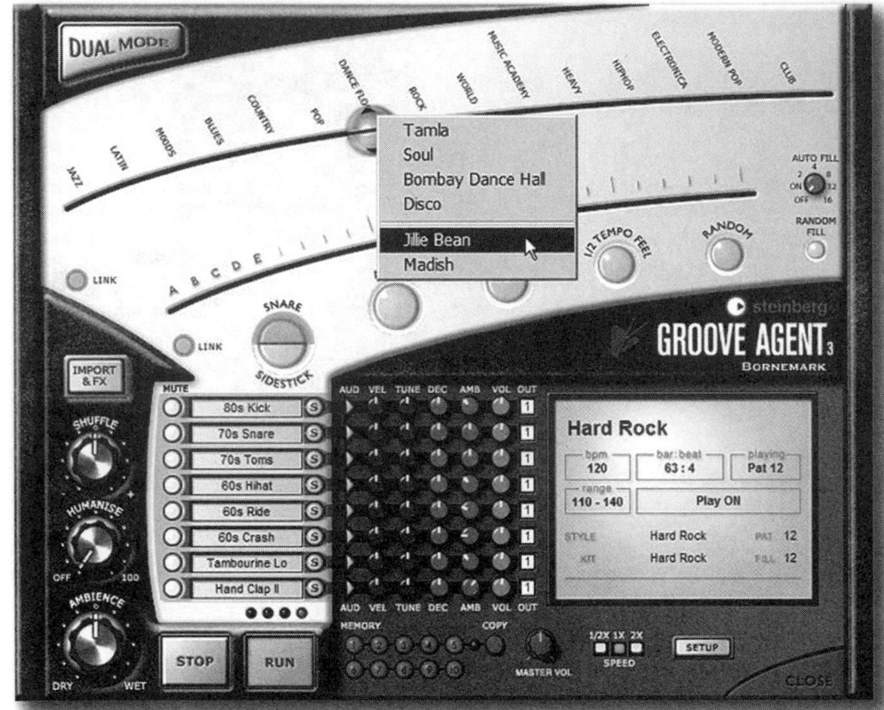

FIGURE 8.15 Steinberg's Groove Agent 3 drum plug-in. (Courtesy of Steinberg Media Technologies GmbH, a division of Yamaha Corporation, www.steinberg.net)

FIGURE 8.16 BFD2 acoustic drum library module. (Courtesy of FXpansion, www.fxpansion.com)

processed (using equalization, effects, etc.) into a session in new and interesting ways.

Pulling loops into a DAW session

When dealing with loops in modern-day production, one concept that needs to be discussed is that of importing loops into a DAW session. As we've seen, it's certainly possible to use the ReWire application to run a supporting client program in conjunction with the host program or even to use an instrument or groove-based plug-in within a compatible session. However, it should be noted that often (but not always) some sort of setup might be required in order to properly configure the application. Or at worst, these applications might use up valuable DSP resources that could otherwise be used for effects and signal processing.

One of the best options for freeing up these resources is to export the instrument or groove to a new audio track. This can be done in several ways (although with forethought, you might be able to come up with new ways of your own):

- The instrument track can be soloed and exported to an audio track in a contiguous fashion from the beginning "00:00:00:00" of the session to the end of the instrument's performance.
- In the case of a repetitive loop, a defined segment (often of a precise length of 4, 8, or more bars that occurs on the metric boundaries) can be selected for export as a sound file loop. Once exported to the file, it can be imported back into the session and looped into the session.
- In the case of an instrument that has multiple parts or voices, each part can be soloed and exported to its own track—thus giving a greater degree of control during a mixdown session.

DJ SOFTWARE

In addition to music production software, there are a growing number of software players, loopers, groovers, effects and digital devices on the market for the 21st century digital DJ. These hardware/software devices make it possible for digital grooves to be created from a laptop, iPad, controller, specially fitted turntable or digital turntable (jog/scratch CD player). Using such hardware-/software-based systems, it's possible to sync, scratch and perform with vinyl within a synchronized digital environment with an unprecedented amount of live performance and wireless interactivity that can be used on the floor, onstage or in the studio (Figures 8.17 and 8.18).

OBTAINING LOOP FILES FROM THE GREAT DIGITAL WELLSPRING

In this day and age, there's absolutely no shortage of preprogrammed loops that will work with a number of DAWs and groove editors. Some sound files will need

FIGURE 8.17
Torq DJ performance production software with Conectiv USB DJ interface. (Courtesy of M-Audio, a division of Avid Technology, Inc., www.m-audio.com)

FIGURE 8.18
Traktor portable laptop DJ rig. (Courtesy of Native Instruments GmbH, www.native-instruments.com)

to be manually edited, shaped or programmed to work with your DAW system, while others can be automatically entered into a loop-based program. Either way, these files can be easily obtained from any number of sources, such as:

- Those that are included for free with newly purchased software
- The Web (both free and for purchase)
- Commercial media
- Free files within websites, promotions and CDs that are loaded into demo content
- Rolling your own (creating your own loops can add a satisfying and personal touch)

It's important to note that at any point during the creation of a composition, audio and MIDI tracks (such as vocals or played instruments) can often be easily recorded into a loop session in order to give the performance a fluid, altered and interesting feel. It's even possible to record a live instrument into a session with a defined tempo and then edit these tracks into defined loops that can be dropped into the current and future sessions to add a live performance touch.

As with most music technologies, the field of looping and laying tracks into a groove continues to advance and evolve at an alarming rate. It's almost a sure bet that your current system will support looping, one way or another. Take time to read the manuals, gather up some loops that fit your style and particular interests and start working them into a session. It might take you some time to master the art of looping... or, then again, you might be a natural Zen master. Either way, the journey is usually educational and always lots of fun!

CHAPTER 9
MIDI and Electronic Music Technology

Today, professional and nonprofessional musicians are using the language of the *Musical Instrument Digital Interface* (MIDI) to perform an expanding range of music and automation tasks within audio production, audio for video, film post, stage production, etc. This industry-wide acceptance can, in large part, be attributed to the cost effectiveness, power and general speed of MIDI production. Once a MIDI instrument or device comes into the production picture, there's often less need (if any at all) to hire outside musicians for a project. This alluring factor allows a musician to compose, edit and arrange a piece in an electronic music environment that's extremely flexible. By this, I'm not saying that MIDI replaces—or should replace—the need for acoustic instruments, microphones and the traditional performance setting. In fact, it's a powerful production tool that assists countless musicians in creating music and audio productions in ways that are innovative and highly personal. In short, MIDI is all about control, repeatability, flexibility, cost-effective production power and fun.

The affordable potential for future expansion and increased control over an integrated production system has spawned the growth of a production industry that allows an individual to cost effectively realize a full-scale sound production, not only in his or her own lifetime … but in a relatively short time. For example, much of modern-day film music owes its very existence to MIDI. Before this technology, composers were forced to create without the benefits of hearing their composition or by creating a reduction score that could only be played on a piano or small ensemble (due to the cost and politics of hiring an orchestra). With the help of MIDI, composers can hear their work in real time, make any necessary changes, print out the scores … and take a full orchestra into the studio to record the final score version. At the other end of the spectrum, MIDI can be an extremely personal tool that lets us perform, edit and layer synthesized and/or sampled instruments to create a song that helps us to express ourselves to the masses—all within the comfort of the home or personal project studio. The moral of this story is that today's music industry would look and sound very different if it weren't for this powerful, four-letter production word.

MIDI PRODUCTION ENVIRONMENTS

One of the more powerful aspects of MIDI production is that a system can be designed to handle a wide range of tasks with a degree of flexibility and ease that best suits an artist's main instrument, playing style and even the user's personal working habits. By opening up almost any industry-related magazine, you'll easily see that a vast number of electronic musical instruments, effects devices, computer systems and other MIDI-related devices are currently available on the new and used electronic music market. MIDI production systems exist in all kinds of shapes and sizes and can be incorporated to match a wide range of production and budget needs. For example, working and aspiring musicians commonly install digital audio and MIDI systems in their homes (Figure 9.1). These production environments can range from ones that take up a corner of an artist's bedroom to larger systems that are integrated into a dedicated project studio. Systems such as these can be specially designed to handle a multitude of applications and have the important advantage of letting artists produce their music in a comfortable environment ... whenever the creative mood hits. Newer, laptop-based systems allow us to make music "wherever and whenever" from the comfort of your trusty backpack. Such production luxuries, that would have literally cost an artist a fortune in the not-too-distant past, are now within the reach of almost every musician.

Once MIDI has been mastered, its repeatability and edit control offer production challenges and possibilities that can stretch beyond the capabilities and cost effectiveness of the traditional multitrack recording environment. When combined with digital audio workstations (DAWs) and modern-day recording technology, much of the music production process can be preplanned and rehearsed before you step into the studio. In fact, it's not uncommon for recorded tracks to be laid down before they ever see the hallowed halls of the professional studio (if they see them at all). In business jargon, this luxury has reduced the number of billable hours to the artist or label to a cost-effective minimum.

FIGURE 9.1
Gettin' it all going in the bedroom studio.

FIGURE 9.2
Art'n Audio Studio, Tailand. (Courtesy of Solid State Logic, www.solid-state-logic.com).

Since its inception, electronic music has been an indispensable tool for the scoring and audio postproduction of television and radio commercials, industrial videos and full-feature motion picture sound tracks (Figure 9.2). For productions that are on a budget, an entire score can be created in the artist's project studio using MIDI, hard-disk tracks and digital recorders—all at a mere fraction of what it might otherwise cost to hire the musicians and rent a studio.

Electronic music production and MIDI are also very much at home on the stage. In addition to using synths, samplers, DAWs and drum machines on the stage, most or all of a MIDI instrument and effects device parameters can be controlled from a pre-sequenced or real-time controller source. This means that all the necessary settings for the next song (or section of a song) can be automatically called up before being played. Once under way, various instrument patch and controller parameters can also be changed during a live performance from a stomp box controller, keyboard or other hardware controller.

One of the media types that can be included in the notion of multimedia is definitely MIDI. With the advent of General MIDI (GM, which is discussed in greater detail within Chapter 10, is a standardized spec that allows any soundcard or GM-compatible device to play back a score using the originally intended sounds and program settings), it's possible (and common) for MIDI scores to be integrated into multimedia games and websites.

With the integration of the General MIDI standard into various media devices, one of the fastest growing MIDI applications, surprisingly, is probably comfortably resting in your pocket or purse right now—the ring tone on your cell phone (Figure 9.3). The ability to use MIDI (and often digital sound files) to let you know who is calling has spawned an industry that allows your cell to be personalized and super fun. One of my favorite ring tone stories happened on Hollywood Boulevard in L.A. This tall, lanky man was sitting at a café when his cell phone started blaring out an "If I Only Had a Brain" MIDI sequence from

FIGURE 9.3
One ringy-dingy ... MIDI (as well as digital audio waveforms) helps us to reach out and touch someone through phone ring tones.

The Wizard of Oz. It wouldn't have been nearly as funny if the guy didn't look *A LOT* like the scarecrow character. Of course, everyone laughed.

WHAT IS MIDI?

Simply stated, the Musical Instrument Digital Interface (MIDI) is a digital communications language and compatible specification that allows multiple hardware and software electronic instruments, performance controllers, computers and other related devices to communicate with each other over a connected network (Figure 9.4). MIDI is used to translate performance- or control-related events (such as playing a keyboard, selecting a patch number, varying a modulation wheel, triggering a staged visual effect, etc.) into equivalent digital messages and then transmit these messages to other MIDI devices where they can be used to control sound generators and other performance/control parameters. The beauty of MIDI is that its data can be recorded into a hardware device or software program (known as a sequencer), where it can then be edited and transmitted to electronic instruments or other devices to create music or control any number of parameters in a performance- or post-performance setting.

In addition to composing and performing a song, musicians can also act as techno-conductors, having complete control over a wide palette of sounds, their timbre (sound and tonal quality), overall blend (level, panning) and other real-time controls. MIDI can also be used to vary the performance and control parameters of electronic instruments, recording devices, control devices and signal processors in the studio, on the road or on the stage.

The term *interface* refers to the actual data communications link and software/hardware systems in a connected MIDI network. Through the use of MIDI, it's possible for all of the electronic instruments and devices within a network to be addressed through the transmission of real-time performance and control-related

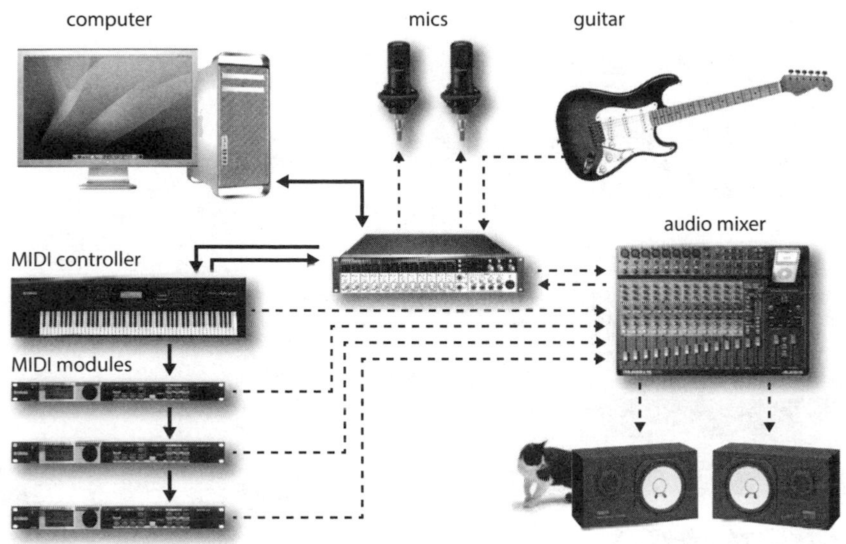

FIGURE 9.4
Example of a typical MIDI system with the MIDI network connections highlighted in solid lines.

MIDI data messages throughout a system to multiple instruments and devices through one or more data lines (which can be chained from device to device). This is possible because a single data cable is capable of transmitting performance and control messages over 16 discrete channels. This simple fact allows electronic musicians to record, overdub, mix and play back their performances in a working environment that loosely resembles the multitrack recording process. Once mastered, MIDI surpasses this analogy by allowing a composition to be edited, controlled, altered and called up with complete automation and repeatability—all of this providing production challenges and possibilities that are well beyond the capabilities of the traditional tape-based multitrack recording process.

What MIDI isn't

For starters, let's dispel one of MIDI's greatest myths: MIDI DOESN'T communicate audio, nor can it create sounds! It is strictly a digital language that instructs a device or program to create, play back or alter the parameters of sound or control function. It is a data protocol that communicates on/off triggering and a wide range of parameters to instruct an instrument or device to generate, reproduce or control audio or production-related functions. Because of these differences, the MIDI data path and the audio routing paths are entirely separate from each another (Figure 9.5). Even when they digitally share the same transmission cable (such as through FireWire or USB), the actual data paths and formats are completely separate.

In short, MIDI communicates information that instructs an instrument to play or a device to carry out a function. It can be likened to the dots on a

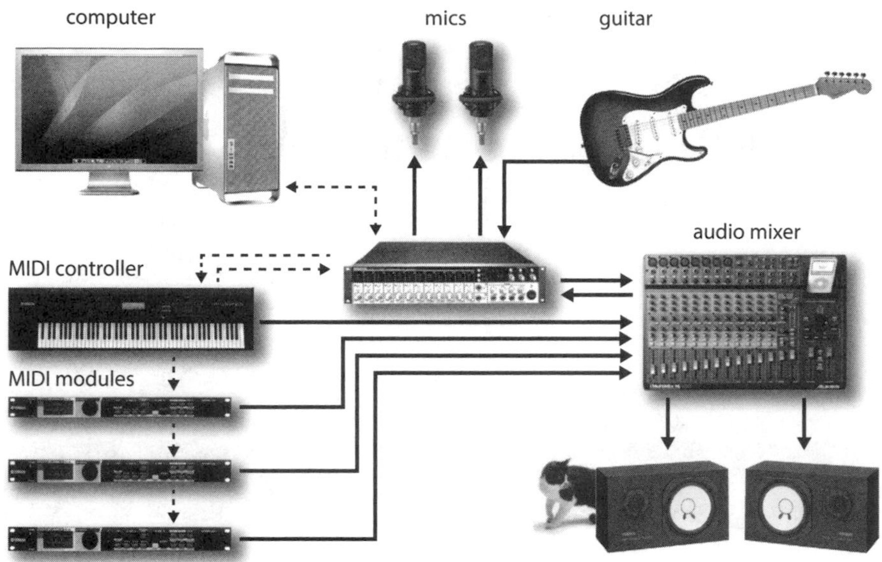

FIGURE 9.5
Example of a typical MIDI system with the audio connections highlighted in solid lines.

player-piano roll; when we put the paper roll up to our ears, we hear nothing, but when the cut-out dots pass over the sensors on a player piano, the instrument itself begins to make beautiful music. It's exactly the same with MIDI. A MIDI file or datastream is simply a set of instructions that pass down a wire in a serial fashion, but when an electronic instrument interprets the data, we begin to hear sound.

SYSTEM INTERCONNECTIONS

As a data transmission medium, MIDI is unique in the world of sound production in that it's able to pack 16 discrete channels of performance, controller and timing information and then transmit it in one direction, using data densities that are economically small and easy to manage. In this way, it's possible for MIDI messages to be communicated from a specific source (such as a keyboard or MIDI sequencer) to any number of devices within a connected network over a single MIDI data chain. In addition, MIDI is flexible enough that multiple MIDI data lines can be used to interconnect devices in a wide range of possible system configurations; for example, multiple MIDI lines can be used to transmit data to instruments and devices over 32, 48, 128 or more discrete MIDI channels!

Of course, those of you who are familiar with the concept of MIDI know that, over the years, the concept of interconnecting electronic instruments and other devices together has changed. These days, you're more likely to connect a device to a computer by using a USB, Firewire or Thunderbolt cable than the older standard cabling systems. In fact, often these interconnections are made "under the

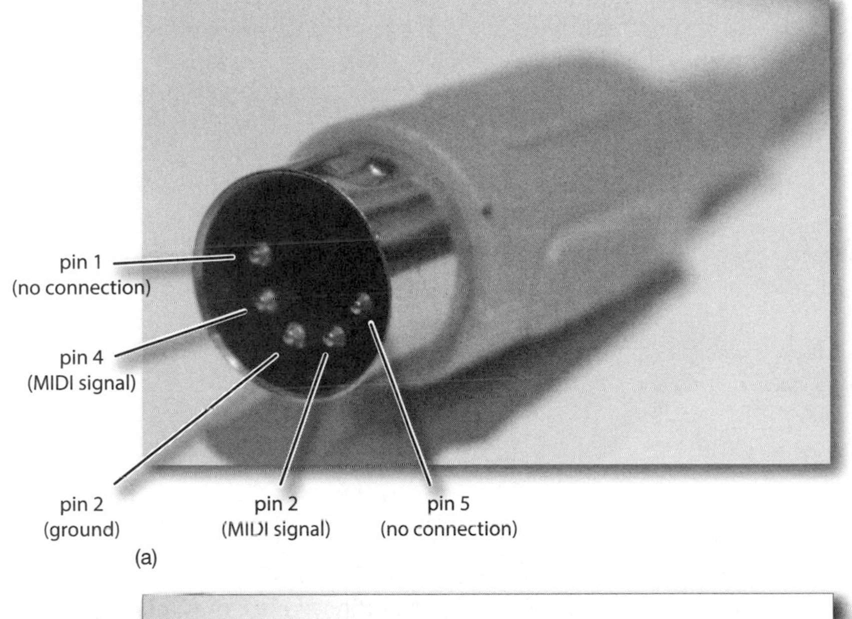

(a)

(b)

FIGURE 9.6
The MIDI cable:
(a) wiring diagram;
(b) the cable itself.

virtual hood," within the overall system and cable connectivity limitations aren't usually an issue. It's for this reason that it's important that we have an understanding of how these data connections are made "at a basic level" … therefore, I present to you, the MIDI cabling system.

The MIDI cable

A MIDI cable (Figure 9.6) consists of a shielded, twisted pair of conductor wires that has a male 5-pin DIN plug located at each of its ends. The MIDI specification currently uses only three of the five pins, with pins 4 and 5 being used as conductors for MIDI data; pin 2 is used to connect the cable's shield to equipment ground. Pins 1 and 3 are currently not in use, although the next section describes an ingenious system for powering devices through these pins that's known as MIDI phantom power. The cables use twisted cable and metal shield groundings to reduce outside interference, such as radio-frequency interference (RFI) or electrostatic interference, both of which can serve to distort or disrupt the transmission of MIDI messages.

MIDI Pin Description

- Pin 1 is not used in most cases; however, it can be used to provide the V– (ground return) of a MIDI phantom power supply.
- Pin 2 is connected to the shield or ground cable, which protects the signal from radio and electromagnetic interference.
- Pin 3 is not used in most cases; however, it can be used to provide the +V (+9 to +15V) of a MIDI phantom power supply.
- Pin 4 is a MIDI data line.
- Pin 5 is a MIDI data line.

MIDI cables come prefabricated in lengths of 2, 6, 10, 20 and 50 feet and can commonly be obtained from music stores that specialize in MIDI equipment. To reduce signal degradations and external interference that tends to occur over extended cable runs, 50 feet is the maximum length specified by the MIDI spec.

It should be noted, however, that in modern-day MIDI production, it's become increasingly common for MIDI data to be transmitted throughout a system network through USB, Firewire and wireless (pad) interconnections. Although the data isn't transmitted through traditional MIDI cabling, the data format adheres to the MIDI protocol.

MIDI PHANTOM POWER

In December 1989, Craig Anderton (musician, audio guru and editor of *Electronic Musician* magazine) submitted an article to *EM* proposing an idea that provides a standardized 12-V DC power supply to instruments and MIDI devices directly through pins 1 and 3 of a basic MIDI cable. Although pins 1 and 3 are technically reserved for possible changes in future MIDI applications (which never really came about), over the years several forward-thinking manufacturers (and project enthusiasts) have begun to implement MIDI phantom power (Figure 9.7) directly into their studio and on-stage systems.

For the more adventurous types who aren't afraid to get their soldering irons hot, I've included a basic schematic for getting rid of your wall wart and powering a system through the MIDI cable itself. It should be noted that not all MIDI cables connect all five pins. Some manufacturers use only three wires to save on manufacturing costs, thus it's best to make sure what you've got before you delve into your own personal project. Of course, neither the publisher nor I

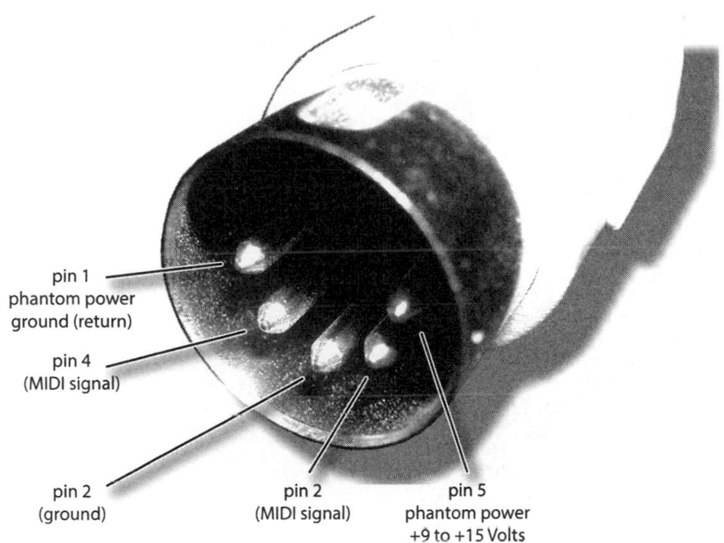

FIGURE 9.7
MIDI phantom power wiring diagram.

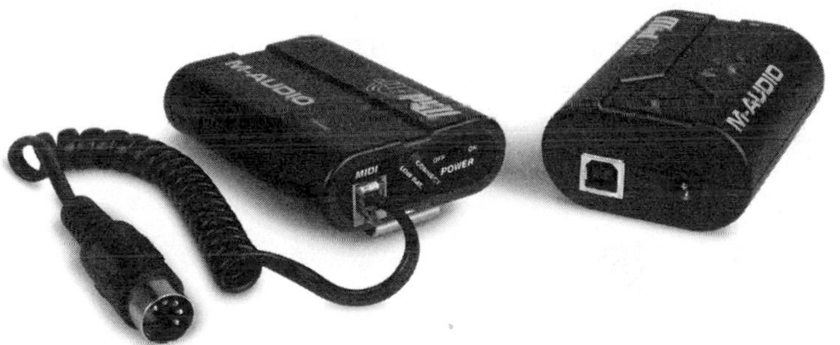

FIGURE 9.8
MidAir wireless USB MIDI transmitter and receiver system. (Courtesy of M-Audio, a division of Avid Technology, Inc., www.m-audio.com)

extend any warranty to anyone regarding damage that might occur as a result of this modification. Think it out carefully, have a steady hand and don't blow anything up! (On a personal note, I've retrofitted my own MIDI guitar rig, so the wall wart's off-stage, eliminating the need for a nearby power plug … which I'd probably kick into the next state in the middle of a gig.)

WIRELESS MIDI

Several companies have begun to manufacture wireless MIDI transmitters (Figure 9.8) that allow a battery-operated MIDI guitar, wind controller, etc., to be footloose and fancy free on-stage and in the studio. Working at distances of up to 500 feet, these battery-powered transmitter/receiver systems introduce very low delay latencies and can be switched over a number of radio channel frequencies.

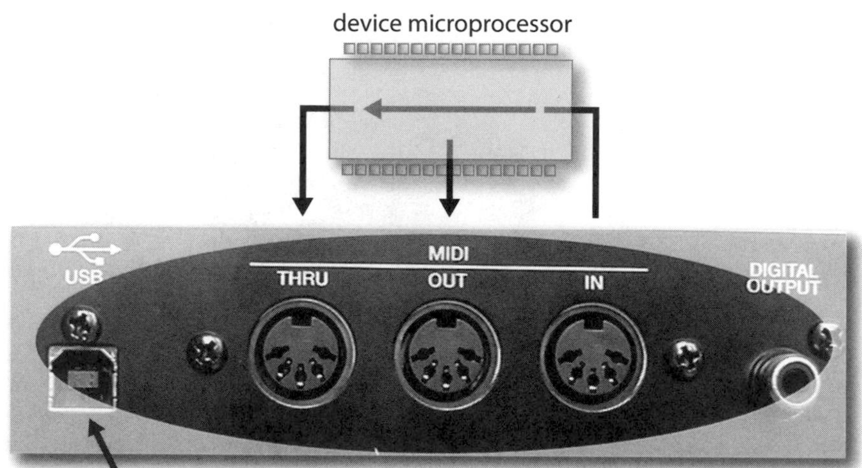

FIGURE 9.9
MIDI In, Out and Thru ports, showing the device's signal path routing.

MIDI JACKS

MIDI is distributed from device to device using three types of MIDI jacks: MIDI In, MIDI Out and MIDI Thru (Figure 9.9). These three connectors use 5-pin DIN jacks as a way to connect MIDI instruments, devices and computers into a music or production network system. As a side note, it's nice to know that these ports (as strictly defined by the MIDI 1.0 spec) are optically isolated to eliminate possible ground loops that might occur when connecting numerous devices together.

- *MIDI In jack:* The MIDI In jack receives messages from an external source and communicates this performance, control and timing data to the device's internal microprocessor, allowing an instrument to be played or a device to be controlled. More than one MIDI In jack can be designed into a system to provide for MIDI merging functions or for devices that can support more than 16 channels (such as a MIDI interface). Other devices (such as a controller) might not have a MIDI In jack at all.
- *MIDI Out jack:* The MIDI Out jack is used to transmit MIDI performance, control messages or SysEx from one device to another MIDI instrument or device. More than one MIDI Out jack can be designed into a system, giving it the advantage of controlling and distributing data over multiple MIDI paths using more than just 16 channels (i.e., 16 channels × *N* MIDI port paths).
- *MIDI Thru jack:* The MIDI Thru jack retransmits an exact copy of the data that's being received at the MIDI In jack. This process is important, because it allows data to pass directly through an instrument or device to the next device in the MIDI chain. Keep in mind that this jack is used to relay an exact copy of the MIDI In datastream, which isn't merged with data being transmitted from the MIDI Out jack.

MIDI ECHO

Certain MIDI devices may not include a MIDI Thru jack at all. Some of these devices, however, may give the option of switching the MIDI Out between being an actual MIDI Out jack and a MIDI Echo jack (Figure 9.10). As with the MIDI Thru jack, a MIDI Echo option can be used to retransmit an exact copy of any information that's received at the MIDI In port and route this data to the MIDI Out/Echo jack. Unlike a dedicated MIDI Out jack, the MIDI Echo function can often be selected to merge incoming data with performance data that's being generated by the device itself. In this way, more than one controller can be placed in a MIDI system at one time. Note that, although performance and timing data can be echoed to a MIDI Out/Echo jack, not all devices are capable of echoing SysEx data.

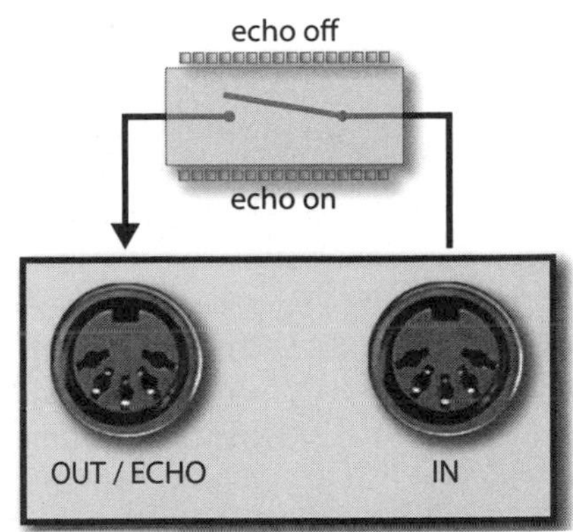

FIGURE 9.10
MIDI echo configuration.

TYPICAL CONFIGURATIONS

Although electronic studio production equipment and setups are rarely alike (or even similar), there are a number of general rules that make it easy for MIDI devices to be connected to a functional network. These common configurations allow MIDI data to be distributed in the most efficient and understandable manner possible.

As a primary rule, there are only two valid ways to connect one MIDI device to another within a MIDI cable chain (Figure 9.11):

- The MIDI Out jack of a source device (controller or sequencer/computer) must be connected to the MIDI In of a second device in the chain.
- The MIDI Thru jack of the second device must be connected to the MIDI In jack of the third device in the chain ... following this same Thru-to-In convention until the end of the chain is reached.

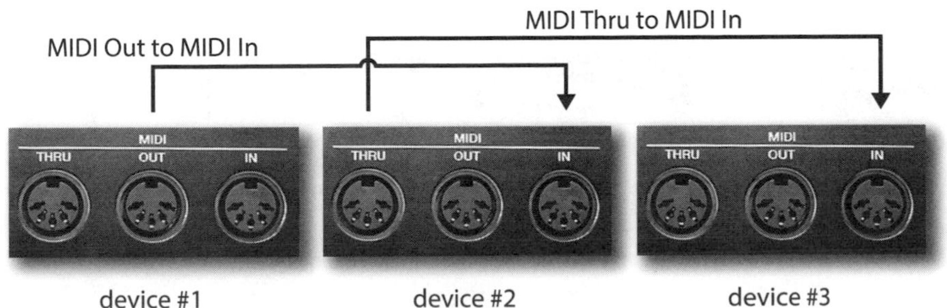

FIGURE 9.11
The two valid means of connecting one MIDI device to another.

The daisy chain

One of the simplest and most common ways to distribute data throughout a MIDI system is the daisy chain. This method relays MIDI data from a source device (controller or sequencer/computer) to the MIDI In jack of the next device in the chain (which receives and acts on this data). This next device then relays an exact copy of the incoming data out to its MIDI Thru jack, which is then relayed to the next device in the chain, and so on through the successive devices. In this way, up to 16 channels of MIDI data can be chained from one device to the next within a connected data network—and it's precisely this concept of transmitting multiple channels through a single MIDI line that makes the whole concept work! Let's try to understand this system better by looking at a few examples.

Figure 9.12a shows a simple (and common) example of a MIDI daisy chain whereby data flows from a controller (MIDI Out jack of the source device) to a synth module (MIDI In jack of the second device in the chain). An exact copy of the data that flows into the second device is then relayed to its MIDI Thru jack to another synth (MIDI In jack of the third device in the chain). If our controller is set to transmit on MIDI channel 3, the second synth in the chain (which is set to channel 2) will ignore the messages and not play, while the third synth (which is set to channel 3) will be playing its heart out. The moral of this story is that, although there's only one connected data line, a wide range of instruments and channel voices can be played in a surprisingly large number of combinations—all by using individual channel assignments along a daisy chain.

Another example (Figure 9.12b) shows how a computer can easily be designated as the master source within a daisy chain so a sequencing program can be used

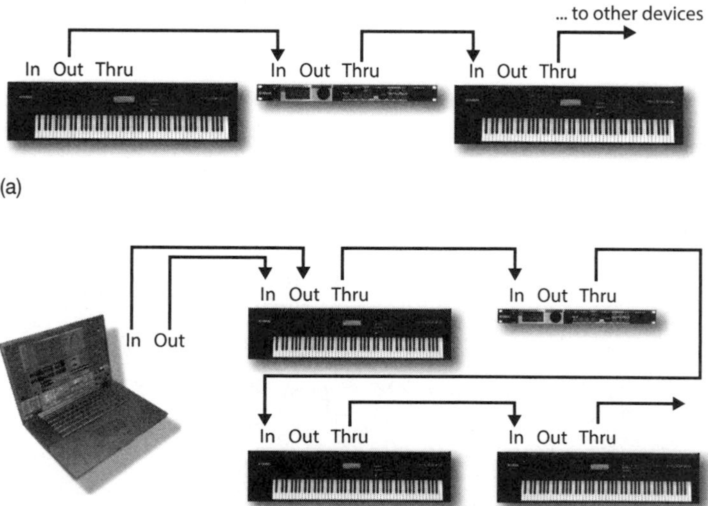

FIGURE 9.12
Example of a connected MIDI system using a daisy chain: (a) typical daisy chain hookup; (b) example of how a computer can be connected into a daisy chain.

to control the entire playback and channel routing functions of a daisy-chained system. In this situation, the MIDI data flows from a master controller/synth to the MIDI In jack of a computer's MIDI interface (where the data can be played into, processed and rerouted through a MIDI sequencer). The MIDI Out of the interface is then routed back to the MIDI In jack of the master controller/synth (which receives and acts on this data). The controller then relays an exact copy of this incoming data out to its MIDI Thru jack (which is then relayed to the next device in the chain) and so on, until the end of the chain is reached. When we stop and think about it, we can see that the controller is essentially used as a "performance tool" for entering data into the MIDI sequencer, which is then used to communicate this data out to the various instruments throughout the connected MIDI chain.

The multiport network

Another common approach to routing MIDI throughout a production system involves distributing MIDI data through the multiple 2, 4 and 8 In/Out ports that are available on the newer multiport MIDI interfaces or through the use of multiple MIDI USB interface devices.

NOTE

Although the distinction isn't overly important, you might want to keep in mind that a MIDI "port" is a virtual data path that's processed through a computer, whereas a MIDI "jack" is the physical connection on a device itself.

In larger, more complex MIDI systems, a multiport MIDI network (Figure 9.13) offers several advantages over a single daisy chain path. One of the most important is its ability to address devices within a complex setup that requires more than 16 MIDI channels. For example, a 2 x 2 MIDI interface that has two independent In/Out paths is capable of simultaneously addressing up to 32 channels (i.e., port A 1–16 and port B 1–16), whereas an 8 x 8 port is capable of addressing up to 128 individual MIDI channels.

This type of network of independent MIDI chains has a number of advantages. As an example, port A might be dedicated to three instruments that are set to respond to MIDI channels 1–6, 7 and finally channel 11, whereas port B might be transmitting data to two instruments that are responding to channels 1–4 and 5–10 and port C might be communicating SysEx MIDI data to and from a MIDI remote controller for a digital audio workstation (DAW). In this modern age of audio interfaces, multiport MIDI interfaces and controller devices that are each fitted with MIDI ports, it's a simple matter for a computer to route and synchronously communicate MIDI data throughout the studio in lots of ingenious and cost-effective ways.

iConnections

In addition to wired connections via standard MIDI, USB and Firewire cables, MIDI can also be connected to many pad and phone apps and devices using a wide variety of wireless and wired methods for connecting portable on-the-go

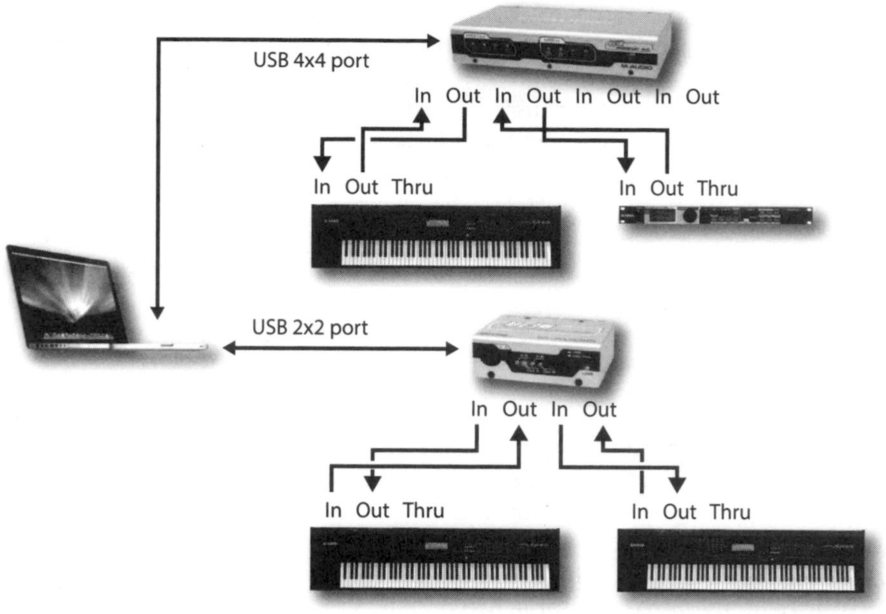

FIGURE 9.13
Example of a multiport network using two MIDI interfaces.

apps to your music production system. Any number of youtube videos and online resources (i.e. www.iosmidi.com) can be found on this ever-expanding subject.

EXPLORING THE SPEC

MIDI is a specified data format that must be strictly adhered to by those who design and manufacture MIDI-equipped instruments and devices. Because the format is standardized, you don't have to worry about whether the MIDI output of one device will be understood by the MIDI in port of a device that's made by another manufacturer. As long as the data ports say and/or communicate MIDI, you can be assured that the data (at least the basic performance functions) will be transmitted and understood by all devices within the connected system. In this way, the user need only consider the day-to-day dealings that go hand-in-hand with using electronic instruments, without having to be concerned with data compatibility between devices.

The MIDI message

MIDI digitally communicates musical performance data between devices as a string of MIDI messages. These messages are traditionally transmitted through a standard MIDI line in a serial fashion at a speed of 31,250 bits/sec. Within a serial data transmission line, data travels in a single-file fashion through a single conductor cable (Figure 9.14a); a parallel data connection, on the other hand, is able to simultaneously transmit digital bits in a synchronous fashion over a number of wires (Figure 9.14b).

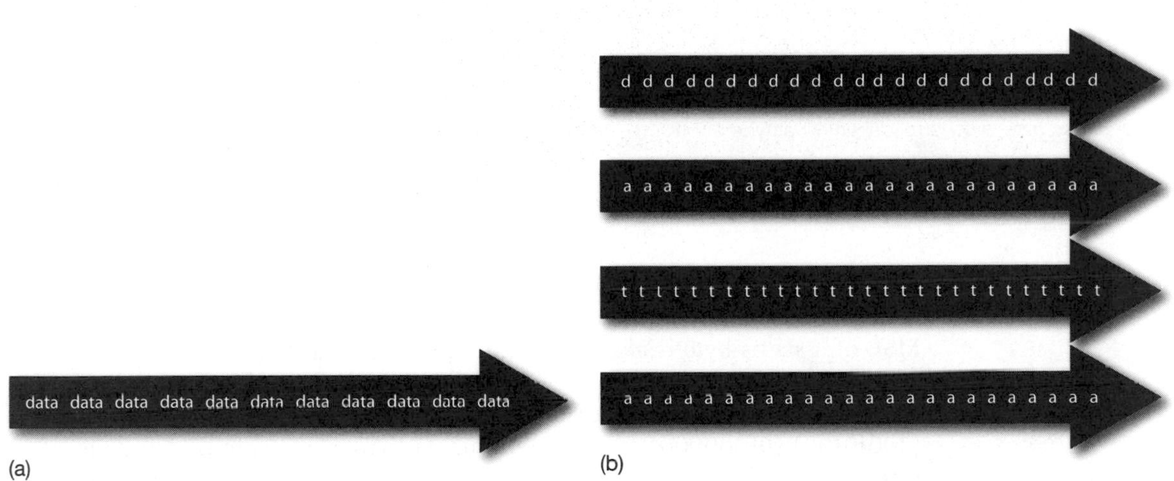

FIGURE 9.14
Serial versus parallel data transmission: (a) Serial data must be transmitted in a single-file fashion over a serial data line. (b) Multiple bits of data can be synchronously transmitted over a number of parallel lines.

When using a standard MIDI cable, it's important to remember that data can only travel in one direction from a single source to a destination (Figure 9.15a). To make two-way communication possible, a second MIDI data line must be used to communicate data back to the device, either directly or through the MIDI chain (Figure 9.15b).

MIDI messages are made up of groups of 8-bit words (known as bytes), which are transmitted in a serial fashion to convey a series of instructions to one or all MIDI devices within a system.

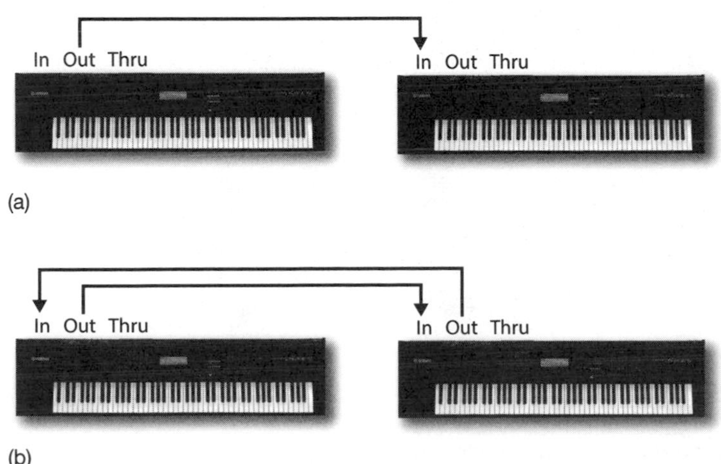

FIGURE 9.15
MIDI data can only travel in one direction through a single MIDI cable: (a) data transmission from a single source to a destination; (b) two-way data communication using two cables.

Only two types of bytes are defined by the MIDI specification: the status byte and the data byte.

- A *status byte* is used to identify what type of MIDI function is to be performed by a device or program. It is also used to encode channel data (allowing the instruction to be received by a device that's set to respond to the selected channel).
- A *data byte* is used to associate a value to the event that's given by the accompanying status byte.

Although a byte is made up of 8 bits, the most significant bit (MSB; the leftmost binary bit within a digital word) is used solely to identify the byte type. The MSB of a status byte is always 1, while the MSB of a data byte is always 0. For example, a 3-byte MIDI Note-On message (which is used to signal the beginning of a MIDI note) might read in binary form as a 3-byte Note-On message of (10010100) (01000000) (01011001). This particular example transmits instructions that would be read as: "Transmitting a Note-On message over MIDI channel #5, using keynote #64, with an attack velocity [volume level of a note] of 89."

FIGURE 9.16 Up to 16 channels can be communicated through a single MIDI cable or data port.

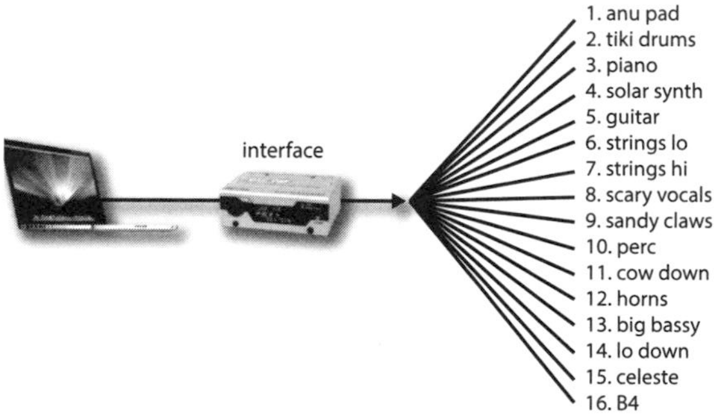

MIDI channels

Just as a public speaker might single out and communicate a message to one individual in a crowd, MIDI messages can be directed to communicate information to a specific device or range of devices within a MIDI system. This is done by embedding a channel-related nibble (4 bits) within the status/channel number byte. This process makes it possible for up to 16 channels of performance or control information to be communicated to a specific device, or a sound generator through a single MIDI data cable (Figure 9.16).

Since this nibble is 4 bits wide, up to 16 discrete MIDI channels can be transmitted through a single MIDI cable or designated port.

0000 = CH#1	0100 = CH#5	1000 = CH#9	1100 = CH#13
0001 = CH#2	0101 = CH#6	1001 = CH#10	1101 = CH#14
0010 = CH#3	0110 = CH#7	1010 = CH#11	1110 = CH#15
0011 = CH#4	0111 = CH#8	1011 = CH#12	1111 = CH#16

Whenever a MIDI device, sound generator within a device or program function is instructed to respond to a specific channel number, it will only react to messages that are transmitted on that channel (i.e., it ignores channel messages

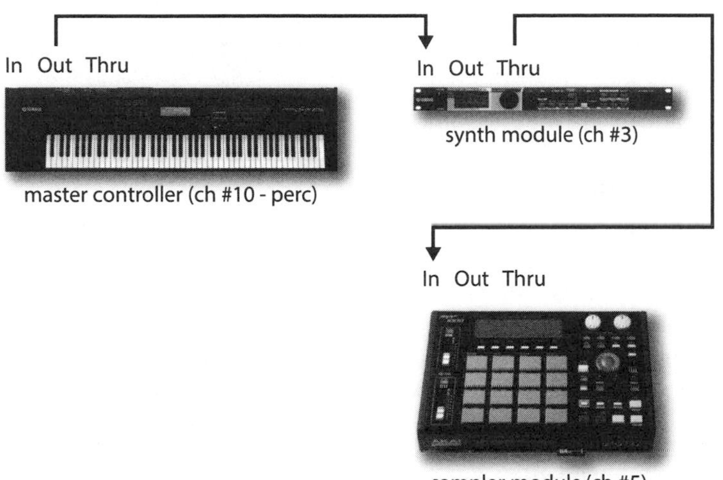

FIGURE 9.17
MIDI setup showing a set of MIDI channel assignments.

that are transmitted on any other channel). For example, let's assume that we're going to create a short song using a synthesizer that has a built-in sequencer (a device or program that's capable of recording, editing and playing back MIDI data) and two other "synths" (Figure 9.17):

1. We could easily start off by recording a simple drum pattern track into the master synth on channel 10 (numerous synths are pre-assigned to output drum/percussion sounds on this channel).
2. Once recorded, the sequence will thereafter transmit the notes and data over channel 10, allowing the synth's percussion section to be played.
3. Next, we could set a synth module to channel 3 and instruct the master synth to transmit on the same channel (since the synth module is set to respond to data on channel 3, its generators will sound whenever the master keyboard is played). We can now begin recording a melody line into the sequencer's next track.
4. Playing back the sequence will then transmit data to both the master synth (perc section 10) and the module (melody line 3) over their respective channels. At this point, our song is beginning to take shape.
5. Now we can set the other synth (or other instrument type) to respond to channel 5 and instruct the master synth to transmit on the same channel, allowing us to further embellish the song.
6. Now that the song's beginning to take shape, the sequencer can play the musical parts to the synths on their respective MIDI channels—all in an environment that gives us complete control over voicing, volume, panning and a wide range of edit functions over each instrument. In short, we've created a true multichannel production environment.

It goes without saying that the above example is just one of the infinite setup and channel possibilities that can be encountered in a production environment. It's often true, however, that even the most complex MIDI and production rooms will have a system—a basic channel and overall layout that makes the day-to-day operation of making music easier. This layout and the basic decisions that you might make in your own room are, of course, up to you. Streamlining a system to work both efficiently and easily will come with time, experience and practice.

MIDI modes

Electronic instruments often vary in the number of sounds and notes that can be simultaneously produced by their internal sound-generating circuitry. For example, certain instruments can only produce one note at a single time (known as a monophonic instrument), while others can generate 16, 32 and even 64 notes at once (these are known as polyphonic instruments). The latter type can easily play chords or more than one musical line on a single instrument at a time.

In addition, some instruments are only capable of producing a single generated sound patch (often referred to as a "voice") at any one time. Its generating circuitry could be polyphonic, allowing the player to lay down chords and bass or melody lines, but it can only produce these notes using a single, characteristic sound at any one time (e.g., an electric piano, a synth bass or a string patch). However, the vast majority of newer synths differ from this in that they're multitimbral in nature, meaning that they can generate numerous sound patches at any one time (e.g., an electric piano, a synth bass and a string patch, as can be seen in Figure 9.18). That's to say that it's common to run across electronic instruments that can simultaneously generate a number of voices, each offering its own control over a wide range of parameters. Best of all, it's also common for different sounds to be assigned to their own MIDI channels, allowing multiple patches to be internally mixed within the device to a stereo output bus or independent outputs.

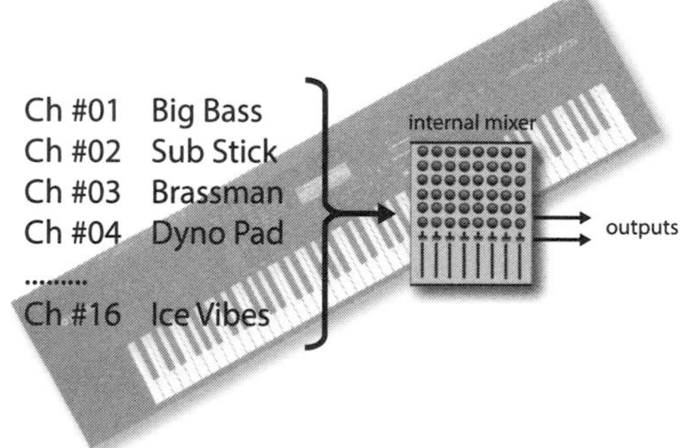

FIGURE 9.18
Multitimbral instruments are virtual bands-in-a-box that can simultaneously generate multiple patches, each of which can be assigned to its own MIDI channel.

The following list and figures explain the four modes that are supported by the MIDI spec:

- *Mode 1 (Omni On/Poly):* In this mode, an instrument will respond to data that's being received on any MIDI channel and then redirect this data to the instrument's base channel. In essence, the device will play back everything that's presented at its input in a polyphonic fashion, regardless of the incoming channel designations. As you might guess, this mode is rarely used.
- *Mode 2 (Omni On/Mono):* As in Mode 1, an instrument will respond to all data that's being received at its input, without regard to channel designations; however, this device will only be able to play one note at a time. Mode 2 is used even more rarely than Mode 1, as the device can't discriminate channel designations and can only play one note at a time.
- *Mode 3 (Omni Off/Poly):* In this mode, an instrument will only respond to data that matches its assigned base channel in a polyphonic fashion.

Data that's assigned to any other channel will be ignored. This mode is by far the most commonly used, as it allows the voices within a multi-timbral instrument to be individually controlled by messages that are being received on their assigned MIDI channels. For example, each of the 16 channels in a MIDI line could be used to independently play each of the parts in a 16-voice, multitimbral synth.

- *Mode 4 (Omni Off/Mono):* As with Mode 3, an instrument will be able to respond to performance data that's transmitted over a single, dedicated channel; however, each voice will only be able to generate one MIDI note at a time. A practical example of this mode is often used in MIDI guitar systems, where MIDI data is monophonically transmitted over six consecutive channels (one channel/voice per string).

Channel voice messages

Channel Voice messages are used to transmit real-time performance data throughout a connected MIDI system. They're generated whenever a MIDI instrument's controller is played, selected or varied by the performer. Examples of such control changes could be the playing of a keyboard, pressing of program selection buttons or movement of modulation or pitch wheels. Each Channel Voice message contains a MIDI channel number within its status byte, meaning that only devices that are assigned to the same channel number will respond to these commands. There are seven Channel Voice message types: Note-On, Note-Off, Polyphonic Key Pressure, Channel Pressure, Program Change, Pitch Bend Change and Control Change:

- *Note-On* messages (Figure 9.19): Indicate the beginning of a MIDI note. This message is generated each time a note is triggered on a keyboard, drum machine or other MIDI instrument (by pressing a key, striking a drum pad, etc.). A Note-On message consists of three bytes of information: a MIDI channel number, a MIDI pitch number, and an attack velocity value (messages that are used to transmit the individually played volume levels [0–127] of each note).
- *Note-Off* messages: Indicate the release (end) of a MIDI note. Each note played through a Note-On message is sustained until a corresponding Note-Off message is received. A Note-Off message doesn't cut off a sound; it merely stops playing it. If the patch being played has a release (or final decay) stage, it begins that stage upon receiving this message. It should be noted that many systems will actually use a Note-On message with a velocity 0 to denote a Note-Off message.
- *Polyphonic Key Pressure* messages (Figure 9.20): Transmitted by instruments that can respond to pressure changes applied to the

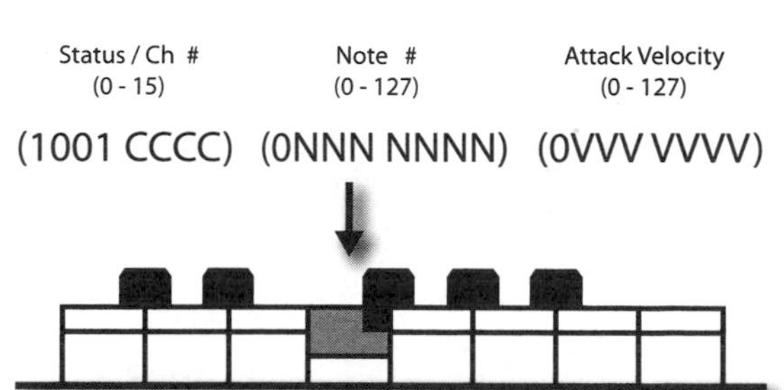

FIGURE 9.19
Byte structure of a MIDI Note-On message.

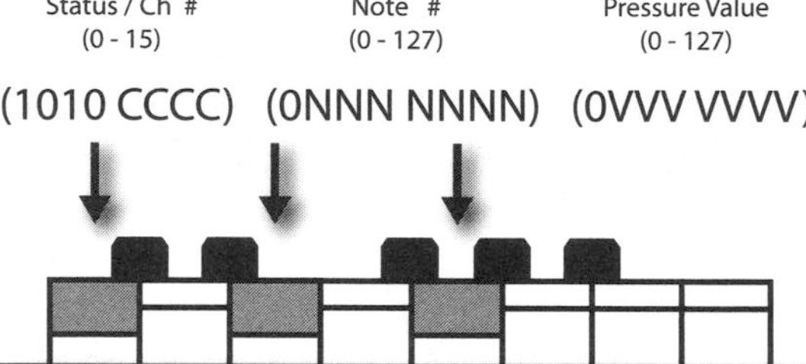

FIGURE 9.20 Byte structure of a MIDI Polyphonic Key Pressure message (generated when additional pressure is applied to each key that's played).

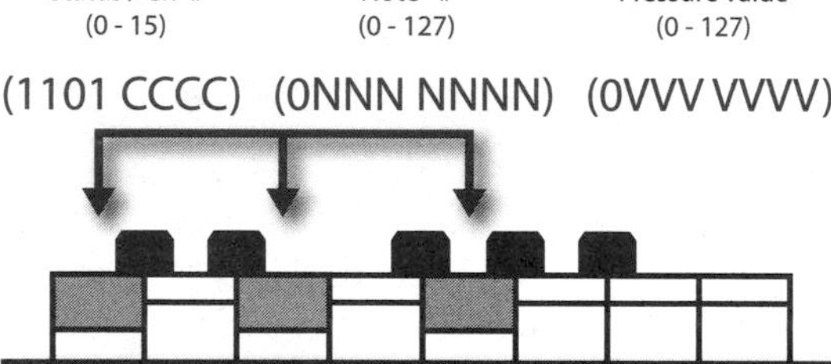

FIGURE 9.21 Byte structure of a MIDI Channel Pressure message (simultaneously affect all notes that are transmitted over a MIDI channel).

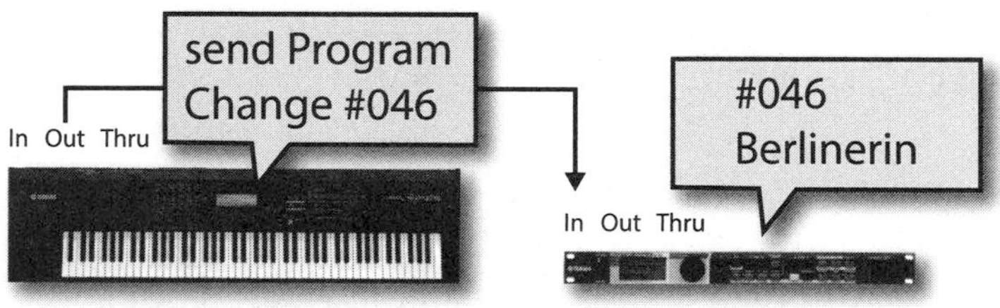

FIGURE 9.22 Program Change messages can be used to change sound patches from a sequencer or from a remote controller.

individual keys of a keyboard. A Polyphonic Key Pressure message consists of three bytes of information: a MIDI channel number, a MIDI pitch number and a pressure value.

- *Channel Pressure* (or *Aftertouch*) messages (Figure 9.21): Transmitted and received by instruments that respond to a single, overall pressure applied to the keys. In this way, additional pressure on the keys can be assigned to control such variables as pitch bend, modulation and panning.
- *Program Change* messages (Figures 9.22 and 9.23): Change the active voice (generated sound) or preset program number in a MIDI instrument or device. Using this message format, up to 128 presets (a user- or factory-defined number that activates a specific sound-generating patch or system setup) can be selected. A Program Change message consists of two bytes of information: a MIDI channel number (1–16) and a program ID number (0–127).

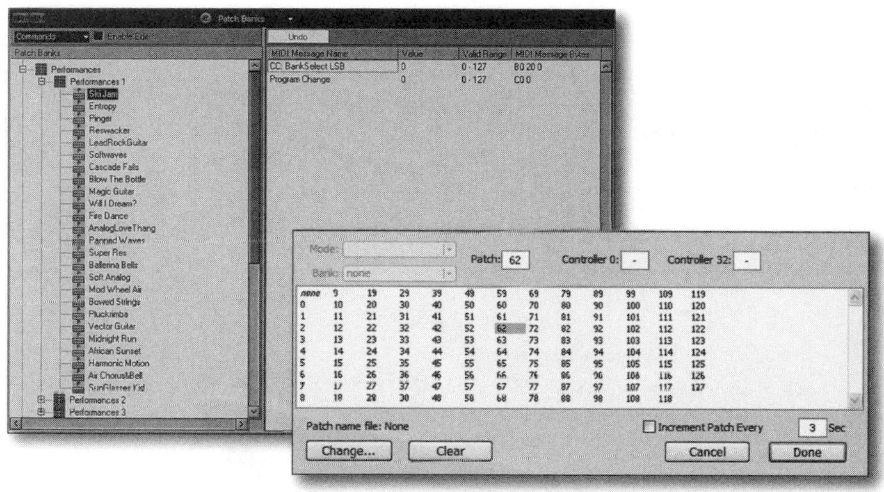

FIGURE 9.23
Workstations and sequencer software systems will often allow patches to be recalled via Program Change messages. (Courtesy of Steinberg Media Technologies GmbH, a division of Yamaha Corporation, www.steinberg.net, and Digidesign, a division of Avid Technology, www.digidesign.com)

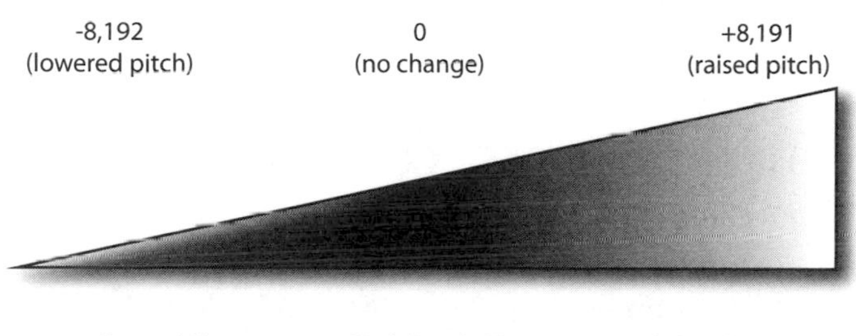

Status / Ch #	Pitch Bend LSB	Pitch Bend MSB
(1111 NNNN)	(0LLL LLLL)	(0MMM MMMM)

FIGURE 9.24
Byte structure of a Pitch Bend Change message.

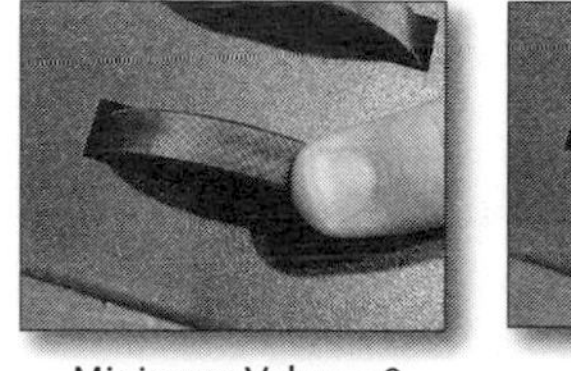

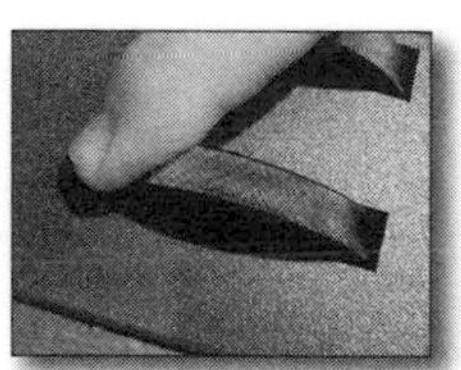

FIGURE 9.25
Pitch bend wheel data value ranges.

- *Pitch Bend Change* messages (Figures 9.24 and 9.25): Transmitted by an instrument whenever its pitch bend wheel is moved in either the positive (raise pitch) or negative (lower pitch) direction from its central (no pitch bend) position.
- *Control Change* messages (Figures 9.26 and 9.27): Transmit information that relates to real-time control over a MIDI instrument's performance

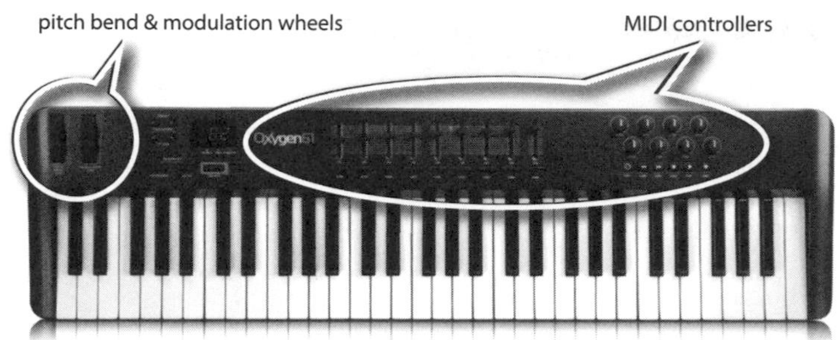

FIGURE 9.26
MIDI keyboard controller. (Courtesy of M-Audio, a division of Avid Technology, Inc., www.m-audio.com).

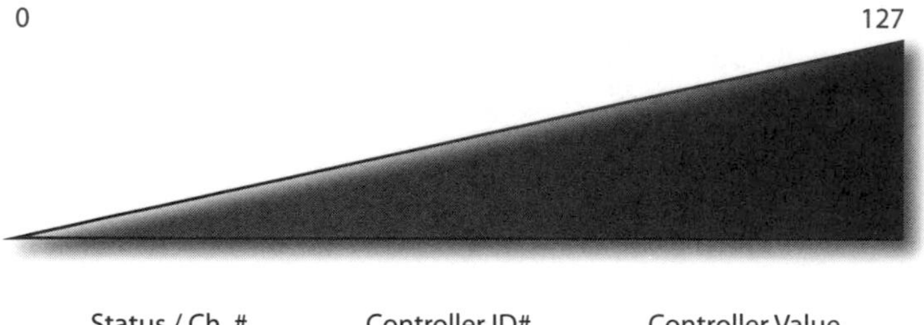

Status / Ch # (0 - 15) | Controller ID# (0 - 127) | Controller Value (0 - 127)

(a) (1011 NNNN) (0CCC CCCC) (0VVV VVVV)

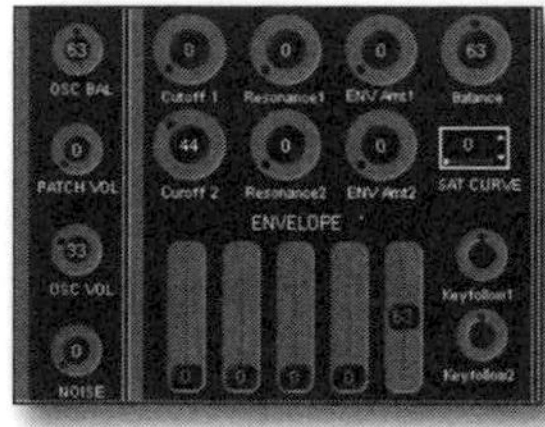
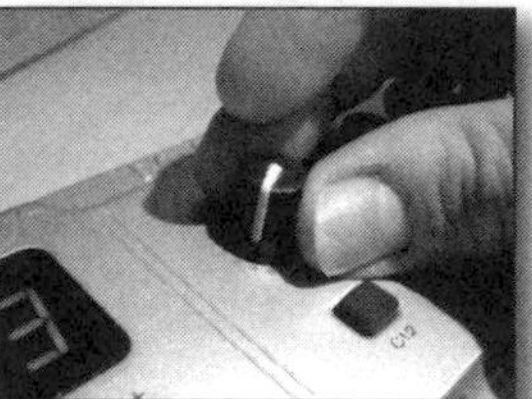
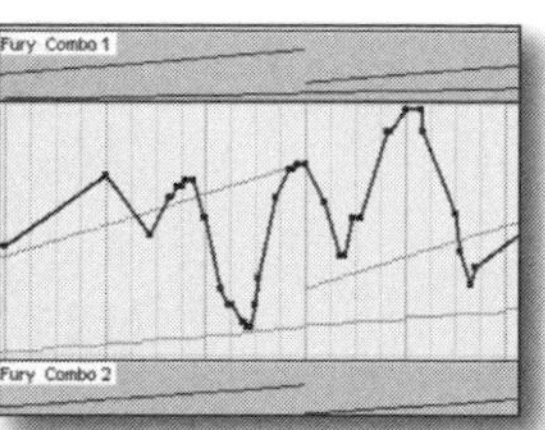

(b)

FIGURE 9.27
Control Change message: (a) byte structure; (b) control messages can be varied in real time or under automation using a number of input methods.

parameters (such as modulation, main volume, balance and panning). Three types of real-time controls can be communicated through control change messages: continuous controllers, which communicate a continuous range of control settings, generally with values ranging from 0–127; switches (controls having an ON or OFF state with no intermediate settings); and data controllers, which enter data either through numerical keypads or stepped up/down entry buttons.

Explanation of controller ID parameters

As you can see in Figure 9.27a, the 2nd byte of the Control Change message is used to denote the *controller ID number*. This all-important value is used to specify which of the device's program or performance parameters are to be addressed.

The following section details the general categories and conventions for assigning controller ID numbers to an associated parameter (as specified by the 1995 update of the MMA (MIDI Manufacturers Association, www.midi.org). An overview of these controllers can be seen in Table 9.1. This is definitely an important table to earmark, because these numbers will be an important guide toward knowing and/or finding the right ID number that can help you on your path toward finding that perfect variable for making it sound right.

System messages

As the name implies, System messages are globally transmitted to every MIDI device in the MIDI chain. This is accomplished because MIDI channel numbers aren't addressed within the byte structure of a System message. Thus, any device will respond to these messages, regardless of its MIDI channel assignment. The three System message types are System-Common messages, System Real-Time messages, and System-Exclusive messages.

System-Common messages are used to transmit MIDI timecode, song position pointer, song select, tune request and end-of-exclusive data messages throughout the MIDI system or 16 channels of a specified MIDI port.

- *MIDI timecode (MTC)* messages: Provide a cost-effective and easily implemented way to translate SMPTE (a standardized synchronization timecode) into an equivalent code that conforms to the MIDI 1.0 spec. It allows time-based codes and commands to be distributed throughout the MIDI chain in a cheap, stable and easy-to-implement way. MTC Quarter-Frame messages are transmitted and recognized by MIDI devices that can understand and execute MTC commands. A grouping of eight quarter frames is used to denote a complete timecode address (in hours, minutes, seconds, and frames), allowing the SMPTE address to be updated every two frames. More in-depth coverage of MIDI timecode can be found in Chapter 11.
- *Song Position Pointer (SPP)* messages: Allow a sequencer or drum machine to be synchronized to an external source (such as a tape machine) from any measure position within a song. This complex timing protocol isn't commonly used, because most users and design layouts currently favor MTC.
- *Song Select* messages: Use an identifying song ID number to request a specific song from a sequence or controller source. After being selected, the song responds to MIDI Start, Stop and Continue messages.
- *Tune Request* messages: Used to request that an equipped MIDI instrument initiate its internal tuning routine.
- *End of Exclusive (EOX)* messages: Indicate the end of a System-Exclusive message.

System Real-Time messages provide the precise timing element required to synchronize all of the MIDI devices in a connected system. To avoid timing delays, the MIDI specification allows System Real-Time messages to be inserted at any point in the data stream, even between other MIDI messages.

Table 9.1 Listing of Controller ID Numbers, Outlining Both the Defined Format and Conventional Controller Assignments

Control #	Parameter
14-Bit Controllers Coarse/MSB (Most Significant Bit)	
0	Bank Select 0-127 MSB
1	Modulation Wheel or Lever 0-127 MSB
2	Breath Controller 0-127 MSB
3	Undefined 0-127 MSB
4	Foot Controller 0-127 MSB
5	Portamento Time 0-127 MSB
6	Data Entry MSB 0-127 MSB
7	Channel Volume (formerly Main Volume) 0-127 MSB
8	Balance 0-127 MSB
9	Undefined 0-127 MSB
10	Pan 0-127 MSB
11	Expression Controller 0-127 MSB
12	Effect Control 1 0-127 MSB
13	Effect Control 2 0-127 MSB
14	Undefined 0-127 MSB
15	Undefined 0-127 MSB
16–19	General Purpose Controllers 1-4 0-127 MSB
20–31	Undefined 0-127 MSB
14-Bit Controllers Fine/LSB (Least Significant Bit)	
32	LSB for Control 0 (Bank Select) 0-127 LSB
33	LSB for Control 1 (Modulation Wheel or Lever) 0-127 LSB
34	LSB for Control 2 (Breath Controller) 0-127 LSB
35	LSB for Control 3 (Undefined) 0-127 LSB
36	LSB for Control 4 (Foot Controller) 0-127 LSB
37	LSB for Control 5 (Portamento Time) 0-127 LSB
38	LSB for Control 6 (Data Entry) 0-127 LSB
39	LSB for Control 7 (Channel Volume, formerly Main Volume) 0-127 LSB
40	LSB for Control 8 (Balance) 0-127 LSB
41	LSB for Control 9 (Undefined) 0-127 LSB

Continued …

Table 9.1 continued

Control #	Parameter
42	LSB for Control 10 (Pan) 0-127 LSB
43	LSB for Control 11 (Expression Controller) 0-127 LSB
44	LSB for Control 12 (Effect control 1) 0-127 LSB
45	LSB for Control 13 (Effect control 2) 0-127 LSB
46–47	LSB for Control 14-15 (Undefined) 0-127 LSB
48–51	LSB for Control 16-19 (General Purpose Controllers 1-4) 0-127 LSB
52–63	LSB for Control 20-31 (Undefined) 0-127 LSB
7-Bit Controllers	
64	Damper Pedal on/off (Sustain) <63 off, >64 on
65	Portamento On/Off <63 off, >64 on
66	Sustenuto On/Off <63 off, >64 on
67	Soft Pedal On/Off <63 off, >64 on
68	Legato Footswitch <63 Normal, >64 Legato
69	Hold 2 <63 off, >64 on
70	Sound Controller 1 (default: Sound Variation) 0-127 LSB
71	Sound Controller 2 (default: Timbre/Harmonic Intensity) 0-127 LSB
72	Sound Controller 3 (default: Release Time) 0-127 LSB
73	Sound Controller 4 (default: Attack Time) 0-127 LSB
74	Sound Controller 5 (default: Brightness) 0-127 LSB
75	Sound Controller 6 (default: Decay Time: see MMA RP-021) 0-127 LSB
76	Sound Controller 7 (default: Vibrato Rate: see MMA RP-021) 0-127 LSB
77	Sound Controller 8 (default: Vibrato Depth: see MMA RP-021) 0-127 LSB
78	Sound Controller 9 (default: Vibrato Delay: see MMA RP-021) 0-127 LSB
79	Sound Controller 10 (default undefined: see MMA RP-021) 0-127 LSB
80–83	General Purpose Controller 5-8 0-127 LSB
84	Portamento Control 0-127 LSB

Continued ...

Table 9.1 continued

Control #	Parameter
85–90	Undefined
91	Effects 1 Depth (default: Reverb Send Level) 0–127 LSB
92	Effects 2 Depth (default: Tremolo Level) 0–127 LSB
93	Effects 3 Depth (default: Chorus Send Level) 0–127 LSB
94	Effects 4 Depth (default: Celesta [Detune] Depth) 0–127 LSB
95	Effects 5 Depth (default: Phaser Depth) 0–127 LSB
Parameter Value Controllers	
96	Data Increment (Data Entry +1)
97	Data Decrement (Data Entry –1)
98	Non-Registered Parameter Number (NRPN): LSB 0–127 LSB
99	Non-Registered Parameter Number (NRPN): MSB 0–127 MSB
100	Registered Parameter Number (RPN): LSB* 0–127 LSB
101	Registered Parameter Number (RPN): MSB* 0–127 MSB
102–119	Undefined
Reserved for Channel Mode Messages	
120	All Sound Off 0
121	Reset All Controllers
122	Local Control On/Off 0 off, 127 on
123	All Notes Off
124	Omni Mode Off (+ all notes off)
125	Omni Mode On (+ all notes off)
126	Poly Mode On/Off (+ all notes off)
127	Poly Mode On (+ mono off + all notes off)

- *Timing Clock* messages: The MIDI Timing Clock message is transmitted within the MIDI datastream at various resolution rates. It is used to synchronize the internal timing clocks of each MIDI device within the system and is transmitted in both the start and stop modes at the currently defined tempo rate. In the early days of MIDI, these rates (which are measured in pulses per quarter note [ppq]) ranged from 24 to 128 ppq; however, continued advances in technology have brought these rates up to 240, 480 or even 960 ppq.
- *Start* messages: Upon receipt of a timing clock message, the MIDI Start command instructs all connected MIDI devices to begin playing from

their internal sequences' initial start point. Should a program be in midsequence, the start command will reposition the sequence back to its beginning, at which point it will begin to play.

- *Stop* messages: Upon receipt of a MIDI Stop command, all devices within the system will stop playing at their current position point.
- *Continue* messages: After receiving a MIDI Stop command, a MIDI Continue message will instruct all connected devices to resume playing their internal sequences from the precise point at which it was stopped.
- *Active Sensing* messages: When in the Stop mode, an optional Active Sensing message can be transmitted throughout the MIDI datastream every 300 milliseconds. This instructs devices that can recognize this message that they're still connected to an active MIDI data stream.
- *System Reset* messages: A System Reset message is manually transmitted in order to reset a MIDI device or instrument back to its initial power-up default settings (commonly mode 1, local control on and all notes off).

System-exclusive messages

The system-exclusive (sys-ex) message allows MIDI manufacturers, programmers and designers to communicate customized MIDI messages between MIDI devices. The purpose of these messages is to give manufacturers, programmers and designers the freedom to communicate any device-specific data of an unrestricted length, as they see fit. Most commonly, sys-ex data are used for the bulk transmission and reception of program/patch data and sample data, as well as real-time control over a device's parameters. The transmission format of a sys-ex message (Figure 9.28), as defined by the MIDI standard, includes a sys-ex status header, manufacturer's ID number, any number of sys-ex data bytes and an EOX byte. When a sys-ex message is received, the identification number is read by a MIDI device to determine whether or not the following messages are relevant. This is easily accomplished by the assignment of a unique 1- or 3-byte ID number to each registered MIDI manufacturer and make. If this number doesn't match the receiving MIDI device, the subsequent data bytes will be ignored. Once a valid stream of sys-ex data has been transmitted, a final EOX message is sent, after which the device will again begin to respond normally to incoming MIDI performance messages.

In actual practice, the general idea behind sys-ex is that it uses MIDI messages to transmit and receive program, patch and sample data or real-time parameter

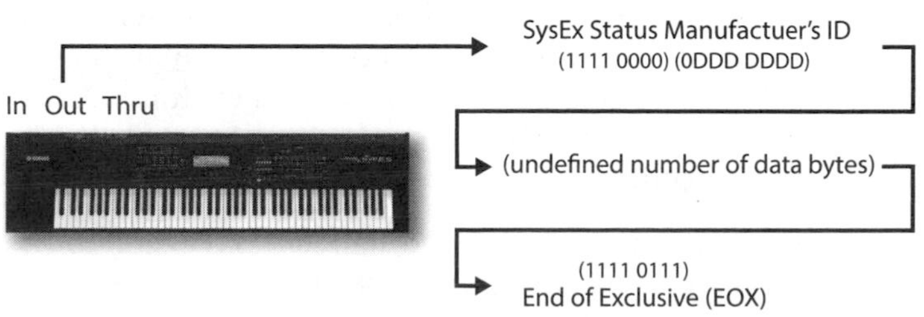

FIGURE 9.28
System-exclusive ID data and controller format.

information between devices. It's sort of like having an instrument or device that's a musical chameleon. One moment it can be configured with a certain set of sound patches and setup data and then, after it receives a new sys-ex data dump, you could easily end up with an instrument that's literally full of new and exciting (or not-so-exciting) sounds and settings. Here are a few examples of how sys-ex can be put to good use:

- *Transmitting patch data between synths:* Sys-ex can be used to transmit patch and overall setup data between synths of identical make and (most often) model. Let's say that we have a Brand X Model Z synthesizer, and, as it turns out, you have a buddy across town that also has a Brand X Model Z. That's cool, except your buddy has a completely different set of sound patches in her synth … and you want them! Sys-ex to the rescue! All you need to do is go over and transfer your buddy's patch data into your synth (to make life easier, make sure you take the instruction manual along).
- *Backing up your current patch data:* This can be done by transmitting a sys-ex dump of your synth's entire patch and setup data to disk, to a sys-ex utility program (often shareware) or to your DAW/MIDI sequencer. This is important: *Back up your factory preset or current patch data before attempting a sys-ex dump!* If you forget and download a sys-ex dump, your previous settings will be lost until you contact the manufacturer, download the dump from their website or take your synth back to your favorite music store to reload the data.
- *Getting patch data from the Web:* One of the biggest repositories of sys-ex data is the Internet. To surf the Web for sys-ex patch data, all you need to do is log on to your favorite search engine and enter the name of your synth. You'll probably be amazed at how many hits will come across the screen, many of which are chock-full of sys-ex dumps that can be downloaded into your synth.
- *Varying sys-ex controller or patch data in real time:* Patch editors or hardware MIDI controllers can be used to vary system and sound-generating parameters, in real time. Both of these controller types can ease the job of experimenting with parameter values or changing mix moves by giving you physical or on-screen controls that are often more intuitive and easier to deal with than programming electronic instruments that'll often leave you dangling in cursor and 3-inch LCD screen hell.

Before moving on, I should also point out that sys-ex data grabbed from the Web, disk, disc or any other medium will often be encoded using several sys-ex file format styles (unfortunately, none of which are standardized). Unfortunately, sequencer Y might not recognize a sys-ex dump that was encoded using sequencer Z. For this reason, dumps are often encoded using easily available, standard sys-ex utility programs for the Mac or PC.

At last, it seems that a single unified standard has begun to emerge from the fray that's so simple that it's amazing it wasn't universally adopted from the start. This system simply records a sys-ex dump as data on a single MIDI track file. Before recording a dump to a sequencer track, you may need to consult the manual to make sure that sys-ex filtering is turned off. Once this is done, simply place the track into record mode, initiate the dump and save the track in an all-important sys-ex dump directory. Using this approach, it would also be possible to:

- Import the appropriate sys-ex dump track (or set of tracks) into the current working session so as to automatically program the instruments before the sequence is played back.
- Import the appropriate sys-ex dump track (or set of tracks) into separate MIDI tracks that can be muted or unassigned. Should the need arise, the track(s) can be activated and/or assigned in order to dump the data into the appropriate instruments.

MIDI AND THE COMPUTER

Besides the coveted place of honor in which most electronic musicians hold their instruments, the most important device in a MIDI system is undoubtedly the personal computer. Through the use of software programs and peripheral hardware, the computer is often used to control, process and distribute information relating to music performance and production from a centralized, integrated control position.

Of course, two computer types dominate modern-day music production: the PC and the Mac. In truth, each brings its own particular set of advantages and disadvantages to personal computing, although their differences have greatly dwindled over the years. My personal take on the matter (a subject that's not even important enough to debate) is that it's a dual-platform world. The choice is yours and yours alone to make. … Many professional software and hardware systems can work on either platform. As I write this, some of my music collaborators are fully Mac, some are PC and some (like me) use both, and it doesn't affect our production styles at all. Coexistence isn't much of a problem, either. Living a dual-platform existence can give you the edge of being familiar with both systems, which can be downright handy in a sticky production pinch.

Connecting to the peripheral world

An important event in the evolution of personal computing has been the maturation of hardware and processing peripherals. With the development of the USB (www.usb.org) and FireWire (www.1394ta.org) protocols, hardware devices such as mice, keyboards, cameras, audio interfaces, MIDI interfaces, CD and hard drives, MP3 players and even portable fans can be plugged into an available port without any need to change frustrating hardware settings or open up the box. External peripherals are generally hardware devices that are designed to do a specific task or range of production tasks. For example, an audio interface is capable of translating analog audio (and often MIDI, control and other media) into digital data that can be understood by the computer. Other peripheral devices can perform such useful functions as printing, media interfacing (video and MIDI),

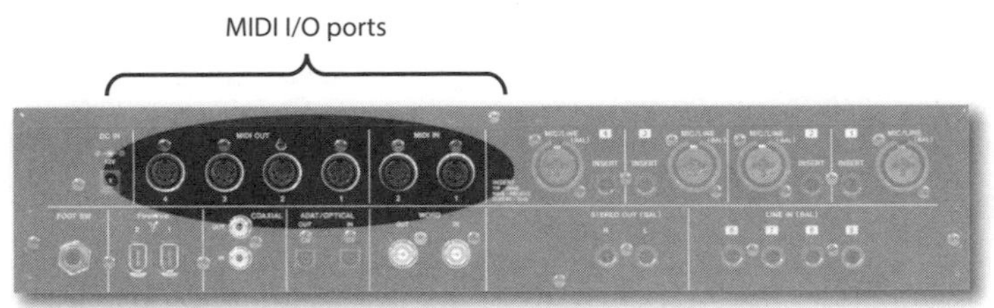

FIGURE 9.29
Many audio interface devices include MIDI I/O ports. (Courtesy of Tascam, www.tascam.com)

scanning, memory card interfacing, portable hard disk storage … the list could literally fill pages.

The MIDI interface

Although computers and electronic instruments both communicate using the digital language of 1's and 0's, computers simply can't understand the language of MIDI without the use of a device that translates the serial messages into a data structure that computers can comprehend. Such a device is known as the *MIDI interface*. A wide range of MIDI interfaces currently exist that can be used with most computer system and OS platforms. For the casual and professional musician, interfacing MIDI into a production system can be done in a number of ways. Probably the most common way to access MIDI In, Out and Thru jacks is on modern-day USB or FireWire audio interface or controller surface (Figure 9.29). It's become a common matter for portable devices to offer 16 channels of I/O (on one port), while multichannel interfaces often include multiple MIDI I/O ports that can give you access to 32 or more channels.

The next option is to choose a USB MIDI interface that can range from simpler devices that include a single port to multiple-port systems that can easily handle up to 64 channels over four I/O ports. The multiport MIDI interface (Figure 9.30) is often the device of choice for most professional electronic musicians who require added routing and synchronization capabilities. These USB devices can easily be ganged together to provide eight or more independent MIDI Ins and Outs to distribute MIDI data through separate lines over a connected network.

In addition to distributing MIDI data, these systems often include driver software that can route and process MIDI data throughout the MIDI network. For example, a multiport interface could be used to merge together several MIDI Ins (or Outs) into a single datastream, filter out specific MIDI message types (used to block out unwanted commands that might adversely change an instrument's sound or performance) or rechannel data being transmitted on one MIDI channel or port to another channel or port (thereby allowing the data to be recognized by an instrument or device).

Another important function that can be handled by most multiport interfaces is synchronization. Synchronization (sync, for short) allows other, external devices (such as DAWs, video decks and other media systems) to be simultaneously

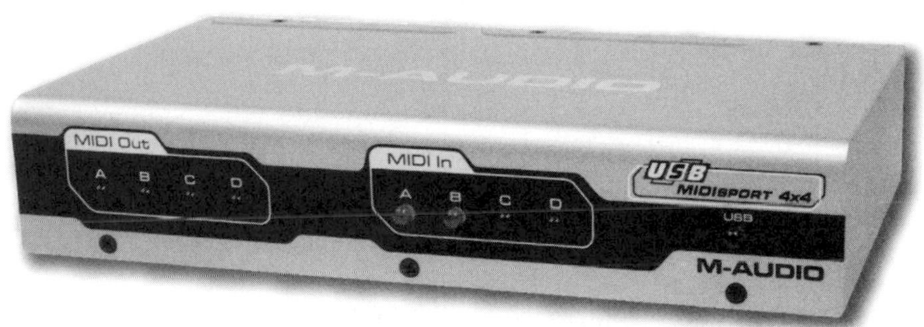

FIGURE 9.30
M-Audio MIDISPORT 4x4 MIDI interface. (Courtesy of M-Audio, a division of Avid Technology, Inc., www.m-audio.com)

played back using the same timing reference. Interfaces that includes sync features will often read and write SMPTE timecode, convert SMPTE to MIDI timecode (MTC) and allow recorded timecode signals to be cleaned up when copying code from one analog device to another (jam sync). Further reading on synchronization can be found in Chapter 11.

In addition to the above interface types, a number of MIDI keyboard controllers and synth instruments have been designed with MIDI ports and jacks built right into them. For those getting started, this useful and cost-saving feature makes it easy to integrate your existing instruments into your DAW and sequencing environment.

ELECTRONIC INSTRUMENTS

Since their inception in the early 1980s, MIDI-based electronic instruments have played a central and important role in the development of music technology and production. These devices (which fall into almost every instrument category), along with the advent of cost-effective analog and digital audio recording systems, have probably been the most important technological advances to shape the industry into what it is today. In fact, the combination of hardware and newer software plug-in technologies has turned the personal project studio into one of the most important driving forces behind modern-day music production.

Inside the toys

Although electronic instruments often differ from one another in looks, form and function, they almost always share a common set of basic building block components (Figure 9.31), including the following:

- *Central processing units (CPUs):* CPUs are one or more dedicated computing devices (often in the form of a specially manufactured microprocessor chip) that contain all of the necessary instructional brains to control the hardware, voice data and sound-generating capabilities of the entire instrument or device.
- *Performance controllers:* These include such interface devices as music keyboards, knobs, buttons, drum pads and/or wind controllers for inputting

performance data directly into the electronic instrument in real time or for transforming a performance into MIDI messages. Not all instruments have a built-in controller. These devices (commonly known as modules) contain all the necessary processing and sound-generating circuitry; however, the idea is to save space in a cramped studio by eliminating redundant keyboards or other controller surfaces.

- *Control panel:* The control panel is the all-important human interface of data entry controls and display panels that let you select and edit sounds and route and mix output signals, as well as control the instrument's basic operating functions.
- *Memory:* Digital memory is used for storing important internal data (such as patch information, setup configurations and/or digital waveform data). This digital data can be encoded in the form of either read-only memory (ROM; data that can only be retrieved from a factory-encoded chip, cartridge, or CD/DVD-ROM) or random access memory (RAM; memory that can be read from and stored to a device's resident memory, cartridge, hard disk or recordable media).
- *Voice circuitry:* Depending on the device type, this section can chain together digital processing "blocks" to either generate sounds (voices) or process and reproduce digital samples that are recorded into memory for playback according to a specific set of parameters. In short, it's used to generate or reproduce a sound patch, which can then be processed, amplified and heard via speakers or headphones.
- *Auxiliary controllers:* These are external controlling devices that can be used in conjunction with an instrument or controller. Examples of these include foot pedals (providing continuous-controller data), breath controllers, and pitch-bend or modulation wheels. Some of these controllers are continuous in nature, while others exist as a switching function that can be turned on and off. Examples of the latter include sustain pedals and vibrato switches.
- *MIDI communications ports:* These data ports and physical jacks are used to transmit and/or receive MIDI data.

Generally, no direct link is made between each of these functional blocks; the data from each of these components is routed and processed through the instrument's CPU. For example, should you wish to select a certain sound patch from the instrument's control panel, the control panel could be used to instruct the CPU to recall all of the waveform and sound-patch parameters from memory that are associated with the particular sound. These instructional parameters would then be used to modify the internal voice circuitry, so that when a key on the keyboard is pressed or a MIDI Note-On message is received, the sound generators will output the desired patch's note and level values.

Instrument plug-ins

In recent years, an almost staggering range of software instruments has come onto the market as instrument plug-ins. These systems, which include all known types of synths, samplers and pitch- and sound-altering devices, are able to

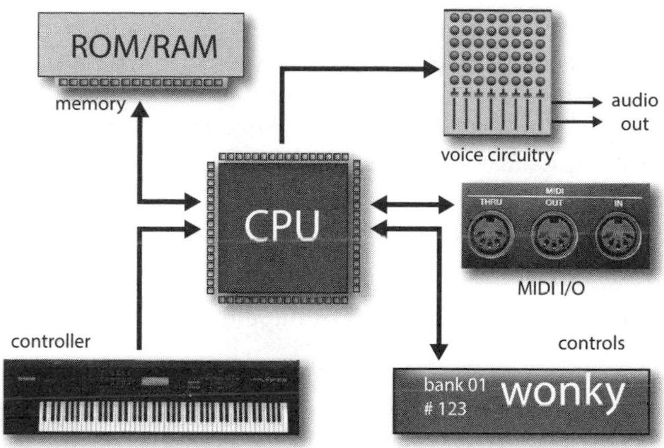

FIGURE 9.31
The basic building blocks of an electronic musical instrument.

communicate MIDI, audio, timing sync and control data between the software instrument (or effect plug-in) and a host DAW program/CPU processor.

Using an established plug-in communications protocol, it's possible for most or all of the audio and timing data to be routed through the host audio application, allowing the instrument or application I/O, timing and control parameters to be seamlessly integrated into the DAW or application. A few of these protocols include:

- Steinberg's VST
- Digidesign's RTAS
- MOTU's MAS

Propellerhead's ReWire is another type of protocol that allows audio, performance and control data of an independent audio program to be wired into a host program (usually a DAW) such that the audio routing and sync timing of the slave program is locked to the host DAW, effectively allowing them to work in tandem as a single production environment. Further reading on plug-in protocols and virtual and plug-in instruments can be found in Chapters 7 and 8.

For the remainder of this section, we'll be discussing the various types of MIDI instruments and controller devices that are currently available on the market. These instruments can be grouped into such categories as keyboards, percussion, controlling devices, MIDI guitars and strings and sequencers.

Keyboards

By far, the most common instruments that you'll encounter in almost any MIDI production facility will probably belong to the keyboard family. This is due, in part, to the fact that keyboards were the first electronic music devices to gain wide acceptance; also, MIDI was initially developed to record and control many of their performance and control parameters. The two basic keyboard-based instruments are the synthesizer and the digital sampler.

THE SYNTH

A synthesizer (or synth) is an electronic instrument that uses multiple sound generators, filters and oscillator blocks to create complex waveforms that can be combined into countless sonic variations. These synthesized sounds have become a basic staple of modern music and range from those that sound "cheesy" to ones that realistically mimic traditional instruments … and all the way to those that generate otherworldly, ethereal sounds that literally defy classification.

Synthesizers generate sounds using a number of different technologies or program algorithms. Examples of these include:

- *FM synthesis:* This technique generally makes use of at least two signal generators (commonly referred to as "operators") to create and modify a voice. It often does this by generating a signal that modulates or changes the tonal and amplitude characteristics of a base carrier signal. More sophisticated FM synths use up to four or six operators per voice, each using filters and variable amplifier types to alter a signal's characteristics.
- *Wavetable synthesis:* This technique works by storing small segments of digitally sampled sound into a memory media. Various sample-based and synthesis techniques make use of looping, mathematical interpolation, pitch shifting and digital filtering to create extended and richly textured sounds that use a surprisingly small amount of sample memory, allowing hundreds if not thousands of samples and sound variations to be stored in a single device or program.
- *Additive synthesis:* This technique makes use of combined waveforms that are generated, mixed and varied in level over time to create new timbres that are composed of multiple and complex harmonics. Subtractive synthesis makes extensive use of filtering to alter and subtract overtones from a generated waveform (or series of waveforms).

Of course, synths come in all shapes and sizes and use a wide range of patented synthesis techniques for generating and shaping complex waveforms, in a polyphonic fashion using 16, 32 or even 64 simultaneous voices (Figures 9.32 and 9.33). In addition, many synths often include a percussion section that can play a full range of drum and "perc" sounds, in a number of styles. Reverb and other basic effects are also commonly built into the architecture of these devices, reducing the need for using extensive outboard effects when being played on-stage or out of the box. Speaking of "out of the box," a number of synth systems are referred to as being "workstations." Such beasties are designed (at least in theory) to handle many of your basic production needs (including basic sound generation, MIDI sequencing, effects, etc.)—all in one neat little package.

SAMPLERS

A sampler (Figures 9.34 through 9.36) is a device that can convert audio into a digital form that is then imported into internal random access memory (RAM).

FIGURE 9.32
Dave Smith Instruments Mopho x4 4-voice analog synth. (Courtesy of Dave Smith Instruments, www.davesmithinstruments.com)

FIGURE 9.33
iZotope Iris Synth Plug-in. (Courtesy of iZotope, Inc, www.izotope.com)

FIGURE 9.34
Akai MPC1000 Music Production Center. (Courtesy of Akai Professional, www.akaipro.com)

FIGURE 9.35 Kontact Virtual Sampler. (Courtesy of Native Instruments GmbH, www.native-instruments.com)

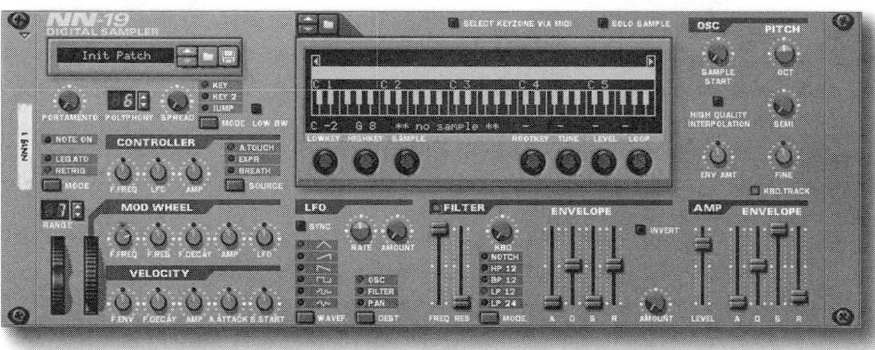

FIGURE 9.36 Reason's NN-19 and NN-XT sample modules. (Courtesy of Propellerhead Software, www.propellerheads.se)

Once audio has been sampled or loaded into RAM (from disk, disc or diskette), segments of sampled audio can then be edited, transposed, processed and played in a polyphonic, musical fashion. In short, a sampler can be thought of as a digital audio memory device that lets you record, edit and reload samples into RAM. Once loaded, these sounds (whose length and complexity are often only limited by memory size and your imagination) can be looped, modulated, filtered and amplified (according to user or factory setup parameters) in a way that allows the waveshapes and envelopes to be modified. Signal processing capabilities, such as basic editing, looping, gain changing, reverse, sample-rate conversion, pitch change and digital mixing can also be easily applied to change the sounds in an almost infinite number of ways.

Samplers derive their sounds from recorded and/or imported audio data stored as digital audio data within a personal computer or other media storage device. Using its digital signal processing (DSP) capabilities, most software samplers are able to store and access samples within the internal memory to:

- Import previously recorded sound files (often in .wav, .aif and other common formats)
- Edit and loop sounds into a usable form
- Vary envelope parameters (e.g., dynamics over time)
- Vary processing parameters
- Save the edited sample performance setup as a file for later recall

A sample can be played according to the standard Western musical scale (or any other scale, for that matter) by altering the playback sample rate over the controller's note range. For example, pressing a low-pitched key on the keyboard will cause the sample to be played back at a lower sample rate, while pressing a high-pitched one will cause the sample to be played back at rates that would put Mickey Mouse to shame. By choosing the proper sample–rate ratios, these sounds can be polyphonically played (whereby multiple notes are sounded at once) at pitches that correspond to standard musical chords and intervals.

A sampler (or synth) with a specific number of voices (e.g., 64 voices) simply means that up to 64 notes can be simultaneously played on a keyboard at any one time. Each sample in a multiple-voice system can be assigned across a performance keyboard, using a process known as splitting or mapping. In this way, a sound can be assigned to play across the performance surface of a controller over a range of notes, known as a zone (Figure 9.37). In addition to grouping samples into various zones, velocity can enter into the equation by allowing

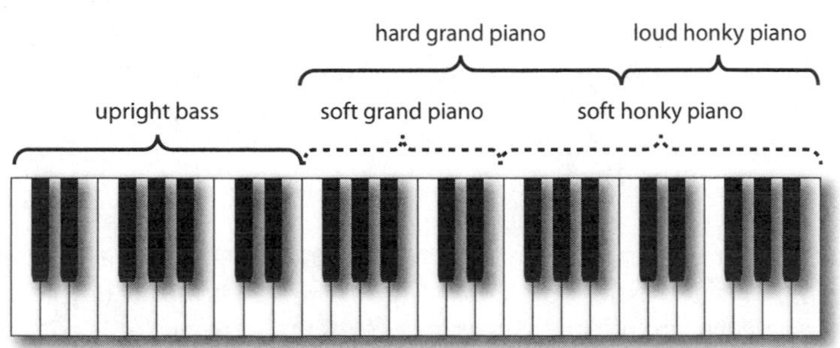

FIGURE 9.37
Example of a sampler's keyboard layout that has been programmed to include zones. Notice that the upper register has been split into several zones that are triggered by varying velocities.

multiple samples to be layered across the same keys of a controller, according to how soft or hard they are played. For example, a single key might be layered so that pressing the key lightly would reproduce a softly recorded sample, while pressing it harder would produce a louder sample with a sharp, percussive attack. In this way, mapping can be used to create a more realistic instrument or wild set of soundscapes that change not only with the played keys but with different velocities as well. Most samplers have extensive edit capabilities that allow the sounds to be modified in much the same way as a synthesizer, using such modifiers as:

- Velocity
- Panning
- Expression (modulation and user control variations)
- Low-frequency oscillation (LFO)
- Attack, delay, sustain and release (ADSR) and other envelope processing parameters
- Keyboard scaling
- Aftertouch

Many sampling systems will often include such features as integrated signal processing, multiple outputs (offering isolated channel outputs for added live mixing and signal processing power or for recording individual voices to a multitrack recording system) and integrated MIDI sequencing capabilities.

Sample DVDs and the Web

Just as patch data in the form of sys-ex dump files can have the effect of breathing new life into your synth, a wide range of free or commercially available samples is commonly available online or from a business entity that lets you experiment with loading new and fresh sounds into your production system. These files can exist as unedited sound file data (which can be imported into any sample system or DAW track), or as data that has been specifically programmed by a professional musician/programmer to contain all the necessary loops, system commands and sound-generating parameters, so that all you ideally need to do is load the sample and start having fun.

The mind boggles at the range of styles and production quality that has gone into producing samples that are just ready and waiting to give your project a boost. The general production level literally runs the entire amateur-to-pro gamut—meaning that, whenever possible, it's wise to listen to examples to determine their quality and to hear how they might fit into your own personal or project style before you buy. As a final caveat … by now, you've probably heard of the legal battles that have been raging over sampled passages that have been "ripped" from recordings of established artists. In the fall of 2004, in the case of *Bridgeport Music et al. v. Dimension Films*, the 6th Circuit U.S. Court of Appeals ruled that the digital sampling of a recording without a license is a violation of copyright, regardless of size or significance. This points to the need for tender loving care when lifting samples off a CD, DVD or the Web.

The MIDI keyboard controller

As computers, sound modules, virtual software instruments and other types of digital devices have come onto the production scene, it's been interesting to note that fewer and fewer devices are being made that include a music keyboard in their design. As a result, the MIDI keyboard controller has gained in popularity as a device that might include a:

- Music keyboard surface
- Variable parameter controls
- Fader, mixing and transport controls
- Switching controls
- Tactile trigger and control surfaces

A Word About Controllers

A MIDI controller is a device that's expressly designed to control other devices (be they for sound, light or mechanical control) within a connected MIDI system. As was previously mentioned, these devices contain no internal tone generators or sound-producing elements but often include a high-quality control surface and a wide range of controls for handling control, trigger and device-switching events. Since controllers have become an integral part of music production and are available in many incarnations to control and emulate many types of musical instrument types, don't be surprised to find controllers of various incarnations popping up all over this book and within electronic music production.

As was stated, these devices contain no internal tone generators or sound-producing elements. Instead they can be used in the studio or on the road as a simple and straightforward surface for handling MIDI performance, control and device-switching events in real time.

As you might imagine, controllers vary widely in the number of features that are offered (Figures 9.38 and 9.39). For starters, the number of keys can vary from the sporty, portable 25-key models to those having 49 and 61 keys and all the way to the full 88-key models that can play the entire range of a full-size grand piano. The keys may be fully or partially weighted and in a number of models the keys might be much smaller than the full piano key size—often making a performance a bit difficult. Beyond the standard pitch and modulation wheels (or similar-type controller), the number of options and general features is up to the manufacturers. With the increased need for control over electronic instruments and music production systems, many model types offer up a wide range of physical controllers for varying an ever-widening range of expressive parameters.

FIGURE 9.38
Novation 25SL USB MIDI Controller. (Courtesy of Novation Digital Music Systems, Ltd., www.novationmusic.com)

FIGURE 9.39
Yamaha DS6IV PRO Disclavier MIDI piano. (Courtesy of Yamaha Corporation of America, www.yamaha.com)

Percussion

Historically speaking, one of the first applications of sample technology made use of digital audio to record and play back drum and percussion sounds. Out of this virtual miracle sprang a major class of sample and synthesis technology that lets an artist (mostly keyboard players) add drum and percussion to their own compositions with relative ease. Over the years, MIDI has brought sampled percussion within the grasp of every electronic musician, whatever skill level, from the frustrated drummer to professional percussionist/programmers—all of whom use their skills to perform live and/or to sequenced compositions.

THE DRUM MACHINE

In its most basic form, the drum machine uses ROM-based, prerecorded waveform samples to reproduce high-quality drum sounds from its internal memory. These factory-loaded sounds often include a wide assortment of drum sets, percussion sets and rare, wacky percussion hits, and effected drum sets (e.g., reverberated, gated). Who knows—you might even encounter scream hits by the venerable King of Soul, James Brown. These prerecorded samples can be assigned to a series of playable keypads that are generally located on the machine's top face, providing a straightforward controller surface that often sports velocity and aftertouch dynamics. Sampled voices can be assigned to each pad and edited using control parameters such as tuning, level, output assignment and panning position.

Because of new cost-effective technology, many drum machines now include basic sampling technology, which allows sounds to be imported, edited and triggered directly from the box (Figures 9.40 and 9.41). As with the traditional "beat box," these samples can be easily mapped and played from the traditionally styled surface trigger pads. Of course, virtual software drum and groove machines are part of the present-day landscape and can be used in a stand-alone, plug-in and rewired production environment.

MIDI DRUM CONTROLLERS

MIDI drum controllers are used to translate the voicing and expressiveness of a percussion performance into MIDI data. These devices are great for capturing

FIGURE 9.40
Alesis SR-18 stereo drum machine. (Courtesy of Alesis, www.alesis.com)

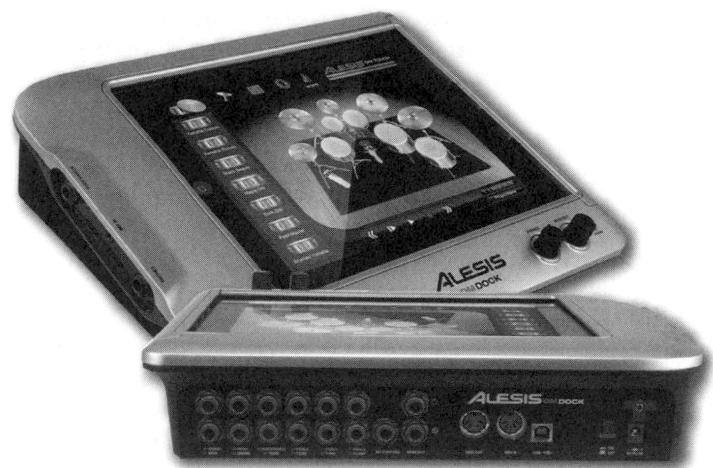

FIGURE 9.41
Alesis DM Dock iPad-based drum module. (Courtesy of Alesis, www.alesis.com)

the feel of a live performance, while giving you the flexibility of automating or sequencing a live event. These devices range from having larger pads and trigger points on a larger performance surface to drum-machine-type pads/buttons. Coming under the "Don't try this at home" category, these controller pads are generally too small and not durable enough to withstand drumsticks or mallets. For this reason, they're generally played with the fingers. It's long been a popular misconception that MIDI drum controllers have to be expensive. This simply isn't true. There are quite a few instruments that are perceived by many to be toys but, in fact, are fully implemented with MIDI and can be easily used as a controller. A few of the ways to perform and sequence percussion include:

- *Drum machine button pads:* One of the most straightforward of all drum controllers is the drum button pad design that's built into most drum machines, portable percussion controllers (Figure 9.42) and certain keyboard controllers. By calling up the desired setup and voice parameters, these small footprint triggers let you go about the business of using your fingers to do the walking through a performance or sequenced track.
- *The keyboard as a percussion controller:* Since drum machines respond to external MIDI data, probably the most commonly used device for triggering percussion and drum voices is a standard MIDI keyboard controller. One advantage of playing percussion sounds from a keyboard is that sounds can be triggered more quickly because the playing surface is designed for fast finger movements and doesn't require full hand/wrist motions. Another advantage is its ability to express velocity over the entire range of possible values (0–127), instead of the limited number of velocity steps that are available on certain drum pad models.
- *Drum pad controllers:* In more advanced MIDI project studios or live stage rigs, it's often necessary for a percussionist to have access to a playing surface that can be played like a real instrument. In these situations, a dedicated drum pad controller would be better for the job. Drum controllers

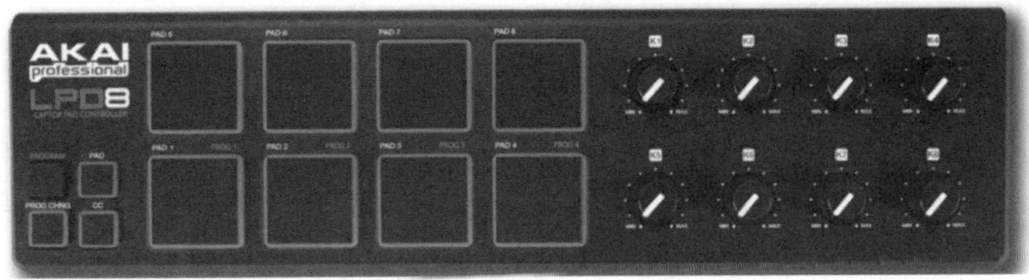

FIGURE 9.42 Akai LPD8 portable USB Drum Controller. (Courtesy Akai Professional, LP. www.akaipro.com)

vary widely in design. They can be built into a single, semiportable case, often having between six and eight playing pads, or the trigger pads can be individual pads that can be fitted onto a special rack, traditional drum floor stand or drum set.

- *MIDI drums:* Another way to MIDI-fy an acoustic drum is through the use of trigger technology. Put simply, triggering is carried out by using a transducer pickup (such as a mic or contract pickup) to change the acoustic energy of a percussion or drum instrument into an electrical voltage. Using a MIDI trigger device (Figure 9.43), a number of pickup inputs can be translated into MIDI so as to trigger programmed sounds or samples from an instrument for use on stage or in the studio.

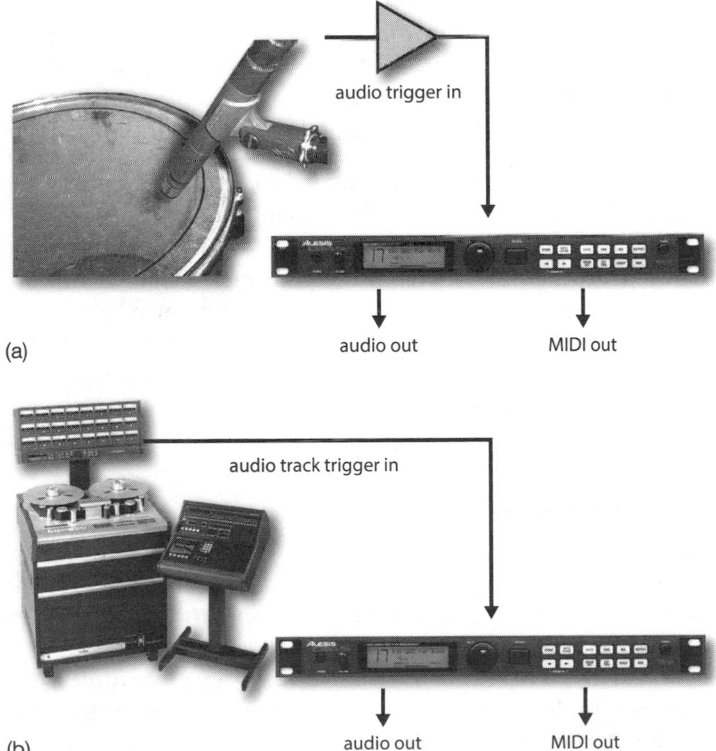

FIGURE 9.43 By using a MIDI trigger device (such as an Alesis DM5), a pickup can be either directly replaced or sent via MIDI to another device or sequenced track (a) using a mic/instrument source or (b) using a recorded track as a source. (Courtesy of Alesis, www.alesis.com).

OTHER MIDI INSTRUMENT AND CONTROLLER TYPES

There are literally tons of instruments and controller types out there that are capable of translating a performance or general body movements into MIDI. You'd be surprised what you'll find searching the Web for wild and wacky controllers—both those that are commercially available and those that are made by soldering iron junkies. A few of the more traditional controllers include MIDI guitars and basses, wind controllers, MIDI vibraphones … the list goes on.

SEQUENCING

Apart from our computers, DAWs and venerable electronic instruments, one of the most important tools that can be found in the modern-day project studio is the MIDI sequencer. Basically, a sequencer is a digital device or software application that's used to record, edit and output MIDI messages in a sequential fashion. These messages are generally arranged in a track-based format that follows the modern production concept of having instruments (and/or instrument voices) located on separate tracks. This traditional interface makes it easy for us humans to view MIDI data as tracks on a digital audio workstation (DAW) or analog tape recorder that follow along a straightforward linear time line.

These tracks contain MIDI-related performance and control events that are made up of such channel and system messages as Note-On, Note-Off, Velocity, Modulation, Aftertouch and Program/Continuous Controller messages. Once a performance has been recorded into a sequencer's memory, these events can be graphically arranged and edited into a musical performance. The data can then be saved as a file or within a DAW session and recalled at any time, allowing the data to be played back in its originally recorded or edited order.

Because most sequencers are designed to function in a way that's similar to the multitrack tape recorder, they give us a familiar operating environment in which each instrument, set of layered instruments or controller data can be recorded onto separate, synchronously arranged tracks. Like its multitrack cousin, each track can be re-recorded, erased, copied and varied in level during playback. However, since the recorded data is digital in nature, a MIDI sequencer is far more flexible in its editing speed and control in that it offers all the cut-and-paste, signal processing and channel-routing features that we've come to expect from a digital production application.

Integrated workstation sequencers

A type of keyboard synth and sampler system known as a keyboard workstation will often include much of the necessary production hardware that's required for music production, including effects and an integrated hardware sequencer. These systems have the advantage of letting you take your instrument and sequencer on the road without having to drag your whole system along. Similar to the stand-alone hardware sequencer, a number of these sequencer systems have the disadvantage of offering few editing tools beyond transport functions,

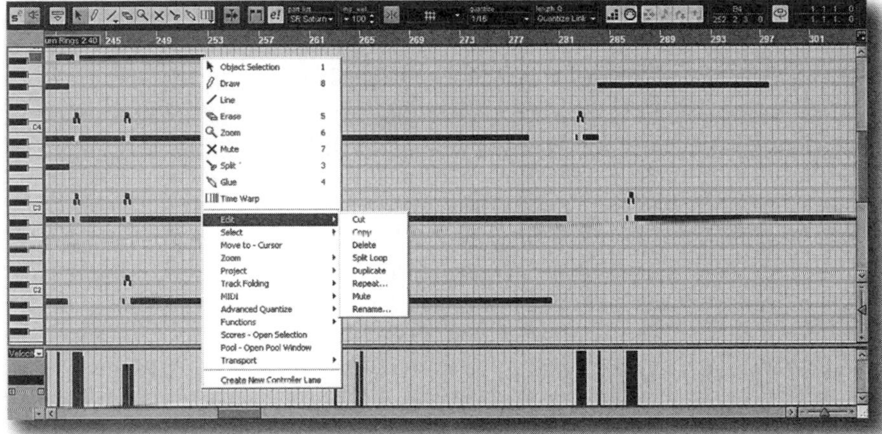

FIGURE 9.44
MIDI edit window within Cubase audio production software. (Courtesy of Steinberg Media Technologies GmbH, a division of Yamaha Corporation, www.steinberg.net)

punch-in/out commands, and other basic edit functions. With the advent of more powerful keyboard systems that include a larger, integrated LCD display, the integrated sequencers within these systems are becoming more powerful, resembling their software sequencing counterparts. In addition, other types of palm-sized sequencers offer such features as polyphonic synth voices, drum machine kits, effects, MIDI sequencing and, in certain cases, facilities for recording multitrack digital audio in an all-in-one package that literally fits in the palm of your hand!

Software sequencers

By far, the most common sequencer type is the software sequencing program (Figure 9.44). These programs run on all types of personal and laptop computers and take advantage of the hardware and software versatility that only a computer can offer in the way of speed, hardware flexibility, memory management, signal routing and digital signal processing. Software sequencers offer several advantages over their hardware counterparts. Here are just a few highlights:

- Increased graphics capabilities (giving us direct control over track and transport-related record, playback, mix and processing functions)
- Standard computer cut-and-paste edit capabilities
- Ability to easily change note and controller values, one note at a time or over a defined range
- A window-based graphic environment (allowing easy manipulation of program and edit-related data)
- Easy adjustment of performance timing and tempo changes within a session
- Powerful MIDI routing to multiple ports within a connected system
- Graphic assignment of instrument voices via Program Change messages
- Ability to save and recall files using standard computer memory media

Basic introduction to sequencing

When dealing with any type of sequencer, one of the most important concepts to grasp is that these devices don't store sound directly; instead, they encode MIDI messages that instruct instruments as to what note is to be played, over what channel, at what velocity and at what, if any, optional controller values. In other words, a sequencer simply stores command instructions that follow in a sequential order. These instructions tell instruments and/or devices how their voices are to be played or controlled. This means that the amount of encoded data is a great deal less memory intensive than its digital audio or digital video recording counterparts. Because of this, the data overhead that's required by MIDI is very small, allowing a computer-based sequencer to work simultaneously with the playback of digital audio tracks, video images, Internet browsing, etc., all without unduly slowing down the computer's CPU. For this reason, MIDI and the MIDI sequencer provide a media environment that plays well with other computer-based production media.

RECORDING

Commonly, a MIDI sequencer is an application within a digital production workspace for creating personal compositions in environments that range from the bedroom to more elaborate professional and project studios. Whether hardware or software based, most sequencers use a working interface that's roughly designed to emulate a traditional multitrack-based environment. A tape-like set of transport controls lets us move from one location to the next using standard Play, Stop, Fast Forward, Rewind and Record command buttons. Beyond using the traditional Record-Enable button to select the track or tracks that we want to record onto, all we need to do is select the MIDI input (source) port, output (destination) port, MIDI channel, instrument patch and other setup requirements. Then press the record button and begin laying down the track.

Tutorial: Setting Up a Session and Laying Down a MIDI Track

1. Pull out a favorite MIDI instrument or call up a favorite plug-in instrument.
2. Route the instrument's MIDI and audio cables (or plug-in routing) to their appropriate workstation and audio mixer devices according to the situation at hand.
3. Call up a MIDI sequencer (either by beginning a DAW session or by using a hardware sequencer) and create a MIDI track that can be recorded to.
4. Set the session to a tempo that feels right for the song.
5. Assign the track's MIDI input to the port that's receiving the incoming MIDI data.
6. Assign the track's MIDI output to the port and proper MIDI channel that'll be receiving the outgoing MIDI data during playback.
7. If a click track is desired, turn it on.
8. Name the track. This will make it easier to identify the MIDI passage in the future.
9. Place the track into the Record Ready mode.
10. Play the instrument or controller. Can you hear the instrument? Do the MIDI activity indicators light up on the sequencer, track and MIDI interface? If not, check your cables and run through the checklist again. If so, press Record and start laying down your first track.

Once you've finished laying down a track, you can jump back to the beginning of the recorded passage and listen to it. From this point, you could then "arm" (a term used to denote placing a track into the record-ready mode) the next track and go about the process of laying down additional tracks until a song begins to form.

Setting the session tempo

When beginning a MIDI session, one of the first aspects to consider is the tempo and time signature. The beats-per-minute (bpm) value will set the general tempo speed for the overall session. This is important to set at the beginning of the session, so as to lock the overall "bars and beats" timing elements to this initial speed that's often essential in electronic music production. This tempo/click element can then be used to lock the timing elements of other instruments and/or rhythm machines to the session (e.g., a drum machine plug-in can be pulled into the session that'll automatically be locked to the session's speed and timing).

> To avoid any number of unforeseen obstacles to a straightforward production, it's often wise to set your session tempo before pressing any record buttons.

Changing tempo

The tempo of a MIDI production can often be easily changed without worrying about changing the program's pitch or real-time control parameters. In short, once you know how to avoid potential conflicts and pitfalls, tempo variations can be made after the fact with relative ease. Basically, all you need to do is alter the tempo of a sequence (or part of a sequence) to best match the overall feel of the song. In addition, the tempo of a session can be dynamically changed over its duration by creating a tempo map that causes the speed to vary by defined amounts at specific points within a song. Care and preplanning should be exercised when a sequence is synced to another media form or device.

Click track

When musical timing is important (as is often the case in modern music and visual media production), a click track can be used as a tempo guide for keeping the performance as accurately on the beat as possible. A click track can be set to make a distinctive sound on the measure boundary or (for a more accurate timing guide) on the first beat boundary and on subsequent meter divisions (e.g., tock, tick, tick, tick, tock, tick, tick, tick, ...). Most sequencers can output a click track by either using a dedicated beep sound (often outputting from the device or main speakers) or by sending Note-On messages to a connected instrument in the MIDI chain. The latter lets you use any sound you want and often at definable velocity levels. For example, a kick could sound on the beat, while a snare sounds out the measure divisions. A strong reason for using a click track (at least initially) is that it serves as a rhythmic guide that can improve the timing accuracy of a performance.

Care should be taken when setting the proper time signature at the session's outset. Listening to a 4/4 click can be disconcerting when the song is being performed in 3/4 time.

The use of a click track is by no means a rule. In certain instances, it can lead to a performance that sounds stiff. For compositions that loosely flow and are legato in nature, a click track can stifle the passage's overall feel and flow. As an alternative, you could turn the metronome down, have it sound only on the measure boundary, and then listen through one headphone. As with most creative decisions, the choice is up to you and your current circumstance.

Multiple track recording

Although only one MIDI track is commonly recorded at a time, most mid- and professional-level sequencers allow us to record multiple tracks at one time. This feature makes it possible for a multi-trigger instrument or for several performers to record to a sequence in one, live pass. For example, such an arrangement would allow for each trigger pad of a MIDI drum controller to be recorded to its own track (with each track being assigned to a different MIDI channel on a single port). Alternatively, several instruments of an on-stage electronic band could be captured to a sequence during a live performance and then laid into a DAW session for the making of an album project.

Punching in and out

Almost all sequencers are capable of punching in and out of record while playing a sequence (Figure 9.45). This commonly used function lets you drop in and out of record on a selected track (or series of tracks) in real time, in a way that mimics the traditional multitrack overdubbing process. Although punch-in and

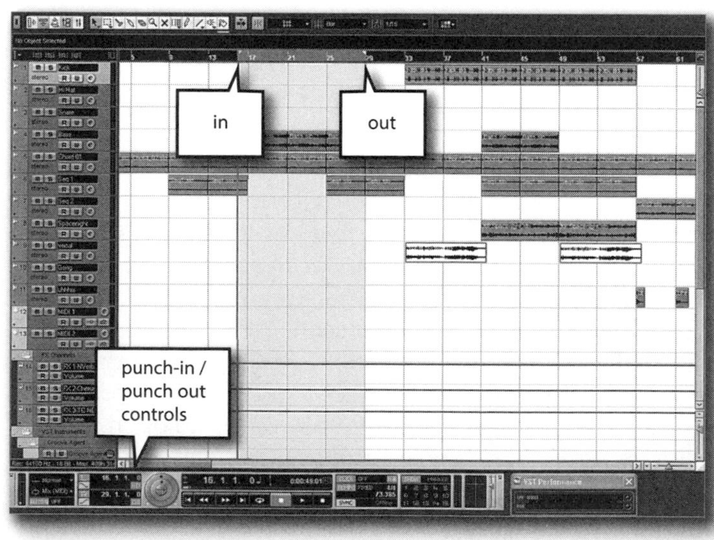

FIGURE 9.45
Steinberg Cubase/Nuendo punch-in and punch-out controls. (Courtesy of Steinberg Media Technologies GmbH, a division of Yamaha Corporation, www.steinberg.net)

punch-out points can often be manually performed on the fly, most sequencers can automatically perform a punch by graphically or numerically entering in the measure/beat points that mark the punch-in and punch-out location points. Once done, the sequence can be rolled back to a point a few measures before the punch-in point and the artist can play along while the sequencer automatically performs the necessary switching functions.

Tutorial: Punching During a Take

1. Set up a track and record a musical passage. Save the session to disk.
2. Roll back to the beginning of the take and play along. Manually punch in and out during a few bars. Was that easy or difficult? Now, undo or revert back to your originally saved session.
3. Read your DAW/sequencer manual and learn how to perform an automated punch (placing your punch-in and punch-out points at the same place as before).
4. Roll back to a point a few bars before the punch and go into the record mode. Did the track automatically place itself into record? Was that easier than doing it manually?
5. Feel free to try other features, such as record looping or stacking.

Step time entry

In addition to laying down a performance track in real time, most sequencers will allow us to enter note values into a sequence one note at a time. This feature (known as step time, step input or pattern sequencing) makes it possible for notes to be entered into a sequence without having to worry about the exact timing. Upon playback, the sequenced pattern will play back at the session's original tempo. Fact is, step entry can be an amazing tool, allowing a difficult or a blazingly fast passage to be meticulously entered into a pattern and then be played out or looped with a degree of technical accuracy that would otherwise be impossible for most of us to play. Quite often, this data entry style is used with fast, high-tech musical styles where real-time entry just isn't possible or accurate enough for the song.

Tutorial: Is It Live or Is It MIDI?

For those who want to dive into a whole new world of experimentation and sonic surprises, here's a new adage: "If the instrument supports MIDI, record the performance data to a DAW track." For example:

- You might record the MIDI out of a keyboard performance to a DAW MIDI track. If there's a problem in the performance you can simply change the note and re-record it—or, if you want to change the sound, simply pick another sound.
- Record the sequenced track from a triggered drum set or controller to a DAW MIDI track. If you do a surround mix at a later time, simply re-record each drum part to new tracks and pan the drums around the soundfield.
- If a MIDI guitar riff needs some tweaking to fill out the sound, double the MIDI track an octave down.
- Acoustic drum recordings can benefit from MIDI by using a trigger device that can accept audio from a mic or recorded track and output the triggered MIDI messages to a sampler or instrument, in order to replace the bad tracks with samples that rock the house. By the way, don't forget to record the trigger outputs to a MIDI track on a DAW or sequencer, just in case you want to edit or change the sounds at a later time. Alternatively, you could even use replacement triggering software to replace or augment poorly recorded acoustic tracks with killer sounds.

It's all up to you ... as you might imagine, surprises can definitely come from experiments like these.

SAVING YOUR MIDI FILES

Just as it's crucial that we carefully and methodically back up our program and production media, it's important that we save our MIDI and session files while we're in production. This can be done in two ways:

- Update your files by periodically saving files over the course of a production. A number of programs can be set up to automatically perform this function at regular intervals.
- At important points throughout a production, you might choose to save your session files under new and unique names, thereby making it possible to easily revert back to a specific point in the production. This can be an important recovery tool should the session take a wrong turn or for re-creating a specific effect and/or mix.

When working with MIDI within a DAW session, it's ALWAYS a good idea to save the original MIDI tracks within the session. This makes it easy to go back and change a note, musical key or sounding voice or to make any other alterations you want. Not saving these files could lead to some major headaches or worse.

MIDI files can also be converted to and saved as a standard MIDI file for use in exporting to and importing from another MIDI program or for distributing MIDI data for use on the Web, in cell phones, etc. These files can be saved in either of two formats:

- *Type 0:* Saves all of the MIDI data within a session as a single MIDI track. The original MIDI channel numbers will be retained. The imported data will simply exist on a single track. Note that if you save a multi-instrument session as a Type 0, you'll loose the ability to save the MIDI data to discrete tracks within the saved sequence.
- *Type 1:* Saves all of the MIDI data within a session onto separate MIDI tracks that can be easily imported into a sequencer in a multitrack fashion.

EDITING

One of the more important features that a sequencer (or MIDI track within a DAW) has to offer is its ability to edit sequenced tracks or blocks within a track. Of course, these editing functions and capabilities often vary from one sequencer to another. The main track window of a sequencer or MIDI track on a DAW is used to display such track information as the existence of track data, track names, MIDI port assignments for each track, program change assignments, volume controller values and other transport commands.

Depending on the sequencer, the existence of MIDI data on a particular track at a particular measure point (or over a range of measures) is indicated by the highlighting of a track range in a way that's extremely visible. For example, in Figure 9.46, you'll notice that the gray MIDI tracks contain graphical bar display

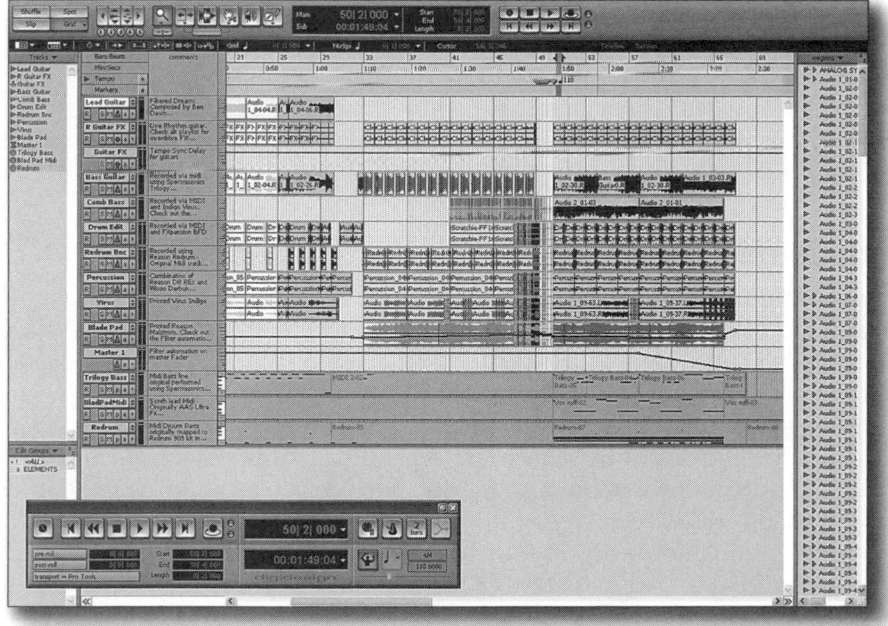

FIGURE 9.46 ProTools showing the main edit screen. (Courtesy of Avid Technology, Inc., www.avid.com)

information. This means that these measures contain MIDI messages, while the non-highlighted areas don't.

By navigating around the various data display and parameter boxes, it's possible to use cut-and-paste and/or edit techniques to vary note values and parameters for almost every facet of a section or musical composition. For example, let's say that we really screwed up a few notes when laying down an otherwise killer bass riff. With MIDI, fixing the problem is totally a no-brainer. Simply highlight each fudged note and drag it to its proper note location—we can even change the beginning and endpoints in the process. In addition, tons of other parameters can be changed, including velocity, modulation and pitch bend, note and song transposition, quantization and humanizing (factors that eliminate or introduce human timing errors that are generally present in a live performance), in addition to full control over program and continuous controller messages ... the list goes on.

Practical editing techniques

When it comes to learning the Ins, Outs and Thrus of basic sequencing, absolutely nothing can take the place of diving in and experimenting with your setup. Here, I'd like to quote Craig Anderton, who said: "Read the manual through once when you get the program (or device), then play with the software and get to know it before you need it. Afterwards, reread the manual to pick up the system's finer operating points." Wise words!

In this section, we'll be covering some of the basic techniques that'll speed you on your way to sequencing your own music. Note that there are no rules to sequencing MIDI. As with all of music production (and the arts, for that matter), there are as many right ways to perform and sequence music as there are musicians. Just remember that there are no hard and steadfast rules to music production—but there are definitely guidelines, tools and tips that can speed and improve the process.

TRANSPOSITION

As was mentioned earlier, a sequencer app is capable of altering individual notes in a number of ways including pitch, start time, length and controller values. In addition, it's generally a simple matter for a defined range of notes in a passage to be altered as a single entity in ways that could alter the overall key, timing and controller processes. Changing the pitch of a note or the entire key of a song is extremely easy to do with a sequencer. Depending on the system, a song can be transposed up or down in pitch at the global level, thereby affecting the musical key of a song. Likewise, a segment can be shifted in pitch from the main edit, piano roll or notation edit windows by simply highlighting the bars and tracks that are to be changed and then calling up the transpose function from the program menu.

QUANTIZATION

By far, most common timing errors begin with the performer. Fortunately, "to err is human," and standard performance timing errors often give a piece a live and natural feel. However, for those times when timing goes beyond the bounds of nature, an important sequencing feature known as quantization can help correct these timing errors. Quantization allows timing inaccuracies to be adjusted to the nearest desired musical time division (such as a quarter, eighth, or sixteenth note). For example, when performing a passage where all involved notes must fall exactly on the quarter-note beat, it's often easy to make timing mistakes (even on a good day). Once the track has been recorded, the problematic passage can be highlighted and the sequencer can recalculate each note's start and stop times so they fall precisely on the boundary of the closest time division.

HUMANIZING

The humanization process is used to randomly alter all of the notes in a selected segment according to such parameters as timing, velocity and note duration. The amount of randomization can often be limited to a user-specified value or percentage range, and parameters and can be individually selected or fine-tuned for greater control. Beyond the obvious advantages of reintroducing human-like timing variations back into a track, this process can help add expression by randomizing the velocity values of a track or selected tracks. For example, humanizing the velocity values of a percussion track that has a limited dynamic range can help bring it to life. The same type of life and human swing can be effective on almost any type of instrument.

SLIPPING IN TIME

Another timing variable that can be introduced into a sequence to help change the overall feel of a track is the slip time feature. Slip time is used to move a selected range of notes either forward or backward in time by a defined number of clock pulses. This has the obvious effect of changing the start times for these notes, relative to the other notes or timing elements in a sequence.

Editing controller values

Almost every sequencer package allows controller message values to be edited or changed, and they often provide a simple, graphic window whereby a line or freeform curve can be drawn that graphically represents the effect that relevant controller messages will have on an instrument or voice. By using a mouse or other input device, it becomes a simple matter to draw a continuous stream of controller values that correspondingly change such variables as velocity, modulation, pan, etc. To physically change parameters, all you need to do is twiddle a knob, move a fader or graphically draw the variables on-screen in a WYSIWYG ("what you see is what you get") fashion.

Some of the more common controller values that can affect a sequence and/or MIDI mix values include (for the full listing of controller ID numbers, see Table 9.1 earlier in this chapter):

Control #	Parameter
1	Modulation Wheel
2	Breath Controller
4	Foot Controller
7	Channel Volume (formerly Main Volume)
8	Balance
10	Pan
11	Expression Controller
64	Damper Pedal on/off (Sustain) <63 off, >64 on

It almost goes without saying that a range of controller events can be altered on one or more tracks by allowing a range of MIDI events to be highlighted and then altered by entering in a parameter or processing function from an edit dialog box. This ability to define a range of events often comes in handy for making changes in pitch/key, velocity, main volume and modulation (to name a few).

Tutorial: Changing Controller Values

1. Read your DAW/sequencer manual and learn its basic controller editing features.
2. Open or create a MIDI track.
3. Select a range of measures and change their Channel Volume settings (controller 7) over time. Does the output level change over the segment?
4. Highlight the segment and reduce the overall Channel Volume levels by 25%.
5. Take the segment's varying controller settings and set them to an absolute value of 95. Did that eliminate the level differences?
6. Now, refer to your DAW/sequencer manual for how to scale MIDI controller events over time.
7. Again, rescale the Channel Volume settings so they vary widely over the course of the segment.
8. Highlight the segment and scale the velocity values so they have a minimum value of 64 and a maximum of 96. Could you see and hear the changes?
9. Again, highlight the segment and instruct the software to fade it from its current value to an ending value of 0. Hopefully, you've just created a Channel Volume fade. Did you see the MIDI channel fader move?
10. Undo and start the fade with an initial value of 0 and a current value of 100% towards the end of the section. Did the segment fade in?

PLAYBACK

Once a sequence is composed and saved to disk, all of the sequence tracks can be transmitted through the various MIDI ports and channels to the instruments or devices to make music, create sound effects for film tracks or control device parameters in real time. Because MIDI data exists as encoded real-time control commands and not as audio, you can listen to the sequence and make changes at any time. You could change the patch voices, alter the final mix or change and experiment with such controllers as pitch bend or modulation—even change the tempo and key signature. In short, this medium is infinitely flexible in the number of versions that can be created, saved, folded, spindled and mutilated until you've arrived at the overall sound and feel you want. Once done, you'll have the option of using the data for live performance or mixing the tracks down to a final recorded media, either in the studio or at home.

During the summer, in a wonderful small-town tavern in the city where I live, there's a frequent performer who'll wail the night away with his voice, trusty guitar and a backup band that consists of several electronic synth modules and a laptop PC/sequencer that's just chock-full of country-'n'-western sequences. His set of songs for the night is loaded into a song playlist feature that's programmed into his sequencer. Using this playlist, he queues his sequences so that when he's finished one song, taken his bow, introduced the next song and complimented the lady in the red dress, all he needs to do is press the space bar and begin playing the next song. Such is the life of an on-the-road sequencer.

In the studio, it's become more the rule than the exception that MIDI tracks will be recorded and played back in sync with a DAW, analog multitrack machine or digital multitrack. As you're aware, using a DAW-based sequencer automatically integrates the world of audio and video with the MIDI production environment. Whenever more than one playback medium is involved in a production, a process known as synchronization is required to make sure that events in the MIDI, analog, digital and video media occur at the same point in time. Locking events together in time can be accomplished in various ways (depending on the equipment and media type used).

MIXING A SEQUENCE

Almost all DAW and sequencer types will let you mix a sequence in the MIDI domain using various controller message types. This is usually done by creating a software interface that incorporates these controls into a virtual on-screen mixer environment that often integrates with the main DAW mix screen. Instead of directly mixing the audio signals that make up a sequence, these controls are able to directly access such track controllers as Main Volume (controller 7), Pan (controller 10), and Balance (controller 8), most often in an environment that completely integrates into the workstation's overall mix controls. Since the mix-related data is simply MIDI controller messages, an entire mix can be easily stored within a sequence file. Therefore, even with the most basic sequencer,

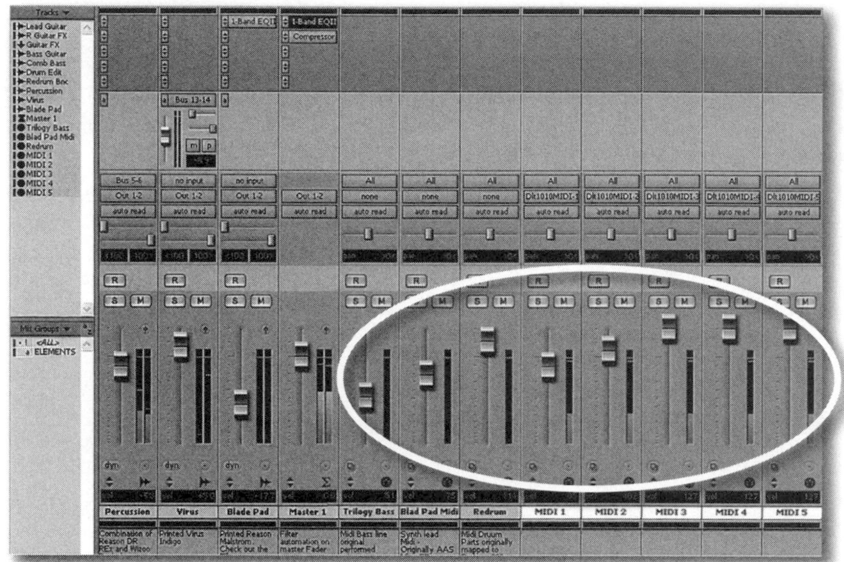

FIGURE 9.47
MIDI tracks can be added into the mix for real-time parameter control over hardware and/or software devices. (Courtesy of Avid Technology, Inc., www.avid.com)

you'll be able to mix and remix your sequences with complete automation and total settings recall whenever a new sequence is opened. As is almost always the case with a DAW's audio and MIDI graphical user interface (GUI), the controller and mix interface will almost always have moving faders and variable controls (Figure 9.47).

TRANSFERRING MIDI TO AUDIO TRACKS

When mixing down a session that contains MIDI tracks, many folks prefer not to mix the sequenced tracks in the MIDI domain. Instead, they'll often export it to an audio track within the project. Here are a few helpful hints that can make this process go more smoothly:

- Set the main volume and velocities to a "reasonably" high output level, much as you would with any recorded audio track.
- Solo the MIDI and instrument's audio input track and take a listen, making sure to turn off any reverb or other effects that might be on that instrument track (you might need to pull out your instrument's manual to flip through these settings). If you really like the instrument effect, of course, go ahead and record it; however, you might consider recording the track both with and without effects.
- If any mix-related moves have been programmed into the sequence, you "might" want to strip out volume, pan and other controller messages before exporting the track. This can be done manually; however, most DAW/sequencer packages include features for stripping controller data from a MIDI track.

- If the instrument has an acoustic (such as a MIDI grand piano or MIDI guitar room/amp setup) or multichannel output component, you might consider recording the instrument to stereo or surround tracks. This will allow for greater flexibility in creating ambient "space" within the final mix.

Important Note

It's always wise to save the original MIDI track or file within the session. This makes future changes in the composition infinitely easier. Failure to save your MIDI files will often limit your future options or result in major production headaches down the road.

MUSIC PRINTING PROGRAMS

Over the past few decades, the field of transcribing musical scores and arrangements has been strongly affected by both the computer and MIDI technology. This process has been greatly enhanced through the use of newer generations of computer software that make it possible for music notation data to be entered into a computer either manually (by placing the notes onto the screen via keyboard or mouse movements), by direct MIDI input or by sheet music scanning technology. Once entered, these notes can be edited in an on-screen environment that lets you change and configure a musical score or lead sheet using standard cut-and-paste editing techniques. In addition, most programs allow the score data to be played directly from the score by electronic instruments via MIDI. A final and important program feature is their ability to quickly print out hard copies of a score or lead sheets in a wide number of print formats and styles.

A music printing program (also known as a music notation program) lets you enter musical data into a computerized score in a number of manual and automated ways (often with varying degrees of complexity and ease). Programs of this type (Figure 9.48) offer a wide range of notation symbols and type styles that can be entered either from a computer keyboard or mouse. In addition to entering a score manually, most music transcription programs will generally accept MIDI input, allowing a part to be played directly into a score. This can be done in real time (by playing a MIDI instrument/controller or finished sequence into the program) or in step time (by entering the notes of a score one note at a time from a MIDI controller) or by entering a standard MIDI file into the program (which uses a sequenced file as the notation source).

In addition to dedicated music printing programs, most DAW or sequencer packages will often include a basic music notation application that allows the sequenced data within a track or defined region to be displayed and edited

FIGURE 9.48
Finale music composition, arranging and printing program. (Courtesy of MakeMusic, Inc., www.finalemusic.com)

directly within the program (Figure 9.49), from which it can be printed in a limited score-like fashion. However, a number of high-level workstations offer scoring features that allow sequenced track data to be notated and edited in a professional fashion into a fully printable music score.

As you might expect, music printing programs will often vary widely in their capabilities, ease of use and offered features. These differences often center around the graphical user interface (GUI), methods for inputting and editing data, the number of instrumental parts that can be placed into a score, the overall selection of musical symbols, the number of musical staves (the lines that music notes are placed onto) that can be entered into a single page or overall score, the ability to enter text or lyrics into a score, etc. As with most programs that deal with artistic production, the range of choices and general functionality reflect the style and viewpoints of the manufacturer, so care should be taken when choosing a professional music notation program, to see which one would be right for your personal working style.

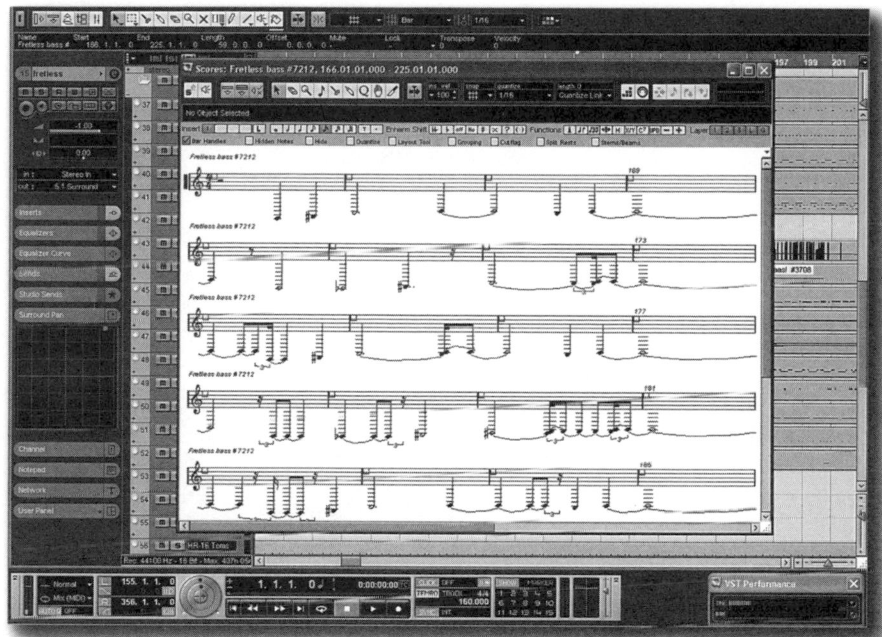

FIGURE 9.49
Steinberg Cubase/Nuendo score window. (Courtesy of Steinberg Media Technologies GmbH, a division of Yamaha Corporation, www.steinberg.net)

CHAPTER 10
Multimedia and the Web

It's no secret that modern-day computers, portable devices, game-stations and even televisions have gotten faster, sleeker, more touchable and sexier in their overall design. In addition to their ability to act as a multifunctional production workhorse, one of the crowning achievements of modern work and entertainment devices is the degree of media and networking integration that has made its way into our collective consciousness in a way that has come to be universally known as the household buzzword *multimedia*.

The combination of working and playing with multimedia has found its way into modern media and computer culture through the use of various hardware and software systems that work in a multitasking environment and combine to bring you an experience that seamlessly involves such media types as:

- Audio and music
- Video and video streaming
- Graphics
- Musical instrument digital interface (MIDI)
- Text

The obvious reason for creating and integrating these media types is the human desire to share and communicate one's experiences with others. This has been done for centuries in the form of books and, in relatively more recent decades, through movies and television. In the here and now, the amazingly powerful and versatile presence of the web can be added to this communications list. Nothing allows individuals and corporate entities alike to reach millions so easily. Perhaps most importantly, the web is a multimedia experience that each individual can manipulate, learn from and even respond to in an interactive fashion. The web has indeed unlocked the potential for experiencing multimedia events and information in a way that makes each of us a participant—not just a passive spectator. To me, this is the true revolution occurring at the dawn of the 21st century!

THE MULTIMEDIA ENVIRONMENT

Although much of recording technology has matured into a relatively stable industry, the web and multimedia is in a full-speed-ahead tailspin of change. With online social media, on-demand video and audio streaming, communications, computer gaming and hundreds of other media option entering onto the marketplace on a daily basis … it's no wonder that things are changing, and fast!

As with all things tech today, I would say that the most important word in multimedia technology today is "integration." In fact, the perfect example of multimedia today is in your pocket or your purse. Your cell phone (handy, mobile or whatever you call it) is a marvel of multimedia technology that:

- keeps you in touch with your friends
- surfs the web to find the best restaurant in the area
- has a web-based or actual GPS to help you find your way
- plays or streams your favorite music
- lets you watch the latest movie or youtube video
- lights your way to the bathroom late at night

… what more could you ever want from a mobile on-the-go device. I don't know … but we're surely to find out in the not-too-distant-future … and, because of these everyday tools, we've come to expect the same or similar experience from our other media devices:

- the computer
- television

The cell phone, handy or whatever you want to call it

Tell me, is there a more socially integrated, music-playing, photo-taking, alarm-waking and egg-cooking device on the planet? In a society where you see folks with their faces buried in that small screen, this device is definitely part of our consciousness … maybe it's your best friend.

The computer

Obviously, the toy that started the multimedia revolution is the computer. The fact that it eats and breaths in a multitasking environment makes it ideal for delivering all of the media that we want, all of the time. From multimedia, to gaming, to music, to video … if you have a good internet connection … you can pretty much hold the media world in the palm of your hands (or at least your lap).

Television and the home theater

As you might expect, newer generations of video and home theater systems incorporate more and more options for offering up "a rich multimedia experience" (to use a favorite marketing phrase). Newer TVs are able to directly connect to the web, allowing us to stream our favorite movies or listen to internet radio "on demand."

So, what holds all of these various media devices together? Media and data distribution/transmission formats!

DELIVERY MEDIA

Although media data can be stored and/or transmitted over a wide range of media storage devices, the most commonly found delivery media at the time of this writing are the:

- networks
- the web
- physical media

Networking

At its most basic level, a shared data *network* is a collection of computers and other data hardware devices that are connected by communications channels that allow for the communication of shared resources and/or data. The most well-known network communications protocols are the Ethernet (a standard that's widely in creating Local Area Networks), and the Internet Protocol Suite (otherwise known as the www).

Within media production, a Local Area Network (or LAN) is a very powerful tool allowing multiple computers to be linked within a production facility. For example, a central server (a dedicated data delivery and shared storage device) can be used in a facility to store large amounts of data that relate to everyday production and data backup. Simpler LAN connections can be made in a project studio to link various computers and to share media and data backup in a simple and cost-effective manner. For example, a single, remote computer within a connected facility could be used to store the large amounts of media data that required for video production, music soundfiles and sample libraries and overall system backups … a RAID (redundant array of independent disks) backup system could also be installed, allowing the data and backups to be duplicated and stored on multiple hard drives (greatly reducing the chance of system data loss).

In short, it's always a good idea to become familiar with the strength, protection and power that a properly designed network can offer a production facility, no matter its size.

The Web

One of the most powerful aspects of multimedia is the ability to communicate experiences either to another individual or to the masses. For this, you need a very large network connection. The largest and most common network of all is the internet (World Wide Web). Here's the basic gist of how this beast works:

FIGURE 10.1
The internet works by communicating requests and data from a user's computer to a single server that's connected to other servers around the world, which are likewise connected to other users' computers.

- The internet (Figure 10.1) can be thought of as a communications network that allows your computer (or connected network) to be connected to an Internet Service Provider (ISP) server (a specialized computer or cluster of ISP computers that are designed to handle, pass and route data between large numbers of user connections).
- These ISPs are connected (through specialized high-speed connections) to a series of network access points (NAPs), which essentially form the connected infrastructure of the World Wide Web (www).

Therefore, in its most basic form the www can be simply thought of as a unified array of connected networks.

Internet browsers transmit and receive information on the web via a Uniform Resource Locator (URL) address. This address is then broken down into three parts: the protocol (e.g., http), the server name (e.g., www.modrec.com) and the requested page or file name (e.g., /index.htm). The connected server is able to translate the server name into a specific Internet Provider (IP) address, which is used to connect your computer to the desired server, after which the requests to receive or send data are communicated and the information is passed to your desktop.

Email works in a similar data transfer fashion, with the exception that an email isn't sent to or requested from a specific server; rather, it's communicated through a worldwide server network from one specific email address (e.g., myname@myprovider.com) directly to a destination email address (e.g., yourname@yourprovider.com).

The Cloud

One of the biggest buzz terms on the web is "The Cloud" or "Cloud Computing." Put simply, storing data in or on the cloud refers to data that is networked with and stored on a remote server. Most commonly, that server would be operated and maintained by a company that will store limited amounts of your personal data for free … or larger amounts of data (with more service options) at a cost. For example, cloud companies can:

- Specialize in promoting artist and dj soundfiles to the industry in a social media form (i.e. www.soundcloud.com).
- Store backup media online in an automated way that keeps your files secure.
- Store program-related data online, reducing the need for installing programs directly onto your computer.
- Store huge amounts of uploadable video data online in a social media context (i.e. www.youtube.com).

Physical media

Although online data distribution and management is increasingly becoming the pre-eminent method for getting information from the distributor to the consumer, physical media (you know, the kind that we can hold in our hands) still has a very important role for delivering media to the consumer masses. Beyond the fun stuff—like vinyl records—the most common delivery media are the CD, DVD and Blu-ray discs.

The flash memory card

In our on-the-go world, one of the more useful media devices is the flash memory card. More specifically, the SD (secure digital) card typically ranges in size from 4GB to 64GB in capacity and comes in various physical sizes. These media cards can be used for storing audio, video, photo, app and any other type of digital info that you care to throw in your pocket.

The CD

Of course, one of the first and most important developments in the mass marketing and distribution of large amounts of digital media was the compact disc (CD), both in the form of the CD-Audio and the CD-ROM. As most are aware, the CD-Audio disc is capable of storing up to 74 minutes of audio at a rate of 44.1 kHz/16 bits. Its close optical cousin, the CD-ROM, is most often capable of storing 700 MB of graphics, video, digital audio, MIDI, text, and raw data. Consequently, these pre-manufactured and user-encoded media are able to store large amounts of music, text, video, graphics, etc., to such an interactive extent that this medium has become a driving force among all communications media. Table 10.1 details the various CD standards (often affectionately called the "rainbow book") that are currently in use.

It's important to note that, in addition to audio, music CDs are capable of encoding small amounts of user-data that can be used to encode imbedded metadata (user information). This metadata (called CD-Text), allows the media creator to enter and encode such information as musician, titles, song number, track artist/title, etc. within the disc itself. Since recently-made CD players, computers and media players are capable of reading this information, it's always helpful to give the listener as much information as possible.

Another system for identifying CD and disc-related data is provided by Gracenote (formerly CDDB or Compact Disc Data Base). In short, Gracenote maintains and licences an internet database that contains CD info, text and images. Many media devices that are connected to the web are able to access this database and display the information.

The DVD

Similar to their optical cousins, DVDs (which, after a great deal of industry deliberation, simply stands for "DVD") can contain any form of data. Unlike CDs, these discs are capable of storing a whopping 4.7 gigabytes (GB) within a single-sided disc and 8.5 GB on a double-layered disc. This capacity makes the DVD the

Table 10.1	CD Format Standards
Format	**Description**
Red Book	Audio-only standard; also called CD-A (Compact Disc Audio)
Yellow Book	Data-only format; used to write/read CD-ROM data
Green Book	CD-I (Compact Disc Interactive) format; never gained mass popularity
Orange Book	CD-R (Compact Disc Recordable) format
White Book	VCD (Video Compact Disc) format for encoding CD-A audio and MPEG-1 or MPEG-2 video data; used for home video and karaoke
Blue Book	Enhanced Music CD format (also known as CD Extra or CD+) can contain both CD-A and data
ISO-9660	A data file format that's used for encoding and reading data from CDs of all types across platforms
Joliet	Extension of the ISO-9660 format that allows for up to 64 characters in its file name (as opposed to the 8 file + 3 extension characters allowed by MS-DOS)
Romeo	Extension of the ISO-9660 format that allows for up to 128 characters in the file name
Rock Ridge	Unix-style extension of the ISO-9660 format that allows for long file names
CD-ROM/XA	Allows for extended usage for the CD-ROM format—Mode-1 is strictly Yellow Book, while Mode-2 Form-1 includes error correction and Mode-2 Form-2 doesn't allow for error correction; often used for audio and video data
CD-RFS	Incremental packet writing system from Sony that allows data to be written and rewritten to a CD or CD-RW (in a way that appears to the user much like the writing/retrieval of data from a hard drive)
CD-UDF	UDF (Universal Disc Format) is an open incremental packet writing system that allows data to be written and rewritten to a CD or CD-RW (in a way that appears to the user much like the writing/retrieval of data from a hard drive) according to the ISO-13346 standard
HDCD	The High-Definition Compatible Digital system adds 6 dB of gain to a Red Book CD (when played back on an HDCD-compatible player) through the use of a special companion mastering technique
Macintosh HFS	An Apple file system that supports up to 31 characters in a file name; includes a data fork and a resource fork that identify which application should be used to open the file

Table 10.2 DVD Video/Audio Formats

Format	Sample Rate (kHz)	Bit Rate	Bit/s	Ch	Common Format	Compression
PCM	48, 96	16, 20, 24	Up to 6.144 Mbps	1 to 8	48 kHz, 16 bit	None
AC3	48	16, 20, 24	64 to 448 kbps	1 to 6.1	192 kbps, stereo	AC3 and 384 kbps, 448 kbps
DTS	48, 96	16, 20, 24	64 to 1536 kbps	1 to 7.1	377 or 754 kbps for stereo and 754.5 or 1509.25 kbps for 5.1	DTS coherent acoustics
MPEG-2	48	16, 20	32 to 912 kbps	1 to 7.1	Seldom used	MPEG
MPEG-1	48	16, 20	384 kbps	2	Seldom used	MPEG
SDDS	48	16	Up to 1289 kbps	5.1, 7.1	Seldom used	ATRAC

perfect delivery medium for encoding video in the MPEG-2 encoding format, data-intensive games, DVD-Audio and numerous DVD-ROM titles. The increased demand for multimedia games, educational products, etc., has spawned the computer-related industry of CD and DVD-ROM authoring. The term *authoring* refers to the creative, design, and programming aspects of putting together a CD/DVD project. At its most basic level, a project can be authored, mastered and burned to disc from a single commercial authoring program. Whenever the stakes are higher, trained professionals and expensive systems are often called in to assemble, master and produce the final disc for mass duplication and sales. Table 10.2 details the various DVD video/audio formats currently in use.

Blu-ray

Although similar in size to the CD and DVD, Blu-ray is able to store up to 25GB of media-related data onto each data layer (50GB for a dual-layer disc). In addition to most of the standard video formats that are commonly encoded onto a DVD, the Blu-ray format is able to playback both compressed and non-compressed PCM audio (Table 10.3) in a multi-channel environment.

DELIVERY FORMATS

Now that we've taken a look at the various delivery media, the next most important aspect of delivering the multimedia experience rests with the data distribution formats (the coding and technical aspects of data delivery) themselves.

Table 10.3 Specification of BD-ROM Primary audio streams

	LPCM	Dolby Digital	Dolby Digital Plus	Dolby TrueHD (Lossless)	DTS Digital Surround	DTS-HD Master Audio (Lossless)	DRA	DRA Extension
Max. Bitrate	27.648 Mbit/s	640 kbit/s	4.736 Mbit/s	18.64 Mbit/s	1.524 Mbit/s	24.5 Mbit/s	1.5 Mbit/s	3.0 Mbit/s
Max. Channel	**8** (48 kHz, 96 kHz) **6** (192 kHz)	5.1	7.1	**8** (48 kHz, 96 kHz), **6** (192 kHz)	5.1	**8** (48 kHz, 96 kHz), **6** (192 kHz)	5.1	7.1
Bits/ sample	16, 20, 24	16, 24	16, 24	16, 24	16, 20, 24	16, 24	16	16
Sample frequency	48 kHz, 96 kHz, 192 kHz	48 kHz	48 kHz	48 kHz, 96 kHz, 192 kHz	48 kHz	48 kHz, 96 kHz, 192 kHz	48 kHz	48 kHz, 96 kHz

When creating content for the various media systems, it's extremely important that you match the media format and bandwidth requirements to the content delivery system that's being used. In other words, it's always smart to maximize the efficiency of the message (media format and required bandwidth) to match (and not alienate) your intended audience. The following section outlines many standard and/or popular formats for delivering media to a target audience.

Digital audio

Digital audio is obviously a component that adds greatly to the multimedia experience. It can augment a presentation by adding a dramatic music soundtrack, help us to communicate through speech or give realism to a soundtrack by adding sound effects. Because of the large amounts of data required to pass video, graphics and audio from a CD-ROM, the internet or other media, the bit- and sample-rate structure of an uncompressed audio file is usually limited compared to that of a professional-quality soundfile. At the "lo-fi" range, the generally accepted soundfile standard for older multimedia production is either 8-bit or 16-bit audio at a sample rate of 11.025 or 22.050 kHz. This standard had come about mostly because older CD drive and processor systems generally couldn't pass the professional 44.1-kHz rate. With the introduction of faster processing systems and better hardware, these limitations have generally been lifted to include 44.1-kHz/16-bit audio (or higher) and compressed data formats that offer CD quality and discrete surround-sound playback capabilities. In addition, there are obvious limitations to communicating uncompressed professional-rate soundfiles over the internet or from a CD or DVD disc that's also streaming full-motion video. Fortunately, with improvements in codec (encode/decode) techniques, hardware speed and design, the overall sonic and production quality of compressed audio data has greatly improved in audience acceptance.

Uncompressed soundfile formats

Although several formats exist for encoding and storing soundfile data, only a few have been universally adopted by the industry. These standardized formats make it easier for files to be exchanged between compatible media devices. Probably the most common file type is the Wave (or .wav) format. Developed for the Microsoft Windows format, this universal file type supports both mono and stereo files at a wde range of uncompressed resolutions and sample rates. Wave files contain pulse-code modulation (PCM) coded audio that follows the Resource Information File Format (RIFF) spec, which allows extra user information to be embedded and saved within the file itself. The newly adopted Broadcast Wave format, which has been adopted by the NARAS Producers and Engineers wing (www.grammy.com/Recording_Academy/Producers_and_Engineers) as the preferred soundfile format for DAW production and music archiving, allows for timecode-related positioning information to be directly embedded within the soundfile's datastream.

In addition to the .wav format, another file format that's commonly used in Apple computers is the Audio Interchange File (AIFF; .aif) format. Like Wave files, AIFF files support mono or stereo, 8-bit, 16-bit and 24-bit audio at a wide range of sample rates—and like Broadcast Wave files, AIFF files can also contain embedded text strings. Table 10.4 details the differences between uncompressed file sizes as they range from the 32-bit/192-kHz rates, all the way down to lo-voice quality 8-bit/10-kHz files.

COMPRESSED CODEC SOUNDFILE FORMATS

As was mentioned earlier, high-quality uncompressed soundfiles often present severe challenges to media delivery systems that are restricted in terms of bandwidth, download times or memory storage. Although the streaming of audio data from various media and high-bandwidth networks (including the web) has improved over the years, memory storage space and other bandwidth limitations have led to the popular acceptance of audio-related data formats known as codecs. These formats can encode audio in a manner that reduces data file size and bandwidth requirements and then decode the information upon playback using a system known as perceptual coding.

Perceptual coding

The central concept behind *perceptual coding* is the psychoacoustic principle that the human ear will not always be able to hear all of the information that's present in a recording. This is largely due to the fact that louder sounds will often mask sounds that are both lower in level and relatively close to another louder signal. These perceptual coding schemes take advantage of this masking effect by filtering out noises and sounds that can't be detected and removing them from the encoded audiostream.

Table 10.4 Audio Bit Rate and File Sizes

Sample Rate	Word Length	No. of Channels	Data Rate (kbps)	MB/min	MB/hour
192	24	2	1152	69.12	4147.2
192	24	1	576	34.56	2073.6
96	32	2	768	46.08	2764.8
96	32	1	384	23.04	1382.4
96	24	2	576	34.56	2073.6
96	24	1	288	17.28	1036.8
48	32	2	384	23.04	1382.4
48	32	1	192	11.52	691.2
48	24	2	288	17.28	1036.8
48	24	1	144	8.64	518.4
48	16	2	192	11.52	691.2
48	16	1	96	5.76	345.6
44.1	32	2	352	21.12	1267.2
44.1	32	1	176	10.56	633.6
44.1	24	2	264	15.84	950.4
44.1	24	1	132	7.92	475.2
44.1	16	2	176	10.56	633.6
44.1	16	1	88	5.28	316.8
32	16	2	128	7.68	460.8
32	16	1	64	3.84	230.4
22	16	2	88	5.28	316.8
22	16	1	44	2.64	158.4
22	8	1	22	1.32	79.2
11	16	2	44	2.64	158.4
11	16	1	22	1.32	79.2
11	8	1	11	0.66	39.6

The perceptual encoding process is said to be lossy or destructive, because once the filtered data has been eliminated it can't be replaced or introduced back into the file. For the purposes of audio quality, the amount of audio that's to be removed from the data can be selected by the user during the encoding process. Higher bandwidth compression rates will remove less data from a stream (resulting in a reduced amount of filtering and higher audio quality), while low bandwidth rates will greatly reduce the datastream (resulting in smaller file sizes, increased filtering, increased artifacts and lower audio quality). The amount of filtering that's to be applied to a file will depend on the intended audio quality and the delivery medium's bandwidth limitations. Due to the lossy character of these encoded files, it's always a good idea to keep a copy of the original, uncompressed soundfile as a data archive backup should changes in content or future technologies occur (never underestimate Murphy's law of technology).

The perceptual coding schemes in most common use are:

- MP3
- MP4
- WMA
- AAC
- RealAudio
- FLAC

Many of the listed codecs are capable of encoding and decoding audio using a constant bit rate (CBR) and variable bit rate (VBR) structure:

- CBR encoding is designed to work effectively in a streaming scenario where the end user's bandwidth is a consideration. With CBR encoding, the chosen bit rate will remain constant over the course of the file or stream.
- VBR encoding is designed for use when you want to create a downloadable file that has a smaller file size and bit rate without sacrificing sound and video quality. This is carried out by detecting which sections will need the highest bandwidth and adjusting the encode process accordingly. When lower rates will suffice, the encoder adjusts the process to match the content. Under optimum conditions, you might end up with a VBR-encoded file that has the same quality as a CBR-encoded file, but is only half the file size.

MP3

MPEG (which is pronounced "M-peg" and stands for the Moving Picture Experts Group; www.mpeg.org) is a standardized format for encoding digital audio into a compressed format for the storage and transmission of various media over the web. As of this writing, the most popular format is the ISO-MPEG Audio Level-2 Layer-3, commonly referred to as MP3. Developed by the Fraunhofer Institute (www.iis.fraunhofer.de) and Thomson Multimedia in Europe, MP3 has advanced the public awareness and acceptance of compressing and distributing digital audio by creating a codec that can compress audio by a substantial factor while still maintaining quality levels that approach those of a CD (depending on which compression levels are used). Although a wide range of compression rates can be chosen to encode/decode an MP3 file, the most common rate for the general masses is 128kbps (kilo bits per second). Although this rate is definitely "lossy" (containing increased distortion, reduced bandwidth and sideband artifacts), it lets us literally put thousands of songs on an on-the-go player. Higher rates of 160, 192 and 320 kbps offer higher "near CD sound quality" with the obvious tradeoff being a larger filesize.

Although faster web connections are capable of streaming MP3 in real time, this format is most often downloaded to the end consumer for storage to disk, disc and memory media for the storage and playback of downloaded songs. Once saved, the data can then be transferred to solid-state playback devices (such as portable MP3 players, PDAs, cell phone players—you name it!). In fact, billions of music tracks are currently being downloaded every month on the internet using MP3, practically every personal computer contains licensed MP3 software, virtually every song has been MP3 encoded and an astounding number of MP3 players are on the global market. This makes it the web's most popular audio compression format by far.

MP4

Like MP3, the MPEG-4 (MP4) codec is largely used for streaming data over the web or for use with portable media devices. Largely based on Apple's QuickTime "MOV" format, MP4 is used to encode both audio and video media data at various quality bit depths and with multichannel (surround) capabilities.

WMA

Developed by Microsoft as their corporate response to MP3, Windows Media Audio (WMA) allows for compression rates that can encode high-quality audio at low bit-rate and file-size settings. Designed for ripping (encoding audio from audio CDs) and soundfile encoding/playback from within the popular Window's Media Player (Figure 10.2), this format has grown in general acceptance and popularity. In addition to its high quality at low bit rates, WMA has the advantage of allowing for real-time streaming over the internet (as witnessed by the large amount of radio and internet stations that stream to the Windows Media Player at various bit-rate qualities). Finally, content providers often favor MWA over MP3, because it is able to provide for a degree of content copy protection through its incorporation of Digital Rights Management (DRM) coding. With the introduction of Windows Media Player Version 9 and later versions, WMA is able to encode and deliver audio in discrete surround sound.

AAC

Jointly developed by Dolby Labs, Sony, ATT, and the Fraunhofer Institute, the Advanced Audio Coding (AAC) scheme is touted as a multichannel-friendly format for secure digital music distribution over the internet. Stated as having the ability to encode CD-quality audio at lower bit rates than other coding formats, AAC not only is capable of encoding 1, 2 and 5.1 surround-soundfiles but can also encode up to 48 channels within a single bitstream at bit/sample rates of up to 24/96. This format is also SDMI (Secure Digital Music Initiative) compliant, allowing copyrighted material to be protected against unauthorized copying and distribution.

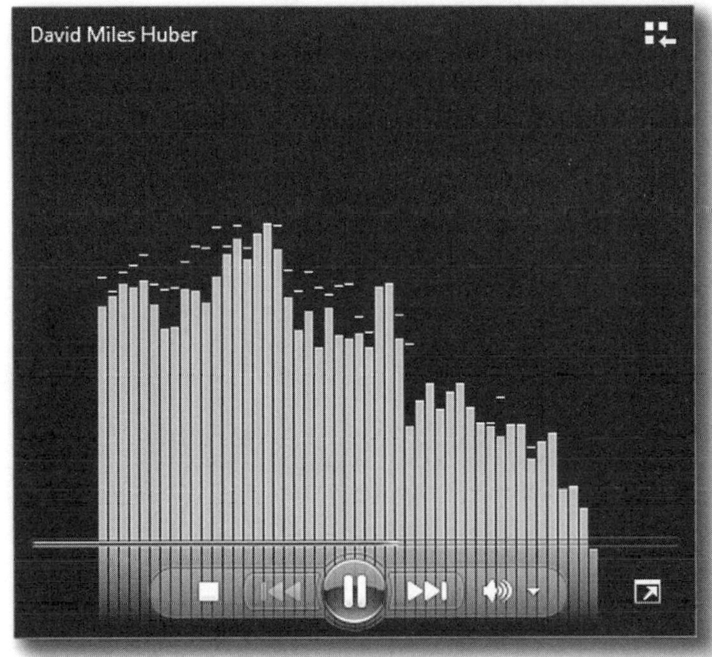

FIGURE 10.2
Window's Media Player—notice that the "Bars and Waves" visualization theme option can act as a basic spectral display.

FLAC

FLAC (Free Lossless Audio Codec) is a format that makes use of a data compression scheme that's capable of reducing an audio file's data size by 40% to 50%, while playing back in a lossless fashion that maintains the sonic integrity of the original stereo and multichannel source audio. As the name suggests, FLAC is a free, open-source codec that can be used by software developers in a royalty-free fashion.

With the increase in memory storage size and higher download speeds, many enthusiasts in audio are beginning to demand higher playback quality. As such, FLAC is growing in popularity as a medium for playing back high-quality, lossless audio, both in stereo and in various surround formats.

Tagged metadata

Within most types of multimedia file formats it's possible to embed a wide range of identifier data directly into the file itself or within a web-related page or event. This "tagged" data (also known as metadata) can identify and provide extensive and extremely important information that relates to the content of the file. For example, let's say that little Sally is looking to find a song that she downloaded from her favorite artist. Now, Sally consumes a lot of music and her iPod is nearly full of thousands of songs. So how can she find her favorite needle in a digital haystack? By searching for songs under "Mr. Right," the player is instantly able to find all of his songs and BOOM … she's groovin' to the beat … all thanks to metadata. On the flip-side, if the song name hasn't been entered (or was incorrectly entered), poor Sally's song would literally get lost in the shuffle.

On another day, let's say that Sally *really* wanted to buy a new song. She could enter her favorite music genre into the field and look through the songs that have been properly tagged with that genre. Maybe she really wants to be more specific and enters the name of her favorite artist … but she already bought the song. Well, she could check the "sounds like" box, and lo and behold, a song pops up that completely flips her out … a sale and (even better) a new fan is born … all because the metadata was properly filled out by her new, favorite band. Are you getting the idea that "tagging" a song, project or artist band data with the proper metadata can be a make or break deal in the new digital age?

Metadata in all its media and website glory tells the world who it is, the song title, what genre type, etc … It's the gateway to telling the world "Hey, I'm here … Listen to me!" Without the proper metadata … it's just another unknown song out in the pouring rain.

Metadata can be entered into an MP3 or FLAC in numerous ways. It can be automatically detected from a CD using a central music database and automatically entered into a music copy (ripping) program or it can be entered manually by the user (Figure 10.3).

Due to the fact that there is no existing set of rules for filling out metadata, folks who make extensive use of iTunes and other music playlist services often go nuts when the tags that are incorrect, conflicting or non-standard. For example, a user might download one album that might go by the artist name of "XYZ Band" and then download another album might have it listed as "The Band XYZ." One would show up in a player under "X" while the same band would also show up under "T" … and it can get A LOT worse than that. In fact, many people actually go so far as to go into their music and manually fix the metadata to their liking. The moral of the story is to research your metadata and stick to a single naming scheme.

It's worth mentioning here that the Producers and Engineers wing of the Grammys have been doing extensive work on the subject of metadata (www.grammy.com/credits), so that tags can be more uniform and more extensive, allowing producer, engineer and other project-related data can be entered into the official metadata. It is hoped that this documentation will provide tracking info for those who legally deserve to be recognized and paid for their services … in the here-and-now, or in the future (when new payment legislations might be passed).

MIDI

One of the unique advantages of MIDI as it applies to multimedia is the rich diversity of musical instruments and program styles that can be played back in real time, while requiring almost no overhead processing from the computer's CPU. This makes MIDI a perfect candidate for playing back soundtracks from multimedia games or over the phone, internet, gaming devices, etc. Still, as one

FIGURE 10.3
Embedded metadata file tags can be added to a media file via various media libraries, rippers or editors and then viewed by a media player or file manager.

might expect, MIDI has taken a back seat to digital audio as a serious music playback format for multimedia.

Most likely, this is due to several factors, including:

- A basic misunderstanding of the medium
- The fact that producing MIDI content requires a fundamental knowledge of music
- The frequent difficulty of synchronizing digital audio to MIDI in a multimedia environment
- The fact that soundcards, phones, etc. often include poorly designed FM synthesizers (although most operating systems now include higher quality software synths).

Fortunately, a number of companies have taken up the banner of embedding MIDI within their media projects and have helped push MIDI a bit more into the web mainstream. As a result, it's becoming more common for your PC to begin playing back a MIDI score on its own or perhaps in conjunction with a game or more data-intensive program.

Standard MIDI files

The accepted format for transmitting files or real-time MIDI information in multimedia (or between sequencers from different manufacturers) is the standard MIDI file. This file type (which is stored with a .mid or .smf extension) is used to distribute MIDI data, song, track, time signature and tempo information to the general masses. Standard MIDI files can support both single and multichannel sequence data and can be loaded into, edited and then directly saved from almost any sequencer package. When exporting a standard MIDI file, keep in mind that they come in two basic flavors—type 0 and type 1:

- Type 0 is used whenever all of the tracks in a sequence need to be merged into a single MIDI track. All of the notes will have a channel number attached to them (i.e., will play various instruments within a sequence); however, the data will have no definitive track assignments. This type might be the best choice when creating a MIDI sequence for the internet (where the sequencer or MIDI player application might not know or care about dealing with multiple tracks).
- Type 1, on the other hand, will retain its original track information structure and can be imported into another sequencer type with its basic track information and assignments left intact.

General MIDI

One of the most interesting aspects of MIDI production is the absolute uniqueness of each professional and even semi-pro project studio. In fact, no two studios will be alike (unless they've been specifically designed to be the same or there's some unlikely coincidence). Each artist will be unique as to his or her own favorite equipment, supporting hardware, way of routing channels and tracks and assigning patches. The fact that each system setup is unique and personal has placed MIDI at odds with the need for system and setup compatibility in the world of multimedia. For example, after importing a MIDI file over the Net that's been created in another studio, the song will most likely attempt to play with a totally irrelevant set of sound patches (it might sound interesting, but it won't sound anything like what was originally intended). If the MIDI file is loaded into another setup, the sequence will again sound completely different, with patches that are so irrelevant that the guitar track might sound like a bunch of machine-gun shots from the planet Gloop.

To eliminate (or at least reduce) the basic differences that exist between systems, a standardized set of patch settings, known as General MIDI (GM), was created. In short, General MIDI assigns a specific instrument patch to each of the 128 available program change numbers. Since all electronic instruments that conform to the GM format must use these patch assignments, placing GM program change commands at the header of each track will automatically instruct the sequence to play with its originally intended sounds and general song settings. In this way, no matter what sequencer and system setup is used to play the file back, as long as the receiving instrument conforms to the GM spec, the sequence will be heard using its intended instrumentation.

Tables 10.5 and 10.6 detail the program numbers and patch names that conform to the GM format. These patches include sounds that imitate synthesizers, ethnic instruments or sound effects that have been derived from early Roland synth patch maps. Although the GM spec states that a synth must respond to all 16 MIDI channels, the first 9 channels are reserved for instruments, while GM restricts the percussion track to MIDI channel 10.

GRAPHICS

Graphic imaging occurs on the computer screen in the form of pixels. These are basically tiny dots that blend together to create color images in much the same way that dots are combined to give color and form to your favorite comic strip. Just as word length affects the overall amplitude range of a digital audio signal, the number of bits in a pixel's word will affect the range of colors that can be displayed in a graphic image. For example, a 4-bit word only has 16 possible combinations. Thus, a 4-bit word will allow your screen to have a total of 16 possible colors; an 8-bit word will yield 256 colors; a 16-bit word will give you 65,536 colors; and a 24-bit word will yield a whopping total of 16.7 million colors! These methods of displaying graphics onto a screen can be broken down into several categories:

- *Raster graphics:* In raster graphics, each image is displayed as a series of pixels. This image type is what is used utilized when a single graphic image is used (i.e., bitmap, JPEG, GIF or TIFF format). The sense of motion can come from raster images only by successively stepping through a number of changing images every second (in the same way that standard video images create the sense of motion).
- *Vector graphics:* This method displays still graphic drawings using geometric shapes (lines, curves and other shapes) that are placed at specific coordinates on the screen. Being assigned coordinates, shape, thickness and fill, these shapes combine together to create an image that can be simple or complex in nature. This script form reduces a file's data size dramatically and is used with several image programs.
- *Vector animation:* Much like vector graphics, vector animation can make use to the above shapes, thickness, fills, shading and computer-generated lighting to create a sense of complex motion that moves from frame to frame (often with a staggering degree of realism). Obviously, with the increased power of modern computers and supercomputers, this graphic art form has attained higher degrees of artistry or realism within modern-day film, video and desktop visual production and design.

Table 10.5 GM Nonpercussion Instrument (Program Change) Patch Map

1. Acoustic Grand Piano	44. Contrabass	87. Lead 7 (fifths)
2. Bright Acoustic Piano	45. Tremolo Strings	88. Lead 8 (bass + lead)
3. Electric Grand Piano	46. Pizzicato Strings	89. Pad 1 (new age)
4. Honky-tonk Piano	47. Orchestral Harp	90. Pad 2 (warm)
5. Electric Piano 1	48. Timpani	91. Pad 3 (polysynth)
6. Electric Piano 2	49. String Ensemble 1	92. Pad 4 (choir)
7. Harpsichord	50. String Ensemble 2	93. Pad 5 (bowed)
8. Clavichord	51. SynthStrings 1	94. Pad 6 (metallic)
9. Celesta	52. SynthStrings 2	95. Pad 7 (halo)
10. Glockenspiel	53. Choir Aahs	96. Pad 8 (sweep)
11. Music Box	54. Voice Oohs	97. FX 1 (rain)
12. Vibraphone	55. Synth Voice	98. FX 2 (soundtrack)
13. Marimba	56. Orchestra Hit	99. FX 3 (crystal)
14. Xylophone	57. Trumpet	100. FX 4 (atmosphere)
15. Tubular Bells	58. Trombone	101. FX 5 (brightness)
16. Dulcimer	59. Tuba	102. FX 6 (goblins)
17. Drawbar Organ	60. Muted Trumpet	103. FX 7 (echoes)
18. Percussive Organ	61. French Horn	104. FX 8 (sci-fi)
19. Rock Organ	62. Brass Section	105. Sitar
20. Church Organ	63. SynthBrass 1	106. Banjo
21. Reed Organ	64. SynthBrass 2	107. Shamisen
22. Accordion	65. Soprano Sax	108. Koto
23. Harmonica	66. Alto Sax	109. Kalimba
24. Tango Accordion	67. Tenor Sax	110. Bag pipe
25. Acoustic Guitar (nylon)	68. Baritone Sax	111. Fiddle
26. Acoustic Guitar (steel)	69. Oboe	112. Shanai
27. Electric Guitar (jazz)	70. English Horn	113. Tinkle Bell
28. Electric Guitar (clean)	71. Bassoon	114. Agogo
29. Electric Guitar (muted)	72. Clarinet	115. Steel Drums
30. Overdriven Guitar	73. Piccolo	116. Woodblock
31. Distortion Guitar	74. Flute	117. Taiko Drum
32. Guitar Harmonics	75. Recorder	118. Melodic Tom

Continued …

Table 10.5 continued

33. Acoustic Bass	76. Pan Flute	119. Synth Drum
34. Electric Bass (finger)	77. Blown Bottle	120. Reverse Cymbal
35. Electric Bass (pick)	78. Shakuhachi	121. Guitar Fret Noise
36. Fretless Bass	79. Whistle	122. Breath Noise
37. Slap Bass 1	80. Ocarina	123. Seashore
38. Slap Bass 2	81. Lead 1 (square)	124. Bird Tweet
39. Synth Bass 1	82. Lead 2 (sawtooth)	125. Telephone Ring
40. Synth Bass 2	83. Lead 3 (calliope)	126. Helicopter
41. Violin	84. Lead 4 (chiff)	127. Applause
42. Viola	85. Lead 5 (charang)	128. Gunshot
43. Cello	86. Lead 6 (voice)	

Table 10.6 GM Percussion Instrument (Program Key Number) Patch Map (Channel 10)

35. Acoustic Bass Drum	51. Ride Cymbal 1	67. High Agogo
36. Bass Drum 1	52. Chinese Cymbal	68. Low Agogo
37. Side Stick	53. Ride Bell	69. Cabasa
38. Acoustic Snare	54. Tambourine	70. Maracas
39. Hand Clap	55. Splash Cymbal	71. Short Whistle
40. Electric Snare	56. Cowbell	72. Long Whistle
41. Low Floor Tom	57. Crash Cymbal 2	73. Short Guiro
42. Closed Hi-Hat	58. Vibraslap	74. Long Guiro
43. High Floor Tom	59. Ride Cymbal 2	75. Claves
44. Pedal Hi-Hat	60. Hi Bongo	76. Hi Wood Block
45. Low Tom	61. Low Bongo	77. Low Wood Block
46. Open Hi-Hat	62. Mute Hi Conga	78. Mute Cuica
47. Low-Mid Tom	63. Open Hi Conga	79. Open Cuica
48. Hi-Mid Tom	64. Low Conga	80. Mute Triangle
49. Crash Cymbal 1	65. High Timbale	81. Open Triangle
50. High Tom	66. Low Timbale	

Note: In contrast to Table 10.5, the numbers in Table 10.6 represent the percussion keynote numbers on a MIDI keyboard, not program change numbers.

DESKTOP VIDEO

With the proliferation of computers, DVD/Blu-ray players, video interface hardware and video editing software systems, desktop and laptop video has begun to play an increasingly important role in home and corporate multimedia production and content. Video is encoded into a datastream as a continuous series of successive frames, which are refreshed at rates that vary from 12 or fewer frames/second (fr/sec) to the standard broadcast rates of 29.97 and 30 fr/sec (or higher for 3D applications). As with graphic files, a single full-sized video frame can be made up of a gazillion pixels, which are themselves encoded as a digital word of *n* bits. Multiply these figures by nearly 30 frames and you'll come up with a rather impressive data file size.

Obviously, it's more common to find such file sizes and data throughput rates on higher-end desktop systems and professional video editing workstations; however, several options are available to help bring video down to data rates that are suitable for multimedia and even the internet:

- *Window size:* The basics of making the viewable picture smaller are simple enough: Reducing the frame size will reduce the number of pixels in a video frame, thereby reducing the overall data requirements during playback.
- *Frame rate:* Although standard video frame rates run at around 30 fr/sec (United States and Japan) and 25 fr/sec (Europe), these rates can be lowered to 12 fr/sec in order to reduce the encoded file size or throughput.
- *Compression:* In a manner similar to that which is used for audio, codecs can be applied to a video frame to reduce the amount of data that's necessary to encode the file by filtering out and smoothing over pixel areas that consume data or by encoding data that doesn't change from frame to frame into a shorthand that reduces data throughput. In situations where high levels of compression are needed, it's common to accept degradations in the video's resolution in order to reduce the file size and/or data throughput to levels that are acceptable to a restrictive medium (e.g., the web).

From all of this, it's clear that there are many options for encoding a desktop video file. When dealing with video clips, tutorials and the like, it's common for the viewing window to be medium in size and encoded at a medium to lower frame rate. This middle ground is often chosen in order to accommodate the standard data throughput that can be streamed off of most CD-ROMs and the web. These files are commonly encoded using Microsoft's Audio-Video Interleave (AVI) format, QuickTime (a common codec that was developed by Apple and can be played by either a Mac or PC) or MPEG 1, 2 or 4 (codecs that vary from lower multimedia resolutions to higher ones that are used to encode DVD movies). Both the Microsoft Windows and Apple OS platforms include built-in or easily obtained applications that allow all or most of these file types to be played without additional hardware or software.

Table 10.7 Internet Connection Speeds

Connection	Speed (bps)	Description
56k dial-up	56 Kbps (usually less)	Common modem connection
ISDN ISDN PRI/E1	128 Kbps (older technology) 1.5 Mbps/1.9 Mbps	
DSL	256 Kbps to 20 Mbps	
Cable	Up to 85 Mbps	
OC-1	52 Mbps	Optical fiber
OC-3	155 Mbps	Optical fiber
OC-12	622 Mbps	Optical fiber
OC-48	2.5 Gbps	Optical fiber
Ethernet	10 Mbps (older technology) up to 100 Gbps	Local-area network (LAN), not an internet connection

MULTIMEDIA AND THE WIN THE "NEED FOR SPEED" ERA

The household phrase "surfin' the web" has become synonymous with jumping onto the Net, browsing the sites and grabbin' onto all of those hot songs, videos, and graphics that might wash your way. Dude, with improved audio and video codecs and faster data connections (Table 10.7), the ability to search on any subject, download files, and stream audio or radio stations from any point in the world has definitely changed our perception of modern-day communications.

Thoughts on being (and getting heard) in cyberspace

Most of us have grown up in this age of the supermarket, where everything is wholesaled, processed, packaged and distributed to a single clearinghouse that's gotten so big that older folks can only shop there with the aid of a motorized shopping cart. For more than six decades, the music industry has largely worked on a similar principle: Find artists who'll fit into an existing marketing formula (or, more rarely, create a new marketing image), produce and package them according to that formula and put tons of bucks behind them to get them heard and distributed. Not a bad thing in and of itself; however, for independent artists (Figure 10.4) the struggle has been, and continues to be, one of getting themselves heard, seen and noticed—without the aid of the well-oiled megamachine. With the creation of cyberspace, not only are established record industry forces able to work their way onto your desktop screen (and into your multimedia speakers), but independent artists now also have a huge medium for getting heard. Through the creation of dedicated websites, search engines, links from other sites, and

FIGURE 10.4
Getting the artist out to the masses is hard work ... Here's my chance to put my best foot (or music site... www.davidmileshuber.com) forward. Moral of the story? Never pass up an opportunity!

independent music dot-coms, as well as through creative gigging and marketing, new avenues have begun to open up for the web-savvy independent artist (Figure 10.4).

Uploading to stardom!

If you build it, they will come! This overly simplistic concept definitely doesn't apply to the web. With an ever-increasing number of dot-whatevers going online every month, expecting people to come to your site just because it's there simply isn't realistic. Like anything that's worthwhile, it takes connections, persistence, a good product and good ol'-fashioned dumb luck to be seen as well as heard! If you're selling your music, T-shirts or whatever at gigs, on the streets and to family and friends, cyberspace can help increase sales by making it possible (and even easy) to get your band, music or clients onto several independent music websites that offer up descriptions, downloadable samples, direct sales and a link that goes directly to your main website. Such a site could definitely help to get the word out to a potentially new public—and help clue your audience in to what you and your music are all about.

Of course, artists aren't the only ones that have websites. Most folks in fields involving the crafts have a site; heck—even some studios have sites for their studio pets or mascots. Here are a few general guidelines that help your website work its magic for you and/or your organization.

- *Make your site a central hub*: It's always a good idea to direct your social network friends and fans to your personal page. This helps to inform them about your upcoming projects, older ones that are for sale, current goings-on in your personal and professional life, etc.

- *Give it a strong, uncluttered main page*: Give your site a simple front page that makes a statement about you and your work. From there, they can dig deeper and be drawn into "all things you."
- *Keep your site simple and straightforward*: Try to make your site navigation easy to use and understand. Don't use problematic animations or programming that can easily confuse your audience.
- *Make your site personal*: Speak directly to your fans: Let your fans know what makes you tick, what you're up to, as well as what's happening in your professional life.
- *Keep it up-to-date*: Nothings worse than going to a site, only to see that there's been no activity on it. Nothing says "I don't have the time or care to keep connected" like a neglected site.

Selling your wares

So many music-sales, direct distribution and direct sales sites have sprung up over the years that the traditional model for music sales has come under attack to the point of being broken. Depending upon your viewpoint, this can be a bad thing or a good thing. As a result of web-based downloads (or even direct physical internet sales), artists are able to sell their projects, remixes, demos, etc. directly to their fans, while retaining the rights to their own work. Cyber distribution sites such as iTunes, CDBaby and countless others offer up secure credit card authorization, billing and artist payment plans for an overall percentage of sales (Figure 10.5).

Of course, the above music sales sites let the prospective buyer listen to music samples before buying, however, it's also fairly simple to put your own music

FIGURE 10.5
iTunes® media player and music website. (Courtesy of Apple Computers, Inc., www.apple.com)

samples on your site that can be listened to, linked to social networks and commented on by the adoring fans. Sites such as www.soundcloud.com let you upload your tracks in a way that can be professionally presented within a site, followed by fans and linked in any number of ways.

No matter what or how many cyber distribution methods you choose for getting your music out, it's always wise to take the time to read through the contractual fine print. Although most are above board and offer to get your music out on a nonexclusive basis ... caveat emptor (let the buyer, and the content provider, beware)! In your excitement to get your stuff out there, you might not realize that you are signing away the rights for free distribution of that particular song (or worse). This hints at the fact that you're dealing with the music business, and, as with any business, you should always tread carefully in the cyber jungle.

Internet radio

Due to the increased bandwidth of many internet connections and improvements in audio streaming technology, many of the world's radio stations have begun to broadcast on the web. In addition to offering a worldwide platform for traditional radio listening audiences, a large number of corporate and independent web radio stations have begun to spring up that can help to increase the fan and listener base of musicians and record labels. Go ahead, get on the web and listen to your favorite Mexican station en vivo ... catch the latest dance craze from London ... or chill to reggae rhythms streaming in from the islands.

ON A FINAL NOTE

One of the most amazing things about multimedia, cyberspace and their related technologies is the fact that they're ever-changing. By the time you read this book, many changes will have occurred. Old concepts will have faded away ... and new and possibly better ones will take over and then begin to take on a new life of their own. Although I've always had a fascination with crystal balls and have often had a decent sense about new trends in technology, there's simply no way to foretell the many amazing things that lie ahead in the fields of music, music technology, gaming, visual media, multimedia and especially cyberspace. As with everything techno, I encourage you to read the trades and surf the web to keep abreast of the latest and greatest tools that have recently arrived—or are about to rise—on the horizon.

CHAPTER 11
Synchronization

Of course, it's a safe bet to say that music itself, in all its various forms, is an indispensable production part of almost all types of media production. In video postproduction, digital video editors, digital audio workstations (DAWs), audio and video transports, automated console systems and electronic musical instruments routinely work together to help create a soundtrack and then refine it into its finished form (Figure 11.1). The underlying technology that allows multiple audio and visual media to operate in tandem (so as to maintain a direct time relationship) is known as *synchronization* or *sync*.

Strictly speaking, synchronization occurs when two or more related events happen at precisely the same relative time. With respect to analog audio and video systems, sync is achieved by interlocking the transport speeds of two or more machines. For computer-related systems (such as digital video, digital audio and

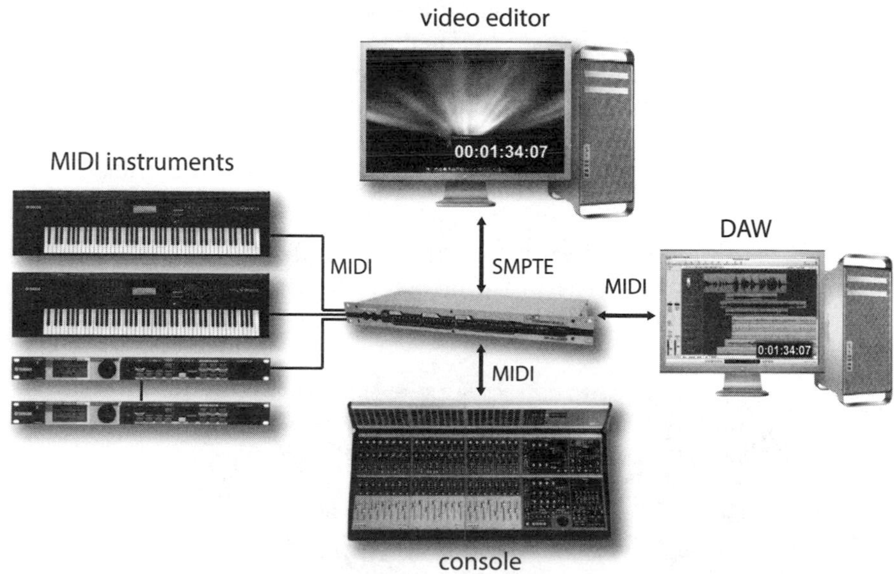

FIGURE 11.1
Example of an integrated audio production system.

MIDI), synchronization between devices is often achieved through the use of a timing clock that can be fed through a separate line or is directly embedded within the digital data line itself. Within such an environment, it's often necessary for analog and digital devices to be synchronized together; resulting in a number of ingenious forms of communication and data translation systems. In this chapter, we'll explore the various forms of synchronization used for both digital and analog devices, as well as current methods for maintaining sync between media types.

TIMECODE

Maintaining relative sync between media devices doesn't require that all transport speeds involved in the process be constant; however, it's important that they maintain the same relative speed and position over the course of a program. Physical analog devices, for example, have a particularly difficult time achieving this. Due to differences in mechanical design, voltage fluctuations and tape slippage, it's a simple fact of life that analog tape devices aren't able to maintain a constant playback speed, even over relatively short durations. For this reason, accurate sync between analog and digital machines would be nearly impossible to achieve over any reasonable program length without some form of timing lock. It therefore quickly becomes clear that if production is to utilize, multiple forms of media and record/playback sync is essential.

The standard method of interlocking audio, video and film transports makes use of a code that was developed by the Society of Motion Picture and Television Engineers (SMPTE; www.smpte.org). This timecode (or SMPTE timecode) identifies an exact position within recorded media or onto tape by assigning a digital address that increments over the course of a program's duration. This address code can't slip in time and always retains its original location, allowing for the continuous monitoring of tape position to an accuracy of between 1/24th and 1/30th of a second (depending on the media type and frame rates being used). These divisional segments are called *frames*, a term taken from film production. Each audio or video frame is tagged with a unique identifying number, known as a timecode address. This eight-digit address is displayed in the form 00:00:00:00, whereby the successive pairs of digits represent hours:minutes:seconds:frames or HH:MM:SS:FF (Figure 11.2).

The recorded timecode address is then used to locate a position on hard disk, magnetic tape or any other recorded media, in much the same way that a letter carrier uses a written address to match up, locate and deliver a letter to a

FIGURE 11.2
Readout of a SMPTE timecode address in HH:MM:SS:FF.

FIGURE 11.3
Location of relative addresses: (a) postal address analogy; (b) timecode addresses and a cue point on longitudinal tape.

specific, physical residence (i.e., by matching up the address, you can then find the desired physical location point, as shown in Figure 11.3a). For example, let's suppose that a time-encoded analog multitrack tape begins at time 00:01:00:00, ends at 00:28:19:00 and contains a specific cue point (such as a glass shattering) at 00:12:53:19 (Figure 11.3b). By monitoring the timecode readout, it's a simple matter to locate the precise position that corresponds to the cue point on the tape and then perform whatever function is necessary, such as inserting an effect into the sound track at that specific point … CRASH!

Timecode word

The total of all time-encoded information that's encoded into each audio or video sync frame is known as a timecode word. Each word is divided into 80 equal segments, which are numbered consecutively from 0 to 79. One word covers an entire audio or video frame, such that for every frame there is a unique and corresponding timecode address. Address information is contained in the digital word as a series of bits that are made up of binary 1's and 0's, which (in the case of an analog SMPTE signal) are electronically encoded in the form of a modulated square wave.

This method of encoding information is known as biphase modulation. Using this code type, a voltage or bit transition in the middle of a half-cycle of a square

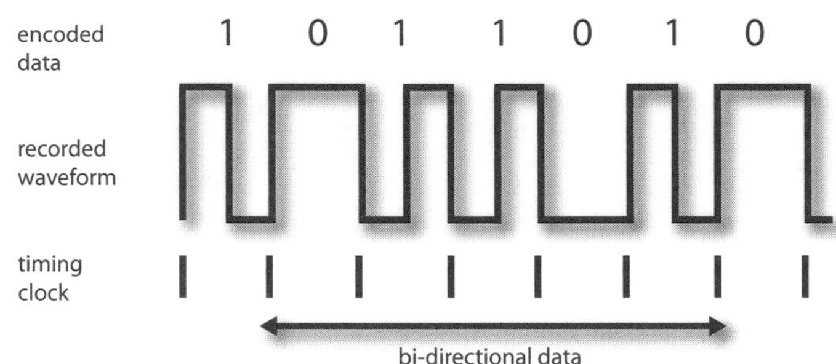

FIGURE 11.4
Biphase modulation encoding.

wave represents a bit value of 1, while no transition within this same period signifies a bit value of 0 (Figure 11.4). The most important feature about this system is that detection relies on shifts within the pulse and not on the pulse's polarity. Consequently, timecode can be read in either the forward or reverse direction, as well as at fast or slow shuttle speeds.

Tutorial: SMPTE Timecode

1. Go to the Tutorial section of www.modrec.com, click on SMPTE Audio Example and play the timecode soundfile. Not my favorite tune, but it's a useful one!
2. The 80-bit timecode word is subdivided into groups of 4 bits (Figure 11.5), whereby each grouping represents a specific coded piece of information. Each 4-bit segment represents a binary-coded decimal (BCD) number that ranges from 0 to 9. When the full frame is scanned, all eight of these 4-bit groupings are read out as a single SMPTE frame number (in hours, minutes, seconds and frames).

Sync information data

An additional form of information that's encoded into the timecode word is sync data. This information exists as 16 bits at the end of each timecode address word. These bits are used to define the end of each frame. Because timecode can be read in either direction, sync data is also used to tell the device which direction the tape or digital device is moving.

Timecode frame standards

In productions using timecode, it's important that the readout display be directly related to the actual elapsed time of a program, particularly when dealing with the exacting time requirements of broadcasting. For this reason, timecode frame rates may vary from one medium, production house or country of origin to another.

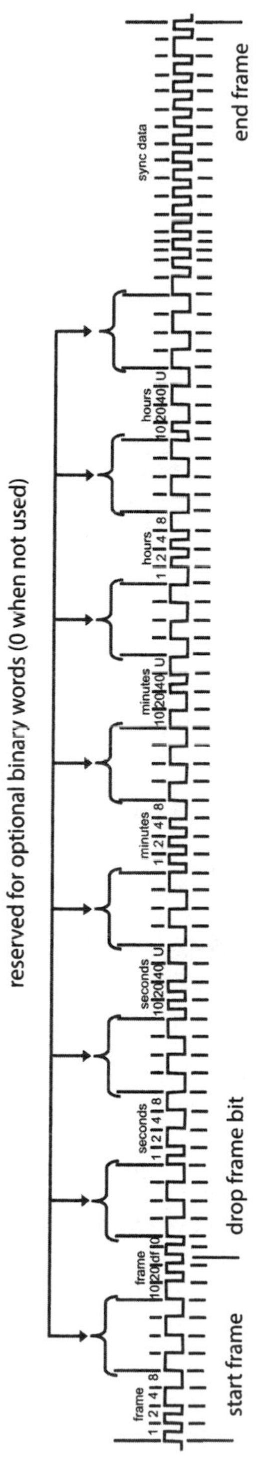

FIGURE 11.5
Biphase representation of the SMPTE timecode word.

- *30 fr/sec (monochrome U.S. video):* In the case of a black-and-white (monochrome) video signal, a rate of exactly 30 frames per second (fr/sec) is used. If this rate (often referred to as non-drop code) is used on a black-and-white program, the timecode display, program length and actual clock-on-the-wall time would all be in agreement.
- *29.97 fr/sec (drop-frame timecode for color NTSC video):* The simplicity of 30 fr/sec was eliminated, however, when the National Television Standards Committee (NTSC) set the frame rate for the color video signal in the United States and Japan at 29.97 fr/sec. Thus, if a timecode reader that's set up to read the monochrome rate of 30 fr/sec were used to read a color program, the timecode readout would pick up an extra 0.03 frame for every second that passes. Over the duration of an hour, the timecode readout would differ from the actual elapsed time by a total of 108 frames (or 3.6 seconds). To correct this difference and bring the timecode readout and the actual elapsed time back into agreement, a series of frame adjustments was introduced into the code. Because the goal is to drop 108 frames over the course of an hour, the code used for color has come to be known as drop-frame code. In this system, two frame counts for every minute of operation are omitted from the code, with the exception of minutes 00, 10, 20, 30, 40 and 50. This has the effect of adjusting the frame count so that it agrees with the actual elapsed duration of a program.
- *29.97 fr/sec (non-drop-frame code):* In addition to the color 29.97 drop-frame code, a 29.97 non-drop-frame color standard can also be found in video production. When using non-drop timecode, the frame count will always advance one count per frame, without any drops. As you might expect, this mode will result in a disagreement between the frame count and the actual clock-on-the-wall time over the course of the program. Non-drop, however, has the distinct advantage of easing the time calculations that are often required in the video editing process (because no frame compensations need to be taken into account).
- *25 fr/sec EBU (standard rate for PAL video):* Another frame rate format that's used throughout Europe is the European Broadcast Union (EBU) timecode. EBU utilizes SMPTE's 80-bit code word but differs in that it uses a 25 fr/sec frame rate. Because both monochrome and color video EBU signals run at exactly 25 fr/sec, an EBU drop-frame code isn't necessary.
- *24 fr/sec (standard rate for film work):* The medium of film differs from all of these in that it makes use of an SMPTE timecode format that runs at 24 fr/sec.

From the above, it's easy to understand why confusion often exists as to which frame rate should be used on a project. Basically, if you are working on an in-house project that doesn't incorporate time-encoded material that comes from the outside world, you should choose a rate that both makes sense for you and is likely to be compatible with an outside facility (should the need arise).

For example, electronic musicians who are working in-house in the US will often choose to work at 30 fr/sec. Those in Europe have it easy, because on that continent 25 fr/sec is the logical choice for all music and video productions. On the other hand, those who work with projects that come through the door from other production houses will need to take special care to reference their timecode rates to those used by the originating media house. This can't be stressed enough: If care isn't taken to keep your timecode references at the proper rate, while keeping degradation to a minimum from one generation to the next, the various media might have trouble syncing up when it comes time to put the final master together—and that could spell BIG trouble.

TIMECODE WITHIN DIGITAL MEDIA PRODUCTION

Given that SMPTE exists as a digitally encoded data form, current-day digital professional media devices are able to accept and communicate SMPTE directly without too much trouble. Professional camera, film, controllers and editing systems are able to directly chain and synchronize SMPTE using a multitude of complicated, yet standardized methods that make use of both digital- and analog-style timecode data streams.

Of course, there are a wide range of approaches that can be taken when approaching the subject of synchronizing video devices (cameras, video editing software and field audio recorders). These can range from "shoots" that make use of multiple cameras and separate field recorders, which are "locked" to a single timecode source on a set … all the way down to a simple camera and digital hand recorder, where the audio is manually synced up within the digital editor, without the use of timecode at all. The types of equipment and the ways that they deal with the technology of sync are ever-changing. It's simply important to keep abreast of current technology, read the manuals (about how connections and settings can best be made) and dive into the field of visual media production.

Broadcast wave file format

Although digital media devices and software are able to import, convert and communicate using the language of SMPTE timecode (in all its various flavors), the folks at the EBU (European Broadcast Union) saw the need to create a universal audio file format that would include data that's important to all forms of audio and visual media production. The result was the Broadcast Wave Format (BWF). Broadcast Wave is in most ways compatible with its Microsoft Wave counterpart, with the exception that it is able to embed metadata (information about the recorded content … take#, date, technical data, etc.) as well as SMPTE timecode data.

The inclusion of such important content and timecode information means that the time-related information will actually be imbedded within the file itself, allowing soundfiles that are imported into a video or audio editor to automatically snap to their appropriate timecode position. Obviously, BWF can be a huge time saver within the production and post-production process.

FIGURE 11.6
Many time-based media devices in the studio can be cost effectively connected via MIDI timecode (MTC).

MIDI TIMECODE

In earlier times, the synchronizing of audio devices to other video and/or audio devices was a very expensive proposition, far beyond the budget of most project or independent production houses. Today, however, an easy-to-use and inexpensive standard makes use of MIDI to transmit sync and timecode data throughout a connected production system (Figure 11.6). This has made it possible for even the most budget-minded bedrooms to be able to synchronize media devices and software using timecode.

MIDI timecode (MTC) was developed to allow electronic musicians, project studios, video facilities and virtually all other production environments to cost effectively and easily translate timecode into time-stamped messages that can be transmitted over MIDI data lines. Created by Chris Meyer and Evan Brooks, MIDI timecode allows SMPTE-based timecode to be distributed throughout the MIDI chain to devices or instruments that are capable of synchronizing to and executing MTC commands. MIDI timecode is an extension of MIDI 1.0, which makes use of existing sys-ex message types that were either previously undefined or were being used for other, non-conflicting purposes.

Since most modern recording systems include MIDI in their design, there's often no need for external hardware when making direct connections. Simply chain the MIDI data lines from the master to the appropriate slaves within the system (via physical cables, USB or virtual internal routing). Although MTC uses a reasonably small percentage of MIDI's available bandwidth (about 7.68% at 30 fr/sec), it's customary (but not necessary) to separate these lines from those that are communicating performance data when using physical MIDI cables. As with conventional SMPTE, only one master can exist within an MTC system, while any number of slaves can be assigned to follow, locate and chase to the master's speed and position. Because MTC is easy to use and is often included free in many system and program designs, this technology has grown to become the most straightforward and commonly used way to lock together such devices as DAWs, modular digital multitracks and MIDI sequencers, as well as analog and videotape machines (by using a MIDI interface that includes a SMPTE-to-MTC converter).

MIDI timecode messages

The MIDI timecode format can be divided into two parts:

- Timecode
- MIDI cueing

The timecode capabilities of MTC are relatively straightforward and allow devices to be synchronously locked or triggered to SMPTE timecode. MIDI cueing is a format that informs a MIDI device of an upcoming event that's to be performed at a specific time (such as load, play, stop, punch in/out, reset). This protocol envisions the use of intelligent MIDI devices that can prepare for a specific event in advance and then execute the command on cue.

MIDI timecode is made up of three message types:

- *Quarter-frame messages:* These are transmitted only while the system is running in real or variable speed time, in either forward or reverse direction. True to its name, four quarter-frame messages are generated for each timecode frame. Since 8 quarter-frame messages are required to encode a full SMPTE address (in hours, minutes, seconds and frames: 00:00:00:00), the complete SMPTE address time is updated once every two frames (In other words, MIDI timecode actually has half the resolution accuracy of its SMPTE timecode counterpart). Each quarter-frame message contains 2 bytes. The first byte is F1, the quarter-frame common header; the second byte contains a nibble (four hits) that represents the message number (0 through 7) and a nibble for encoding the time field digit.
- *Full messages:* Quarter-frame messages are not sent in the fast-forward, rewind or locate modes, because this would unnecessarily clog a MIDI data line. When the system is in any of these shuttle modes, a full message is used to encode a complete timecode address. After a fast shuttle mode is entered, the system generates a full address message and then places itself in a pause mode until the time-encoded slaves have located to the correct position. Once playback has resumed, MTC will again begin sending incremental quarter-frame messages.
- *MIDI cueing messages:* MIDI cueing messages are designed to address individual devices or programs within a system. These 13-bit messages can be used to compile a cue or edit decision list, which in turn instructs one or more devices to play, punch in, load, stop, and so on, at a specific time. Each instruction within a cueing message contains a unique number, time, name, type and space for additional information. At the present time, only a small percentage of the possible 128 cueing event types has been defined.

SMPTE/MTC conversion

Although MIDI timecode connections can be directly made between compatible MIDI devices, a SMPTE-to-MIDI converter is required to read incoming LTC SMPTE timecode and convert it into MIDI timecode (and vice versa) for other device types. These conversion systems are available as a stand-alone device or as an integrated part of an audio interface or multiport MIDI interface/patch bay/ synchronizer system (Figure 11.7).

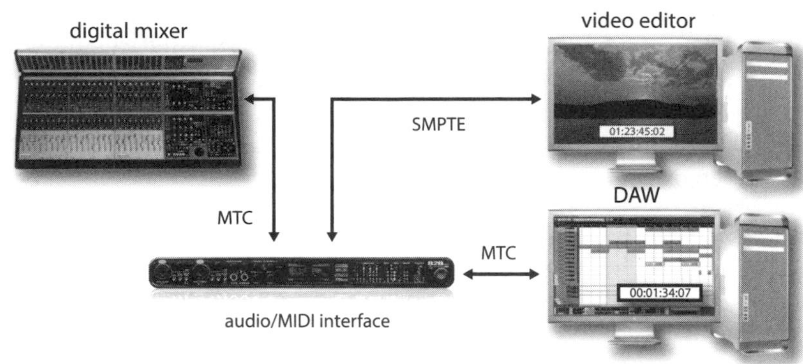

FIGURE 11.7
SMPTE timecode can often be generated throughout a production system, possibly as either LTC or as MTC via a capable MIDI or audio interface.

Timecode production in the analog audio and video worlds

Fortunately for us, most digital editing systems (such as digital video and audio workstations) are able to communicate timecode in a relatively seamless and straightforward manner (at least at a basic level, often requiring that the various systems be set to the same framerates, etc). Synching analog to analog or analog to digital devices, on the other hand, is often far less straightforward and should be understood … at least at its basic level.

Timecode that's recorded onto an analog audio or video cue track of an older-style video tape recorder is known as longitudinal timecode (LTC). LTC encodes a biphase timecode signal onto an analog track in the form of a modulated square wave at a bit rate of 2400 bits/sec. This recording of a perfect square wave onto a magnetic audio track is difficult, even under the best of conditions. For this reason, the SMPTE standard has set forth an allowable rise time of 25 ±5 microseconds for the recording and reproduction of valid code. This tolerance requires a signal bandwidth of 15 kHz, which is well within the range of most professional audio recording devices. Variable-speed timecode readers are often able to decode timecode information at shuttle rates ranging from 1/10th to 100 times normal playing speed. This is effective for most audio applications; however, in video postproduction it's often necessary to monitor videotape at slow or still speeds.

Because LTC can't be read at speeds slower than 1/10th to 1/20th normal play speed, two methods can be used for reading timecode from an analog playback device. The first of these uses a character generator to burn timecode addresses directly into the video image of a worktape copy. This superimposed readout allows the timecode to be easily identified, even at very slow or still picture shuttle speeds. As was mentioned, in most modern-day situations, LTC code will not be necessary whenever digital video and audio devices are involved.

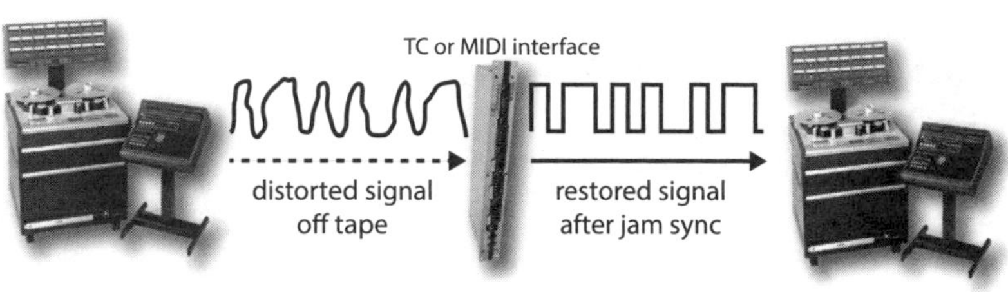

FIGURE 11.8
Jam sync is used to restore distorted SMPTE when copying code from one machine to another.

Timecode refresh and jam sync

Longitudinal timecode operates by recording a series of square-wave pulses onto magnetic tape. As you now know, it's somewhat difficult to record a square waveform onto analog magnetic tape without having the signal suffer moderate to severe waveform distortion (Figure 11.8). Although timecode readers are designed to be relatively tolerant of waveform amplitude fluctuations, such distortions are severely compounded when code is copied from one analog recorder to another by one or more generations. For this reason, a timecode refresher has been incorporated into most timecode synchronizers and MIDI interface devices that have sync capabilities. Basically, this process reads the degraded timecode information from a previously recorded track and then amplifies and regenerates the square wave back into its original shape so it can be freshly recorded to a new track or read by another device.

Should the quality of a SMPTE signal degrade to the point where the synchronizer can't differentiate between the pulses, the code will disappear and the slaves will stop unless the system includes a feature known as jam sync. Jam sync also refers to the synchronizer's ability to output the next timecode value, even though one has not appeared at its input. The generator is said to be working in a freewheeling fashion, since the generated code may not agree with the actual recorded address values; however, if the dropout occurs for only a short period, jam syncing works well to detect or refresh the signal. (This process is often useful when dealing with dropouts or undependable code from VHS audio tracks.) Two forms of jam sync options are available:

- Freewheeling
- Continuous

In the freewheeling mode, the receipt of timecode causes the generator's output to initialize when a valid address number is detected. The generator then begins to count in an ascending order on its own, ignoring any deterioration or discontinuity in code and producing fresh, uninterrupted SMPTE address numbers. Continuous jam sync is used in cases where the original address numbers must remain intact and shouldn't be regenerated as a continuously ascending count. After the reader has been activated, the generator updates the address count for

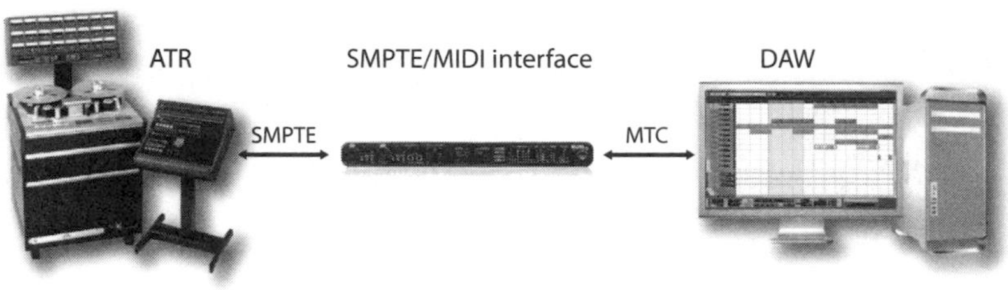

FIGURE 11.9 Example of timecode sync production using a simple MIDI interface synchronizer (possibly one that's designed into an audio interface) within a project studio.

each frame in accordance with incoming address numbers and outputs an identical, regenerated copy.

Synchronization using SMPTE timecode

To achieve a frame-by-frame timecode lock between multiple audio, video and film analog transports, it's necessary to use a device or integrated system that's known as a synchronizer. The basic function of a synchronizer is to control one or more tape, computer-based or film transports (designated as slave machines) so their speeds and relative positions are made to accurately follow one specific transport (designated as the master).

The use of a synchronizer within a project studio environment (Figure 11.9) often involves a multiport MIDI interface that includes provisions for locking an analog audio or video transport to a digital audio, MIDI or electronic music system by translating LTC SMPTE code into MIDI timecode (more on this later in the chapter). In this way, one simple device can cost effectively serve multiple purposes to achieve lock with a high degree of accuracy. Systems that are used in video production and in higher levels of production will often require a greater degree of control and remote-control functions throughout the studio or production facility. Such a setup will often require a more sophisticated device, such as a control synchronizer or an edit decision list (EDL) controller.

SMPTE OFFSET TIMES

In the real world of audio production, programs or songs don't always begin at 00:00:00:00 (as might easily happen when using a video or audio workstation). Let's say that you were handed a recording that needed a synth track to be laid down onto track 7 of a song that goes from 00:11:24:03 to 00:16:09:21. Instead of inserting more than 11 minutes of empty bars into a MIDI track on your synched DAW, you could simply insert an offset start time of 00:11:24:03. This means that the sequenced track will begin to increment from measure 1 at 00:11:24:03 and will maintain relative offset sync throughout the program.

Offset start times are also useful when synchronizing devices to an analog or videotape source that doesn't begin at 00:00:00:00. As you're probably aware, it takes a bit of time for an analog audio transport to settle down and begin playing (this wait time often quadruples whenever a videotape transport is involved). If a program's timecode were to begin at the head of the tape, it's extremely unlikely that you would want to start a program at 00:00:00:00, since playback would be delayed and extremely unstable at this time. Instead, most programming involving an analog audio or video media is striped with an appropriate pre-roll of anywhere from 10 seconds to 2 minutes. Such a pre-roll gives any analog transports ample time to begin playback and sync up to the master timecode source.

In addition, it's often wise to start the actual production or first song at an offset time of 01:00:00:00 (some facilities begin at 00:01:00:00). This minimizes the possibility that the synchronizer will become confused by rolling over at midnight; that is, if the content starts at 00:00:00:00, the pre-roll would be in the 23:59:00:00 range and the synchronizer would try to rewind to zero (rolling the tape off the reel) instead of rolling forward. Not always fun in the heat of a production!

DISTRIBUTION OF SMPTE SIGNALS

In a basic audio production system, the only connection that's usually required between the master machine and a synchronizer is the LTC timecode track. Generally, when connecting analog slave devices, two connections will need to be made between each transport and the synchronizer. These include lines for the timecode reproduce track and the control interface (which often uses the Sony 9-pin remote protocol for giving the synchronizer full logic transport and speed-related feedback information). LTC signal lines can be distributed throughout the production system in the same way that any other audio signal is distributed. They can be routed directly from machine to machine or patched through audio switching systems via balanced, shielded cables or unbalanced cables, or a combination of both. Because the timecode signal is biphase or symmetrical, it's immune to cable polarity problems.

TIMECODE LEVELS

One problem that can plague systems using timecode is crosstalk. This happens when a high-level signal leaks into adjacent signal paths or analog tape tracks. Currently, no industry standard levels exist for the recording of timecode onto magnetic tape or digital tape track; however, the levels shown in Table 11.1 can help you get a good signal level while keeping distortion and analog crosstalk to a minimum.

Table 11.1 Optimum Timecode Recording Levels

Tape Format	Track Format	Optimum Recording Level
ATR	Edge track (highest number)	–5 to –10 VU
Digital device	Highest number track or dedicated timecode I/O ports	–20 dB

Note: If the VTR is equipped with automatic gain compensation (AGC), override the AGC and adjust the signal gain controls manually.

REAL-WORLD APPLICATIONS USING TIMECODE AND MIDI TIMECODE

Before we delve into the many possible ways that a system can be set up to work in a timecode environment, it needs to be understood that each system will often have its own particular personality and that the connections, software and operation of one system might differ from those of another. This is often due to factors such as system complexity and the basic hardware types that are involved, as well as the type of hardware and software systems that are installed in a DAW. Larger, more expensive setups that are used to create television and film soundtracks will often involve extensive timecode and system interconnections that can easily get complicated.

Fortunately, the use of MIDI timecode and digital systems has greatly reduced the cost and complexity of connecting and controlling a synchronous pro and project studio system down to levels that can be easily managed by both experienced and novice users. Having said these things, I'd still like to stress that solving synchronization problems will often require as much intuition, perseverance, insight and art as it will technical skill. For the remainder of this chapter, we'll be looking into some of the basic concepts and connections that can be used to get your system up and running. Beyond this, the next best course of action will be to consult your manuals, seek help from an experienced friend or call the tech department about the particular hardware or software that's giving both you and your system the willies.

Master/slave relationship

Since synchronization is based on the timing relationship between two or more devices, it follows that the logical way to achieve sync is to have one or more devices (known as slaves) follow the relative movements of a single transport or device (known as the master). The basic rule to keep in mind is that there can be only one master in a connected system; however, any number of slaves can be set to follow the relative movements of a master transport or device (Figure 11.10).

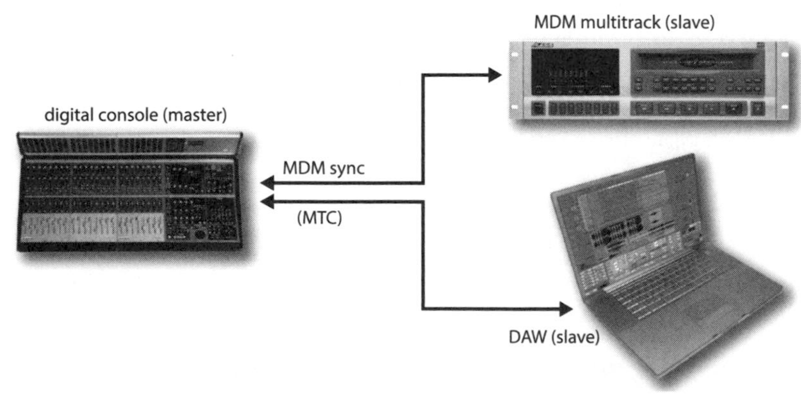

FIGURE 11.10 There can be only one master in a synchronized system; however, there can be any number of slaves.

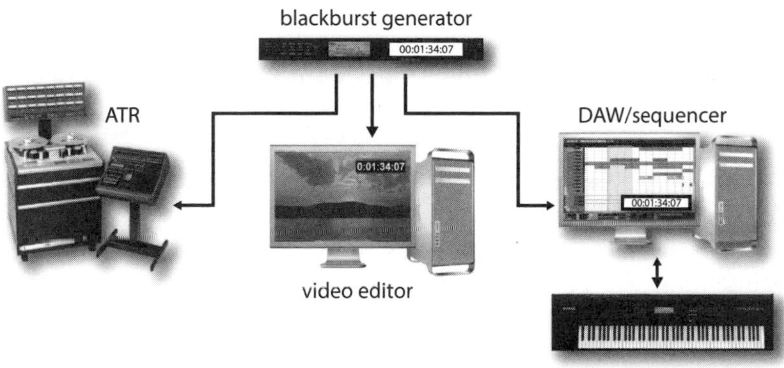

FIGURE 11.11
Example of a system whose overall timing elements are locked to a black burst reference signal.

Generally, the rule for deciding which device will be the master in a production system can best be determined by asking a few questions:

- What type of media is the master timecode media recorded on?
- Which device will provide the most stable timing reference?
- Which device will most easily serve as the master?

If the master comes to you from an outside source, asking lots of questions about the source specs will most likely solve many of your problems. If the project is in-house and you have total say in the matter, you might want to research your options more fully, to make the best choice for your facility. The following sections can help give you insights into which devices will best serve as the master within a particular system.

Video's need for a stable timing reference

Whenever a video signal is copied from one machine to another, it's essential that the scanned data (containing timing, video and user information) be copied in perfect sync from one frame to the next. Failure to do so will result in severe picture breakup or, at best, the vertical rolling of a black line over the visible picture area. Copying video from one machine to another generally isn't a problem, because the video recording device that's doing the copying can obtain its sync source from the playback machine. Video postproduction houses, however, often simultaneously use any number of video recorders, switchers and edit controllers during the production and editing of a single program. Mixing and switching between these combined sources without a stable sync source would almost certainly result in chaos … with the end result being a very unhappy client.

Fortunately, referencing all of the video, audio and timing elements to an extremely stable timing source (called a black burst or house sync generator) will generally resolve this sync nightmare. This reference clock serves to synchronize the video frames and timecode addresses that are received or transmitted by every video-related device in a production facility, so the leading frame edge of every video signal occurs at exactly the same instant in time (Figure 11.11). By resolving all video and audio devices to a single black burst reference, you're assured that relative frame transitions and speeds throughout the system will be consistent and stable. This even holds true for slaved analog machines, because their transport's wow and flutter can be smoothed out when locked to such a stable timing reference.

Digital audio's need for a stable timing reference

The process of maintaining a synchronous lock between digital audio devices or between digital and analog systems differs fundamentally from the process of maintaining relative speed between analog transports. This is due to the fact that a digital system generally achieves synchronous lock by adjusting its playback sample rate (and thus its speed and pitch ratio) so as to precisely match the relative playback speed of the master transport. Therefore, whenever a digital system is synchronized to a time-encoded master, a stable timing source is extremely important in order to keep jitter (in this case, an increased distortion due to rapid pitch shifts) to a minimum. In other words, the source's program speed should vary as little as possible to prevent any degradation in the digital signal's quality. As such, a digital audio system that's working within a video production environment would also benefit from the above mentioned house sync timing source.

Video workstation or recorder

Since video is often an extremely stable timing source, a digital video editor (or even analog video device) would be the next best stable timing source. This process shouldn't be taken lightly, because the timecode must conform to the timecode addresses on the original video master or working copy (see the earlier discussion of jam sync). Basically, the rule of thumb is: If you're working on a project that was created out of house, always use the code that was provided by the original production team. Striping your own code or erasing over the original code with your own would render the original timing elements useless, because the new code wouldn't relate to the original addresses or include any timing variations that might be a part of the original master source. In short, make sure that your working copy includes SMPTE that's a regenerated copy of the original code! Should you overlook this, you might run into timing and sync troubles, either immediately or later in the postproduction phase, while putting the music or dialog back together with the final video master—factors that will definitely lead to premature hair and client loss.

Digital audio workstations

A computer-based DAW can often be set to act as either a master or slave. This will ultimately depend on the software and the situation, because most professional workstations can be set to chase (or be triggered by) a master timecode source, as well as generate timecode (often in the form of MIDI timecode or SMPTE within a higher-end system).

Most modern DAWs include support for displaying a video track within a session (both as a separate video screen that can be displayed on the monitor

FIGURE 11.12
Most high-end DAW systems are capable of importing a videofile directly into the project session window. (Courtesy of Steinberg Media Technologies GMBH, www.steinberg.net)

desktop and in the form of a video thumbnail track that appears within the track view). Of course, the video track provides important visual cues for tracking live music, accurately placing automation moves and effects (sfx) at specific hitpoints within the scene or for adding music sweetening (Figure 11.12). This feature allows audio to be built up within a DAW environment without the need for synching to an external device at all. As you might expect, the use of recorded tracks, software instruments and internal mixing capabilities, tracks can easily be built up, spotted and mixed—all inside the box.

Routing timecode to and from your computer

From a connections standpoint, most DAW, MIDI and audio application software packages are flexible enough to let you choose from any number of available sync sources (whether connected to a hardware port, MIDI interface port or virtual sync driver). All you have to do is assign all of the slaves within the system to the device driver that's generating the system's master code (Figure 11.13). In many cases, it's best to have your DAW or editor generate the master code for the system with the appropriate settings and timecode address.

Analog audio recorders

In many audio production situations, whenever an analog tape recorder is connected in a timecode environment, this machine will most often want to act as the master in an LTC environment. Although this might be counter-intuitive, it's far easier and less expensive for an analog recorder to generate a master SMPTE

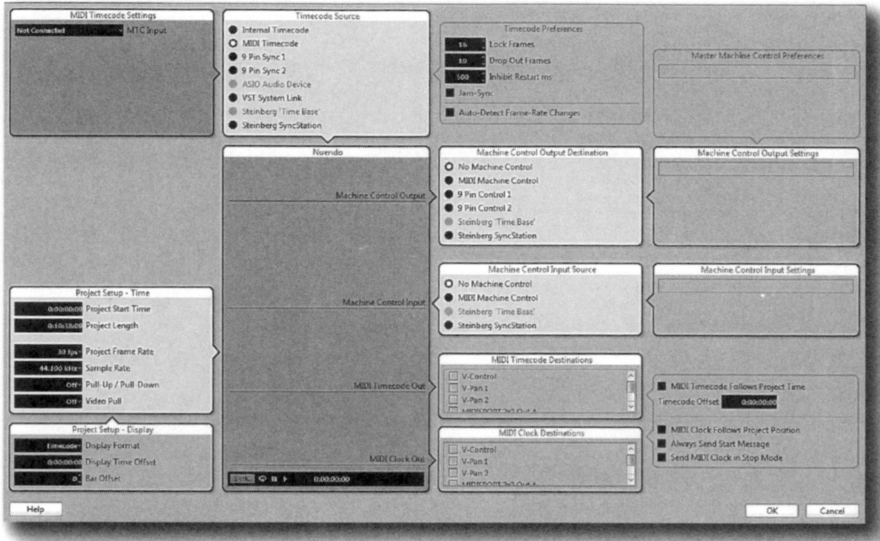

FIGURE 11.13 Cubase/Nuendo Sync Preferences dialog box. (Courtesy of Steinberg Media Technologies GMBH, www.steinberg.net)

code, than to be controlled in an external slave relationship. This is because of the special equipment that's generally required to continuously adjust the regulator's speed (using a DC capstan servo), so as to maintain a synchronous relationship to the master SMPTE address.

A simple caveat

The above guidelines are just that—guidelines. As you might expect, each and every setup will be slightly different, and might require that you come up with a novel solution to a quirky problem. Again, the internet is full of insights and solutions from those who have already gone down that long and treacherous path. Be warned though. It's important that you make your decisions wisely, lest a problem raise its ugly head at a later and crucial time.

KEEPING OUT OF TROUBLE

Here are a few guidelines that can help save your butt when using SMPTE and other timecode translations during a project:

- Familiarize yourself with the hardware and software involved in a project BEFORE the session starts.
- When in doubt about frame rates, special requirements or anything else, for that matter … ask! You (and your client) will be glad you did.
- Fully document your timecode settings, offsets, start times, etc.
- If the project isn't to be used in-house, ask the producer what the proper frame rate should be. Don't assume or guess it.

- When beginning a new session (when using a tape-based device), always stripe the master contiguously from the beginning to end before the session begins. It never hurts to stripe an extra tape, just in case. This goes for both analog and digital devices.
- Whenever analog machines are involved, start generating new code at a point before midnight (i.e., 01:00:00:00 or 00:01:00:00 to allow for a pre-roll). If the project isn't to be used in-house, ask the producer what the start times should be. Don't assume or guess it.
- Never dub (copy) timecode directly. Always make a refreshed (jam synched) copy of the original timecode (from an analog master) before the session begins.
- Never use slow analog videotape speeds. In EP mode, a VHS deck runs too slowly to record or reproduce a reliable code.
- Disable noise reduction on analog audio tracks (on both audio and video decks).
- Work with copies of the original production video, and make a new one when sync troubles appear.
- It's not unusual for the timecode to be read incorrectly (when short drop-outs occur on the track, usually on videotape). When this happens, you might set the synchronizer to freewheel once the transports have initially locked.

In closing, I'd like to point out that synchronization can be a simple procedure or it can be a fairly complex one, depending on your requirements and the type of equipment that's involved. A number of books and articles have been written on this subject. If you're serious about production, I suggest that you do your best to keep up on it. Although the fundamentals stay the same, new technologies and techniques are constantly emerging. As always, the best way to learn is simply by reading and then jumping in and doing it.

CHAPTER 12
Amplifiers

In the world of audio, amplifiers have many applications. They can be designed to amplify, equalize, combine, distribute or isolate a signal. They can even be used to match signal impedances between devices. At the heart of any *amplifier* (*amp*) system is either a vacuum tube or a semiconductor-type transistor series of devices. Everyone has heard of these regulating devices, but few have a grasp of how they operate, so let's have a basic look into these electronic wonders.

AMPLIFICATION

To best understand how the theoretical process of amplification works, let's draw on an analogy. The original term for the tube used in early amplifiers is *valve* (a term that's still used in England and other Commonwealth countries). If we hook up a physical valve to a high-pressure water hose, large amounts of water pressure can be controlled with very little effort, simply by turning the valve (Figures 12.1a and 12.1b). Using a small amount of expended energy, a trickle of water can be turned into a high-powered gusher and back again. In practice, both the vacuum tube and the transistor work much like this valve. For example, a vacuum tube operates by placing a DC current across its plate and a heated cathode element (Figure 12.2). A wire mesh grid separating these two elements acts like a control valve, allowing electrons to pass from the plate to the cathode. By introducing a small and varying signal at the input onto the tube's grid, a much larger electrical signal can be used to correspondingly regulate the flow of electrons between the plate and the cathode (Figure 12.3).

Although the transistor (a term originally derived from "trans-resistor" meaning a device that can easily change resistance) operates under a different electrical principle than a tube-based amp, the valve analogy still applies. Figure 12.4 shows a basic amplifier schematic with a DC power source that's placed across the transistor's collector and emitter points. As with the valve analogy, by presenting a small control signal at the transistor's base, the resistance between the collector and emitter will correspondingly change. This allows a much larger analogous signal to be passed to the device's output.

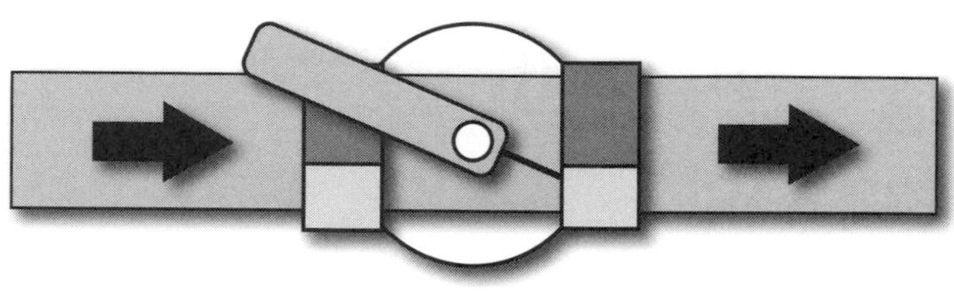

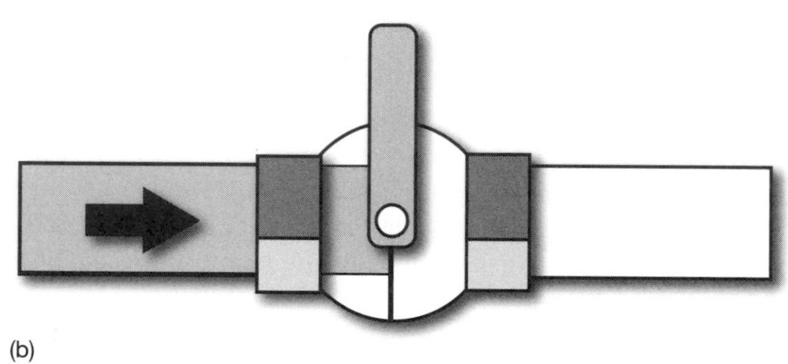

FIGURE 12.1
The current through a vacuum tube or transistor is controlled in a manner that's similar to the way that a valve tap can control water pressure through a water pipe: (a) open valve; (b) closed valve.

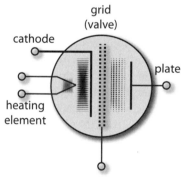

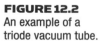

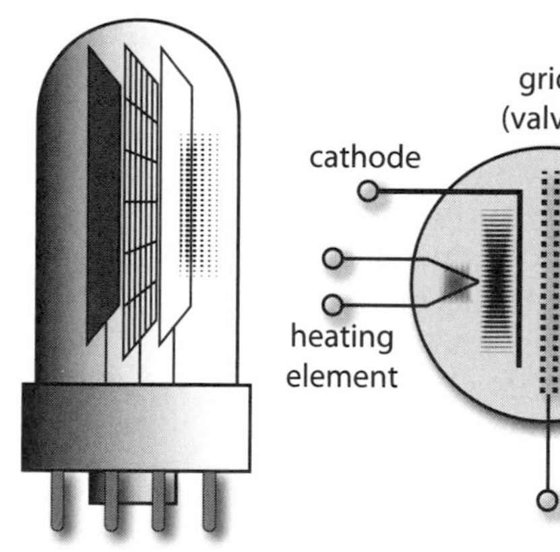

FIGURE 12.2
An example of a triode vacuum tube.

As a device, the transistor isn't inherently linear; that is, applying an input signal to the base won't always produce a corresponding output change. The linear operating region of a transistor lies between the device's lower-end cutoff region and an upper saturation point (Figure 12.5). Within this operating region, however, changes at the input will produce a corresponding (linear) change in the collector's output signal. When operating near these cutoff or saturation points, the base current lines won't be linear and the output will be distorted. In order to keep the signal within this linear operating range, a DC bias voltage signal is applied to the base of the transistor (for much the same reason a high-frequency bias signal is applied to an analog recording head). After a corrective voltage has been applied and sufficient amplifier design characteristics have been met, the amp's dynamic range will be limited by only two factors: noise (which results from thermal electron movement within the transistor and other circuitry) and saturation.

Amplifier *saturation* results when the input signal is so large that its DC output supply isn't large enough to produce the required, corresponding output signal. Overdriving an amp in such a way will cause a mild to severe waveform distortion effect known as *clipping* (Figure 12.6). For example, if an amp having a

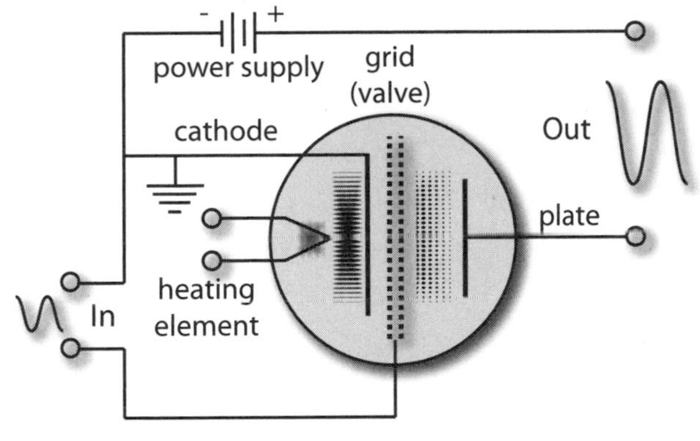

FIGURE 12.3
A schematic showing how small changes in voltage at the tube's grid can produce much larger, corresponding amplitude changes between its cathode and plate.

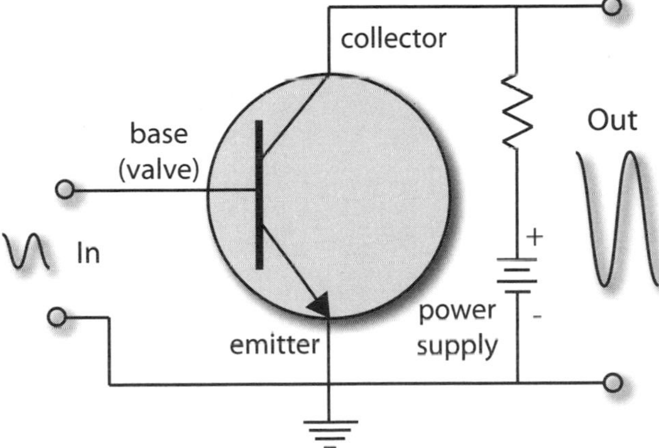

FIGURE 12.4
A simple schematic showing how small changes in current at the transistor's base can produce much larger, corresponding amplitude changes through the emitter and collector to the output.

supply voltage of +24 volts (V) is operating at a gain ratio of 30:1, an input signal of 0.5 V will produce an output of 15 V. Should the input be raised to 1 V, the required output level would have to be increased to 30 V. However, since the maximum output voltage is limited to 24 V, levels above this point will be chopped off or "clipped" at the upper and lower edges of the waveform. Whenever a transistor and integrated circuit design clips, severe odd-order harmonics are often introduced that are immediately audible. Tube amp designs, on the other hand, tend to lend a more musical sounding, even-order harmonic aspect to a clipped signal. I'm sure you're aware that clipping distortion can be a sought-after part of an instrument's sound (electric guitars thrive on it); however, it's rarely a desirable effect in quality studio and monitoring gear. The best way to avoid undesirable distortion is

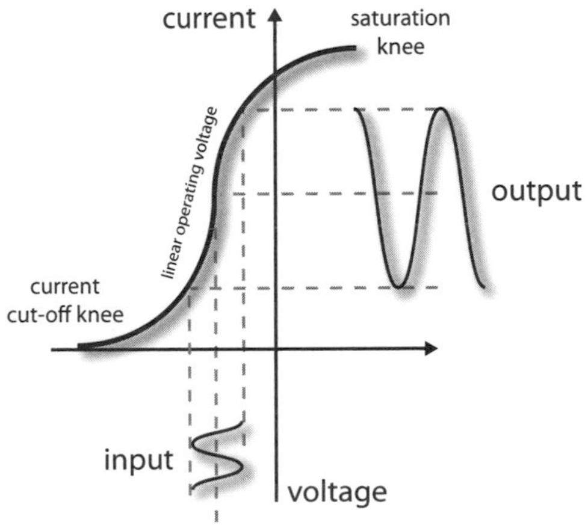

FIGURE 12.5
Operating region of a transistor.

to be aware of the various amp and device gain stages throughout the studio's signal chains.

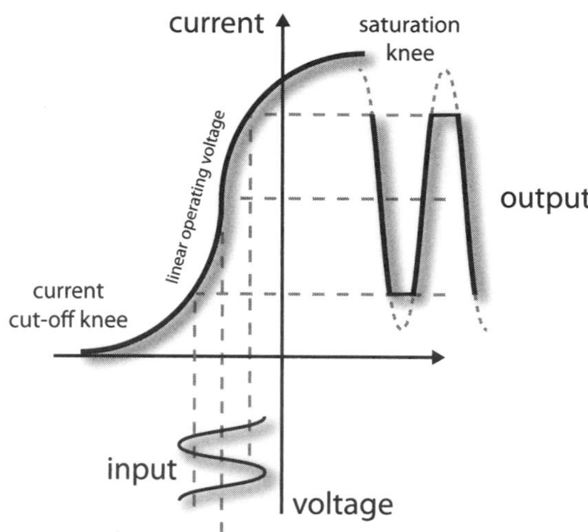

FIGURE 12.6
A clipped waveform.

THE OPERATIONAL AMPLIFIER

An *operational amplifier* (*op-amp*) is a stable, high-gain, high-bandwidth amp that has a high-input impedance and a low-output impedance. These qualities allow op-amps (Figure 12.7) to be used as a basic building block for a wide variety of audio and video applications, simply by adding components onto the basic circuit in a building-block fashion to fit the design's needs. To reduce an op-amp's output gain to more stable, workable levels, a negative feedback loop is often required. *Negative feedback* is a technique that applies a portion of the output signal through a limiting resistor back into the negative or phase-inverted input terminal. By feeding a portion of the amp's output back into the input out of phase, the device's output signal level is reduced. This has the effect of controlling the gain (by varying the negative resistor value) in a way that also serves to stabilize the amp and further reduce distortion.

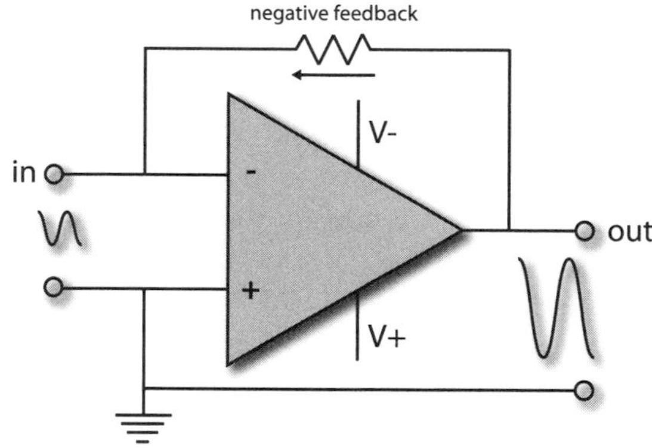

FIGURE 12.7
Basic op-amp configuration.

Preamplifiers

One of the mainstay amplifier types found at the input section of most professional mixer, console and outboard devices is the *preamplifier* (*preamp*). This amp type is often used in a wide range of applications, such as boosting a mic's signal to line level, providing variable gain for various signal types and isolating input signals and equalization, just to name a few. Preamps are an important component in audio engineering because they often set the "tone" of how a device or system will sound. Just as a microphone has its own sonic character, a preamp design will often have its own "sound." Questions such as "Are the op-amps designed from quality components?" "Do they use tubes or transistors?" and "Are they quiet or noisy?" are all-important considerations that can greatly affect the overall sound of a device.

Equalizers

Basically, an *equalizer* is nothing more than a frequency-discriminating amplifier. In most analog designs, equalization (EQ) is achieved through the use of

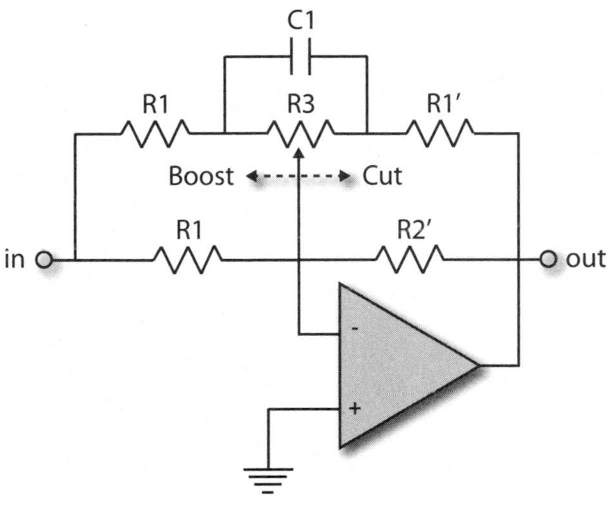

FIGURE 12.8
Low-frequency equalizer circuit.

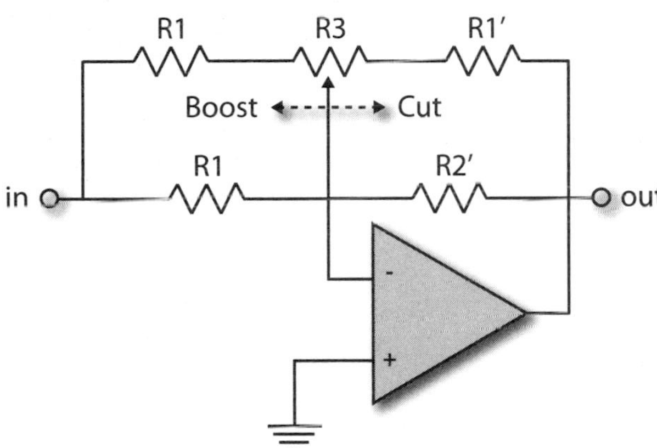

FIGURE 12.9
High-frequency equalizer circuit.

resistor/capacitor networks that are located in an op-amp's negative feedback loop (Figures 12.8 and 12.9) in order to boost (amplify) or cut (attenuate) certain frequencies in the audible spectrum. By changing the circuit design, complexity and parameters, any number of EQ curves can be achieved.

SUMMING AMPLIFIERS

A *summing amp* (also known as an active combining amplifier) is designed to combine any number of discrete inputs into a single output signal bus, while providing a high degree of isolation between them (Figure 12.10). The summing amplifier is an important component in analog console/mixer design because the large number of internal signal paths require a high-degree of isolation in order to prevent signals from inadvertently leaking into other audio paths.

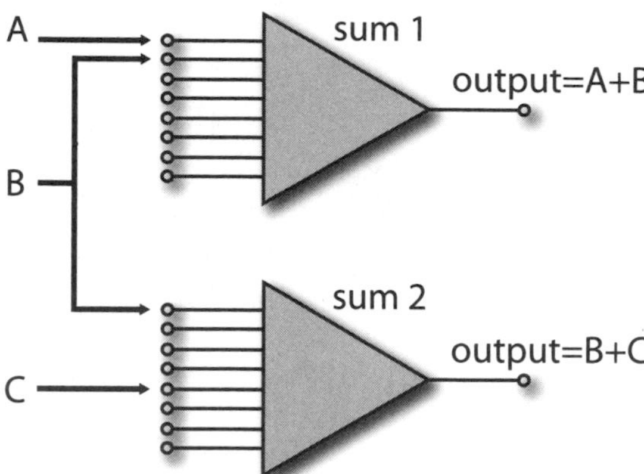

FIGURE 12.10
A summing amp is used to provide isolation between various inputs and/or outputs in a signal chain.

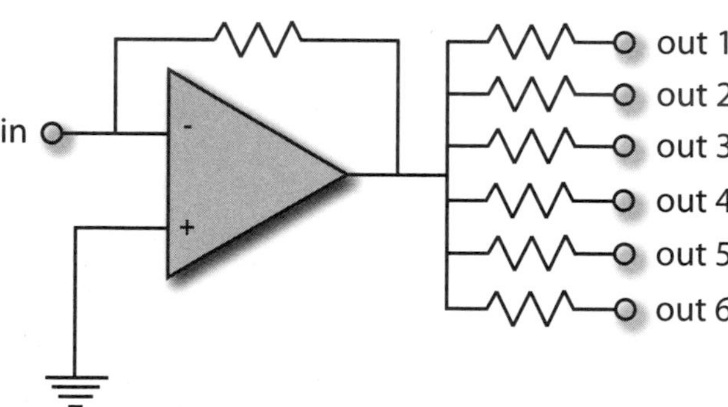

FIGURE 12.11
Distribution amp.

DISTRIBUTION AMPLIFIERS

Often, it's necessary for audio signals to be distributed from one device to several other devices or signal paths within a recording console or music studio. In this situation, a *distribution amp* isn't used to provide gain but instead will amplify the signal's current (power) that's being delivered to one or more loads (Figure 12.11). Such an amp, for example, could be used to boost the overall signal power so that a single feed could be distributed to a large number of headphones during a string or ensemble session.

POWER AMPLIFIERS

As you might expect, *power amplifiers* (Figures 12.12 and 12.13) are used to boost the audio output to a level that can drive one or more loudspeakers at their rated volume levels. Although these are often reliable devices, power amp designs have their own special set of problems. These include the fact that transistors don't like to work at the high temperatures that can be generated during

FIGURE 12.12
QSC Audio's RMX 1850HD professional power amplifier. (Courtesy of QSC Audio Products, Inc., www.qscaudio.com.)

FIGURE 12.13
Bryston 9B SST professional 5-channel amplifier. (Courtesy of Bryston Ltd., www.bryston.ca.)

continuous, high-level operation. Such temperatures can also result in changes in the unit's response and distortion characteristics … or outright failure. This often requires that protective measures (such as fuse and thermal protection) be taken. Fortunately, many of the newer amplifier models offer protection under a wide range of circuit conditions (such as load shorts, mismatched loads and even open "no-load' circuits) and are usually designed to work with speaker impedance loads ranging between 4 and 16 ohms (most speaker models are designed to present a nominal load of 8 ohms). When matching an amp to a speaker set, the amp should be capable of delivering sufficient power to properly drive the speakers. If the speaker's sensitivity rating is too low or the power rating too high for what the amp can deliver, there could be a tendency to "overdrive" the amp at levels that could cause the signal to be clipped. In addition to sounding distorted, clipped signals can contain a high-level DC component that could potentially damage the speaker's voice coil drivers.

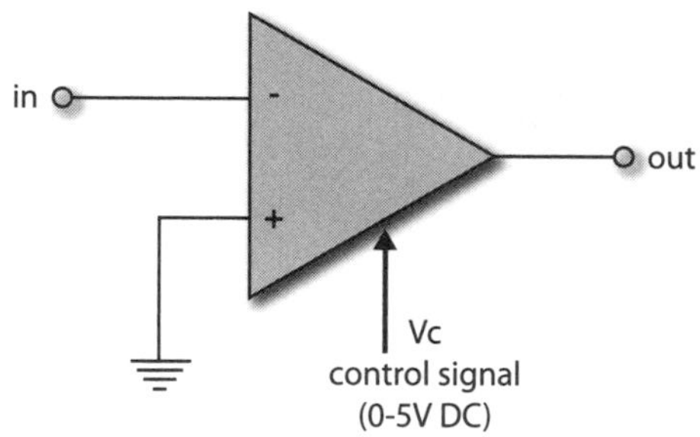

FIGURE 12.14
Simplified example of a voltage-controlled amplifier.

VOLTAGE- AND DIGITALLY CONTROLLED AMPLIFIERS

Up to this point, our discussion has largely focused on analog amps whose output levels are directly proportional to the signal level that's present at its input. Several exceptions to this principle are the *voltage-controlled amplifier* (*VCA*) and the *digitally-controlled amplifier* (*DCA*). In the case of the VCA, the overall output gain is a function of an external DC voltage (generally ranging from 0 to 5 V) that's applied to the device's control input (Figure 12.14). As the control voltage is increased, the analog signal will be proportionately attenuated. Likewise, a digitally controlled external voltage can be used to control the amp's overall gain. Certain older console automation systems, automated analog signal processors and even newer digital console designs make use of VCA technology to digitally store and automate levels.

With the wide acceptance of digital technology in the production studio, it's now far more common to find devices that use digitally controlled amplifiers to control the gain of an audio signal. Although most digital devices change the gain of a signal directly within the digital domain, it's also possible to change the gain of an analog signal using an external digital source. Much like the VCA, the overall gain of an analog amp can be altered by placing a series of digitally controlled step resistors into its negative feedback loop and digitally varying the amount of resistance that's required to achieve the desired gain.

CHAPTER 13

Power- and Ground-Related Issues

Throughout this book, we've seen how a production facility, in all its forms, involves the interconnection of various digital and analog devices to create a common task—to capture and produce good music without adding clicks, pops and spurious noises to our tracks. This brings us to two aspects that are often overlooked in the overall design of a facility:

1. The need for proper grounding techniques (the way that devices interconnect without introducing outside electrical noises)
2. The need for proper power conditioning (the purity and isolation of a room's power from the big, bad outside world)

GROUNDING CONSIDERATIONS

Proper grounding is essential to maintaining equipment safety; however, within an audio facility, small AC voltage potentials between various devices in an audio system can leak into a system's grounding circuit. Although these potentials are small, they are sometimes large enough to induce noise in the form of hums, buzzes or radio-frequency (RF) reception that can be injected (and amplified) directly into the audio signal path. These unwanted signals generally occur whenever improper grounding allows a piece of audio equipment to detect two or more different paths to ground.

Because grounding problems arise as a result of electrical interactions between any number of equipment combinations, the use of proper grounding techniques and troubleshooting within an audio production facility are by their very nature situational and often frustrating. As such, the following procedures are simply a set of introductory guidelines for dealing with this age-old problem. There are a great number of technical papers, books, methods and philosophies on grounding, and it's recommended that you carefully research the subject further before tackling any major ground-related problems. When in doubt, an experienced professional should be contacted, and care should be taken not to sacrifice safety.

- Keep all studio electronics on the same AC electrical circuit—most stray hums and buzzes occur whenever parts of the sound system are plugged into outlets from different AC circuits. Plugging into a circuit that's connected to such noise-generating devices as air conditioners, refrigerators, light dimmers, neon lights, etc., will definitely invite stray noise problems. Because most project studio devices don't require a lot of current (with the possible exception of power amplifiers), it's often safe to run all of the devices from a single, properly grounded line from the electrical circuit panel.
- Keep audio wiring away from AC wiring—whenever AC and audio cables are laid side-by-side, portions of the 60-Hz signal might be induced into a high-gain, unbalanced circuit as hum. If this occurs, check to see if separating or shielding the lines helps reduce the noise.
- When all else fails: If you only hear hum coming from a particular input channel, check that device for ground-related problems. If the noise still exists when the console or mixer inputs are turned down, check the amp circuit or any device that follows the mixer. If the problem continues to elude you, then …
- Disconnect all of the devices (both power and audio) from the console, mixer or audio interface, then methodically plug them back in one at a time (it's often helpful to monitor through a pair of headphones).
- Check the cables for bad connections or improper polarity. It's also wise to keep the cables as short as possible (especially in an unbalanced circuit).
- Another common path for ground loops is through a chassis into a 19-inch rack that not only provides a mount for multiple devices, but a common ground path. Test this by removing devices from the rack, one at a time. If needed, a device can be isolated from the rack by special nylon mounting washers.
- Investigate the use of a balanced power source, if traditional grounding methods don't work.

Troubleshooting a ground-related problem can be tricky and finding the problem's source might be a needle-in-a-haystack situation. When putting on your troubleshooting hat, it's usually best to remain calm, be methodical and consult with others who might know more than you do (or might simply have a fresh perspective).

POWER CONDITIONING

Whether your facility is a project studio or full-sized professional facility, it's often wise to regulate and/or isolate the voltage supply that's feeding one of your studio's most precious investments (besides you and your staff) … the equipment! This discussion of power conditioning can basically be broken down into three topics:

- Voltage regulation
- Eliminating power interruptions
- Keeping the lines quiet

In an ideal world, the power that's being fed to your studio outlets should be very close to the standard reference voltage of the country you are working in (e.g., 120V, 220V, 240V). The real fact of the matter is that these line voltages regularly fluctuate from this standard level, resulting in voltage sags (a condition that can seriously under-power your equipment), surges (rises in voltage that can harm or reduce the working life of your equipment), transient spikes (sharp, high-level energy surges from lightning and other sources that can do serious damage) and brown-outs (long-term sags in the voltage lines). Through the use of a voltage regulator, high-level, short-term spikes and surge conditions can be clamped, thereby reducing or eliminating the chance that the main voltage will rise above a standard, predetermined level.

Certain devices that are equipped with voltage regulation circuitry are able to deal with power sags, long-term surges and brown-outs by electronically switching between the multiple voltage level taps of a transformer so as to match the output voltage to the ideal main level (or as close to it as possible). One of the best approaches for regulating voltage fluctuations both above and below nominal power levels is to use an adequately powered uninterruptible power supply (UPS). In short, a quality UPS works by using a regulated power supply to constantly charge a rechargeable battery or bank of batteries. This battery supply is again regulated and used to feed sensitive studio equipment (such as a computer, bank of effects devices, etc.) with a clean and constant voltage supply.

Multiple-Phase power

One of the best defenses against noise, power drops and other equipment-related power problems is to make use of multiple circuits in your studio power design. This approach reduces the amount of potential interference between power systems by physically placing the major system groups on their own power circuit. For example, an ideal scenario would place each of the following groups on their own power phase (separate circuit):

Phase 1: Equipment, computer (UPS isolated) and studio power

Phase 2: Lighting

Phase 3: Air Conditioning and heating

The use of such a separate circuit design can greatly help to guard against the everyday surges, noise and any other number of gremlins that might creep into your beloved production system.

Balanced power

For those facilities that are located in areas where power lines are overtasked by heavy machinery, air conditioners and the like, a balanced power source might be considered. Such a device makes use of a power transformer (Figure 13.1) that has two secondary windings, with a potential to ground on each side of 60V. Because each side of the circuit is 180° out of phase with the other, a 120V

FIGURE 13.1
Furman IT-1220 balanced-output power conditioner. (Courtesy of Furman Sound, Inc., www.furmansound.com)

supply is maintained. Also, since the two 60V legs are out of phase, any hum, noise or RF that's present at the device's input will be canceled at the transformer's center tap (a null point that's tied to ground).

A few important points relating to a balanced power circuit include:

- A balanced power circuit is able to reduce line noise if all of the system's gear is plugged into it. As a result, the device must be able to deliver adequate power.
- Balanced power will not eliminate noise from gear that's already sensitive to hums and buzzes.
- When to use balanced power is often open to interpretation, depending on who you talk to. For example, some feel that a balanced power conditioner should be used only after all other options to eliminate noise have been explored, while others believe it is a starting point from which to build a noise-free environment.

CHAPTER 14

The Art and Technology of Mixing

In the past, almost all commercial music was mixed by a professional recording engineer under the supervision of a producer and/or artist. Although this is still true on many levels, with the emergence of the project studio, the vast majority of production facilities have become much more personal and cost effective in nature. Additionally, with the maturation of the digital revolution, artists, individuals, labels and enthusiasts are taking the time to become experienced in the common sense rules of creative and commercial mixing in their own production workspaces.

Within the music production industry, it's a well-known fact that most professional mixers have to earn their "ears" by logging countless hours behind the console. Although there's no substitute for this experience, the mixing abilities and ears of producers and musicians outside of the pro studio environment are also steadily improving as equipment quality gets better and as practitioners become more knowledgeable about proper mixing environments and techniques, often by mixing their own projects and compositions.

In this chapter, we'll be taking a basic look at the technology and art of mixing, and will gain insights into how the console, mixer, digital audio workstation and modern production equipment can work together to improve both your techniques and your sound.

The basic purpose of an *audio production console* or mixer (Figures 14.1 through 14.4) is to give us full control over volume, tone, blending and spatial positioning for any or all signals that are applied to its inputs from microphones, electronic instruments, effects devices, recording systems and other audio devices. An audio production console (which also goes by the name of *board, desk* or *mixer*) should also provide a straightforward way to quickly and reliably route these signals to any appropriate device in the studio or control room so they can be recorded, monitored and/or mixed into a final product. A console or mixer can be likened to an artist's palette in that it provides a creative control surface that allows an engineer to experiment and blend all the possible variables onto a sonic canvas.

FIGURE 14.1
DMH and Martin Skibba at a Solid State Logic Duality console at nhow hotel, Berlin. (Courtesy of nhow Berlin, www.nhow-hotels.com/berlin/en, Photo courtesy of Sash)

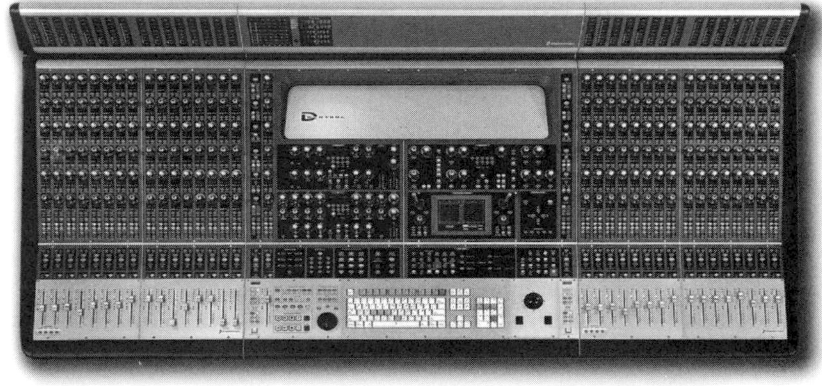

FIGURE 14.2
Digidesign ICON D-Command Integrated console. (Courtesy of Avid Technology, Inc., www.avid.com)

FIGURE 14.3
API Model 1608 analog mixing console. (Courtesy of Automated Processes, Inc., www.apiaudio.com)

FIGURE 14.4
Onyx 4-bus analog mixing console. (Courtesy of Loud Technologies, Inc., www.mackie.com)

THE RECORDING PROCESS

Before multitrack recorders came onto the production scene, all the sounds and effects of a recording had to be mixed together at one time during a live performance. If the recorded blend (or *mix*, as it's called) wasn't satisfactory or if one musician made a mistake, the selection had to be performed over until the desired balance and performance was obtained. However, with the introduction of multitrack recording, the production phase of a modern recording session has radically changed into one that generally involves three stages:

- Recording
- Overdubbing
- Mixdown

Recording

The *recording* phase involves the physical process of capturing live or sequenced instruments onto a recorded medium (disk, tape or whatever). Logistically, this process can be carried out in a number of ways:

- All the instruments to be used in a song can be recorded in a single live pass.
- Live musicians can be used to lay down the basic tracks (usually rhythm) of a song, thereby forming the basic foundation tracks that can be added to at a later time.
- Electronic or groove instruments can be performed into the MIDI sequenced tracks of a DAW in such a way as to build up the foundations of a song.

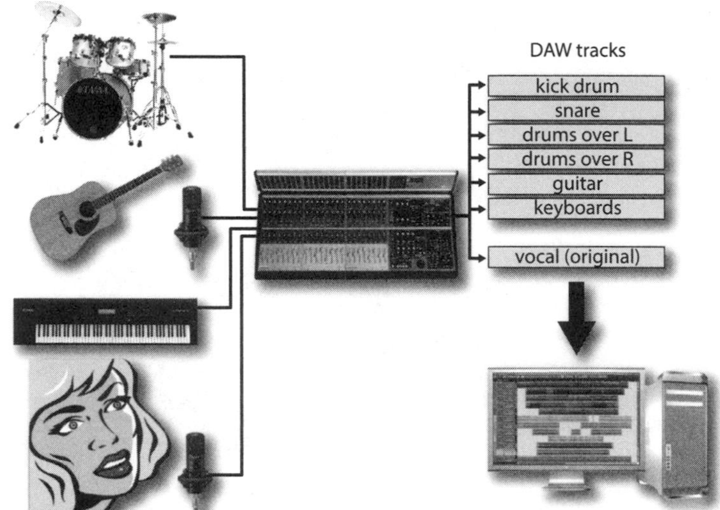

FIGURE 14.5
When recording popular-styled music, each instrument is generally recorded onto a separate track (or stereo tracks) of a recording device.

The last two of these procedures are the most commonly encountered in the recording of popular music. The resulting foundation tracks (to which other tracks can be added at a later time) are called *basic, rhythm* or *bed* tracks. These consist of instruments that provide the rhythmic foundations of a song and often include drums, bass, rhythm guitar and keyboards (or any combination thereof). An optional vocal guide (scratch track) can also be recorded at this time to help the musicians and vocalists capture the proper tempo and that all-important feel of a song.

When recording popular music, each instrument is generally recorded onto separate tracks of a tape or DAW recorder (Figure 14.5). This is accomplished by plugging each mic into an input strip on the console (or mic panel that's located in the studio), an input on an audio interface or an available input on a mixer. Once the input is turned on, the input strip/channel gain can be set to its optimum level and is then assigned to a track on a DAW or tape machine.

The beauty behind the modern recording process is that the various instrument mics can be recorded separately onto a track of a DAW or recorder, while having as much isolation between the recorded tracks as possible. This process is vitally important to the outcome of the overall project. The name of the game is to capture the best performance with the highest possible quality, best track isolation, and at optimum signal levels (often without regard to level balances on the other tracks).

For example, a soft vocal whisper might easily be boosted to recorded levels that are equal to those of a neighboring electric guitar track … or a toy piano might be boosted to the same track level as a full grand piano. In this conventional "one-instrument-per-track" setting, each signal should be recorded at a reasonably high level without overloading the digital or analog track. When recording to analog tape, recording at an optimal level will result in a good signal-to-noise ratio, so that tape hiss, preamp noise or other artifacts won't impair the final product. Digital tracks, on the other hand, are more forgiving (due to the increased headroom, especially at higher bit rates), so it's always a good idea to record signals to disk at recommended levels (often peaking at 12 dB below the maximum overload point).

Of course, the beauty behind this process is that the final volume, effects and placement changes can be made at a later time, during the mixdown stage. However, since levels aren't recorded with respect to balance, it's important that a monitor mix be created in order for the engineer and musicians properly hear what's being recorded to the various tracks.

MONITORING

Since the instruments have been recorded at levels that probably won't relate to the program's final balance, a separate mix must be made in order for the artists, producer and engineer to hear the instruments in their proper musical perspective; for this, a separate mix is often set up for *monitoring*. As you'll learn later in this chapter, a multitrack performance can be monitored in several ways. No particular method is right or wrong; rather, it's best to choose a method that matches your own personal production style. No matter which monitoring style is chosen, the overall result will generally be as follows:

- Commonly, when using a DAW, the actual session mix will often serve as the actual monitor (and possibly even studio monitor/headphone mix.
- When using a DAW, a separate monitor mix might need be created, so that the musicians can better hear in the studio over headphones. In fact, multiple separate "cue" mixes are often available (depending on each musician's individual listening needs).
- When using a console or hardware mixer during the recording process, each signal being fed into a channel will be fed to a monitor mix section (Figure 14.6). This sub-mixer section is used to mix the individual inputs and instrument groups (with regard to level, panning, effects, etc.) into a musical balance that's then fed to the control room's main monitor speakers.

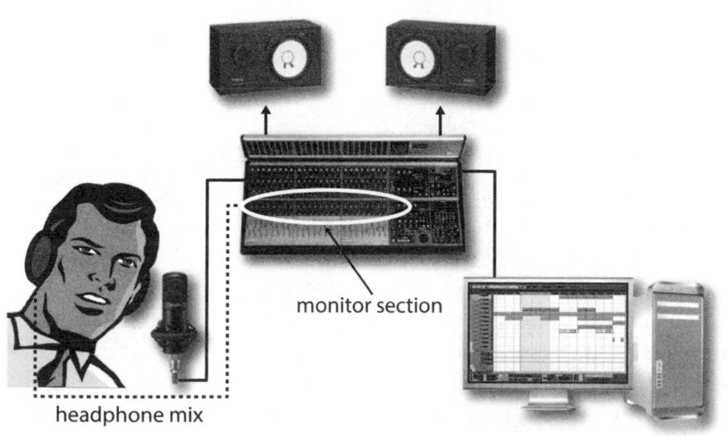

FIGURE 14.6
During recording, each signal can be fed to the monitor mix section, where the various instruments can be mixed and then fed to the control room's main speakers and/or the performer's headphones.

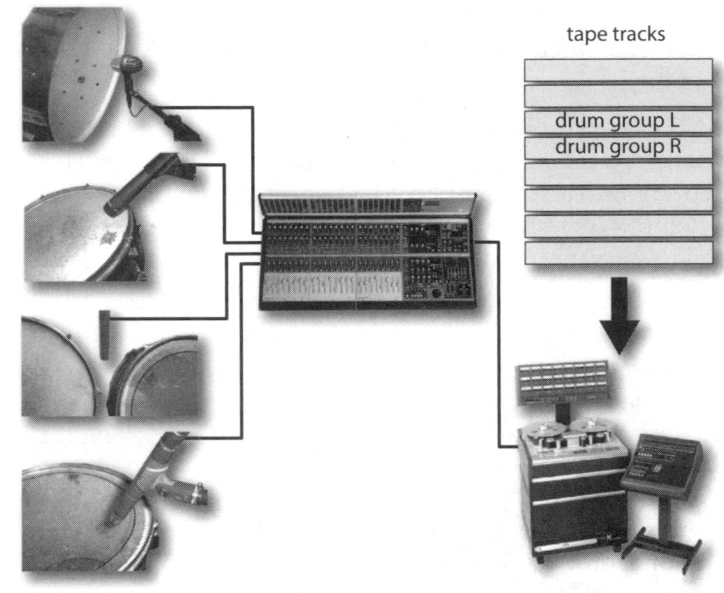

FIGURE 14.7
Several instruments can be "grouped" onto a single track (or set of tracks) by assigning each signal's input strip to the same console output bus.

GROUPING RECORDED TRACKS

When recording to an analog recorder (as most DAWs have few or no track limitations), an instrument or group of instruments that require multiple mics can be recorded onto a single track (or a stereo pair of tracks) by assigning the various input strips on a console or mixer to the same output bus (in a process known as *grouping*). These combined signals can then be balanced in level, panned, equalized and processed by monitoring the grouped outputs at the console or recording device (Figure 14.7). Whenever multiple sources are recorded onto a single track or tracks, a great deal of care should be taken when setting the various volumes, tonalities and placements. As a general rule, it's far more difficult to make changes to instruments that have been combined to one track, because changes to one instrument will almost always directly affect the overall group mix … so, whenever possible always record to separate tracks.

THE POWER OF PREPARATION

Whenever a producer is asked about the importance of preparation, more often than not they'll most likely place preparation at or near the top of the list for capturing a project's sound, feel and performance. The rhythm tracks (i.e., drums, guitar, bass and possibly piano) are often the driving backbone of a popular song, and recording them improperly can definitely get the project off to a bad start. Beyond making sure that the musicians are properly prepared and that the instruments are tuned and in top form (both of these being primarily the producer and/or band's job); it's the engineer's job to help capture a project's sound to disk or tape during the recording phase. Recording the best possible sound (both musically and technically), without having to excessively rely on the "fix it in the mix" approach will definitely start the project off on the right track.

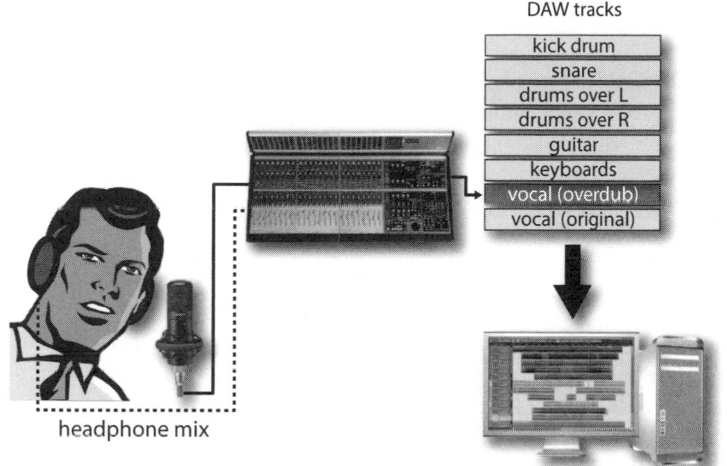

FIGURE 14.8
Once the basic recorded tracks have been "laid down," additional tracks can be added at a later time during the overdub phase.

Overdubbing

Instruments that aren't present during the original performance can be added at a later time to the existing multitrack project during a process known as *overdubbing* (Figure 14.8). At this stage, musicians listen to the previously recorded tracks over headphones and then play along in sync while recording new and separate tracks that are added to the basic tracks in order to fill out and finish the project.

Since the overdub is an isolated take that's being recorded to a new and separate track, it can be laid down to disk or tape in any number of ways. For example, with proper preparation and talent, an overdub take might get recorded in a single pass. However, if the musician has made minor mistakes during an otherwise good performance, sections of a take could be "punched in" on the same track to correct a bad overdub section (As a note: most DAWs can do this "non-destructively" i.e. without losing any of the previously-recorded data and with the ability to make in/out adjustments at a later time). For more information on overdubbing to a DAW track, please refer to Chapters 1 and 7, while info on overdubbing in sync to an analog recorder can be found in Chapter 5.

Mixdown

Once all the musical parts have been performed and recorded to everyone's satisfaction, the *mixdown* or *mix* stage can begin. During this process, the audio is repeatedly played while adjustments in level, panning, EQ, effects, etc., are made for each track and/or track grouping. Throughout this artistic process, the individually recorded signals are blended into a composite stereo, surround or mono signal that's fed to the master mixdown recorder (or internally mixed within a DAW's software mixer). When a number of mixes have been made and a single version has been approved, this recording (called the *master* or *final mix*) can be mastered to its intended medium and/or assembled (along with other songs or scenes in the project) into a final product.

REMEMBER

It's always wise to save your individual mixes under a new and identifiable name. ... You never know when you might want to revert to a previous mix version.

When a recording is made "in-the-box" using a DAW, the mixdown process is often streamlined, since the tracks, mixer and effects are all integrated into the system. In fact, much of the preparation is probably underway, as basic mix moves were probably made and saved into the session during the recording and overdub phases. When working "outside the box," the process can be quite different (depending upon the type of system that you're working with. Newer hardware console systems that include facilities to control and communicate with the DAW recording system can be equally as fluid and straightforward, with all of the console level, routing and effects settings being recalled on the console surface. Fully analog consoles will require that you manually set up and recall all of session settings each and every time that you need to return to the session. At this point, playback outputs of a DAW or multitrack recorder can be fed to the console's line inputs, by switching the console to the mixdown mode or by changing the appropriate input switches to the Line or Tape position.

Of course, there are several ways that a mix can be recorded to its final intended medium.

- Physically routed to and recorded on an ATR
- Recorded onto a new set of tracks on the DAW
- The mix can be exported to a set of mixdown tracks

For example, certain DAWS require that a final mix be "bounced" (another term for exported) to a final stereo or multichannel mix, while others are able to "export" the results of a final mix to a soundfile in non-realtime (without having to play through the mix in real-time). Those who prefer an analog sound can also play the mix out of a console or DAW in real-time to an analog recorder.

UNDERSTANDING THE UNDERLYING CONCEPT OF "THE MIXING SURFACE"

In order to understand the process of mixing, it's important to understand one of the most important concepts in all of audio technology: the *signal chain* (also known as the *signal path*). As is true with literally any audio system, the recording console can be broken down into functional components that are chained together into a larger (and hopefully manageable) number of signal paths. By identifying and examining the individual components that work together to form this chain, it becomes easier to understand the basic layout of any mixing system, no matter how large or complex. To better understand the layout of a mixer, let's start with the concept that it's built of numerous building-block components, each having an output (source) that moves to an input (destination).

In such a signal flow chain, the output of each source device must be connected to the input of the device or section that follows it, and so on and so forth … until the end of the audio path is reached. Whenever a link in this source-to-destination path is broken, no signal will pass. Although this "gozinta-gozouta" approach might seem like a simple concept, keeping it in mind can save your sanity and your butt when paths, devices and cables that look like tangled piles of spaghetti get out of hand.

Let's start our quest for understanding by taking a conceptual look at various mixing systems—from the analog hardware mixer to the general layout within a virtual mixing environment. In a traditional hardware mixer (which also goes by the name of board, desk or console) design, the signal flow for each input travels vertically down a plug-in strip known as an I/O module (Figure 14.9) in a manner that generally flows:

- From the input section
- Through a sends section (which taps off to external processing/monitoring devices)
- Into an equalizer (and other processing functions, such as dynamics)
- Passing through a monitor mix section (which taps off to external monitoring devices)
- To an output fader that includes pan positioning
- Into a routing section that can send signal to selected mix/signal destinations.

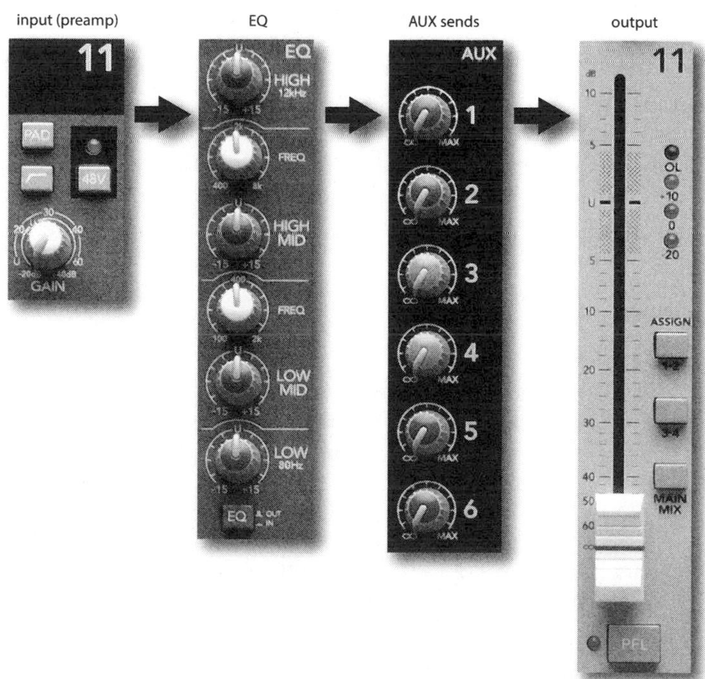

FIGURE 14.9
General anatomy of an input strip on the Onyx 4-bus analog mixing console. (Courtesy of Loud Technologies, Inc., www.mackie.com)

Although the layout of a traditional analog hardware mixer won't always hold true for the graphical user interface (GUI) layout of a virtual DAW mixer, the signal flow will most likely follow along the same or similar paths. Therefore, grasping the concept of an analog console's signal chain is also important for grasping mixers in the virtual world. A DAW's virtual mixer is likewise built from numerous building-block components, each having an input (source) that flows through the signal chain to an output (destination). The output of each source device must be literally or virtually connected to the input of the device that follows it—and so on until the end of the audio path is reached. Again, as with all of audio technology, keeping this simple concept in mind is important when paths, plug-ins and virtual paths seem to meld together into a ball of confusion. When the going gets rough, slow down, take a deep breath, read the manual (if you have the time)—and above all be patient and keep your wits about you.

Figures 14.10 through 14.13 show the general I/O stages of several virtual mixing systems. It's important that you take the time to familiarize yourself with the inner workings of your own DAW (or those that you might come in contact with) by reading the manual, pushing buttons and by diving in and having fun with your own projects.

So, now that we have a fundamental idea of how a hardware and software mixing system is laid out, let's discuss the various stages in greater detail as they flow through the process, starting with a channel's input, on through the various processing and send stages and then out to the final mix destination.

1 Channel input (preamp)

The first link in the input chain is the channel input (Figure 14.14). This serves as a preamp section to optimize the signal gain levels at the input of an I/O module before the signal is processed and routed. On a hardware console, mixer or audio interface that has built-in mic preamps, either a mic or line input can be selected to be the signal source (Figure 14.15). Although these values vary between designs, *mic trims* are typically capable of boosting a signal over a range of +20 to +70 dB, while a *line trim* can be varied in gain over a range of –15 (15-dB pad) to +45 dB or more. Gain trims are a necessary component in the signal path, because the output level of a microphone is typically very low (–45 to –55 dB) and requires that a high-quality, low-noise amp be used to raise and/or match the various mic levels in order for the signal to be passed throughout the console at an optimum level (as determined by the console's design and standard operating levels).

Of course, a mic preamp can take many forms—it might be integrated into a console or mixers input strip (as referred to above), it might exist as an external hardware pre that was carefully chosen for its pristine or special sound character or it might be directly integrated into our workstation's audio interface. Any of these options are a valid way of boosting the mic's signal to a level that can be manipulated, monitored and/or recorded.

Whenever a mic or line signal is boosted to levels that cause the preamp's output to be overdriven, severe clipping distortion will almost certainly occur. To avoid the dreaded LED overload light, the input gain must be reduced (by simply turning down the gain trim or by inserting an attenuation pad into the circuit). Conversely, signals that are too low in level will unnecessarily add noise into the signal path. Finding the right levels is often a matter of knowing your equipment, watching the meter/overload displays and using your experience.

Input attenuation pads that are used to reduce a signal by a specific amount (e.g., −10 or −20 dB) may be inserted ahead of the preamp, in order to prevent input overload. On many consoles, the preamp outputs may be phase-reversed, via the "Ø" button. This is used to change the signal's phase by 180° in order to compensate for polarity problems in the mic cable or during placement. High- and low-pass filters may also follow the preamp, allowing such extraneous signals as amp/tape hiss or subsonic floor rumble to be filtered out.

If the mixer or interface is so equipped, digital input signals can be inserted into the signal path. If this is the case, care must be taken to ensure that word sync issues are addressed so as to reduce possible timing conflicts and distortion between the digital devices. (Further info on wordclock can be found in Chapter 6.)

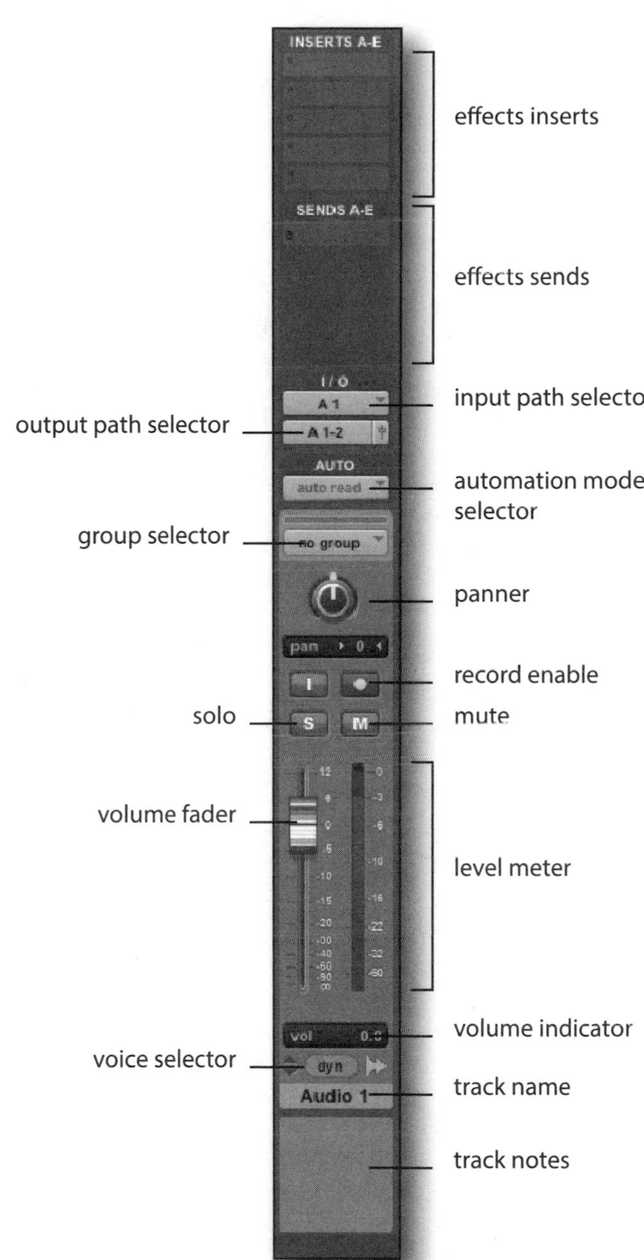

FIGURE 14.10
Virtual mixer strip layout for Pro Tools. (Courtesy of Avid Technology, Inc., www.avid.com)

GAIN LEVEL OPTIMIZATION

As we enter of our discussion on console and mixer layouts, it's extremely important that we touch base on the concept of signal flow or *gain level optimization*. In fact, the idea of optimizing levels as they pass from one functional block in an input strip to the next or from one device to another is one of the more important concepts to be grasped in order to create professional-quality recordings. Although it's possible to go into a great deal of math in this section, I feel that it's

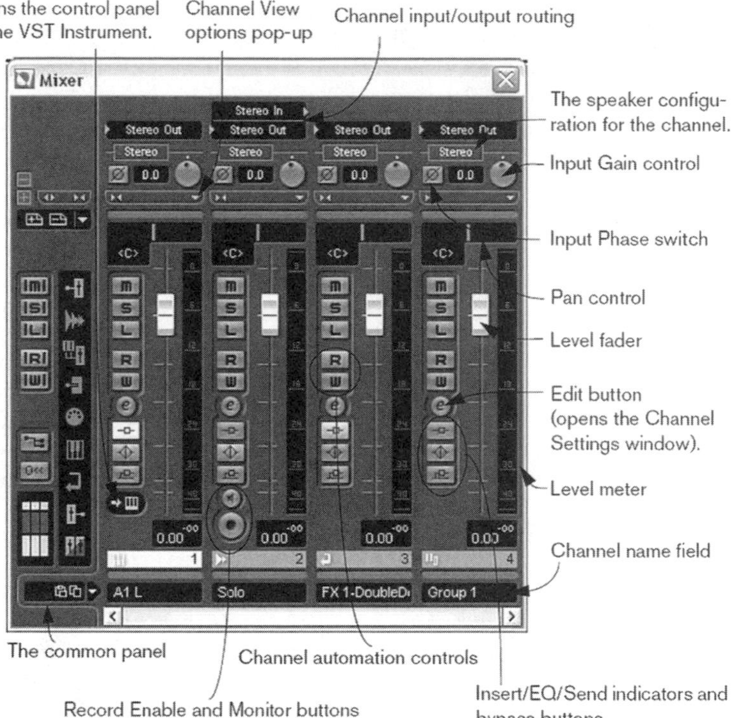

FIGURE 14.11
Steinberg's Cubase and Nuendo virtual mixer. (Courtesy of Steinberg Media Technologies GmbH, a division of Yamaha Corporation, www.steinberg.net)

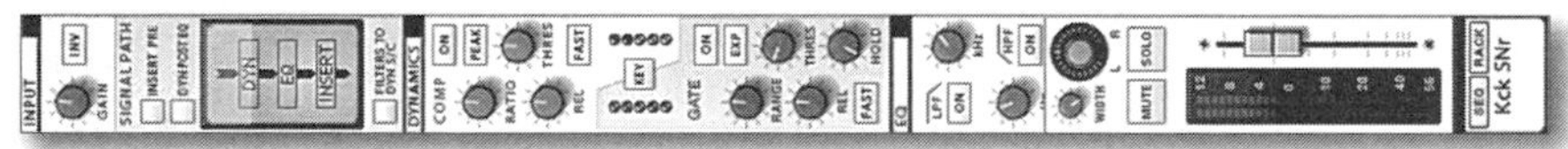

FIGURE 14.12
Input strip for Reason's Mixer (side view). (Courtesy of Propellerhead Software, www.propellerheads.se)

FIGURE 14.13
Apollo FireWire and Thunderbolt audio interface mixer software app. (Courtesy of Universal Audio, www.uaudio.com © 2013 Universal Audio, Inc. All rights reserved. Used with permission)

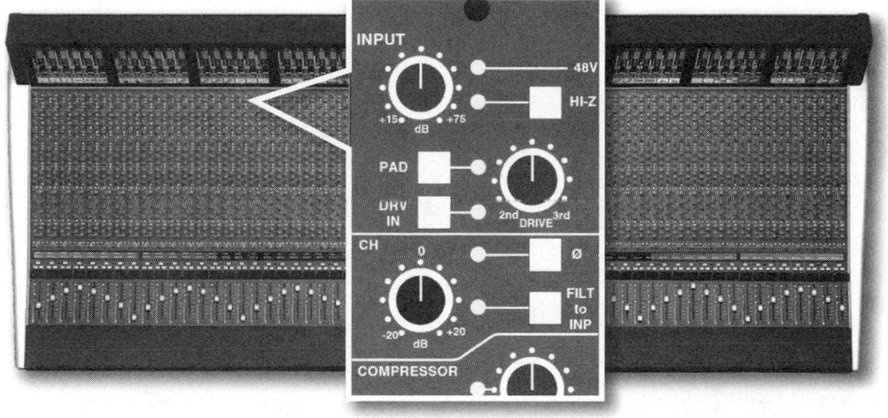

FIGURE 14.14
Channel input section of the Solid State Logic Duality Console (Courtesy of Solid State Logic, www.solid-state-logic.com)

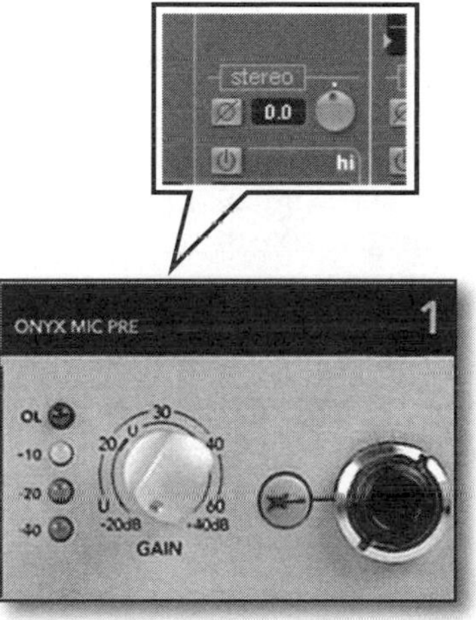

FIGURE 14.15
Analog and DAW interface input sections. (Courtesy of Loud Technologies, Inc., www.mackie.com and Steinberg Media Technologies GmbH, www.steinberg.net)

far more important that you understand the underlying principles of level optimization, internalize them in everyday practice and let common sense be your guide. For example, it's easy to see that, if a mic that's plugged into an input strip is overdriven to the point of distortion, the signal following down the entire path will be distorted. By the same notion, driving the mic preamp at too low a signal will require that it be excessively boosted at a later point in the chain, resulting in increased noise. From this, it follows that the best course of action is to optimize the signal levels at each point in the chain (regardless of whether the signal path is within an input strip, internal to a specific device, or pertains to I/O levels as they pass from one device to another throughout the studio.)

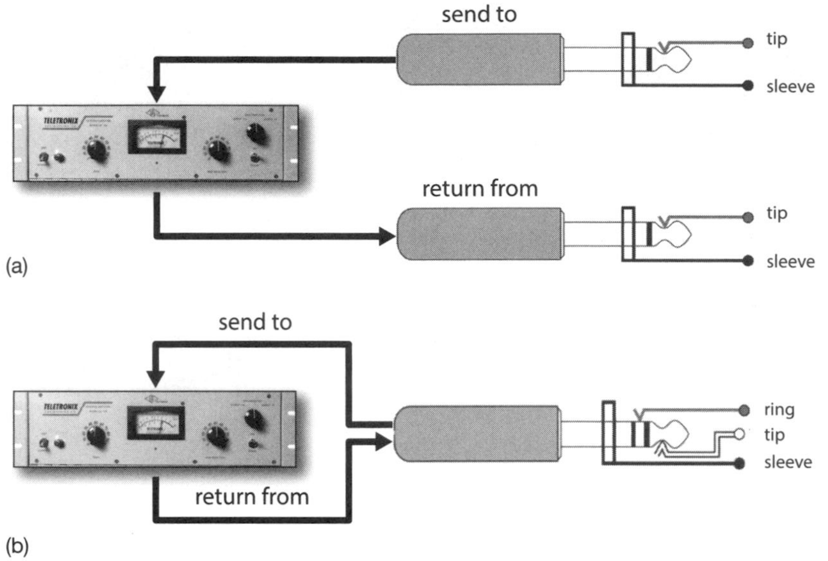

FIGURE 14.16
Direct send/return signal paths. (a) Two jacks can be used to send signals to and return signals from an external device. (b) A single TRS (stereo) jack can be used to insert an external device into an input strip's path.

From a practical standpoint, level adjustments usually begin at the mic preamp. Although many of these points are situational and even debatable, one straightforward approach to setting proper gains on an input strip is to set the gain on the main strip fader (and possibly the master and group output faders) to 0 dB (unity gain). While monitoring levels for that channel or channel grouping, turn the mic preamp up until an acceptable gain is reached. Should the input overload LED light up, back off on the input level and adjust the output gain structure accordingly. Care should be taken when inserting devices into the signal chain at a direct insert point, making sure that the in and out signals are at or near their optimum level. In addition, the EQ section can also cause level overload problems whenever a signal is overly boosted within a frequency range.

INSERT POINT

Many mixer and certain audio interface designs provide a break in the signal chain that occurs after the channel input. A direct send/return or insert point (often referred to simply as direct or insert) can be used to send the strip's line level audio signal out to an external gain or effects processing device. The external device's output signal can then be inserted back into the signal path, where it can be routed to a destination or mixed back into the audio program. It's important to note that plugging a signal processor into an insert point will only affect the audio on that channel.

Should you wish to affect a number of channels, the auxiliary send and group output section can be used to process a combined set of input channels together. Physically, the send and return insert jacks of a hardware mixer or audio interface can be accessed as either two separate jacks on the top or back of a mixer/console (Figure 14.16a), or as a single, stereo jack in the form of a tip-ring-sleeve (TRS) connector that carries the send, return and common signals (as shown in Figure 14.16b) or finally as access points on a console patch bay.

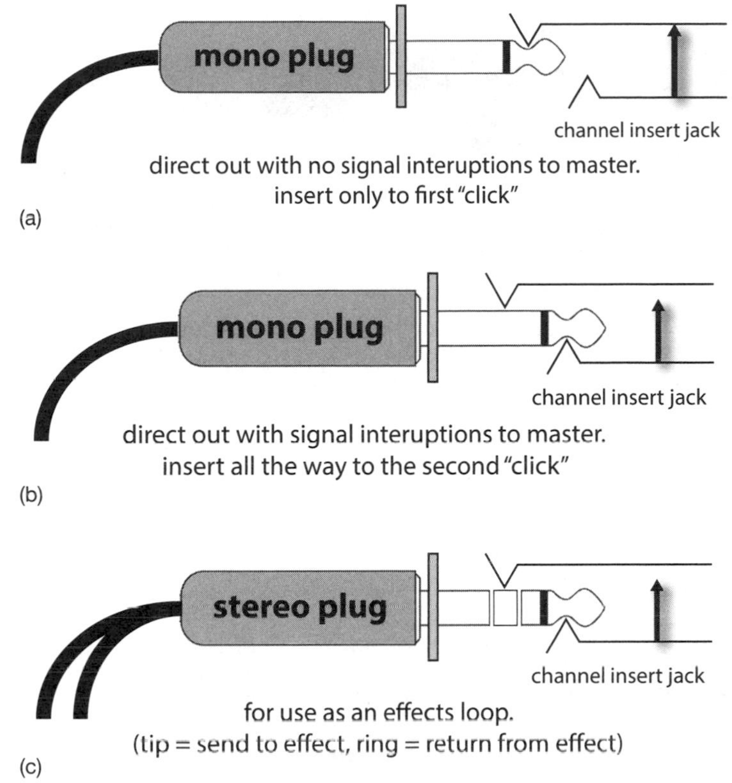
mono plug
channel insert jack
direct out with no signal interuptions to master.
insert only to first "click"
(a)
mono plug
channel insert jack
direct out with signal interuptions to master.
insert all the way to the second "click"
(b)
stereo plug
channel insert jack
for use as an effects loop.
(tip = send to effect, ring = return from effect)
(c)

FIGURE 14.17 Various insert positions for an unbalanced TRS send/return loop: (a) direct out with no signal interruption to a mixer's main output path—insert only to first click; (b) direct out with an interruption to the mixer's main output path—insert all the way to the second click; (c) for use as an insert loop (tip = send to external device, ring = return from external device, sleeve = common circuit ground). (Courtesy of Loud Technologies, Inc., www.mackie.com)

A number of mixers and console manufacturers (such as Mackie and Tascam) use unbalanced, stereo TRS jacks to directly insert a signal in several interesting ways (Figure 14.17; note: please consult the manual for your particular system's jack layout):

- Inserting a mono send plug to the first click will connect the cable to the direct out signal, without interrupting the return signal path.
- Inserting a mono send plug all the way in will connect the cable to the direct out signal while interrupting the return signal path.
- Inserting a stereo TRS plug all the way in allows the cable to be used as its intended send/return loop.

Within a workstation environment, inserts are extremely important in that they allow audio or MIDI processing/effects plug-ins to be directly inserted into the virtual path of that channel (Figure 14.18). Often, a workstation allows multiple plug-ins to be inserted into a channel in a stacked fashion, allowing complex and unique effects to be built up. Of course, keep in mind that the extensive use of insert plug-ins can eat up processing power. Should the stacking of multiple plug-ins become a drain on your CPU (something that can be monitored by watching your processor usage meter—aka "busy bar"), many DAWs allow the track to be frozen, meaning that the total sum of the effects can be written to an audiofile, allowing the effects to be played back without causing undue strain on the CPU.

FIGURE 14.18 An effects plug-in can be inserted into the virtual path of a DAW's channel strip. (Courtesy of Steinberg Media Technologies GmbH, www.steinberg.net, and Universal Audio, www.uaudio.com)

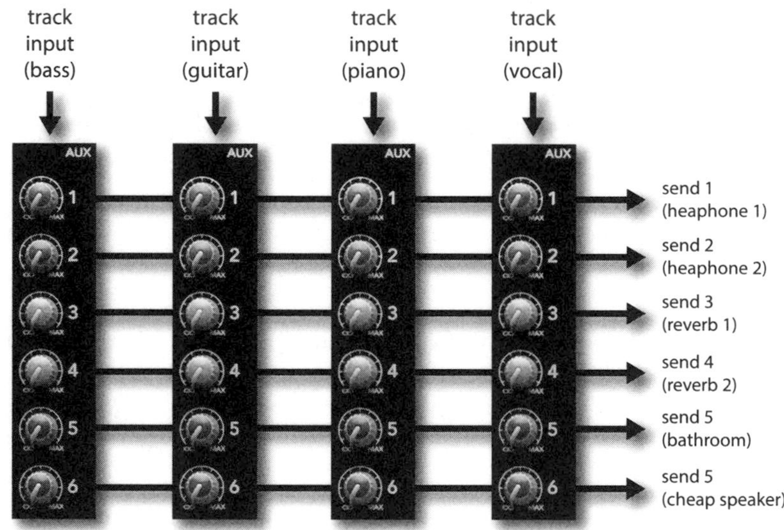

FIGURE 14.19 Although a hardware mixer's signal path generally flows vertically from top to bottom, an aux sends path flows in a horizontal fashion, in that the various channel signals are mixed together to feed a mono or stereo send bus. The combined sum can then be sent to any device (efx, headphone or submix feed during either recording or mixdown. (Courtesy of Loud Technologies, Inc., www.mackie.com)

② Auxiliary send section

In a hardware or virtual setting, the auxiliary (aux) sends are used to route and mix signals from one or more input strips to the various effects output sends and monitor/headphone cue sends of a console. These sections are used to create a submix of any (or all) of the various console input signals to a mono or stereo send, which can then be routed to any signal processing or signal destination (Figure 14.19).

It's not uncommon for up to eight individual aux sends to be found on an input strip. An auxiliary send can serve many purposes. For example, one send could be used to drive a reverb unit, signal processor, etc., while another could be used to drive a speaker that's placed in that great sounding bathroom down the hall. A pair of sends (or a stereo send) could be used to provide a headphone mix

FIGURE 14.20 An effects plug-in can be inserted into an effects send bus, allowing multiple channels to share the same effects processor. (Courtesy of Steinberg Media Technologies GmbH, www.steinberg.net, and Universal Audio, www.uaudio.com)

for several musicians in the studio, while another send could feed a separate mix to the hard-of-hearing drummer. Hopefully, you can see how a send can be used for virtually any signal routing or effects processing task that needs to be handled. How you deal with a send is up to you, your needs and your creativity.

With regard to a workstation, using an aux send is a way of keeping the processing load on the CPU to a minimum (Figure 14.20). For example, let's say that we wanted to make wide use of a reverb plug-in that's generally known to be a CPU hog. Instead of separately plugging the reverb into a number of tracks as inserts, we can greatly save on processing power by plugging the reverb into an aux send bus. This lets us selectively route and mix audio signals from any number of tracks and then send the summed signals to a single plug-in that can then be mixed back into the master out bus. In effect, we've cut down on our power requirements by using a single plug-in to do the job of numerous plug-ins.

③ Equalization

The most common form of signal processing is equalization (EQ). The audio equalizer (Figure 14.21) is a device or circuit that lets us control the relative amplitude of various frequencies within the audible bandwidth. Like the auxiliary sends, it derives its feed directly from the channel input section. Put another way, it exercises tonal control over the harmonic or timbral content of a recorded sound. EQ may need to be applied to a single recorded channel, to a group of channels or to an entire program (often as a step in the mastering process) for any number of other reasons, including:

- To correct for specific problems in an instrument or in the recorded sound (possibly to restore a sound to its natural tone)
- To overcome deficiencies in the frequency response of a mic or in the sound of an instrument
- To allow contrasting sounds from several instruments or recorded tracks to better blend together in a mix
- To alter a sound purely for musical or creative reasons.

FIGURE 14.21
Equalizer examples. (Courtesy of Loud Technologies, Inc., www.mackie.com) and Steinberg Media Technologies GmbH, www.steinberg.net)

THE "GOOD RULE"
Good musician + good instrument + good performance + good acoustics + good mic + good placement = good sound.

When you get right down to it, EQ is all about compensating for deficiencies in a sound pickup, "shaping" the sound of an instrument so that it doesn't interfere with other instruments in a mix or reducing extraneous sounds that make their way into a track. To start our discussion on how to apply EQ, let's take another look at the "Good Rule," (see left).

Let's say that at some point in the "good" chain something falls short—like, a mic was placed in a bad spot for a particular instrument during a session that's still under way. Using this example, we now have two options. We can change the mic position and overdub the track or re-record the entire song—or, we can decide to compensate by applying EQ. These choices represent an important philosophy that's held by many producers and engineers (including myself): Whenever possible, EQ should NOT be used as a bandage to doctor a session after it's been completed. By this I mean that it's often a good idea to correct a problem on the spot (i.e. change the mic or mic position) rather than rely on the hope that you can fix it later in the mix using EQ and other methods.

Although it's usually better to deal with problems as they occur, it isn't always possible. When a track needs fixing after it's completed, EQ is a good option when:

- There's no time or money left to redo the track.
- The existing take was simply magical … and too much feeling would be lost if the track were to be redone.
- You have no control over a track that's already been recorded during a previous session.

When EQ is applied to a track, bus or signal, the whole idea is to take out the bad and leave the good. If you keep adding EQ to the signal, it'll degrade the gain structure and lead to a creeping up of volume. Thus, it's often a good idea to use EQ to take away a deficiency in the signal but not necessarily to boost the desirable part of the track (which would in effect serve to turn up the overall gain). Such a boost will often throw off a mix's overall balance and reduce its overall headroom. A couple of active examples would include:

- Reducing the high end on a bass guitar instead of boosting its primary bass notes
- Using a peak filter to pull out the ring of a snare drum (a perfect example of a problem that should've been corrected during the session).

Using EQ might or might not always be the best course of action; for example, to bring out an upper presence on a recorded vocal track, it might be best to use a peak curve to slightly boost the upper midrange. Just like life, use your best judgment … nothing's ever absolute.

It's always a good idea to be patient with your "EQ style" ... especially when you're just starting out. Learning how to EQ properly (i.e. not over EQ) takes time and practice.

④ Dynamics section

Many top-of-the-line analog consoles offer a *dynamics section* on each of their I/O modules (Figure 14.22). This allows individual signals to be dynamically processed more easily, without the need to scrounge up tons of outboard devices. Often, a full complement of compression, limiting and expansion (including gating) is provided. A complete explanation of dynamic control can be found in Chapter 15 (Signal Processing).

⑤ Monitor section

During the recording phase, since the audio signals are commonly recorded to tape or DAW at their optimum levels (without regard to the relative musical

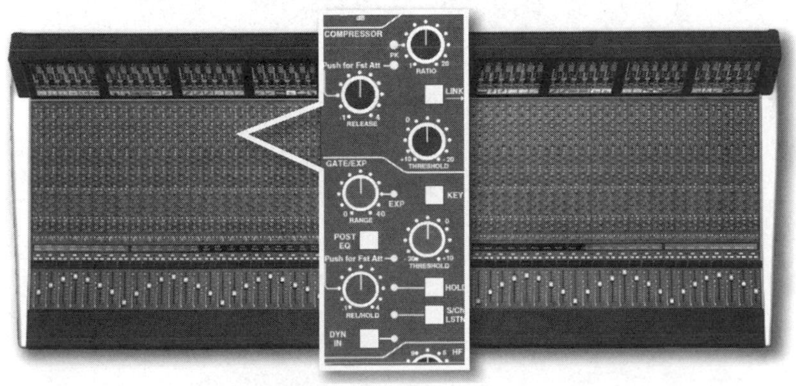

FIGURE 14.22
Dynamics section of a Solid State Logic Duality Console. (Courtesy of Solid State Logic, www.solid-state-logic.com)

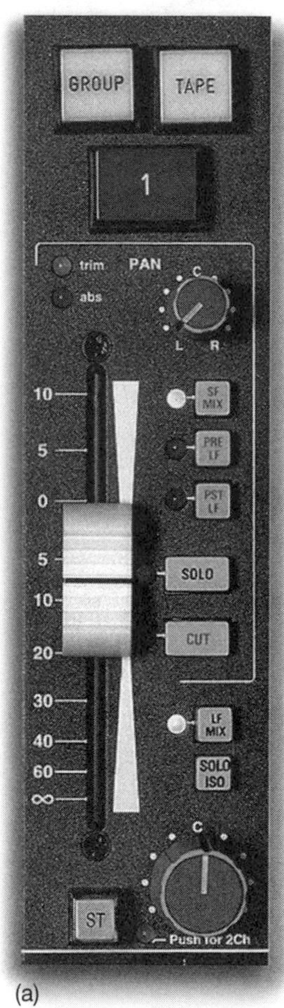

(a)

(b)

FIGURE 14.23
Monitor mix sections: (a) Legacy model SSL XL9000K monitor mix section. (Courtesy of Solid State Logic, www.solid-state-logic.com) (b) Software monitor section (at top) within the Cubase/Nuendo virtual mixer. (Courtesy of Steinberg Media Technologies GmbH, www.steinberg.net)

balance on other tracks), a means for creating a separate monitor mix in the control room is necessary in order to hear a musically balanced version of the production. A separate monitor section (as well as an aux bus) can be used to provide control over each input's level, pan, effects, etc., as well as to route this mix to the control room's mono, stereo or surround speakers (Figure 14.23). The approach and techniques of monitoring tracks during a recording will often vary from mixer to mixer (as well as among individuals). Again, no method is right or wrong compared to another. It simply depends on what type of equipment you're working with, and on your own personal working style.

Note that during the overdub and general production phases, this monitor phase can be easily passed over in favor of mixing the monitor signal levels directly at the main faders in a standard mixdown environment. The straightforward system lets us pass go and collect $200 by setting up a rough mix all through the production phase, allowing us to finesse the mix during production. By the time the final mix rolls around, many or most of the "mix as you go" and automation kinks will have been worked out.

IN-LINE MONITORING

Many hardware console designs incorporate an I/O small fader section that can be used to directly feed the recorded signal that's being fed to either the multitrack recorder or the monitor mixer (depending on its selected operating mode). In the standard monitor mix mode (Figure 14.24a), the small fader is used to adjust the monitor level for the associated recording track. In the *flipped* mode (Figure 14.24b), the small fader is used to control the signal level that's being sent to the recording device, while the larger, main fader is used to control the monitor mix levels. This function allows multitrack record levels (which aren't often changed during a session) to be located out of the way, while the more frequently used monitor levels are assigned to the larger, more accessible master faders.

DIRECT INSERT MONITORING

Another flavor of in-line monitoring that has gained favor over the years is known as *direct insert monitoring*. This method (which is often being used in newer console designs) makes use of the direct send/returns of each input strip to insert the recording device directly into the input strip's signal path (Figure 14.25). Using this approach, which is closely tied to the Insert Point section from earlier in the chapter:

- The direct send for each of the associated tracks (which can be either before or after the EQ section) is routed to the associated interface or track input of a multitrack DAW or ATR.
- The return signal is then routed from the recording device's output back into the same input strip's return path, where it's injected back into the strip (usually before the strip's effects send, pan and main fader path).

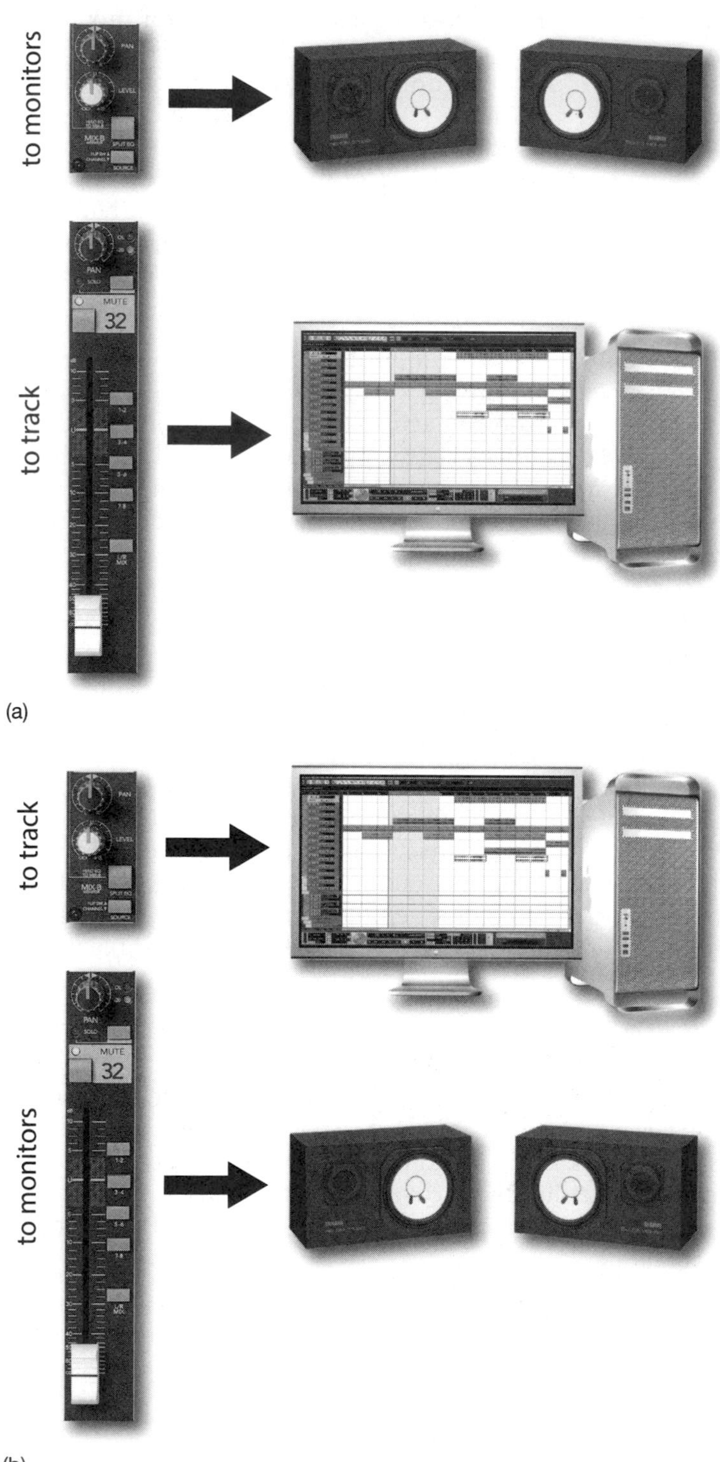

FIGURE 14.24
Flipped-fader monitor modes: (a) standard monitor mode; (b) flipped monitor mode.

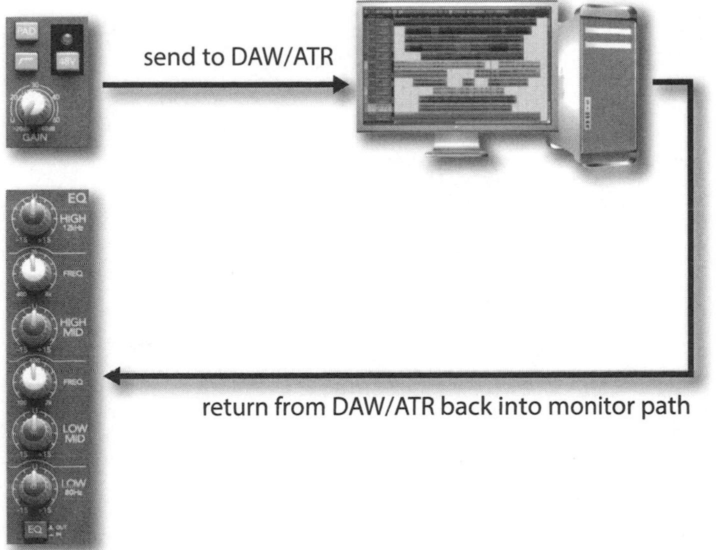

FIGURE 14.25
A track can be easily recorded and monitored by directly inserting the DAW or ATR record track into a strip's insert send and return signal path.

With this approach, the input signal directly following the mic/line preamp will be fed to the DAW or ATR input (with record levels being adjusted by the preamp's gain trim). The return path (from the DAW or ATR) is then fed back into the input strip's signal path so it can be mixed (along with volume, pan, effects, sends, etc.) without regard for the levels that are being recorded to tape, disk or other medium. This system greatly simplifies the process, since playing back the track won't affect the overall monitor mix at all—because the recorder's outputs are already being used to drive the console's monitor mix signals.

SEPARATE MONITOR SECTION

Certain British consoles (particularly those of older design) incorporate a separate mixing section that's dedicated specifically to the task of mixing the monitor feed. Generally located on the console's right-hand side (Figure 14.26), the inputs to this section are driven by the console's multitrack output and tape return buses, and offer level, pan, effects and "foldback" (an older British word for headphone monitor control). During mixdown, this type of design has the distinct advantage of offering a large number of extra inputs that can be assigned to the main output buses during mixdown for use with effects returns, electronic instrument inputs and so on. During a complex recording session, this monitoring approach will often require an extra amount of effort and concentration to avoid confusing the inputs that are being sent to tape or DAW with the corresponding return strips that are being used for monitoring (which is probably why this design style has fallen out of favor in modern console designs).

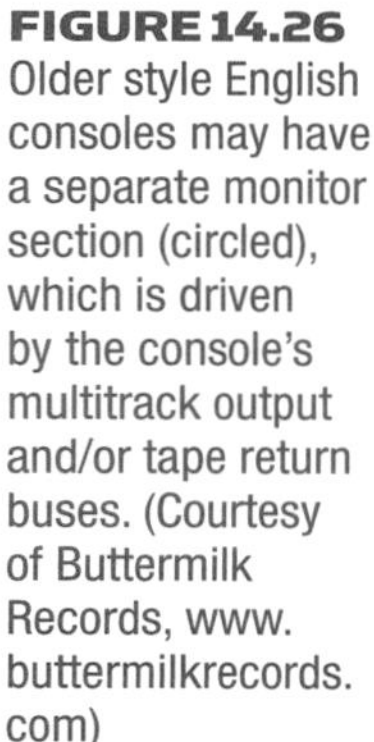

FIGURE 14.26 Older style English consoles may have a separate monitor section (circled), which is driven by the console's multitrack output and/or tape return buses. (Courtesy of Buttermilk Records, www.buttermilkrecords.com)

⑥ Channel fader

Each input strip contains an associated channel fader (which determines the strip's bus output level) (Figures 14.27) and pan pot (and 14.28), which is often designed into or near the fader and determines the signal's left/right placement in the stereo and/or surround field). Generally, this section includes a solo/mute feature, which performs the following functions:

- *Solo:* When pressed, the monitor outputs for all other channels will be muted, allowing the listener to monitor only the selected channel (or soloed channels) without affecting the multitrack or main stereo outputs during the recording or mixdown process.
- *Mute:* This function is basically the opposite of the solo button, in that when it is pressed the selected channel is cut or muted from the main and/or monitor outputs.

Depending on the hardware mixer or controller interface design, the channel fader might be motorized, allowing automation moves to be recorded and played back in the physical motion of moving faders. In the case of some of the high-end console and audio interface/controller designs, a flip fader mode can be called up that literally reassigns the control of the monitor section's fader to that of the main channel fader. This "flip" allows the monitoring of levels during the recording process to be controlled from the larger, long-throw faders. In addition to swapping monitor/channel fader functions, certain audio interface/controller designs allow a number of functions such as panning, EQ, effects

FIGURE 14.27
Output fader section of the Solid State Logic Duality Console. (Courtesy of Solid State Logic, www.solid-state-logic.com)

stereo hardware panner

stereo software panner

surround panner

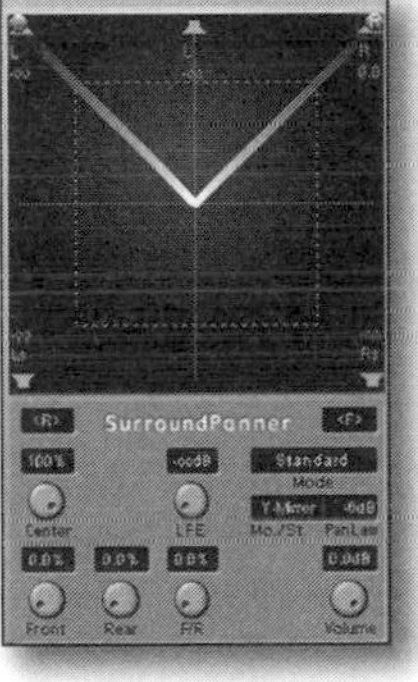

FIGURE 14.28
Example of various hardware and software pan pot configurations.

sends, etc., to be swapped with the main fader, allowing these controls to be finely tuned under motorized control.

7 Output section

In addition to the concept of the signal path as it follows through the chain, there's another important signal path concept that should be understood: output bus. From the above input strip discussion, we've seen that a channel's audio signal by and large follows a downward path from its top to the bottom; however, when we take the time to follow this path, it's easy to spot where audio is routed off the strip and onto a horizontal output path. Conceptually, we can think of this path (or bus) as a single electrical conduit that runs the horizontal length of a console or mixer (Figure 14.29). Signals can be inserted onto or routed off of this bus at multiple points.

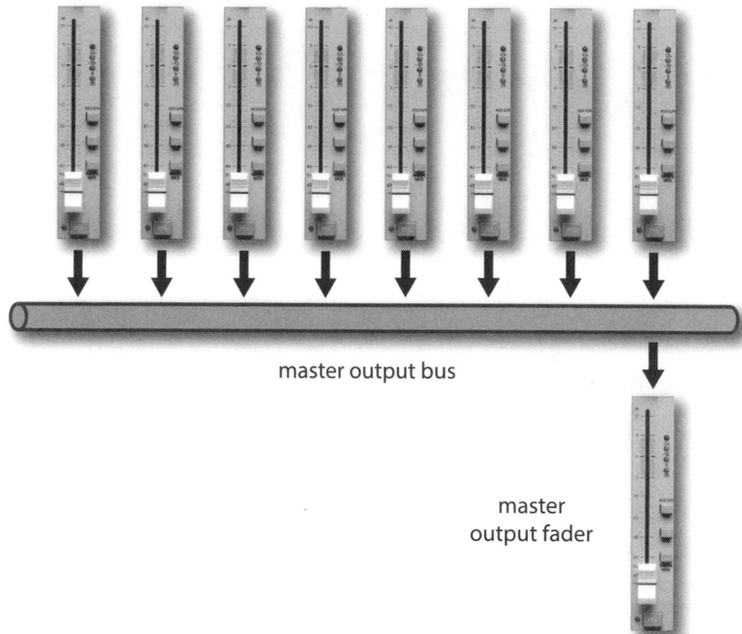

FIGURE 14.29
Example of a master output bus, whereby multiple inputs are mixed and routed to a master output fader.

Much like a city transit bus, this signal path follows a specific route and allows audio signals to get on or off the line at any point along its path. Aux sends, monitor sends, channel assignments, and main outputs are all examples of signals that are injected into buses for routing to one or more output destinations. The main stereo or surround buses, (which are used to feed the channel faders and pan positioners) are then fed to the mixer's main output buses, combined with the various effects return signals and finally routed to the monitor speakers and/or recording device.

As you might expect, a number of newer consoles and mixer designs also include a built-in FireWire, USB and Thunderbolt audio interface that can directly route DAW signals to and from the hardware control/mixer surface in a straightforward manner.

8 Channel assignment

After the channel output fader on a console or larger mixer, the signal is often routed to the strip's track assignment matrix (Figure 14.30), which is capable of distributing the signal to any or all tracks of a connected multitrack recorder. Although this section electrically follows either the main or small fader section (depending on the channel's monitor mode), the track assign buttons will often be located either at the top of the input strip or designed into the main output fader (often being placed at the fader's right-hand side). Functionally, pressing any or all assignment buttons will route the input strip's main signal to the corresponding track output buses. For example, if a vocal mic is plugged into channel 14, the engineer might assign the signal to track 14 by pressing (you guessed it) the "14" button on the matrix. If a quick overdub on track 15 is also needed, all the engineer has to do is unpress the "14" button and reassign the signal to track 15.

Many newer consoles offer only a single button for even- and odd-paired tracks, which can then be individually assigned by using the strip's main output pan pot. For example, pressing the button marked "5/6" and panning to the left routes the signal only to output bus 5, while panning to the right routes it to bus 6. This simple approach accomplishes two things:

- Fewer buttons need to be designed into the input strip (lowering production costs and reducing the number of moving parts).
- Panning instruments within a stereo soundfield and then assigning their outputs to a pair of tracks on the multitrack recorder can easily build up a stereo submix.

Of course, the system for assigning a track or channel on a DAW varies with each design, but is often straightforward. As always, it's best to check with the manual, before beginning a session.

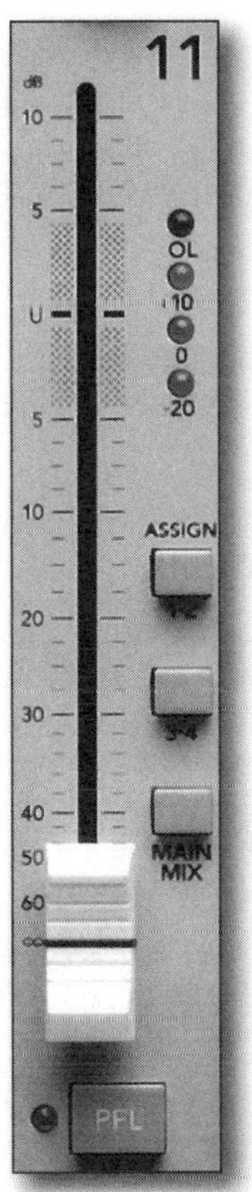

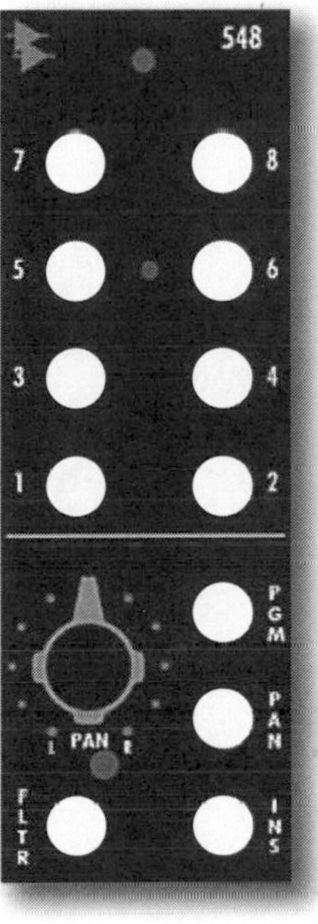

FIGURE 14.30
Channel assignment section of an Onyx 4-bus (Courtesy of Loud Technologies, Inc., www.mackie.com) and API 1608 Console (Courtesy of Automated Processes, Inc., www.apiaudio.com).

9 Grouping

Many DAWs, consoles and professional mixing systems allow any number of input channels to be organized into groups. Such groupings allow the overall relative levels of a series of channels to be interlinked into organized groups according to instrument or scene change type. This important feature makes it possible for multiple instruments to retain their relative level balance while offering control over their overall group level from a single fader or stereo fader pair. Individual group bus faders often have two functions. They:

- Vary the overall level of a grouped signal that's being sent to a recorded track.
- Vary the overall submix level of a grouped signal that's being routed to the master mix bus during mixdown.

The obvious advantage to grouping channels is that it makes it possible to avoid the dreaded and unpredictable need to manually change each channel volume individually. Why try to move 20 faders when you can adjust their overall levels from just one? For example, the numerous tracks of a string ensemble and a drum mix could each be varied in relative level by assigning them to their own stereo or surround sound groupings and moving a single fader … ahhhh … much easier and more accurate!

It's important to realize that there are two methods for grouping signals together, when using either a hardware console or a DAW. These can be done by:

- Grouping the signals together onto the same, combined audio bus.
- Grouping the level "control elements" of the grouped signals, while keeping the signals separate and discrete.

Using the first method, grouped signals are literally assigned to the same audio buss, where they are combined into a composite mono, stereo or multichannel track. Of course, this means that all of your mix decisions have to be made at that time, as all of the signals will be electrically combined together at this time, with little or no recourse for hitting the "undo" button" at a later time. The physical grouping together of channel signals can most be easily done on a console or mixer that has separate "group" outputs by assigning the various channels in the desired group to their own output bus (Figure 14.31). During mixdown, each instrument group bus can be summed into the main stereo or surround output through the use of pan pots or L/R assignment buttons.

The second grouping system doesn't combine the signals, but makes use of a single control voltage or digital assignment to control the "relative" levels of tracks that are assigned to a group number. This method (Figure 14.32), allows the signals to be discretely isolated and separated from each other in a way that varies

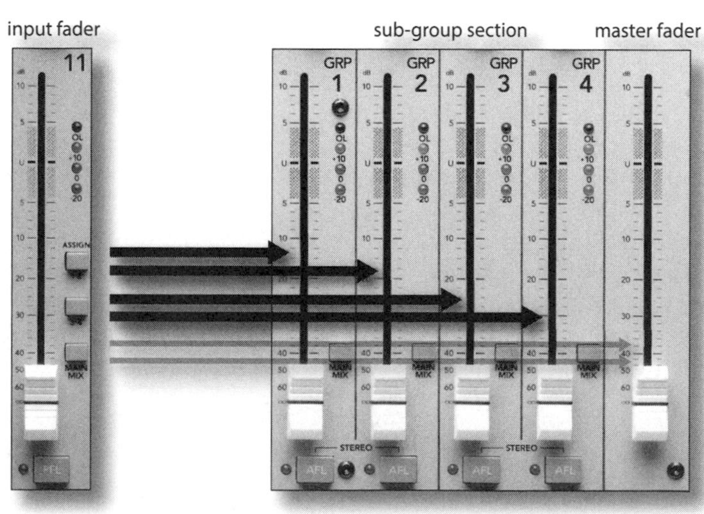

FIGURE 14.31
Simplified anatomy of the output grouping section on the Mackie Onyx 4-bus analog console, whereby the signals can be combined into a physical group output bus. (Courtesy of Loud Technologies, Inc., www.mackie.com)

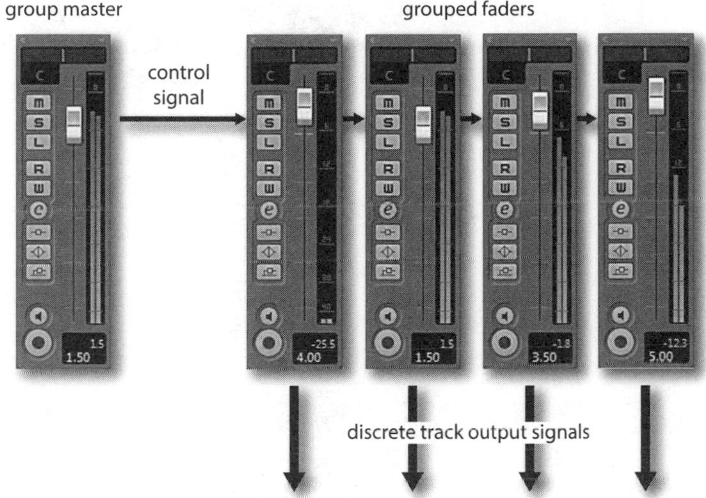

FIGURE 14.32
Tracks can be assigned as a group, so that the relative track levels change, while keeping the track signals separate in a mix.

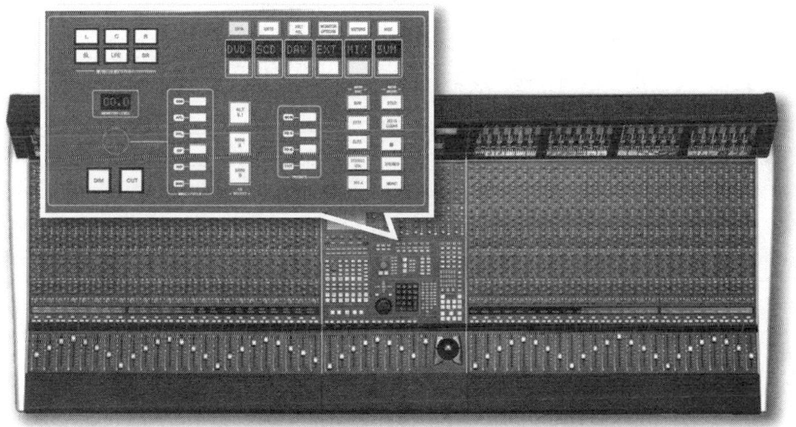

FIGURE 14.33
Monitor level section of the Solid State Logic Duality Console. (Courtesy of Solid State Logic, www.solid-state-logic.com)

the overall group levels in a way that's completely non-destructive and isolated (allowing for changes at any time during a mix or later recall). These group levels can then be controlled from a single fader or via on-screen automation control.

(10) Monitor level section

Most console, mixing and DAW systems include a central *monitor section* that controls levels for the various monitoring functions (such as control room level, studio level, headphone levels and talkback). This section (Figure 14.33) often makes it possible to easily switch between multiple speaker sets and can also provide switching between the various input and recording device sources that are found in the studio (e.g., surround, stereo and mono output buses; tape returns; aux send monitoring; solo monitoring).

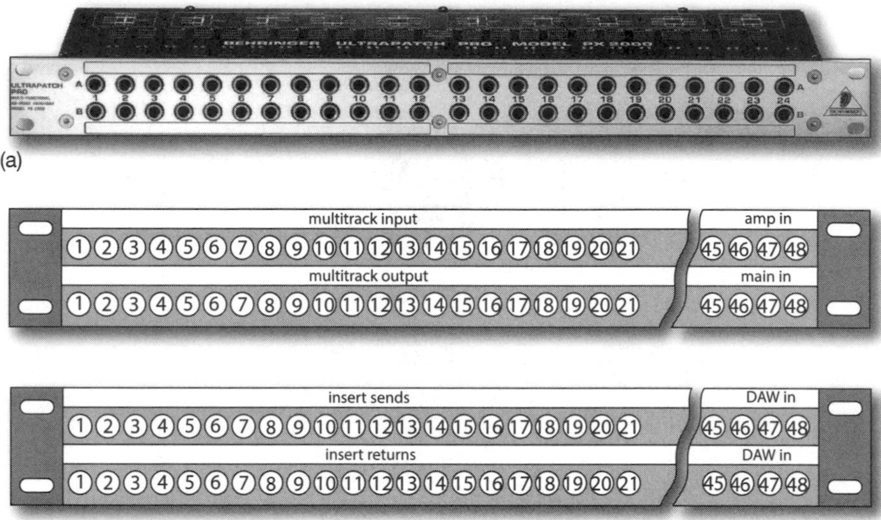

FIGURE 14.34
The patch bay: (a) Ultrapatch PX2000 patch bay (courtesy of Behringer International GmbH, www.behringer.de); (b) rough example of a labeled patch bay layout.

⑪ Patch bay

A *patch bay* (Figure 14.34) is a panel that's found on larger consoles which (under the best of conditions) contains accessible jacks that correspond to the various inputs and outputs of every access point within a mixer or recording console. Most professional patch bays (also known as patch panels) offer centralized I/O access to most of the recording, effects and monitoring devices or system blocks within the production facility (as well as access points that can be used to connect between different production rooms).

Patch bay systems come in a number of plug and jack types as well as wiring layouts. For example, prefabricated patch bays are available using tip–ring–sleeve (balanced), or tip–sleeve (unbalanced) 1/4-inch phone configurations, as well as RCA (phono) connections. These models will often place interconnected jacks at the panel's front and rear so that studio users can reconfigure the panel simply by rearranging the plugs at the rear access points. Other professional systems using the professional telephone-type (TT or mini Bantam-TT) plugs often require that you hand-wire the connections in order to configure or reconfigure a bay (usually an amazing feat of patience, concentration and stamina).

Patch jacks can be configured in a number of ways to allow for several signal connection options among inputs, outputs and external devices (Figure 14.35):

- *Open*: When no plugs are inserted, each I/O connection entering or leaving the panel is independent of the other and has no electrical connection.
- *Half-normaled*: When no plugs are inserted, each I/O connection entering the panel is electrically connected (with the input being routed to the output). When a jack is inserted into the top jack, the in/out connection is still intact, allowing you to tap into the signal path. When a jack is inserted

into the bottom jack, the in/out connection is broken, allowing only the inserted signal to pass to the input.

- *Normaled*: When no plugs are inserted, each I/O connection entering the panel is electrically connected (with the input routing to the output). When a jack is inserted into the top jack, the in/out connection is broken, allowing the output signal to pass to the cable. When a jack is inserted into the bottom jack, the in/out connection is broken, allowing the input signal to pass through the inserted cable connection.
- *Parallel*: In this mode, each I/O connection entering the panel is electrically connected (with the input routing to the output). When a jack is inserted into either the top or bottom jack, the in/out connection will still be intact, allowing you to tap into both the signal path's inputs and outputs.

Breaking a normaled connection allows an engineer to patch different or additional pieces of equipment into a circuit that's normally connected. For example, a limiter might be temporarily patched between a mic preamp output and an equalizer input. The same preamp/EQ patch point could be used to insert an effect or other device type. These useful access points can also be used to bypass a defective component or to change a signal path order …. Versatility is definitely the name of the game here!

12 Metering

The level of a signal's strength, at an input, output bus and other console level point, is often measured by visual meter display (Figure 14.36). Meter and

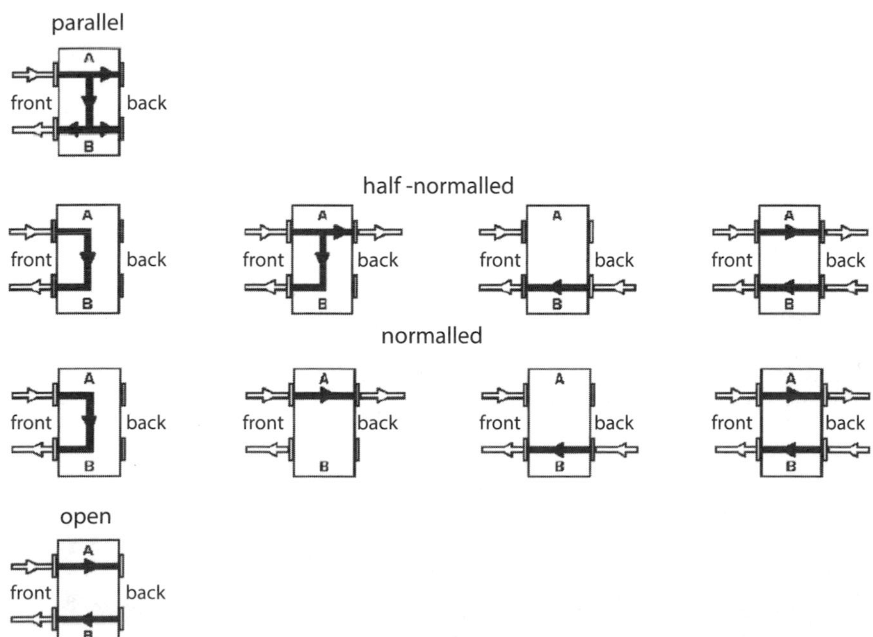

FIGURE 14.35
Typical patch bay signal routing schemes. (Courtesy of Behringer International GmbH, www.behringer.de)

FIGURE 14.36
A set of LED, light-bar and VU meter displays.

indicator types will often vary from system to system. For example, banks of readouts that indicate console bus output and tape return levels might use VU metering, peak program meters (PPMs, found in European designs) or digital/software readouts. It's also not uncommon to find LED overload indicators on an input strip's preamp, which give quick and easy peak indications as to whether you've approached or reached the component's headroom limits (a sure sign to back off your levels).

The basic rule regarding levels isn't nearly as rigid as you might think and will often vary depending on whether the device or recording medium is analog or digital. In short, if the signal level is too low, tape, amp and even digital noise could be a problem, because the levels throughout the signal chain will probably not be optimized. If the level is too high, overloaded preamps, saturated tape or clipped digital converters will often result in a distorted signal. Here are a few rules of thumb:

- In analog recording, the proper recording level is achieved when the highest reading on the meter is near the zero level, although levels slightly above or below this might not be a problem, as shown in Figure 14.37. In fact, (slightly) overdriving some analog devices and tape machines will often result in a sound that's "rough" and "gutsy."
- When recording digitally, noise is often less of a practical concern (especially when higher bit rates are used). It's often a good idea to keep levels at a reasonable level (i.e., peaking at about –12 on the meter), while keeping a respectful distance from the dreaded clip or "over" indicator. Unlike analog, digital is usually very unforgiving of clipped signals and will generate grunge that's guaranteed to make you cringe! Because there is no real standard for digital metering levels beyond these guidelines, many feel that giving a headroom margin that's 12 dB below "0" full scale is usually a wise precaution … especially when recording at a 24-bit depth.

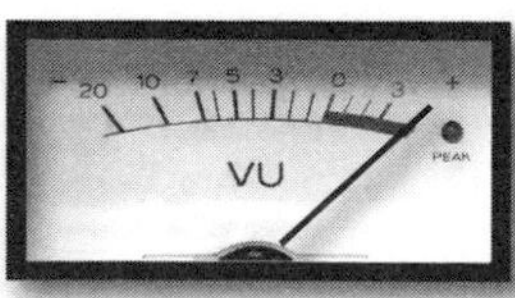

FIGURE 14.37
VU meter readings for analog devices that are too low, too high and just right.

It's important to keep in mind that recorded distortion isn't always easy (or possible) to fix. You might not be able to see the peak levels that could distort your soundfiles, but your ears might hear them after the damage is done. When in doubt, back off … and give your levels some breathing room.

The following sections on readout displays and dynamic range processors can be used in the recording process to tame the audio signal in such a way that instruments, vocals, and the like can be placed squarely in the sonic pocket, at an optimum, undistorted signal level.

THE FINER POINTS OF METERING

Amplifiers, magnetic tape and even digital media are limited in the range of signal levels that they can pass without distortion. As a result, audio engineers need a basic standard to help determine whether the signals they're working with will be stored or transmitted without distortion. The most convenient way to do this is to use a visual level display, such as a *meter*. Two types of metering ballistics (active response times) are encountered in recording sound to either analog or digital media:

- Average (rms)
- Peak

From Chapter 2, we know that the root-mean-square (rms) value was developed to determine a meaningful average level of a waveform over time. Since humans perceive loudness according to a signal's average value (in a way that doesn't bear much relationship to a signal's instantaneous peak level), the displays of many meters will indicate an average signal-level readout. The total amplitude measurement of the positive and negative peak signal levels is called the *peak-to-peak value*. A readout that measures the maximum amplitude fluctuations of a waveform is a *peak-indicating meter*.

One could definitely argue that both average and peak readout levels have their own sets of advantages. For example, the ear's perception of loudness is largely proportional to the rms (average) value of a signal, not its peak value. On the other hand, a peak readout displays the actual amplitude at a particular point in time and not the overall perceived level. For this reason, a peak meter might show readings that are noticeably higher at a particular point in the program than the averaged rms counterpart (Figure 14.38). Such a reading will alert you to the fact that the short-term peaks are at levels that are above the clipping point, while the average signal is below the maximum limits. Under such conditions (where short-duration peaks are above the distortion limit), you might or might not hear distortion as it often depends on the makeup of the signal that's being recorded; for example, the clipped peaks of a bass guitar will not be nearly as noticeable as the clipped high-end peaks of a cymbal. The recording medium often plays a part in how a meter display will relate to sonic reality; for example,

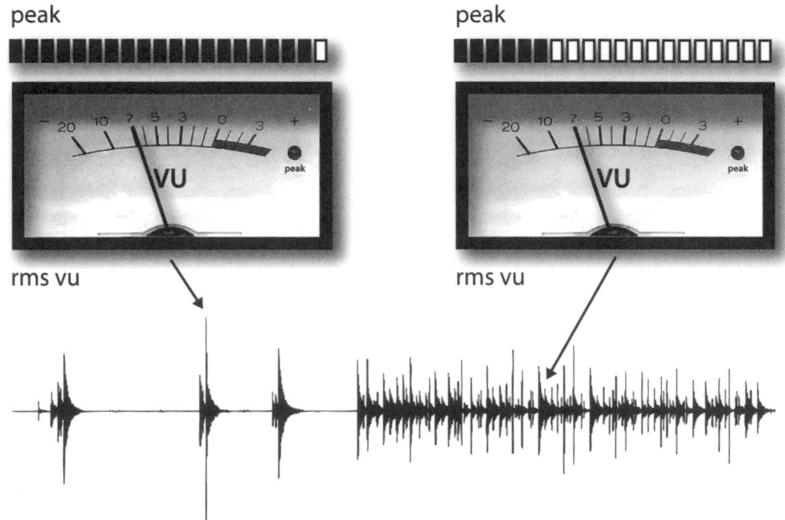

FIGURE 14.38
A peak meter reads higher at point A than at point B, even though the average loudness level is the same.

recording a signal with clipped peaks onto a tube analog tape machine might be barely noticeable (because the tubes and the tape medium act to smooth over these distortions), whereas a DAW or digital recorder might churn out hash that's as ugly as the night (or your current session) is long.

Getting to the heart of the matter, it goes without saying that, whenever the signal is too high (hot), it's an indication for you to grab hold of the channel's mic trim, output fader or whatever level control is the culprit and turn it down. In doing so, you've actually become a dynamic range-changing device. In fact, the main channel fader (which can be controlling an input level during recording or a tape track's level during mixdown) is by far the most intuitive and most often used dynamic gain-changing device in the studio.

In practice, the difference between the maximum level that can be handled without incurring distortion and the average operating level of the system is called *headroom*. Some studio-quality preamplifiers are capable of signal outputs as high as 26 dB above 0 VU and thus are said to have 26 dB of headroom. With regard to analog tape, the 3% distortion level for analog magnetic tape is typically only 8 dB above 0 VU. For this reason, the best recording level for most program material is around 0 VU (although higher levels are possible provided that short-term peak levels aren't excessively high). In some circumstances (i.e., when using higher bias, low-noise/high-output analog tape), it's actually possible to record at higher levels without distortion, because the analog tape formulation is capable of handling higher magnetic flux levels. With regard to digital media, the guidelines are often far less precise and will often depend on your currently chosen bit rate. Since a higher bit rate (e.g., 24 or 32 bits) directly translates into a wider dynamic range, it's often a good idea to back off from the maximum level, because noise generally isn't a problem at these bit rates.

In fact, the optimal average recorded level for a 44.1-k/16-bit recording is often agreed to be around -12 dB, so as to avoid peak clipping. When recording at higher word lengths, the encoded dynamic range is often so wide that the average level could easily be further reduced by 10 or more decibels. Reducing the recorded levels beyond this could have the unintended side effect of unnecessarily increasing distortion due to quantization noise (although this usually isn't an overriding factor at higher bit rates).

… Now that we've gotten a few of the basic concepts out of the way, let's take a brief look at two of the most common meter readout displays.

THE VU METER

The traditional signal-level indicator for analog equipment is the VU meter (see Figure 14.37). The scale chosen for this device is calibrated in volume units (hence its name) and is designed to display a signal's average rms level over time. The standard operating level for most consoles, mixers and analog tape machines is considered to be 0 VU. Although VU meters do the job of indicating rms volume levels, they ignore the short-term peaks that can overload a track. This means that the professional console systems must often be designed so that unacceptable distortion doesn't occur until at least 14 dB above 0 VU. Typical VU meter specifications are listed in Table 14.1.

Since recording is an art form, I have to rise to the defense of those who prefer to record certain instruments (particularly drums and percussion) at levels that bounce or even "pin" VU needles at higher levels than 0 VU. When recording to a professional analog machine, this can actually give a track a "gutsy" feel that can add impact to a performance. This is rarely a good idea when recording instruments that contain high-frequency/high-level signals (such as a snare or cymbals), because the peak transients will probably distort in a way that's hardly pleasing … and it's almost NEVER a good idea when recording to a digital system. Always be aware that you can often add distortion to a track at a later time

Table 14.1 VU Meter Specifications

Characteristic	Specification
Sensitivity	Reads 0 VU when fed a +4-dBm signal (1.228 V into a 600-Ω circuit)
Frequency response	±0.2 dB from 35 Hz to 10 kHz; ±0.5 dB from 25 Hz to 16 kHz
Overload capability	Can withstand 5 times 0-VU level (approximately +18 dBm) continuously and 10 times 0-VU level (+24 dBm) for 0.5 sec

(using any number of ingenious tricks), but you can't remove it from an existing track. As always, it's wise to talk such moves over with the producer and artist beforehand.

THE AVERAGING/PEAK METER

While many analog devices display levels using the traditional VU meter, most digital hardware and software devices will often display a combination of VU and peak program-level metering, using an LCD or on-screen readout. This best-of-both-worlds system makes sense in the digital world as it gives us a traditional readout that visually corresponds to what our ears are hearing, while providing a quick-and-easy display of the peak levels at any point in time. Often, the peak readout is frozen in position for a few seconds before resetting to a new level (making it easier to spot the maximum levels), or it's permanently held at that level until a higher level comes along to bump it up. Of course, should a peak level approach the clipping level, a red "clip" indicator will often light, showing that it's time to back off the levels.

The idea of "pinning" the input levels of any digital recording device is a definite no-no. The dreaded "clip" indicator of a digital meter means that you've reached the saturation point, with no headroom to spare. Unfortunately, digital standard operating levels are far more ambiguous than their analog counterparts, and you should consult the particular device manual for level recommendations (if it says anything about the subject at all). In light of this, I actually encourage you to slightly pin the meters on several digital devices (make sure the monitors aren't turned up too far) just to find out how obnoxious even the smallest amount of clipping can sound … and how those levels might vary from device to device. Pretty harsh, huh?

In recent years, a number of high-quality sound level meters have come onto the market that allow the user to visually inspect the peak, loudness and spectral levels within a recording (see Figures 17.24 and 17.25 in the chapter on monitoring). These tools can be extremely useful for obtaining the best overall level of a mix, as well as the careful balancing of relative song levels within a project.

DIGITAL CONSOLE AND DAW MIXER/CONTROLLER TECHNOLOGY

In the last decade, console design, signal processing and signal routing technology have undergone a tremendous design revolution with the advent of digital technology. DAWs, digital console and mixers, the iPad and controller design have found their ways into professional, project and audio production facilities at an amazing pace. These systems use large-scale integrated circuits and central processors to convert, process, route and interface to external audio and computer-related devices with relative ease. In addition, this technology makes it possible for many of the costly and potentially faulty discrete switches and level controls that are required for such functions as track selection, gain and EQ to be replaced by an assignable digital system. The big bonus, however, is that since routing and other control functions are digitally encoded, it becomes a simple matter for level, patch and automation settings to be saved into memory for instantaneous recall at any time.

From a functional standpoint, the basic signal flow of a digital console, mixer's or even a DAWs virtual mixing environment is conceptually similar to that of an analog console, in that the input must first be boosted in level by a mic/line preamp (where it will be converted into digital data). From this point, the signals can pass through EQ, dynamics and other signal processing blocks; and various effects and monitor sends, volume, routing and other sections that you might expect, such as main and group output fader controls. From a control standpoint, however, sections of a digital console might be laid out in a manner that's completely different. The main difference is a fundamental and philosophical change in the concept of the input strip itself.

The virtual input strip

By the very nature of analog strip design, all of the hundreds of physical controls must be duplicated for each strip along the console. This is the biggest contributing factor to cost and reliability problems in traditional analog design. Since most (if not all) of the functions of a digital console pass through the device's central processor, this duplication is neither necessary nor cost effective. As a result, designers have often opted to keep the most commonly used controls (such as the pan, solo, mute and channel volume fader) in their traditional input strip position. However, controls such as EQ, input signal processing, effects sends, monitor levels and (possibly) track assignment have been designed into a central, virtual control panel (Figure 14.39) that can be used to focus on and vary a particular channel's setting parameters. These controls can be quickly and easily assigned to a particular input strip by pressing the "select" button on the relevant input strip channel.

In certain digital console, mixer and hardware controller designs, each "virtual input strip" may be fitted with a virtual pot (V-pot) that can be assigned to a particular parameter for instant access. In others, the main parameter panel can be multipurpose in its operation, allowing a number of controls and readouts itself to be reconfigured in a chameleon-like fashion to fit the task at hand. In other cases, touch-screen displays can be used to provide for an infinite degree of user control, while some might use software "soft" buttons that can easily reconfigure the form and function of the various buttons and dial controls in the control panel. Finally, console parameters can be controlled by physically placing a readout display at their traditional control points on the input strip. Such a literal "knob per function" control/display system lets you simply grab the desired parameter and alter it (as though it were actually a hardware console). Generally, this system is more expensive than those using a central control panel, because a large number of readout indicators, control knobs and digital control interface systems are required. The obvious advantage to this interface

FIGURE 14.39
Presonus StudioLive 16.0.2 Digital mixing console with channel parameter section highlighted. (Courtesy of Presonus Audio Electronics, Inc., www.presonus.com)

is that you'd have instant access (both physically and visually) to any and all of these parameters at once, in an analog fashion.

From the above, it's easy to see that digital consoles and hardware controllers often vary widely in their layout, ease of operation and cost effectiveness. As with any major system device, it's often a good idea to familiarize yourself with the layout and basic functions before buying or taking on a project that involves such a beastie. As you might expect, operating manuals and basic tutorials are often available on the various manufacturers' websites.

Tutorial: Channel Select

1. Find yourself a digital mixer, console or DAW with a software mixer. (If one's not nearby, you could visit a friend or a friendly studio or music store.)
2. Feed several tracks into it and create a simple mix.
3. Press the channel select for the first input and experiment with the EQ and/or other controls.
4. Move on to the next channel and start building your mix.

(a)

(b)

FIGURE 14.40
Example of virtual, on-screen DAW mixers. (a) ProTools on-screen mixer. (Courtesy of Digidesign, a division of Avid Technology, Inc., www.digidesign.com.), (b) Cubase/Nuendo virtual mixer (Courtesy of Steinberg Media Technologies GmbH, www.steinberg.net)

The DAW software mixer surface

Of course, in this new millennium, the vast majority of mixers that exist in the word are software mixers that are integrated into our DAW software. With the increased power of the computer-based digital audio workstation, mixing "in the box" has become the norm, rather than the exception.

Through the use of a traditional (and sometimes not-so-traditional) user interfacing, these mixers offer functional on-screen control over levels, panning, EQ, effects, DSP, mix automation and a host of functions that are simply too long to list here. Often these software mixers (Figure 14.40) emulate their hardware counterparts by offering basic controls (such as fader, solo, mute, select and

FIGURE 14.41
Mackie Universal Control. (Courtesy of Loud Technologies, Inc., www.mackie.com)

routing) in the virtual input strip. Likewise, selecting a channel track will assign many of the channel's functions to a wide range of virtual parameters that can be controlled from a virtual strip or control panel.

When using a software mixer in conjunction with a DAW's waveform/edit window on a single monitor screen, it's easy to feel squeezed by the lack of visual "real estate." For this reason, many opt for a dual-monitor display arrangement. Whether you are working in a PC or Mac environment, this working arrangement is easier and more cost effective than you might think and the benefits will immediately become evident, no matter what audio, graphics or program environment you're working on.

When using a DAW controller, the flexibility of the DAW's software environment can also be combined with the hands-on control and motorized automation of a hardware or iPad/Android-based controller surface (Figures 14.41 and 14.42). More information on the DAW software mixing environment and hardware controllers can be found in Chapter 7.

Mixdown level and effects automation

One of the great strengths of the digital age is how easily all of the mix and effects parameters can be automated and recalled within a mix. Although many

FIGURE 14.42
A wireless DAW iPad mixing application in action.

large-scale consoles are capable of automation, computer-based DAW systems are particularly strong in this area, allowing complete control of all parameters throughout all phases of a production. The beauty of being able to set basic levels within a mix or to mix under automation is that a mix can be built up over time, and even save multiple mix versions, so that we can go back and explore other production avenues or to correct a potential problem. In short, the job of mixing becomes much less of a chore, giving us the time to pursue less of the technology of mixing … and more of the art of mixing. What could be bad about that?

Although terminologies and functional offerings will differ from one system to the next, control over the basic automation functions will be carried out in one of two operating modes (Figure 14.43):

- Write mode
- Read mode

Write mode

Once the mixdown process has gotten under way, the process of writing the automation data into the system's memory can begin (actually, that's not entirely true, because basic mix moves will often begin during the recording or overdub phase). When in the write mode, the system will begin the process of encoding mix moves for the selected channel or channels in real time. This mode is used to record all of the settings and moves that are made on a selected strip or strips (allowing track mixes to be built up individually) or on all of the input strips (in essence, storing all of the mix moves, live and in one pass). The first approach can help us to focus all of our attention on a difficult or particularly important part or passage. Once that channel has been finessed, another channel or group of channels can then be written into the system's memory … and then another, until an initial mix is built up.

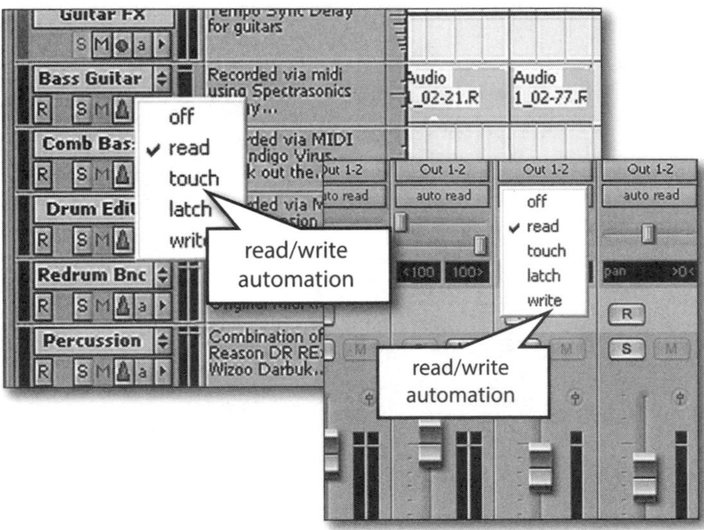

(a)

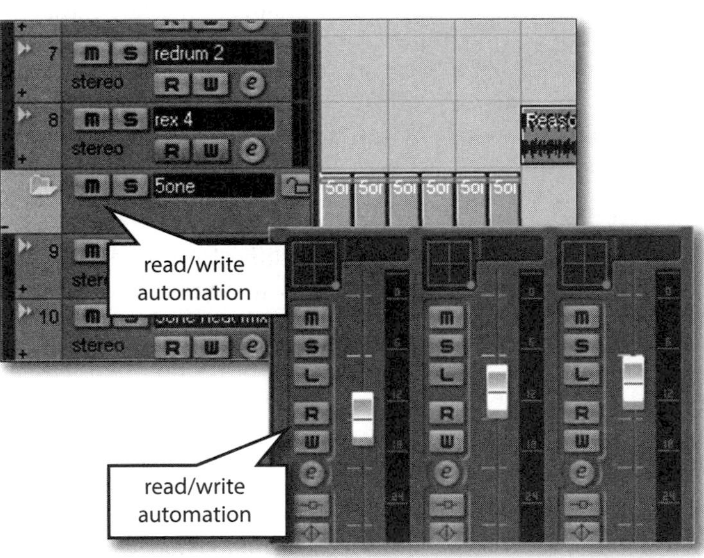

(b)

FIGURE 14.43
Automation mode selections: (a) Pro Tools showing auto selectors within the edit and mixer screens (courtesy of Digidesign, a division of Avid Technology, Inc., www.digidesign.com); (b) Cubase/Nuendo showing auto selectors within the edit and mixer screens (courtesy of Steinberg Media Technologies GmbH, www.steinberg.net.)

Often, modern automation systems will let us update previously written automation data by simply grabbing the fader (either on-screen or on the console/controller) and moving it to the newly desired position. Once the updated move has been made, the automation will remain at that level or position until a previously written automation move is initiated, at which point the values will revert to the existing automation settings.

Whenever data is updated over previously written automation on an older analog mixer or when a DAW controller isn't equipped with moving faders, the concept of matching current fader level or controller positions to their previously written position becomes important. For example, let's say that we needed to redo several track moves that occur in the middle of a song. If the current controller positions don't match up with their previously written points, the mix levels could jump during the updated transition (and thus during playback).

Read mode

An automated console or DAW that has been placed into the read mode will play the mix information from the system's automation data, allowing the on-screen and moving faders to follow or match the written mix moves in real time. Once the final mix has been achieved, all you need to do is press play, sit back and listen to the mix.

Drawn (rubberband) automation

In addition to physically moving on-screen and controller faders under read/write automation control, one of the most accurate ways to control various automation parameters on a DAW is through the drawing and editing of on-screen *rubberbands*. These useful tools offer a simple, graphic form of automation that lets us draw fades and complicated mix automation moves over time. This user interface is so-named because the graphic lines (that represent the relative fade, volume, pan and other parameters) can be bent, stretched, and contorted like a rubberband (Figure 14.44).

Commonly, all that's needed to define a new mix point is to click on a point on the rubberband (at which point a box handle will appear) and drag it to the desired position. You can change a move simply by clicking on an existing handle (or range of handles) and moving it to a new position.

MIXING AND BALANCING BASICS

Once all of the tracks of a project have been recorded, assembled and edited, it's time to put the above technology to use to mix your tracks into their final media forms. The goal of this process is to combine audio, MIDI and effects tracks into a pleasing form that makes use of such traditional tools as:

- Relative level
- Spatial positioning (the physical panned placement of a sound within a stereo or surround field)
- Equalization (affecting the relative frequency balance of a track)
- Dynamics processing (altering the dynamic range of a track, group or output bus to optimize levels or to alter the dynamics of a track so it fits within a mix)
- Effects processing (adding reverb-, delay- or pitch-related effects to a mix in order to augment or alter the piece in a way that is natural, unnatural or just plain interesting.

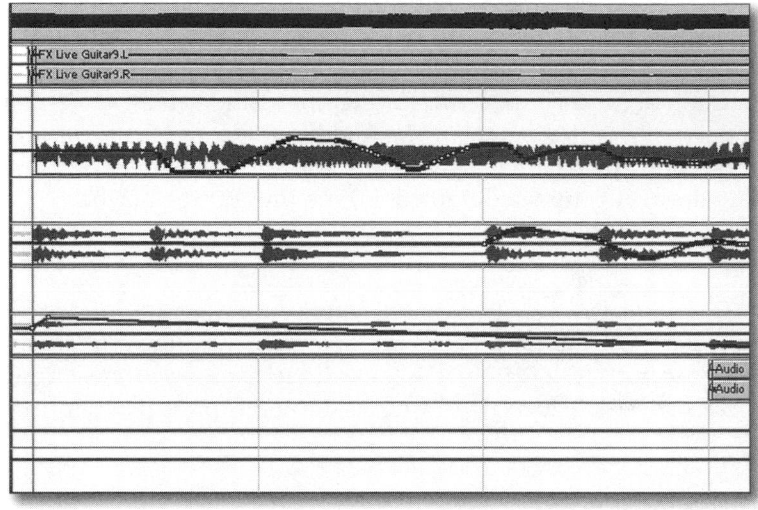

(a)

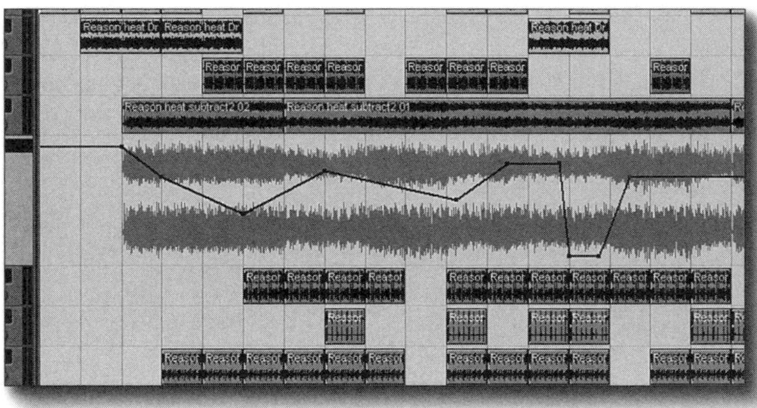

(b)

FIGURE 14.44
Automation rubberbands: (a) Pro Tools (courtesy of Digidesign, a division of Avid Technology, Inc., www.digidesign.com); (b) Cubase/Nuendo (courtesy of Steinberg Media Technologies GmbH, www.steinberg.net.)

Within a well-produced project, sounds can be built up and placed into a sonic stage through the use of natural, psychoacoustic and processed signal cues to create a pleasing, interesting and balanced soundscape. Now, it's pretty evident that volume can be used to move sound forward and backward within the soundfield, and that relative channel levels can be used to position a sound within that field. It's less obvious that changes in timbre (often but not always introduced through EQ), delay and reverb can be used to move sounds within the stereo or surround field. Of course, all of this sounds simple enough;

however, the dedication that's required to hone your skills within this evolving art is what mixing careers are made of.

Before we get started, I'd like to first point you in a direction that just might help your overall mixing technique:

- It's often a good idea to create your own "desert island" favorite mix disc ... basically, a compilation of several of your favorite songs. By putting these on in your mix room, you can get a general idea of how the room sounds. If there's a problem (particularly, if the room's new to you), it can help point it out and help you to avoid certain pitfalls.
- Secondly, you might think about inserting a high-resolution frequency-analyzer into the output mix bus. This device (or plug-in) can literally help you see how your mix is "shaping up" and might help pinpoint possible problem areas.

OK ... now let's take a moment to understand the process better by walking through a fictitious mix. Remember, there's no right or wrong way to mix as long as you watch your levels along the signal path. Also, there's no doubt that, over time, you'll develop your own sense of style. The important thing is to keep your ears open, care about the process and make it sound as good as you can.

Let's begin building the mix by setting the output volumes on each of the instruments to a level that's acceptable to your main mixer or console. From a practical standpoint, you might want to set your tracks to unity gain or to some easily identifiable marking.

The next step is to begin playing the project and changing the fader and pan levels for any instrument or voice tracks until they begin to blend in the mix. Once that's done, you can sit back and listen to how the overall mix levels hold up over the course of the song.

EQ can be applied to any problematic track that doesn't quite fit into the mix properly. You might want to be aware of how a track sounds on it's own, versus how it sounds within the overall mix. Trying to fix the sound of a track is often best done when heard within the overall mix, as you can hear all the basic interactions within the overal blend.

From time-to-time, low frequency energy can get into tracks that can "muddy" your overall mix. Sounds such as room rumble, low-octave rubbish and other gremlins should be "rolled-off" at the low-end, so as not to fight with the bass clarity of your mix.

As with all things sound, letting you ears be your EQ guide takes time and experience... be patient with yourself.

Should the mix need to be changed at any point from its initial settings, you might turn the automation on for that track and begin building up the mix. You might want to save your mix at various stages of its development under a different name 01firstmix, 02mybestmix, 03jimsnewapproach, etc. This makes it easier to return to the point where you began to take a different path.

You might want to group (or link) various instrument sections, so that overall levels can be automated. For example, during the bridge of a song you might want to reduce the volume on several tracks by simply grabbing a single fader … instead of moving each track individually.

This calls to mind a very important aspect of most types of production (actually it's a basic tenant in life): Keep it simple! If there's a trick you can use to make your project go more smoothly, use it. For example, most musicians interact with their equipment in a systematic way. To keep life simple, you might want to explore the possibility of creating a basic mixing template file that has all of the instruments and general assignments already programmed into it.

When the mix comes together, you can then begin to think about inserting some form of EQ or dynamics processing to the master channel output bus. This is not a "required" stage in the process … simply an additional tool that can be used to shape the overall sound of your mix. Of course, this subject starts many a barroom brawl amongst engineers. Adding too much compression to a mix can take the life out of your recording, while adding just the right amount can help tame the mix's overall levels. Likewise, bad EQ decisions at the mix bus stage can wreck a mix like a barrel of monkeys. Applying too much EQ to the master bus means that problems in your mix might exist that need attention.

Once you've built your basic mix, you might want to create a rough mix that can be burned to CD. Take it out to the car, take it to a friend's studio, put it in your best/worst boom box and/or have a friend critique it. Take notes and then revisit the mix after a day. You might want to take the time to compare your new mix to your desert island disc, to hear how it stacks up against your favorite (or the producer's favorite) mixes. It is often surprising what you'll hear.

Need more inputs?

It's no secret that modern multi-miking and electronic music techniques have had a major impact on the physical requirements of mixing hardware and software. This is largely due to the increased need for the large number of physical inputs, outputs and effects that are commonly encountered in the modern project and MIDI production facility. For those who are using a large number of hardware instruments and who are doing their production and mixing work on an analog or digital hardware mixer, it's easy to understand how we might run out of physical inputs when you're faced with synths that have four outputs and a drummer that needs 12 mics. With all of these issues and more, you can see how a system might easily outgrow your connection needs, leaving you with the unpleasant choice of either upgrading or dealing with your present system as best you can. Moral of the story: It's always wise to try to anticipate your future mixing requirements when buying an audio interface, console or mixer.

One way to keep from running out of inputs on your main mixer or console is to use an outboard line mixer. These rack-mountable mixers (also known as submixers) are often equipped with extra line-level inputs, each having equalization, pan and effects send capabilities that can be mixed down to either two or four channels. These channels can then be used to free up a multitude of inputs on your main mixing device.

For those who prefer to mix (or submix) within a DAW, the problems associated with the mixing of virtual instruments basically come down to a need for raw computing speed and power. In a virtual environment, any number of software instruments can be plugged and played into a workstation in real time, or the tracks can be transferred to disk and then mixed into the project as soundfile tracks (making sure to save all of the original instrument settings and MIDI files). These tracks can continue to add up within a session, but as long as the CPU "busy bar" indicator shows that the system is able to keep up with the processing and disk demands, none of this should be a problem.

When using a workstation-based system, there are often a number of ways to tag on additional analog inputs. For example, a number of interface designs offer eight or more analog inputs; however, a number of these systems offer ADAT lightpipe I/O. In short, this means that a number of mic preamp designs (usually having eight mic pres) that include ADAT lightpipe outputs can be plugged into a system, thereby adding eight more analog inputs to the interface.

A FINAL FOOTNOTE ON THE ART OF MIXING

Actually, the topic of the art of mixing could easily fill a book (and I'm sure it has); however, I'd simply like to point out the fact that it is indeed an art form … and as such is a very personal process. I remember the first time I sat down at a console (an older Neve 1604). I was truly petrified and at a loss as to how to approach the making of my first mix. Am I over-equalizing? Does it sound right? Will I ever get used to this sea of knobs? Well, folks, as with all things … the answers come to you when you simply sit down and mix, mix, mix! It's always a good idea to watch others in the process of practicing their art and take the time to listen to the work of others (both the known and the not-so-well known). With practice, it's a foregone conclusion that you'll begin to develop your own sense of the art and style of mixing … which, after all, is what it's all about. It's all up to you now. Just dive into the deep end … and have fun!

CHAPTER 15
Signal Processing

Over the years, signal processing has become an increasingly important part of audio and music production. It's the function of a *signal processor* to change, augment or otherwise modify an audio signal in either the analog or digital domain. This chapter offers some insight into the basics of effects processing and how they can be integrated into a recording or mixdown in ways that sculpt sound using forms that are subtle, lush or just plain whimsical and wacky.

Of course, the processing power of these effects systems can be harnessed in either a hardware or software plug-in form. Regardless of how you choose to work with sound, the important rule to remember is that there are no rules; however, there are a few general guidelines that can help you get the sound that you want. When using effects, the most important asset you can have is experience and your own sense of artistry. The best way to learn the art of processing, shaping and augmenting sound is through experience ... and gaining experience takes time, a willingness to learn and lots of patience.

THE WONDERFUL WORLD OF ANALOG, DIGITAL OR WHATEVER

Signal processing devices and their applied practices come in all sizes, shapes and flavors. These tools and techniques might be analog, digital or even acoustic in nature (Figure 15.1). The very fact that early analog processors have made a serious comeback (in the form of reissued hardware and software plug-in emulations) points to the importance of embracing past tools and techniques while combining them with the technological advances of the day to make the best possible production.

Although they aren't the first thoughts that comes to mind, the use of acoustics and ambient mic techniques are often the first line of defense when dealing with the processing of an audio signal. For example, as we saw in Chapter 4, changing a mic or its position might be the best option for changing the character of a pickup. Placing a pair of mics out into a room or mixing a guitar amp with a second, distant pickup might fill the ambience of an instrument in a way that a device just might not be able to duplicate with signal processing. In short, never underestimate the power of your acoustic environment as an effects tool.

FIGURE 15.1
Side and rear effects racks at Jackpot! Recording Studio in Portland, OR. (Courtesy of Jackpot! Recording Studio, Inc., www.jackpotrecording.com; photo credit: Jason quiqley, www.photojq.com)

For those wishing to work in the world of analog, an enormous variety of devices can be put to use in a production. Although these generally relate to devices that alter a source's relative volume levels (e.g., equalization and dynamics), there are also a number of analog devices that can be used to alter effects that are time based. For example, an analog tape machine can be patched so as to make an analog delay or regenerative echo device. Although they're not commonly used, spring and plate reverb units that can add their own distinctive sound can still be found on the used and sometimes new market.

On the other hand, the world of digital audio has definitely set signal processing on fire by offering an almost unlimited range of effects that are available to the musician, producer and engineer. One of the biggest advantages to working in the digital signal processing (DSP) domain is the fact that software programming can be used to configure a processor in order to achieve an ever-growing range of effects (such as reverb, echo, delay, equalization, dynamics, pitch shifting, gain changing and signal re-synthesis).

The task of processing a signal in the digital domain is accomplished by combining logic or programming circuits in a building-block fashion. These logic blocks follow basic binary computational rules that operate according to a special program algorithm. When combined, they can be used to alter the numeric values of sampled audio in a highly predictable way. After a program has been configured (from either internal ROM, RAM or system software), complete control over a program's setup parameters can be altered and inserted into a chain as an effected digital audio stream. Since the process is fully digital, these settings can be saved and precisely duplicated at any time upon recall. Even more amazing is how the overall quality and functionality have steadily increased while at the same time becoming more cost effective. It has truly brought an overwhelming amount of production power to the beginner, as well as to the pro.

PLUG-INS

In addition to both analog and digital hardware devices, an ever-growing list of signal processors is available for the Mac and PC platforms in the form of software *plug-ins*. These software utilities offer virtually every processing function imaginable (often at a fraction of the price of their hardware counterparts) with little or no reduction in quality, capabilities or automation features. These programs (which are created and marketed by large and small companies alike) are designed to be integrated into an editor or DAW production environment in order to perform a particular real-time or non-real-time processing function.

Currently, several plug-in standards exist, each of which functions as a platform that serves as a bridge to connect the plug-in through the computer's operating system (OS) to the digital audio production software. This means that any plug-in (regardless of its manufacturer) will work with an OS and DAW that's compatible with that platform standard, regardless of its form, function and/or manufacturer. As of this writing, the most popular standards are VST (PC/Mac), AudioSuite (Mac), Audio Units (Mac), MAS (MOTU for PC/Mac), as well as TDM and RTAS (Digidesign for PC/Mac).

By and large, effects plug-ins operate in a native processing environment (Figure 15.2a). This means that the computer's main CPU processor carries out the DAW or other program's required DSP functions. With the ever-increasing speed and power of modern-day CPUs, this has become less and less of a problem; however, hardware cards and external systems can be added to your system to "accelerate" the DSP processing power of your computer (Figure 15.2b) by adding additional CPUs that are dedicated to signal processing plug-in functions.

Plug-in control and automation

One of the more fun and powerful aspects of working with various signal processing plug-ins on a DAW platform is the ability to control and automate many or all of the various effects parameters with relative ease. These controls can be manipulated on-screen (via hands-on or track parameter controls) or from an external hardware controller (Figure 15.3), allowing the parameters to be physically controlled in real time.

SIGNAL PATHS IN EFFECTS PROCESSING

Before delving into the process of effecting and/or altering sound, we should first take a quick look at an important signal path concept—the fact that a signal processing device can be inserted into an analog or digital chain in several ways. The most common of these are:

- Insert routing
- Send routing

(a)

(b)

FIGURE 15.2
DSP signal processing plug-ins. (a) Examples of native effects plug-ins. (Courtesy of Loud Technologies Inc., www.mackie.com). (b) Producer/engineer Billy Bush mixing Garbage's "Not Your Kind of People" with Universal Audio's Apollo and UAD2 accelerated plug-ins. (Photo courtesy of Universal Audio / David Goggin, www.uaudio.com)

Insert routing

Insert routing is often used to alter the sonic or effects characteristics of a particular track or channel signal. It occurs whenever a processor is inserted directly into a signal path in a serial fashion. Using this approach, the path passes in

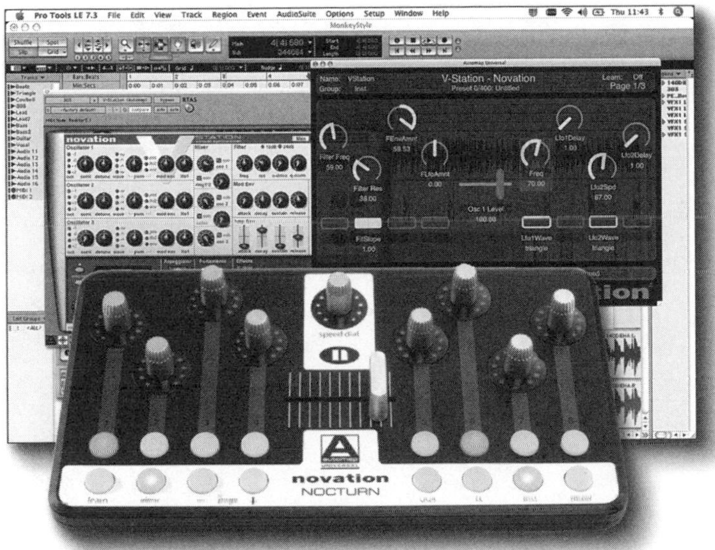

FIGURE 15.3
Novation's Nocturn hardware/plug-in controller with an automap function that automatically maps the controls to various parameters on the target plug-in. (Courtesy of Novation Digital Music Systems, Ltd., www.novationmusic.com)

from an audio source, through the signal processor and directly back out to the mixing section or to another device or point in the signal chain. This method of inserting a device is generally used for the processing of a single instrument, voice or grouped signals that are present on a particular hardware or virtual input strip. Often, but not always, this device tends to be a level-based processor (such as an equalizer, compressor or limiter). In keeping with the "no-rules" concept, time- and pitch-changing devices can also be used to tweak an instrument or voice in the signal chain. Here are but a few examples of how inline routing can be used:

- A device can be plugged into an input strip's insert (direct send/return) point. This approach is often used to insert an outboard device directly into an input strip's signal path (Figure 15.4).
- A console's main output bus could be run through a device (such as a stereo compressor) to affect an overall mix or submix grouping.
- An effects stomp box could be placed between a mic preamp and console input to create a grungy distortion effect.
- A DAW plug-in could be inserted into an input path to process only the signal on that channel (Figure 15.5).

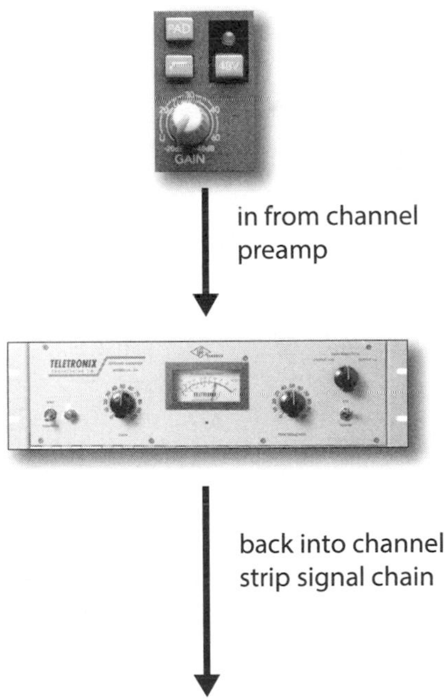

FIGURE 15.4
An example of insert routing, whereby the processed signal is inserted directly into the path of an input strip.

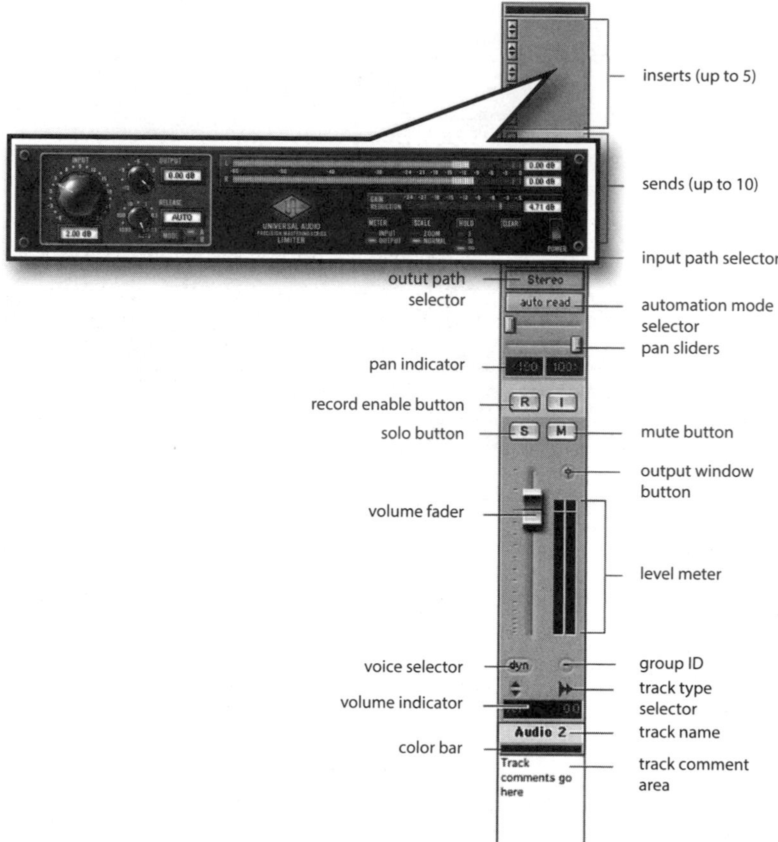

FIGURE 15.5
Example of how an inline effect can be inserted into a single channel on a DAW track. (Courtesy of Digidesign, a division of Avid Technology, Inc., www.digidesign.com)

EXTERNAL CONTROL OVER AN INSERT EFFECTS SIGNAL PATH

Certain insert effects processors allow for an external audio source to act as a control for affecting a signal as it passes from the input to the output of a device (Figure 15.6). Devices that offer an external "key" or "sidechain" input can be quite useful, allowing a signal source to be used as a control for varying another audio path. For example:

- A gate (an infinite expander that can control the passing of audio through a gain device) might take its control input from an external "key" signal that will determine when a signal will or will not pass in order to reduce leakage.
- A vocal track could be inserted into a vocoder's control input, so as to synthetically add a robot-like effect to a track.
- A voice track could be used for vocal ducking at a radio station, to fade out the music when a narrator is speaking.
- An external keyed input can be used to make a mix "pump" or "breathe" in a dance production.

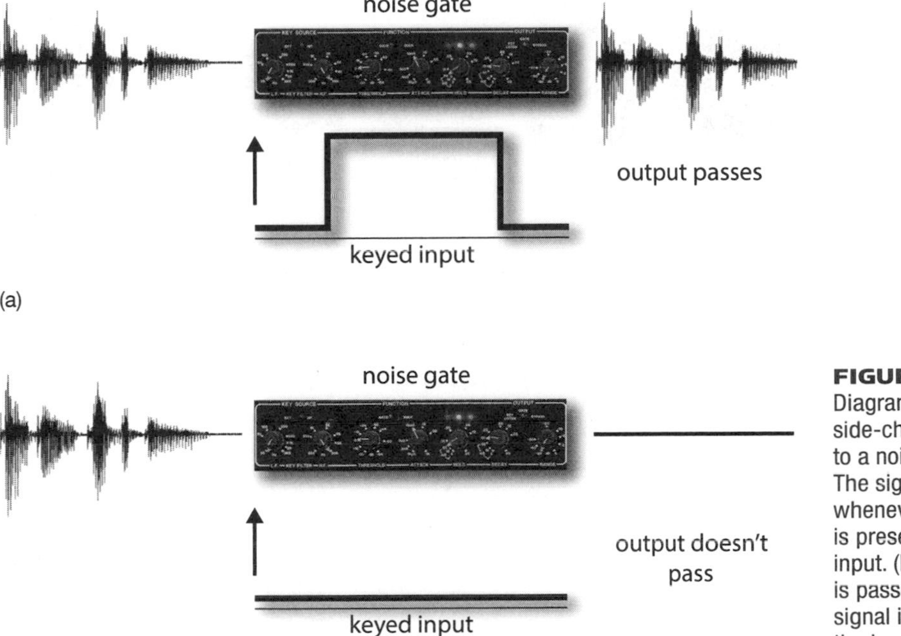

FIGURE 15.6 Diagram of a key side-chain input to a noise gate. (a) The signal is passed whenever a signal is present at the key input. (b) No signal is passed when no signal is present at the key input.

It's important to note that there is no set standard for providing a side-chain key in software. Some software packages provide native side-chain capability, while some support side-chaining via "multiple input" plug-ins and complex signal routing, while others don't support side-chaining at all.

Send routing

Effects "Sends" are often used to augment a signal (generally being used to add reverb, delay or other time-based effects). This type differs from an insert, in that instead of inserting a signal-changing device directly into the signal path, a portion of the signal (which is essentially routed through an effects "aux" mixer path) is "sent" to an effects device. Once effected, the signal can then be proportionately mixed back into the main out signal path, so as to add an effects blend to the track or overall output mix.

As an example, Figure 15.7 shows how sends from any number of channel inputs can be mixed together and then be sent to an effects device. The effected output signal can then be sent back into either the effects return section or to a set of spare mixer inputs … and to the main mixing bus outputs.

Viva la difference

There are a few insights that can help you to understand the difference between insert and send effects routing. Let's start by reviewing the basic ways that their signal paths function:

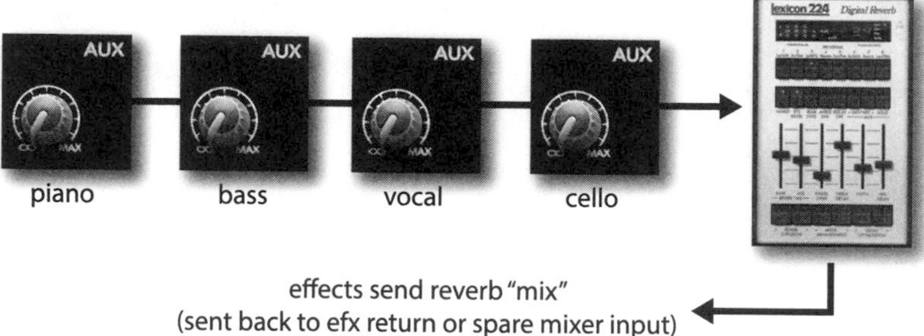

FIGURE 15.7
An "aux" sends path flows in a horizontal fashion to a send bus. The combined sum can then be effected (or sent to another destination) or returned back to the main output bus. (Courtesy of Loud Technologies Inc., www.mackie.com)

- First, from Figures 15.4 and 15.5, we can see that the signal flow of an effects insert is basically "vertical," moving from the channel input through an inserted device and then back into the input strip's signal chain (for mixing, recording or whatever).
- From Figure 15.7 we can see that an effects send basically functions in a "horizontal" direction, allowing portions of the various input signals to be mixed together and sent to an effects device, which can then be routed back into the mixer signal chain for further effects and mix blending.
- Finally, it's important to grasp the idea that when a device is "inserted" into a signal chain, it's usually a single, dedicated hardware/software device (or chain of devices) that is used to perform a specific function (Figure 15.8a). If a large number of track/channels are used in a session or mix, inserting numerous dsp effects into multiple strips could take up too much processing power. On the other hand, setting up an effects send can save a great deal of dsp processing overhead by creating a single effects send "mix" that can then be routed to a single effects device (Figure 15.8b) ... basically, going like this: "why use a whole bunch of the same EFX devices to do a job on every channel, when sending to a single EFX device will do."
- There is one more situation that you should be aware of. Many hard and software devices have an internal "mix" control that serves as a side-chain mix control for varying the amount of "dry" (original) signal to be mixed with the "wet" (effected) signal, as seen in Figure 15.9. This setting can have a profound effect on your settings (when used in either an insert or send setting) ... and a careful awareness is advised.

On a final note, each effects technique has its own set of strengths and weaknesses ... it's important to play with each, so as to know the difference and when to make best use of them.

EFFECT PROCESSING

From this point on, this chapter will be taking an in-depth look at many of the signal processing devices, applications and techniques that have traditionally

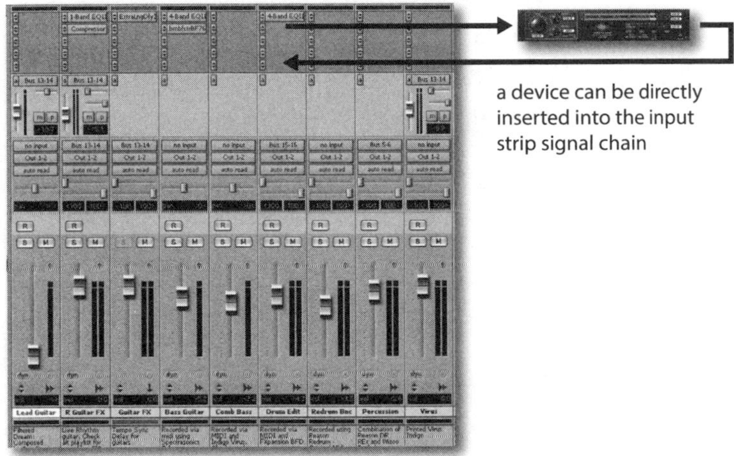
a device can be directly inserted into the input strip signal chain

(a)

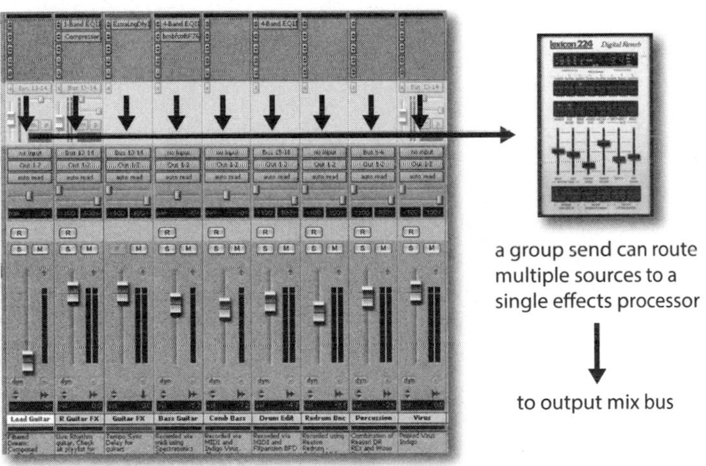
a group send can route multiple sources to a single effects processor
to output mix bus

(b)

FIGURE 15.8
An effect can be: (a) Directly inserted into an input strip signal chain. (b) Created by being sent through an effects send bus, allowing multiple channels to be assigned to a single effects processor. (Courtesy of Digidesign, a division of Avid Technology, Inc., www.digidesign.com, and Universal Audio, www.uaudio.com. © 2013 Universal Audio, Inc. All rights reserved. Used with permission.)

POWER
LEVELS
M•ONE DUAL EFFECTS PROCESSOR
t.c. electronic
MIX
wet/dry mix control

FIGURE 15.9
Example of a commonly found mix control that allows an inserted device to mix the non-effected (dry) with the effected (wet) signals.

been the cornerstone of music and sound production, including systems and techniques that exert an ever-increasing degree of control over:

- *The spectral content of a sound:* in the form of equalization or intelligent equalization and bandpass filtering
- *Amplitude level processing:* in the form of dynamic range processing
- *Time-based effects:* Augmentation or re-creation of room ambience, delay, time/pitch alterations and tons of other special effects that can range from being sublimely subtle to "in yo' face."

Hardware and virtual effects in action

The following sections offer some insight into the basics of effects processing and how they can be integrated into a recording or mixdown. It's a forgone conclusion that the power of these effects can be harnessed in hardware or software plug-in form. The important rule to remember is that there are no rules; however, there are a few general guidelines that can help you get the sound that you want. When using effects, the most important asset you can have is experience and your own sense of artistry. The best way to learn the art of processing, shaping and augmenting sound is through experience; gaining experience takes time, a willingness to learn and lots of patience.

As stated in Chapter 14, an audio equalizer (Figures 15.10 and 15.11) is a circuit, device or plug-in that lets us exert control over the harmonic or timbral content of a recorded sound. EQ may need to be applied to a single recorded channel, to a group of channels or to an entire program.

Equalization refers to the alteration in frequency response of an amplifier so that the relative levels of certain frequencies are more or less pronounced than others. EQ is specified as either plus or minus a certain number of decibels at a certain frequency. For example, you might want to boost a signal by "+4 dB at 5 kHz." Although only one frequency was specified in this example, in reality a range of frequencies above, below and centered around the specified frequency will often be affected. The amount of boost or cut at frequencies other than the one named is determined by whether the curve is peaking or shelving, by the bandwidth of the curve (a factor that's affected by the Q settings and determines how many frequencies will be affected around a chosen centerline), and by the amount of boost or cut at the named frequency. For example, a +4 dB

FIGURE 15.10
Manley Massive Passive Stereo Equalizer (Courtesy of Manley Laboratories, Inc., www.manleylabs.com)

(a)

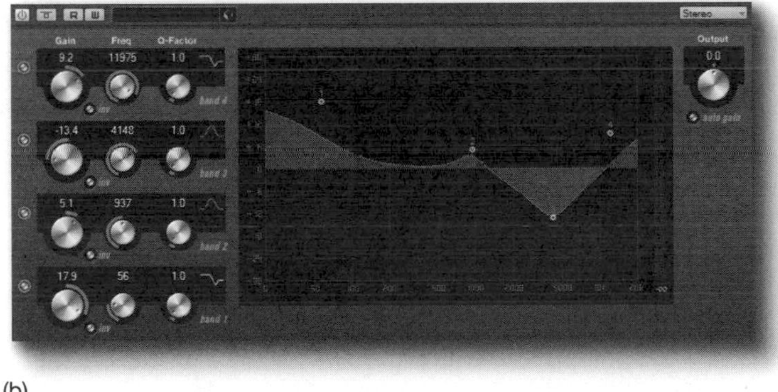

(b)

FIGURE 15.11
EQ plug-ins:
(a) Pro Tools 4-band EQII (courtesy of Digidesign, a division of Avid Technology, Inc., www.digidesign.com); (b) Cubase/Nuendo (Courtesy of Steinberg Media Technologies GmbH, a division of Yamaha Corporation, www.steinberg.net)

boost at 1000 Hz might easily add a degree of boost or cut at 800 and 1200 Hz (Figure 15.12).

Older equalizers and newer "retro" systems often base their design around filters that use passive components (i.e., inductors, capacitors and resistors) and employ amplifiers only to make up for internal losses in level, called *insertion loss*. Most equalization circuits today, however, are of the active filter type that changes their characteristics by altering the feedback loop of an operational amp. This is by far the most common analog EQ type and is generally favored

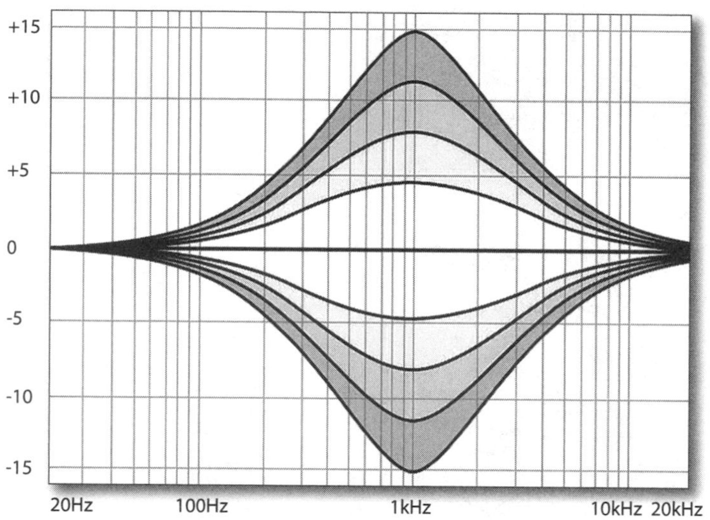

(a)

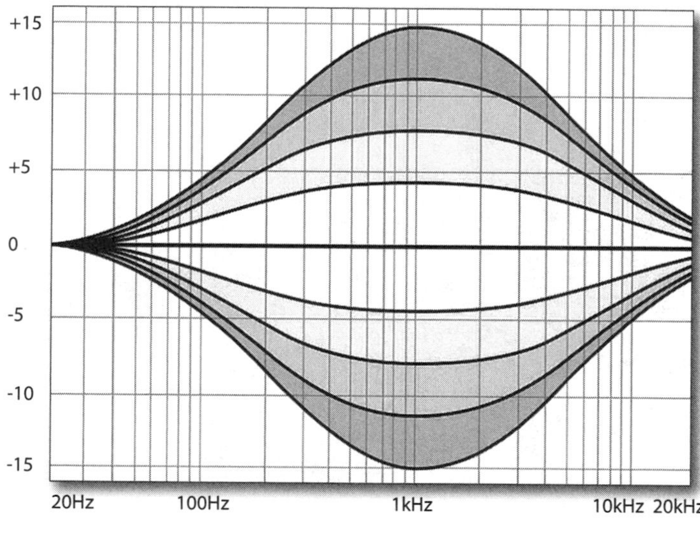

(b)

FIGURE 15.12 Various boost/cut EQ curves centered around 1 kHz: (a) center frequency, 1-kHz bandwidth 1 octave, ±15 dB boost/cut; (b) center frequency, 1kHz bandwidth 3 octaves, ±15 dB boost/cut. (Courtesy of Mackie Designs, www.mackie.com)

over its passive counterpart due to its low cost, size and weight, as well as its wide gain range and line-driving capabilities.

Peaking filters

The most common EQ curve is created by a *peaking filter*. As its name implies, a peak-shaped bell curve can either be boosted or cut around a selected center

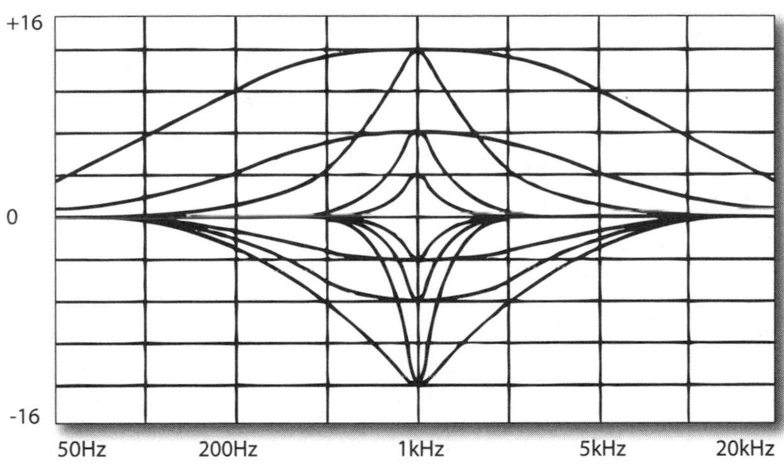

FIGURE 15.13
Peaking equalization curves.

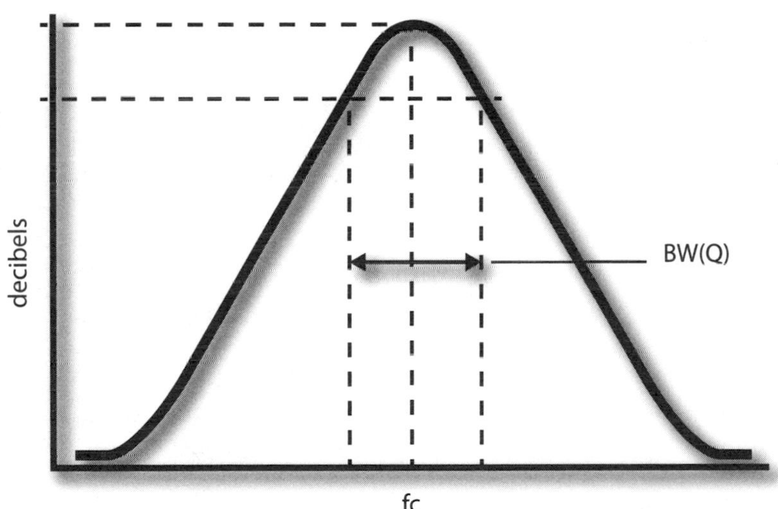

FIGURE 15.14
The number of hertz between the two points that are 3 dB down from the center frequency determines the bandwidth of a peaking filter.

frequency. Figure 15.13 shows the curves for a peak equalizer that's set to boost or cut at 1000 Hz. The *quality factor* (Q) of a peaking equalizer refers to the width of its bell-shaped curve. A curve with a high Q will have a narrow bandwidth with few frequencies outside the selected bandwidth being affected, whereas a curve having a low Q is very broadband and can affect many frequencies (or even octaves) around the center frequency. *Bandwidth* is a measure of the range of frequencies that lie between the upper and lower –3-dB (half-power) points on the curve (Figure 15.14). The Q of a filter is an inverse measure of the bandwidth (such that higher Q values mean that fewer frequencies will be affected, and vice versa). To calculate Q, simply divide the center frequency by

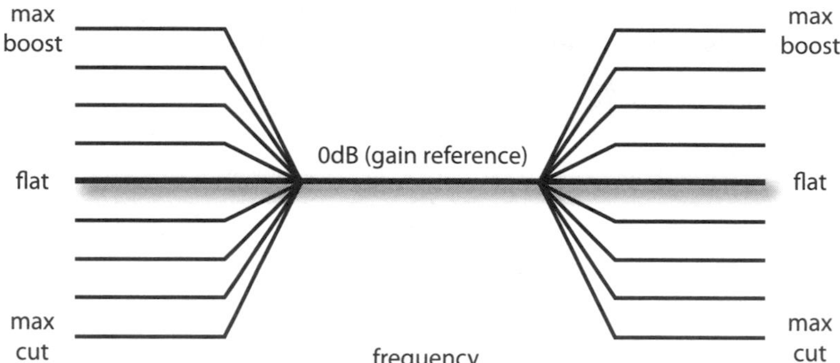

FIGURE 15.15
High/low, boost/cut curves of a shelving equalizer.

the bandwidth. For example, a filter centered at 1 kHz that's a third of an octave wide will have its –3 dB frequency points located at 891 and 1123 Hz, yielding a bandwidth of 232 Hz (1123 – 891). This EQ curve's Q, therefore, will be 1 kHz divided by 232 Hz or 4.31.

Shelving filters

Another type of equalizer is the *shelving filter*. Shelving refers to a rise or drop in frequency response at a selected frequency, which tapers off to a preset level and continues at that level to the end of the audio spectrum. Shelving can be inserted at either the high or low end of the audio range and is the curve type that's commonly found on home stereo bass and treble controls (Figure 15.15).

High-pass and low-pass filters

Equalizer types also include *high-pass* and *low-pass filters*. As their names imply, this EQ type allows certain frequency bandwidths to be passed at full level while other sections of the audible spectrum are attenuated. Frequencies that are attenuated by less than 3 dB are said to be inside the *passband*; those attenuated by more than 3 dB are in the *stopband*. The frequency at which the signal is attenuated by exactly 3 dB is called the *turnover* or *cutoff frequency* and is used to name the filter frequency.

Ideally, attenuation would become infinite immediately outside the passband; however, in practice this isn't always attainable. Commonly, attenuation is carried out at rates of 6, 12 and 18 dB per octave. This rate is called the *slope* of the filter. Figure 15.16a, for example, shows a 700-Hz high-pass filter response curve with a slope of 6 dB per octave, and Figure 15.16b shows a 700-Hz low-pass filter response curve having a slope of 12 dB per octave. High- and low-pass filters differ from shelving EQ in that their attenuation doesn't level off outside the passband. Instead, the cutoff attenuation continues to increase. A high-pass filter in combination with a low-pass filter can be used to create a *bandpass filter*, with the passband being controlled by their respective turnover frequencies and the *Q* by the filter's slope (Figure 15.17).

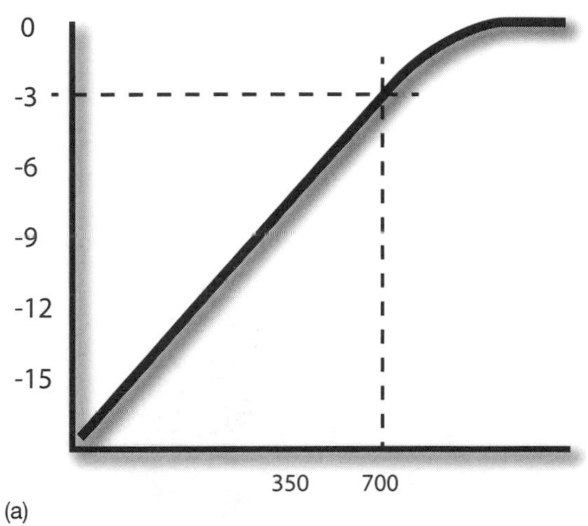

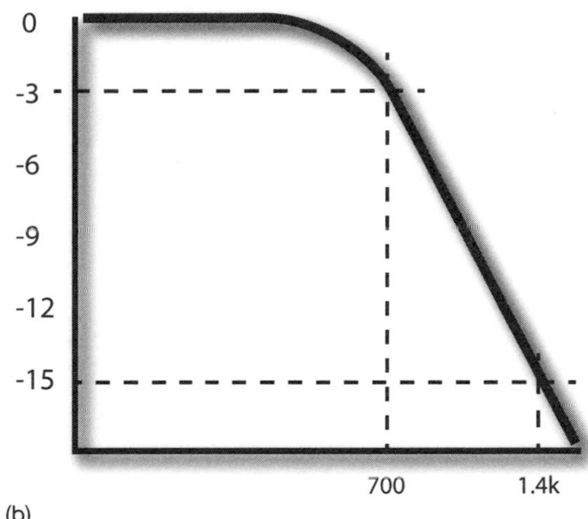

FIGURE 15.16
A 700-Hz filter: (a) high-pass filter with a slope of 6 dB per octave; (b) low-pass filter with a slope of 12 dB per octave.

Equalizer types

The four most commonly used equalizer types that can incorporate one or more of the previously described filter types are the:

- Parametric equalizer
- Selectable frequency equalizer
- Graphic equalizer
- Notch filter

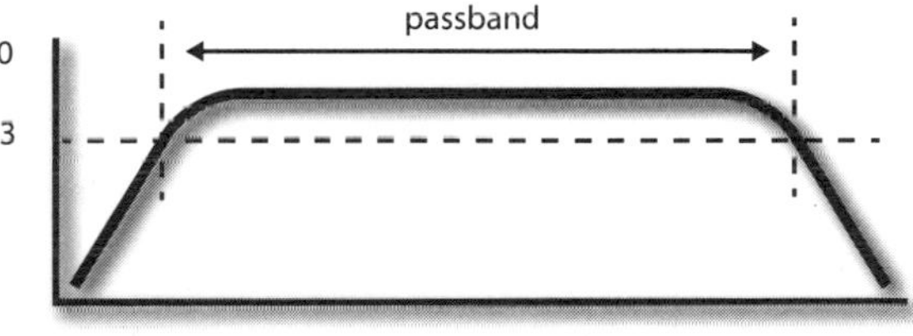

FIGURE 15.17
A bandpass filter is created by combining a high- and low-pass filter with different cutoff frequencies.

The *parametric equalizer* (Figure 15.18) lets you adjust most or all of its frequency parameters in a continuously variable fashion. Although the basic design layout will change from model to model, each band will often have an adjustment for continuously varying the center frequency. The amount of boost or cut is also continuously variable. Control over the center frequency and Q can be either selectable or continuously variable, although certain manufacturers might not have provisions for a variable Q.

Generally, each set of frequency bands will overlap into the next band section, so as to provide smooth transitions between frequency bands or allow for multiple curves to be placed in nearby frequency ranges. Because of its flexibility and performance, the parametric equalizer has become the standard design for most input strips, digital equalizers and workstations.

The *selectable frequency equalizer* (Figure 15.19), as its name implies, has a set number of frequencies from which to choose. These equalizers usually allow a

FIGURE 15.18
The EQF-100 full range, parametric vacuum tube equalizer. (Courtesy of Summit Audio, Inc., www.summitaudio.com)

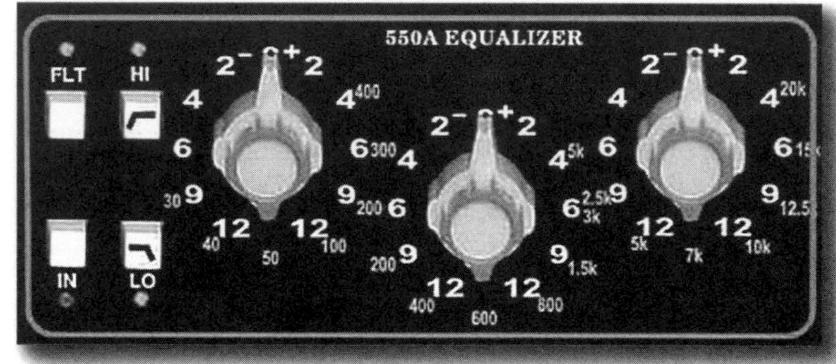

FIGURE 15.19
The API 550A selectable frequency equalizer within the 7600 outboard module. (Courtesy of API Audio, www.apiaudio.com).

FIGURE 15.20
Rane GE 130 single-channel, 30-band, 1/3-octave graphic equalizer. (Courtesy of Rane Corporation, www.rane.com)

boost or cut to be performed at a number of selected frequencies with a predetermined Q. They are most often found on older console designs, certain low-cost production consoles and outboard gear.

A *graphic equalizer* (Figure 15.20) provides boost and cut level control over a series of center frequencies that are equally spaced (ideally according to music intervals). An "octave band" graphic equalizer might, for example, have 12 equalization controls spaced at the octave intervals of 20, 40, 80, 160, 320 and 640 Hz and 1.25, 2.5, 5, 10 and 20 kHz, while 1/3-octave equalizers could have up to 36 center frequency controls. The various EQ band controls generally use vertical sliders that are arranged side by side so that the physical positions of these controls could provide a "graphic" readout of the overall frequency response curve at a glance. This type is often used in applications that can help fine-tune a system to compensate for the acoustics in various types of rooms, auditoriums and studio control rooms.

Notch filters are often used to zero in on and remove 60- or 50-Hz hum or other undesirable discrete-frequency noises. They use a very narrow bandwidth to fine-tune and attenuate a particular frequency in such a way as to have little effect on the rest of the audio program. Notch filters are used more in film location sound and broadcast than in studio recording, because severe narrow-band problems aren't often encountered in a well-designed studio ... hopefully.

Applying equalization

When you get right down to it, EQ is all about compensating for deficiencies in a sound pickup or about reducing extraneous sounds that make their way into a pickup signal. To start our discussion on how to apply EQ, let's again revisit the all-important "Good Rule" from Chapter 4.

Whenever possible, EQ should not be used as a Band-Aid. By this, I mean that it's often a good idea to correct for a problem on the spot rather than to rely on the hope that you can "fix it in the mix" at a later time using EQ and other methods.

THE "GOOD RULE"
Good musician + good instrument + good performance + good acoustics + good mic + good placement = good sound.

When in doubt, it's often better to deal with a problem as it occurs. This isn't always possible, however ... therefore, EQ is best used in situations where:

- There's no time or money left to redo the track.
- The existing take was simply magical and shouldn't be re-recorded.
- The track was already recorded during a previous session.

EQ in action!

Although most equalization is done by ear, it's helpful to have a sense of which frequencies affect an instrument in order to achieve a particular effect. On the whole, the audio spectrum can be divided into four frequency bands: low (20 to 200 Hz), low-mid (200 to 1000 Hz), high-mid (1000 to 5000 Hz) and high (5000 to 20,000 Hz). When the frequencies in the 20- to 200-Hz (low) range are modified, the fundamental and the lower harmonic range of most bass information will be affected. These sounds often are felt as well as heard, so boosting in this range can add a greater sense of power or punch to music. Lowering this range will weaken or thin out the lower frequency range.

The fundamental notes of most instruments lie within the 200- to 1000-Hz (low-mid) range. Changes in this range often result in dramatic variations in the signal's overall energy and add to the overall impact of a program. Because of the ear's sensitivity in this range, a minor change can result in an effect that's very audible. The frequencies around 200 Hz can add a greater feeling of warmth to the bass without loss of definition. Frequencies in the 500- to 1000-Hz range could make an instrument sound hornlike, while too much boost in this range can cause listening fatigue.

Higher-pitched instruments are most often affected in the 1000 to 5000 Hz (high-mid) range. Boosting these frequencies often results in an added sense of clarity, definition and brightness. Too much boost in the 1000 to 2000 Hz range can have a "tinny" effect on the overall sound, while the upper mid-frequency range (2000 to 4000 Hz) affects the intelligibility of speech. Boosting in this range can make music seem closer to the listener, but too much of a boost can also cause listening fatigue.

The 5000-to-20,000 Hz (high-frequency) region is composed almost entirely of instrument harmonics. For example, boosting frequencies in this range will often add sparkle and brilliance to a string or woodwind instrument. Boosting too much might produce sibilance on vocals and make the upper range of certain percussion instruments sound harsh and brittle. Boosting at around 5000 Hz has the effect of making music sound louder. A 6 dB boost at 5000 Hz, for example, can sometimes make the overall program level sound as though it's been doubled in level; conversely, attenuation can make music seem more distant. Table 15.1 provides an

Table 15.1 Instrumental Frequency Ranges of Interest

Instrument	Frequencies of Interest
Kick drum	Bottom depth at 60–80 Hz, slap attack at 2.5 kHz
Snare drum	Fatness at 240 Hz, crispness at 5 kHz
Hi-hat/cymbals	Clank or gong sound at 200 Hz, shimmer at 7.5 kHz to 12 kHz
Rack toms	Fullness at 240 Hz, attack at 5 kHz
Floor toms	Fullness at 80–120 Hz, attack at 5 kHz
Bass guitar	Bottom at 60–80 Hz, attack/pluck at 700–1000 Hz, string noise/pop at 2.5 kHz
Electric guitar	Fullness at 240 Hz, bite at 2.5 kHz
Acoustic guitar	Bottom at 80–120 Hz, body at 240 Hz, clarity at 2.5–5 kHz
Electric organ	Bottom at 80–120 Hz, body at 240 Hz, presence at 2.5 kHz
Acoustic piano	Bottom at 80–120 Hz, presence at 2.5–5 kHz, crisp attack at 10 kHz, honky-tonk sound (sharp Q) at 2.5 kHz
Horns	Fullness at 120–240 Hz, shrill at 5–7.5 kHz
Strings	Fullness at 240 Hz, scratchiness at 7.5–10 kHz
Conga/bongo	Resonance at 200–240 Hz, presence/slap at 5 kHz
Vocals	Fullness at 120 Hz, boominess at 200–240 Hz, presence at 5 kHz, sibilance at 7.5–10 kHz

Note: These frequencies aren't absolute for all instruments, but are meant as a subjective guide.

analysis of how frequencies and EQ settings can interact with various instruments. (For more information, refer to the Microphone Placement Techniques section in Chapter 4.)

Tutorial: Equalization

1. Solo an input strip on a mixer, console or DAW. Experiment with the settings using the previous frequency ranges. Can you improve on the original recorded track or does it take away from the sound?
2. Using the input strip equalizers on a mixer, console or DAW, experiment with the EQ settings and relative instrument levels of an entire mix using the previous frequency ranges as a guide. Can you bring an instrument out without changing the fader gains? Can you alter the settings of two or more instruments to increase the mix's overall clarity?
3. Plug the main output buses of a mixer, console or DAW into an outboard equalizer, and change the program's EQ settings using the previous frequency range discussions as a guide.

One way to zero in on a particular frequency using an equalizer (especially a parametric one) is to accentuate or attenuate the EQ level and then vary the center frequency until the desired range is found. The level should then be scaled back until the desired effect is obtained. If boosting in one-instrument range causes you to want to do the same in other frequency ranges, it's likely that you're simply raising the program's overall level. It's easy to get caught up in the "bigger! better! more!" syndrome of wanting an instrument to sound louder. If this continues to happen on a mix, it's likely that one of the frequency ranges of an instrument or ensemble is too dominant and requires attenuation. On the subject of laying down a recorded track with EQ, there are a number of situations and differing opinions regarding them:

- Some use EQ liberally to make up for placement and mic deficiencies, whereas others might use it sparingly, if at all. One example where EQ is used sparingly is when an engineer knows that someone else will be mixing a particular song or project. In this situation, the engineer who's doing the mix might have a very different idea of how an instrument should sound. If large amounts of EQ were recorded to a track during the session, the mix engineer might have to work very hard to counteract the original EQ settings.
- If everything was recorded flat, the producer and artists might have difficulty passing judgment on a performance or hearing the proper balance during the overdub phase. Such a situation might call for equalization in the monitor mix, while leaving the recorded tracks alone.

- In situations where several mics are to be combined onto a single track or channel, the mics can be individually equalized only during the recording phase. In situations where a project is to be engineered, mixed and possibly even mastered, the engineer might want to know in advance the type and amount of EQ that the producer and/or artist would want.
- Above all, it's wise that any "sound-shaping" should be determined and discussed with the producer and/or artist before the sounds are committed to a track.

In the end, there's no getting around the fact that an equalizer is a powerful tool. When used properly, it can greatly enhance or restore the musical and sonic balance of a signal. Experimentation and experience are the keys to proper EQ usage, and no book can replace the trial-and-error process of "just doing it!"

Before moving on, it's important to keep one age-old viewpoint in mind—that an equalizer shouldn't be regarded as a cure-all for improper mic technique; rather, it should be used as a tool for correcting problems that can't be easily fixed on the spot through mic and/or performance adjustments. If an instrument is poorly recorded during an initial recording session, it's often far more difficult and time consuming to "fix it in the mix" at a later time. Getting the best possible sound down onto tape will definitely improve your chances for attaining a sound and overall mix that you can be proud of.

SOUND-SHAPING EFFECTS PLUG-INS

Another class of effects devices that aren't equalizers, but instead affect the overall tonal character of a track or mix come under the category of sound-shaping devices. These systems can either be hardware or plug-in in nature and are used to alter the tonal and/or overtone balance of a signal. For example, a device that's been around for decades in the Aphex Aural Exiter. This device is able to add a sense of presence to a sound by generating additional overtones that are subdued or not present in the program signal. Another class of sound-shaper comes in the form of virtual tape machine plug-ins that closely model and emulate analog tape recorders … right down to the sonic character, tape noise, distortion, changes with virtual tape formulation and bias settings (Figures 15.21 and 15.22).

DYNAMIC RANGE

Like most things in life that get out of hand from time to time, the level of a signal can vary widely from one moment to the next. For example, if a vocalist gets caught up in the moment and lets out an impassioned scream following a soft whispery passage, you can almost guarantee that the mic's signal will jump from its optimum recording level into severe distortion … OUCH! Conversely, if you set an instrument's mic to accommodate the loudest level, its signal might be buried in the mix during the rest of the song. For these and other reasons, it becomes obvious that it's sometimes necessary to exert some form of control

FIGURE 15.21
The Studer A800 tape emulation plugin plug-in for the Apollo and the UAD effects processing card. (Courtesy of Universal Audio, www.uaudio.com © 2013 Universal Audio, Inc. All rights reserved. Used with permission).

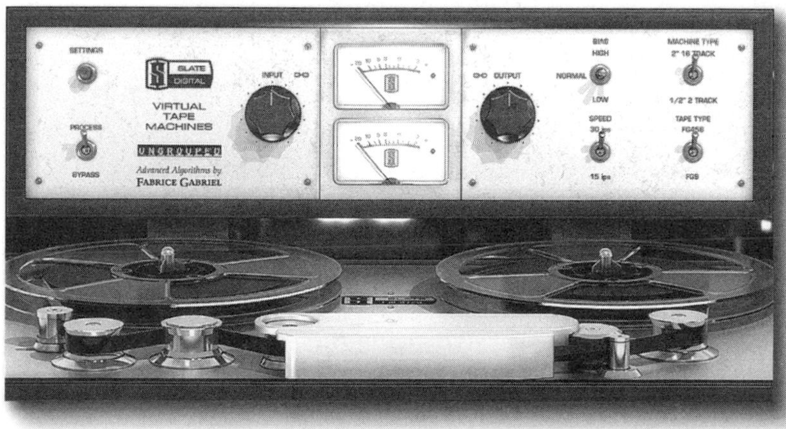

FIGURE 15.22
The Slate Digital Virtual Tape Machines plug-in that's capable of emulating a 16-track 2" and ½" mastering tape deck. (Courtesy of Slate Digital, www.slatedigital.com).

over a signal's dynamic range by using various techniques and dynamic controlling devices. In short, the dynamics of an audio program's signal resides somewhere in a continuously varying realm between three level states:

- Saturation
- Average signal level
- System/ambient noise

As you may remember from various chapters in this book, *saturation* occurs when an input signal is so large that an amp's supply voltage isn't large enough to produce the required output current or is so large that a digital converter reaches full scale (where the A/D output reads as all 1's). In either case, the results generally don't sound pretty and should be avoided in the studio's audio chain. The *average signal level* is where the overall signal level of a mix resides. Logically, if an instrument's level is too low, it can get buried in the mix ... if it's too high, it can unnecessarily stick out and throw the entire balance off. It is here that the art of mixing at an average level that's high enough to stand out in the sonic crowd, while still retaining enough dynamic "life," truly becomes an applied balance of skill and magic.

Dynamic range processors

The overall dynamic range of music is potentially on the order of 120 to 140 dB, whereas the overall dynamic range of a compact disc is often 80 to 90 dB, and analog magnetic tape is on the order of 60 dB (excluding the use of noise-reduction systems, which can improve this figure by 15 to 30 dB). However, when working with 20- and 24-bit digital word lengths, a system, processor or channel's overall dynamic range can actually approach or exceed the full range of hearing. Even with such a wide dynamic range, unless the recorded program is played back in a noise-free environment, either the quiet passages will get lost in the ambient noise of the listening area (35 to 45 dB SPL for the average home and much worse in a car) or the loud passages will simply be too loud to bear. Similarly, if a program of wide dynamic range were to be played through a medium with a limited dynamic range (such as the 20- to 30-dB range of an AM radio or the 40- to 50-dB range of FM), a great deal of information would get lost in the background noise. To prevent these problems, the dynamics of a program can be restricted to a level that's appropriate for the reproduction medium (theater, radio, home system, car, etc.) as shown in Figure 15.23. This gain reduction can be accomplished either by manually riding the fader's gain or through the use of a *dynamic range processor* that can alter the range between the signal's softest and loudest passages.

The concept of automatically changing the gain of an audio signal (through the use of compression, limiting and/or expansion) is perhaps one of the most misunderstood aspects of audio recording. This can be partially attributed to the fact that a well-done job won't be overly obvious to the listener. Changing the dynamics of a track or overall program will often affect the way in which it will be perceived (either unconsciously or consciously) by making it "seem" louder, by reducing its volume range to better suit a particular medium or by making it possible for a particular sound to ride at a better level above other tracks within a mix.

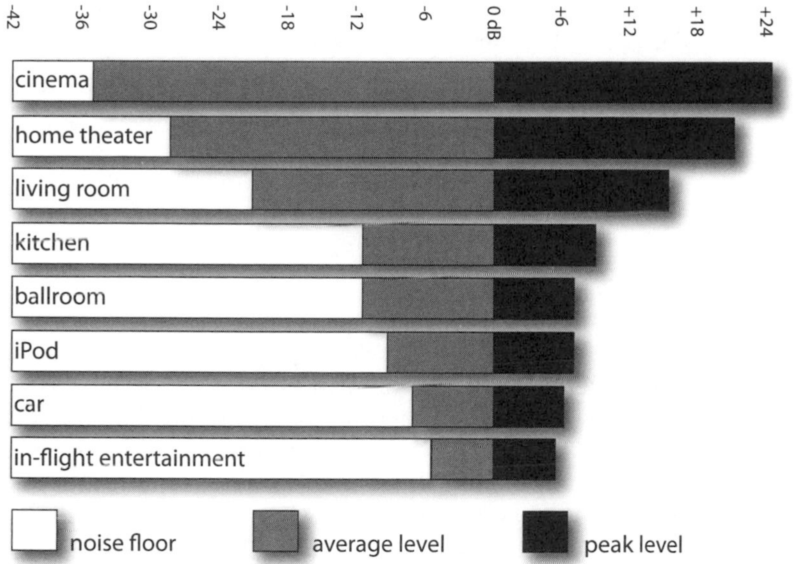

FIGURE 15.23
Dynamic ranges of various audio media, showing the noise floor (black), average level (white) and peak levels (gray). (Courtesy of Thomas Lund, tc electronic, www.tcelectronic.com)

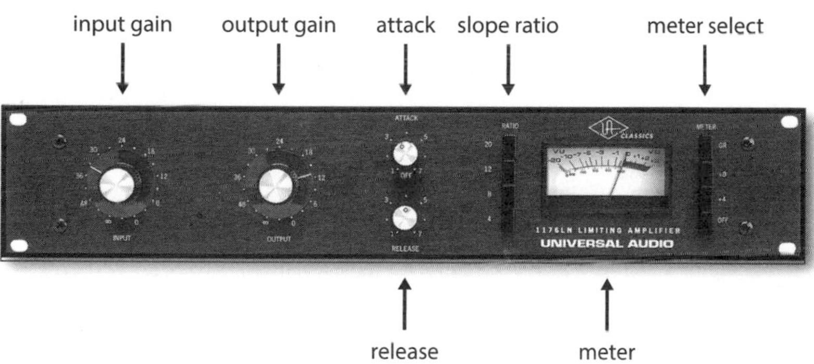

FIGURE 15.24
Universal Audio 1176LN limiting amplifier. (Courtesy of Universal Audio, www.uaudio.com. © 2013 Universal Audio, Inc. All rights reserved. Used with permission.)

Compression

A *compressor* (Figure 15.24), in effect, can be thought of as an automatic fader. It is used to proportionately reduce the dynamics of a signal that rises above a user-definable level (known as the *threshold*) to a lesser volume range. This process is done so that:

- The dynamics can be managed by the electronics and/or amplifiers in the signal chain.
- The range is appropriate to the overall dynamics of a playback or broadcast medium.
- An instrument better matches the dynamics of other recorded tracks within a song or audio program.

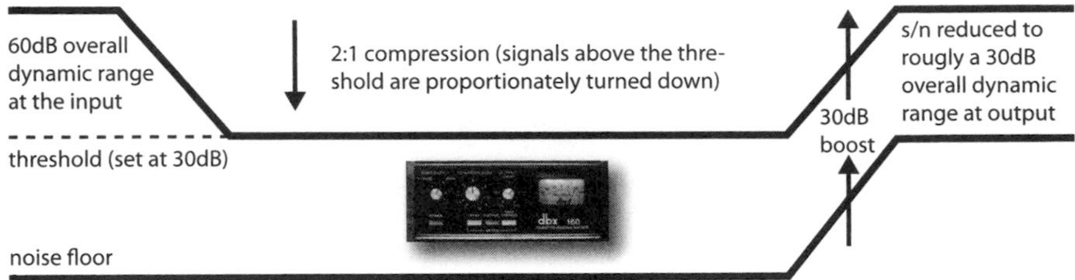

FIGURE 15.25
A compressor reduces input levels that exceed a selected threshold by a specified amount. Once reduced, the overall signal can then be boosted in level, thereby allowing the softer signals to be raised above other program or background sounds.

Since the signals of a track, group or program will be automatically turned down (hence the terms *compressed* or *squashed*) during a loud passage, the overall level of the newly reduced signal can now be amplified. In other words, once the dynamics have been reduced downward, the overall level can be boosted such that the range between the loud and soft levels is less pronounced (Figure 15.25). We've not only restored the louder signals back to a prominent level but we have also turned up the softer signals. In effect, we've turned up the softer signals that would otherwise be buried in the mix or ambient background noise.

The most common controls on a compressor (and most other dynamic range devices) include input gain, threshold, output gain, slope ratio, attack, release and meter display:

- *Input gain*: This control is used to determine how much signal will be sent to the compressor's input stage.
- *Threshold*: This setting determines the level at which the compressor will begin to proportionately reduce the incoming signal. For example, if the threshold is set to –20 dB, all signals that fall below this level will be unaffected, while signals above this level will be proportionately attenuated, thereby reducing the overall dynamics. On some devices, varying the input gain will correspondingly control the threshold level. In this situation, raising the input level will lower the threshold point and thus reduce the overall dynamic range. Most quality compressors offer hard and soft knee threshold options. A *soft knee* widens or broadens the threshold range, making the onset of compression less obtrusive, while the *hard knee* setting causes the effect to kick in quickly above the threshold point.
- *Output gain*: This control is used to determine how much signal will be sent to the device's output. It's used to boost the reduced dynamic signal into a range where it can best match the level of a medium or be better heard in a mix.
- *Slope ratio*: This control determines the slope of the input-to-output gain ratio. In simpler terms, it determines the amount of input signal (in decibels) that's needed to cause a 1 dB increase at the compressor's output (Figure 15.26). For example, with a ratio of 4:1, for every 4 dB increase at the input there will be only a 1 dB increase at the output; an 8-dB input increase will raise the output by 2 dB, while a ratio of 2:1 will produce a 1 dB increase in output for every 2 dB increase at its input. Get the idea?

- *Attack*: This setting (which is calibrated in milliseconds; 1 msec = 1 thousandth of a second) determines how fast or how slowly the device will turn down signals that exceed the threshold. It is defined as the time it takes for the gain to decrease to a percentage (usually 63%) of its final gain value. In certain situations (as might occur with instruments that have a long sustain, such as the bass guitar), setting a compressor to instantly turn down a signal might be audible (possibly creating a sound that pumps the signal's dynamics). In this situation, it would be best to use a slower attack setting. On the other hand, such a setting might not give the compressor time to react to sharp, transient sounds (such as a hi-hat). In this case, a fast attack time would probably work better. As you might expect, you'll need to experiment to arrive at the fastest attack setting that won't audibly color the signal's sound.
- *Release*: Similar to the attack setting, release (which is calibrated in milliseconds) is used to determine how slowly or quickly the device will restore a signal to its original dynamic level once it has fallen below the threshold point (defined as the time required for the gain to return to 63% of its original value). Too fast a setting will cause the compressor to change dynamics too quickly (creating an audible pumping sound), while too slow a setting might affect the dynamics during the transition from a loud to a softer passage. Again, it's best to experiment with this setting to arrive at the slowest possible release that won't color the signal's sound.
- *Meter display*: This control changes the compressor's meter display to read the device's output or gain reduction levels. In some designs, there's no need for a display switch, as readouts are used to simultaneously display output and gain reduction levels.

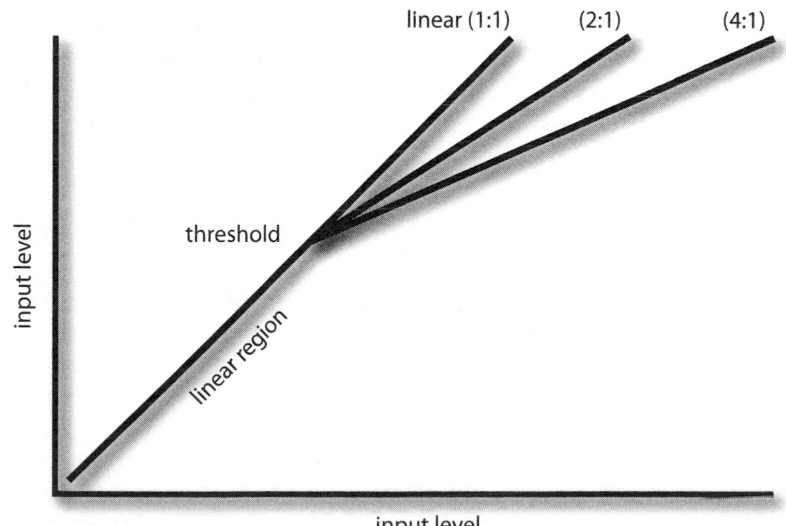

FIGURE 15.26 The output of a compressor is linear below the threshold and follows an input/output gain reduction ratio above this point, while the signal is proportionately reduced for signals that fall above the threshold level.

As was previously stated, the use of compression (and most forms of dynamics processing) is often misunderstood, and compression can easily be abused. Generally, the idea behind these processing systems is to reduce the overall dynamic range of a track, music or sound program or to raise its overall perceived level without adversely affecting the sound of the track itself. It's a well-known fact that over-compression can actually squeeze the life out of a performance by limiting the dynamics and reducing the transient peaks that can give life to a performance. For this reason, it's important to be aware of the general nuances of the controls we've just discussed.

During a recording or mixdown session, compression can be used in order to balance the dynamics of a track to the overall mix or to keep the signals from overloading preamps, the recording medium and your ears. Compression should be used with care for any of the following reasons:

- Minimize changes in volume that occur whenever the dynamics of an instrument or vocal are too great for the mix. As a tip, a good starting point might be a 0-dB threshold setting at a 4:1 ratio, with the attack and release controls set at their middle positions.
- Smooth out momentary changes in source-to-mic distance.
- Balance out the volume ranges of a single instrument. For example, the notes of an electric or upright bass often vary in volume from string to string. Compression can be used to "smooth out" the bass line by matching their relative volumes. In addition, some instruments (such as horns) are louder in certain registers because of the amount of effort that's required to produce the notes. Compression is often useful for equalizing these volume changes. As a tip, you might start with a ratio of 5:1 with a medium-threshold setting, medium attack and slower release time. Over-compression should be avoided to avoid pumping effects.
- Reduce other frequency bands by inserting a filter into the compression chain that causes the circuit to compress frequencies in a specific band (frequency-selective compression). A common example of this is a de-esser, which is used to detect high frequencies in a compressor's circuit so as to suppress those "SSSS," "CHHH" and "FFFF" sounds that can distort or stand out in a recording.
- Reduce the dynamic range and/or boost the average volume of a mix so that it appears to be significantly louder (as occurs when a television commercial seems louder than the show).

Although it may not always be the most important, this last application often gets a great deal of attention, because many producers strive to cut their recordings as "hot" as possible. That is, they want the recorded levels to be as far above the normal operating level as possible without blatantly distorting. In this competitive business, the underlying logic behind the concept is that louder recordings (when broadcast on Top 40 radio or podcast or played on a multiple CD changer or an MP3 player) will stand out from the softer recordings in a playlist. In fact, reducing the dynamic range of a song or program's dynamic range will actually make the overall levels appear to be louder. By using a slight (or not-so-slight) amount of compression and limiting to squeeze an extra 1 or 2 dB gain out of a song, the increased gain will also add to the perceived bass and highs because of our ears' increased sensitivity at louder levels (remember the Fletcher–Munson curve discussed in Chapter 2). To achieve these hot levels without distortion, multiband compressors and limiters often are used during the mastering process to remove peaks and to raise the average level of the program. You'll find more on this subject in Chapter 19 (Mastering).

Compressing a mono mix is done in much the same way as one might compress a single instrument. Adjusting the threshold, attack, release and ratio controls, however, is more critical in order to prevent the pumping of prominent instruments within the mix. Compressing a stereo mix gives rise to an additional problem: If two independent compressors are used, a peak in one channel will only reduce the gain on that channel and will cause sounds that are centered in a stereo image to shift (or jump) toward the channel that's not being compressed (since it will actually be louder). To avoid this center shifting, most compressors (of the same make and model) can be linked as a stereo pair. This procedure of ganging the two channels together interconnects the signal-level sensing circuits in such a way that a gain reduction in one channel will cause an equal reduction in the other (thereby preventing the stereo center information from shifting in the mix).

Before we move on, let's take a look at a few examples of the use of compression in various applications. Keep in mind, these are only beginning suggestions … nothing can substitute for experimenting and finding the settings that work best for you and the situation:

- *Acoustic guitar:* A moderate degree of compression (3 to 8 dB) with a medium compression ratio can help to pull an acoustic forward in a mix. A slower attack time will allow the string's percussive attack to pass through.
- *Bass guitar:* The electric bass is often a foundation instrument in pop and rock music. Due to variations in note levels from one note to another on an electric bass guitar (or upright acoustic, for that matter), a compressor can be used to even out the notes and add a bit of presence and/or punch to the instrument. Since the instrument often (but not always) has a slower attack, it's often a good idea to start with a medium attack (4:1, for example) and threshold setting, along with a slower release time setting. Harder compression of up to 10:1 with gain reductions ranging from 5 to 10 dB can also give a good result.
- *Brass:* The use of a faster attack (1 to 5 ms) with ratios that range from 6:1 to 15:1 and moderate to heavy gain reduction can help keep the brass in line.
- *Electric guitar:* In general, an electric guitar won't need much compression, because the sound is often evened out by the amp, the instrument's natural sustain character and processing pedals. If desired, a heavier compression ratio, with 10 or more decibels of compression can add to the instrument's "bite" in a mix. A faster attack time with a longer release is often a good place to start.

- *Kick drum and snare:* These driving instruments often benefit from added compression. For the kick, a 4:1 ratio with an attack setting of 10 ms or slower can help emphasize the initial attack while adding depth and presence. The snare attack settings might be faster, so as to catch the initial transients. Threshold settings should be set for a minimum amount of reduction during a quiet passage, with larger amounts of gain reduction happening during louder sections.
- *Synths:* These instruments generally don't vary widely in dynamic range, and thus won't require much (or any) compression. If needed, a 4:1 ratio with moderate settings can help keep synth levels in check.
- *Vocals:* Singers (especially inexperienced ones) will often place the mic close to their mouths. This can cause wide volume swings that change with small moves in distance. The singer might also shout out a line just after delivering a much quieter passage. These and other situations lead to the careful need for a compressor, so as to smooth out variations in level. A good starting point would be a threshold setting of 0 dB, with a ratio of 4:1 with attack and release settings set at their midpoints. Gain reductions that fall between 3 and 6 dB will often sit well in a mix (although some rock vocalists will want greater compression) … be careful of over-compression and its adverse pumping artifacts. Given digital's wide dynamic range, you might consider adding compression later in the mixdown phase, rather than during the actual session.
- *Final mix compression:* It's often a common practice to compress an entire mix during mixdown. If the track is to be professionally mastered, you should consult with the mastering engineer before the deed is done (or you might provide him or her with both a compressed and uncompressed version). When applying bus compression, it is usually a good idea to start with medium attack and release settings, with a light compression ratio (say, 4:1). With these or your preferred settings, reduce the threshold detection until a light amount of compression is seen on the meter display. Levels of between 3 and 6 dB will provide a decent amount of compression without audible pumping or artifacts (given that a well-designed unit or plug-in is used).

Tutorial: Compression

1. Go to the Tutorial section of www.modrec.com and download the tutorial soundfiles that relate to compression (which include instrument/music segments in various dynamic states).
2. Listen to the tracks. If you have access to an editor or DAW, import the files and look at the waveform amplitudes for each example. If you'd like to DIY, then …

3. Record or obtain an uncompressed bass guitar track and monitor it through a compressor or compression plug-in. Increase the threshold level until the compressor begins to kick in. Can you hear a difference? Can you see a difference on the console or mixer meters?
4. Set the levels and threshold to a level you like and then set the attack time to a slow setting. Now, select a faster setting and continue until it sounds natural. Try setting the release to its fastest setting. Does it sound better or worse? Now select a slower setting and continue until it sounds natural.
5. Repeat the above routine and settings using a snare drum track. Were your findings different?

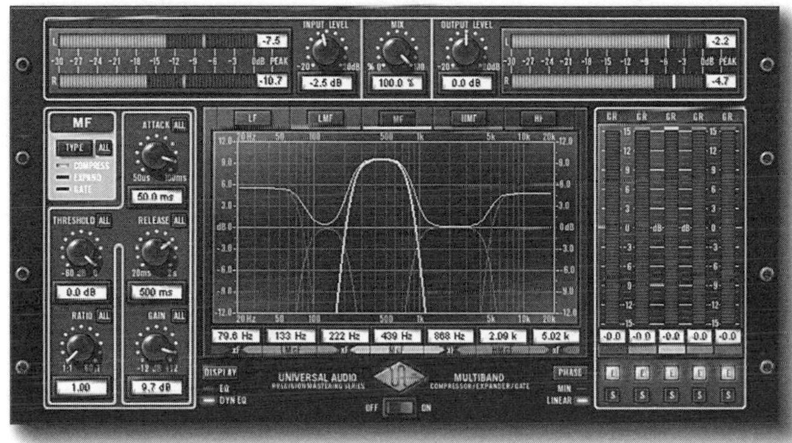

FIGURE 15.27
Universal Audio's UAD Precision Multiband plug-in. (Courtesy of Universal Audio, www.uaudio.com. © 2013 Universal Audio, Inc. All rights reserved. Used with permission.)

Multiband compression

Multiband compression (Figure 15.27) works by breaking up the audible spectrum into various frequency bandwidths through the use of multiple bandpass filters. This allows each of the bands to be isolated and processed in ways that strictly minimize the problems or maximize the benefits in a particular band. Although this process is commonly done in the final mastering stage, multiband techniques can also be used on an instrument or grouping. For example:

- The dynamic upper range of a slap bass could be lightly compressed, while heavier amounts of compression could be applied to the instrument's lower register.
- An instrument's high end can be brightened simply by adding a small amount of compression. This can act as a treble boost while accentuating some of the lower-level high frequencies.

More information on the subject can be found in Chapter 19 (Mastering).

FIGURE 15.28 Waves L1 Ultramaximizer limiting/quantization plug-in. (Courtesy of Waves Ltd., www.waves.com)

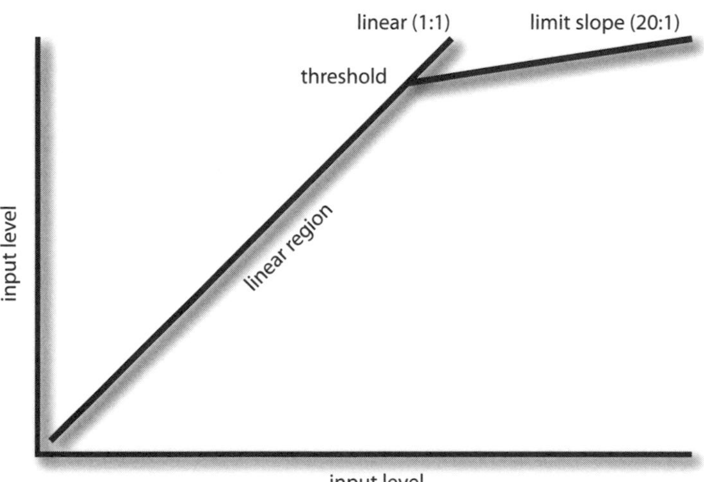

FIGURE 15.29 The output of a limiter is linear below the threshold and follows a high input/output gain reduction ratio (10:1, 20:1 or more) above this point.

Limiting

If the compression ratio is made large enough, the compressor will actually become a limiter. A limiter (Figure 15.28) is used to keep signal peaks from exceeding a certain level in order to prevent the overloading of amplifier signals, recorded signals onto tape or disc, broadcast transmission signals, and so on. Most limiters have ratios of 10:1 (above the threshold, for every 10 dB increase at the input there will be a gain of 1 dB at the output) or 20:1 (Figure 15.29), although some have ratios that can range up to 10:1. Since a large increase above the threshold at the input will result in a very small increase at its output, the likelihood of overloading any equipment that follows the limiter will be greatly reduced. Commonly, limiters have three basic functions:

- *To prevent signal levels from increasing beyond a specified level:* Certain types of audio equipment (often those used in broadcast transmission) are often designed to operate at or near their peak output levels. Significantly increasing these levels beyond 100% would severely distort the signal and possibly damage the equipment. In these cases, a limiter can be used to prevent signals from significantly increasing beyond a specified output level.
- *To prevent short-term peaks from reducing a program's average signal level:* Should even a single high-level peak exist at levels above the program's rms average, the average level can be significantly reduced. This is especially true whenever a digital audio file is normalized at any percentage value, because the peak level will become the normalized maximum value and not the average level. Should only a few peaks exist in the file, they can easily be zoomed in on and manually reduced in level. If multiple peaks exist, then a limiter should be considered.
- *To prevent high-level, high-frequency peaks from distorting analog tape:* When recording to certain media (such as cassette and videotape), high-energy, transient signals actually don't significantly add to the program's level, relative to the distortion that could result from their presence (if they saturated the tape) or from the noise that would be introduced into the program (if the signal was recorded at such a low level that the peaks wouldn't distort).

Unlike the compression process, extremely short attack and release times are often used to quickly limit fast transients and to prevent the signal from being audibly pumped. Limiting a signal during the recording and/or mastering phase should only be used to remove occasional high-level peaks, as excessive use would trigger the process on successive peaks and would be noticeable. If the program contains too many peaks, it's probably a good idea to reduce the level to a point where only occasional extreme peaks can be detected.

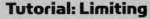

Tutorial: Limiting

1. Go to the Tutorial section of www.modrec.com, click on Limiting and download the soundfiles (which include instrument/music segments in various states of limiting).
2. Listen to the tracks. If you have access to an editor or DAW, import the files and look at the waveform amplitudes for each example. If you'd like to DIY, then …
3. Feed an isolated track or entire mix through a limiter or limiting plug-in.
4. With the limiter switched out, turn the signal up until the meter begins to peg (you might want to turn the monitors down a bit).
5. Now reduce the level and turn it up again … this time with the limiter switched in. Is there a point where the level stops increasing, even though you've increased the input signal? What does the gain reduction meter show? Decrease and increase the threshold level and experiment with the signal's dynamics. What did you find out?

FIGURE 15.30 The Aphex Model 622 Logic-Assisted Expander/Gate. (Courtesy of Aphex Systems, Inc., www.aphex.com)

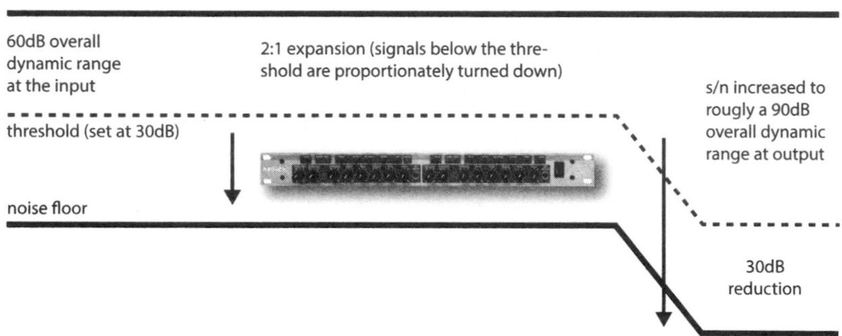

FIGURE 15.31 Commonly, the output of an expander is linear above the threshold and follows a low input/output gain expansion ratio below this point.

Expansion

Expansion is the process by which the dynamic range of a signal is proportionately increased. Depending on the system's design, an *expander* (Figure 15.30) can operate either by decreasing the gain of a signal (as its level falls below the threshold) or by increasing the gain (as the level rises above it). Most expanders are of the first type, in that as the signal level falls below the expansion threshold the gain is proportionately decreased (according to the slope ratio), thereby increasing the signal's overall dynamic range (Figure 15.31). These devices can also be used as noise reducers. You can do this by adjusting the device so that the noise is downwardly expanded during quiet passages, while louder program levels are unaffected or only moderately reduced. As with any dynamics device, the attack and release settings should be carefully set to best match the program material. For example, choosing a fast release time for an instrument that has a long sustain can lead to audible pumping effects. Conversely, slow release times on a fast-paced, transient instrument could cause the dynamics to return to its linear state more slowly than would be natural. As always, the best road toward understanding this and other dynamics processes is through experimentation.

The noise gate

One other type of expansion device is the noise gate (Figure 15.32). This device allows a signal above a selected threshold to pass through to the output at unity gain and without dynamic processing; however, once the input signal falls below this threshold level, the gate acts as an infinite expander and effectively mutes the signal by fully attenuating it. In this way, the desired signal is allowed to

FIGURE 15.32
Noise gates are commonly included within many dynamic plug-in processors. (Courtesy of Steinberg Media Technologies GmbH, a division of Yamaha Corporation, www.steinberg.net)

pass while background sounds, instrument buzzes, leakage or other unwanted noises that occur between pauses in the music aren't. Here are a few reasons why a noise gate might be used:

- To reduce leakage between instruments. Often, parts of a drum kit fall into this category; for example, a gate can be used on a high-tom track in order to reduce excessive leakage from the snare.
- To eliminate noise from an instrument or vocal track during silent passages.

The general rules of attack and release apply to gating as well. Fortunately, these settings are a bit more obvious during the gating process than with any other dynamic tool. Improperly set attack and release times will often be immediately obvious when you're listening to the instrument or vocal track (either on its own or within a mix) because the sound will cut in and out at inappropriate times.

Commonly, a key input (as previously shown in Figure 15.6) is included as a side-chain path to a noise gate. A key input is an external control that allows an external analog signal source (such as a miked instrument, synthesizer or oscillator) to trigger the gate's audio output path. For example, a mic or recorded track of a kick drum could be used to key a low-frequency oscillator. Whenever the kick sounds, the oscillator will be passed through the gate. By combining the two, you can have a deep kick sound that'll make the room shake, rattle and roll.

TIME-BASED EFFECTS

Another important effects category that can be used to alter or augment a signal revolves around delays and regeneration of sound over time. These time-based

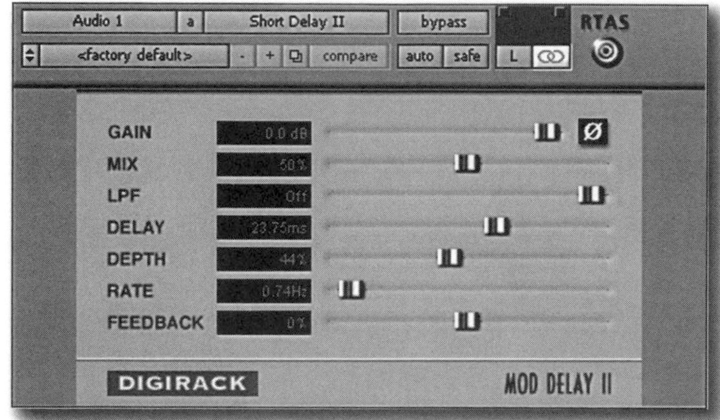

FIGURE 15.33
Pro Tools Mod Delay II. (Courtesy of Avid Technology, Inc., www.avid.com)

effects often add a perceived depth to a signal or change the way we perceive the dimensional space of a recorded sound. Although a wide range of time-based effects exist, they are all based on the use of delay (and/or regenerated delay) to achieve such results as:

- Time-delay or regenerated echoes, chorus and flanging
- Reverb

Delay

One of the most common effects used in audio production today alters the parameter of time by introducing various forms of delay into the signal path. Creating a delay circuit is a relatively simple task to accomplish digitally. Although dedicated delay devices (often referred to as digital delay lines, or DDLs) are readily available on the market, most multifunction signal processors and time-related plug-ins are capable of creating this straightforward effect (Figure 15.33). In its basic form, digital delay is accomplished by storing sampled audio directly into RAM. After a defined length of time (usually measured in milliseconds), the sampled audio can be read out from memory for further processing or direct output (Figure 15.34). Using this basic concept, a wide range of effects can be created simply by assembling circuits and program algorithms into blocks that can introduce delays or regenerated echo loops. Of course, these circuits will vary in complexity as new processing blocks are introduced.

DELAY IN ACTION!

Less than 15 ms

Probably the best place to start looking at the delay process is at the sample level. By introducing delays downward into the microsecond (one millionth of a second) range, control over a signal's phase characteristics can be introduced to the point where selective equalization actually begins to occur. In effect, controlling very short-term delays is actually how EQ is carried out in the digital domain!

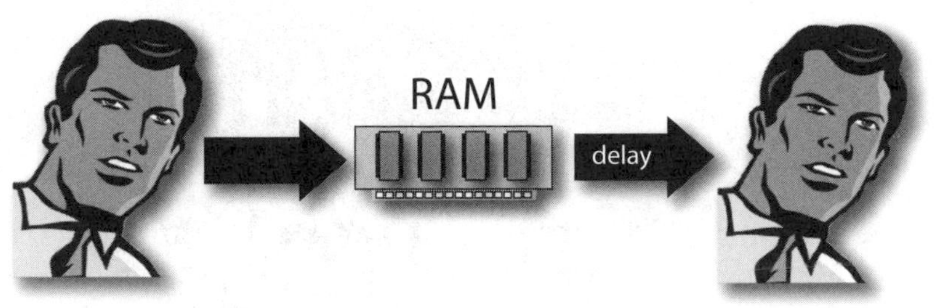

FIGURE 15.34
A digital delay device stores sampled audio into RAM, where it can be read out at a later time.

Whenever delays that fall below the 15-ms range are slowly varied over time and then are mixed with the original undelayed signal, an effect known as *combing* is created. Combing is the result of changes that occur when equalized peaks and dips appear in the signal's frequency response. By either manually or automatically varying the time of one or more of these short-term delays, a constantly shifting series of effects known as flanging can be created. Depending on the application, this effect (which makes a unique "swishing" sound that's often heard on guitars or vocals) can range from being relatively subtle to having moderate to wild shifts in time and pitch. It's interesting to note the differences between the effects of phasing and flanging. Phasing uses all-pass filters to create uneven peaks and notches, whereas flanging uses delay lines to create even peaks and notches … although, the results are somewhat similar.

15 to 35 ms

By combining two identical (and often slightly delayed) signals that are slightly detuned in pitch from one another, an effect known as *chorusing* can be created. Chorusing is an effects tool that's often used by guitarists, vocalists and other musicians to add depth, richness and harmonic structure to their sound. Increasing delay times into the 15- to 35-ms range will create signals that are spaced too closely together to be perceived by the listener as being discrete delays. Instead, these closely spaced delays create a *doubling effect* when mixed with an instrument or group of instruments (Figure 15.35). In this instance, the delays actually fool the brain into thinking that more instruments are playing than actually are … subjectively increasing the sound's density and richness. This effect can be used on background vocals, horns, string sections and other grouped instruments to make the ensemble sound as though it has doubled (or even tripled) its actual size. This effect also can be used on foreground tracks, such as vocals or instrument solos, to create a larger, richer and fuller sound. Some "chorus" delay devices introduce slight changes in delay and pitch shifting, allowing for detunings that can create an interesting, humanized sound.

Should time or budget be an issue, it's also possible to create this doubling effect by actually recording a second pass to a new set of tracks. Using this method, a 10-piece string section could be made to sound like a much larger ensemble. In addition, this process automatically gives vocals, strings, keyboards and other legato instruments a more natural effect than the one you get by using an

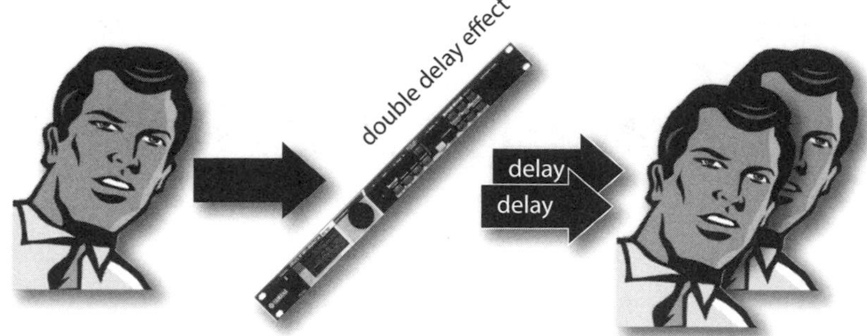

FIGURE 15.35 In certain instances, doubling can fool the brain into thinking that more instruments are playing than actually are.

electronic effects device. This having been said, these devices can actually go a long way toward duplicating the effect. Some delay devices even introduce slight changes in delay times in order to create a more natural, humanized sound. As always, the method you choose will be determined by your style, your budget and the needs of your particular project.

More than 35 ms

When the delay time is increased beyond the 35- to 40-ms point, the listener will begin to perceive the sound as being a discrete echo. When mixed with the original signal, this effect can add depth and richness to an instrument or range of instruments that can really add interest to an instrument within a mix.

Adding delays to an instrument that are tied to the tempo of a song can go even further toward adding a degree of depth and complexity to a mix. Most delay-based plug-ins make it easy to insert tempo-based delays into a track, often by simply pressing a button. For hardware delay devices, it's usually necessary to calculate the tempo math that's required to match the session. Here's the simple math for making the calculations:

60,000/tempo = time (in ms)

For example, if a song's tempo is 100 bpm (beats per minute), then the amount of delay needed to match the tempo at the beat level would be:

60,000/100 = 600 ms

Using divisions of this figure (300, 150, 75, etc.) would insert delays at 1/2, 1/4, 1/8th measure intervals.

Caution should be exercised when adding delay to an entire musical program, because the program could easily begin to sound muddy and unintelligible. By feeding the delayed signal back into the circuit, a repeated series of echo … echo … echoes can be made to simulate the delays of yesteryear—you'll definitely notice that Elvis is still in the house.

Tutorial: Delay

1. Go to the "Tutorial" section of www.modrec.com, and download the delay tutorial soundfiles (which include segments with varying degrees of delay).
2. Listen to the track. If you'd like to DIY, then …
3. Insert a digital delay unit or plug-in into a program channel and balance the dry track's output mix so that the input signal is set equally with the delayed output signal. (*Note:* If there is no mix control, route the delay unit's output to another input strip and combine delayed/undelayed signals at the console.)
4. Listen to the track with the mix set to listen equally to the dry and effected signal.
5. Vary the settings over the 1- to 10-ms range. Can you hear any rough EQ effects?
6. Manually vary the settings over the 10- to 35-ms range. Can you simulate a rough phasing effect?
7. Increase the settings above 35 ms. Can you hear the discrete delays?
8. If the unit has a phaser setting, turn it on. … How does it sound different?
9. Now change the delay settings a little faster to create a wacky flange effect. If the unit has a flange setting, turn it on. Try playing with the time-based settings that affect its sweep rate. … Fun, huh?

Reverb

In professional audio production, natural acoustic reverberation is an extremely important tool for the enhancement of music and sound production. A properly designed acoustical environment can add a sense of space and natural depth to a recorded sound that'll often affect the performance as well as its overall sonic character. In situations where there is little, no or substandard natural ambience, a high-quality reverb device or plug-in (Figure 15.36) can be extremely helpful in filling the production out and giving it a sense of dimensional space and perceived warmth. In fact, reverb consists of closely spaced and random multiple echoes that are reflected from one boundary to another within a determined space (Figure 15.37). This effect helps give us perceptible cues as to the size, density and nature of a space (even though it might have been artificially generated). These cues can be broken down into three subcomponents:

- Direct signal
- Early reflections
- Reverberation

The *direct signal* is heard when the original sound wave travels directly from the source to the listener. *Early reflections* is the term given to those first few reflections that bounce back to the listener from large, primary boundaries in a given space. Generally, these reflections are the ones that give us subconscious cues as to the perception of size and space. The last set of reflections makes up the signal's *reverberation* characteristic. These sounds are comprised of zillions of random reflections that travel from boundary to boundary within the confines of a room. These reflections are so closely spaced in time that the brain can't

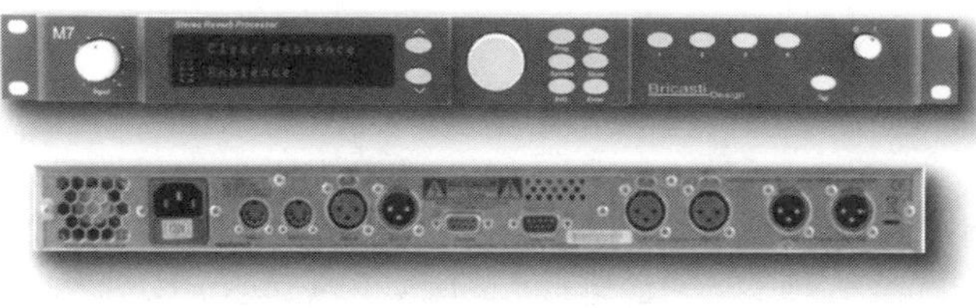
(a)

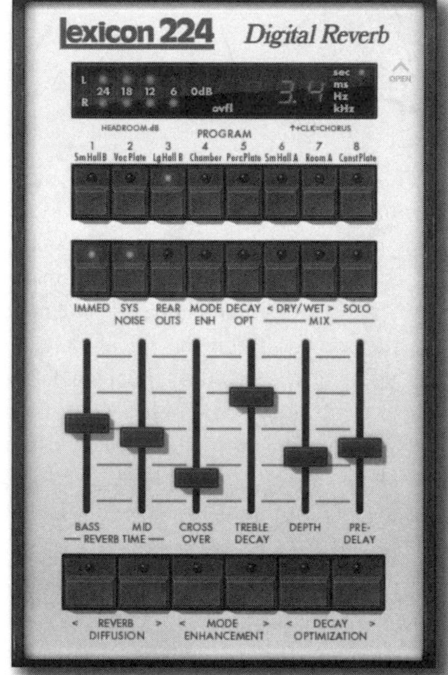

(b)

FIGURE 15.36
Digital reverb effects processors. (a) Bricasti M7 Stereo Reverb Processor. (Courtesy of Bricasti Design Ltd, www.bracasti.com) (b) Lexicon 224 reverb plugin for the Apollo and the UAD effects processing card. (Courtesy of Universal Audio, www.uaudio.com © 2013 Universal Audio, Inc. All rights reserved. Used with permission).

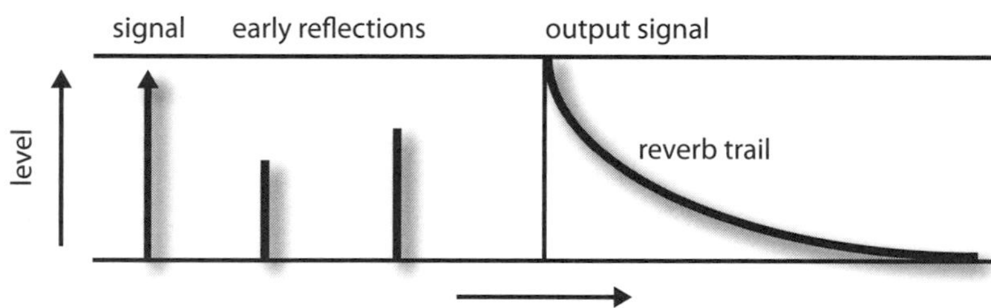

FIGURE 15.37
Signal level versus reverb time.

discern them as individual reflections, so they're perceived as a single, densely decaying signal.

REVERB TYPES

By varying program and setting parameters, a digital reverb device can be used to simulate a wide range of acoustic environments, reverb devices and special effects. A few popular categories include:

- *Hall:* Simulates the acoustics of a concert hall. This is often a diffuse, lush setting with a longer RT60 decay time (the time that's required for a sound to decay by 60 dB).

- *Chamber:* Simulates the acoustics of an echo chamber. Like a live chamber, these settings often simulate the brighter reflectivity of tile or cement surfaces.
- *Room:* As you might expect, these settings simulate the acoustics of a mid- to large-sized room. It's often best suited to intimate solo instruments or a chamber atmosphere.
- *Live (stage):* Simulates a live performance stage. These settings can vary widely but often simulate long early-delay reflections.
- *Spring:* Simulates the low-fidelity "boingyness" of yesteryear's spring reverb devices.
- *Plate:* Simulates the often-bright diffuse character of yesteryear's metallic plate reverb devices. These settings are often used on vocals and percussion instruments.
- *Reverse:* These backward-sounding effects are created by reversing the decay trail's envelope so that the decay increases in level over time and is quickly cut off at the tail end, yielding a sudden break effect. This can also be realistically created in a DAW by reversing a track or segment, applying reverb, and then reversing it again to yield a true backward reverb trail.
- *Gate:* Cuts off the decay trail of a reverb signal. These settings are often used for emphasis on drums and percussion instruments.

Tutorial: Reverb Types

1. Go to the Tutorial section of www.modrec.com, click on Reverb Types and download the soundfile.
2. Listen to the track.

Psychoacoustic enhancement

A number of signal processors rely on psychoacoustic cues in order to fool the brain into perceiving a particular effect. The earliest and most common of these devices are those that enhance the overall presence of a signal or entire recording by synthesizing upper-range frequency harmonics and inserting them into a mix in order to brighten the perceived sound. Although the additional harmonics won't significantly affect the program's overall volume, the effect is a marked increase in its perceived presence. Other psychoacoustic devices that make use of complex harmonic, phase, delay and equalization parameters have become standard production tools in the field of mastering in order to shape the final sound into one that's interesting, with a sonic character all its own.

In addition to synthesizing harmonics in order to change or enhance a recording or track, other digital psychoacoustic processors deal exclusively with the subject of spatialization (the placement of an audio signal within a three-dimensional acoustic field), even though the recording is being played back over stereo speakers. By varying the parameters of a stereo or multiple input source, this processing function creates phase and amplitude paths that can fool the brain into perceiving that the stereo image is actually emanating from a soundfield that's wider than the physical speaker positions. In practice, care should be taken when using these devices, because the effect is often carried off with degrees of success that vary from system to system. In addition, the use of phase relationships to expand the stereo soundfield can actually cause obvious cancellation problems when the program is listened to in mono.

Pitch shifting

Ever had a perfectly good vocal take that was spoiled by just one or two flat notes? Or had a project come in the door with a guitar track that was out of tune? Or needed to change the key on a 30-second radio spot? It's times like these that pitch shifting can save your day! *Pitch shifting* can be used to vary the pitch of a signal or soundfile (either upward or downward) in order to transpose the relative pitch of an audio program without affecting its duration. This process can take place in either real time or non-real time. Pitch shifting works by writing sampled audio data to a temporary memory, where it's resampled to either a higher or a lower sample rate (according to the desired final pitch). Once this is done, the processor either adds interpolated samples to (lowers the pitch) or subtracts them from (raises the pitch) the resampled data to return it back to the original output rate, while keeping the altered pitch intact. Figure 15.38 gives two basic examples of how this is often carried out.

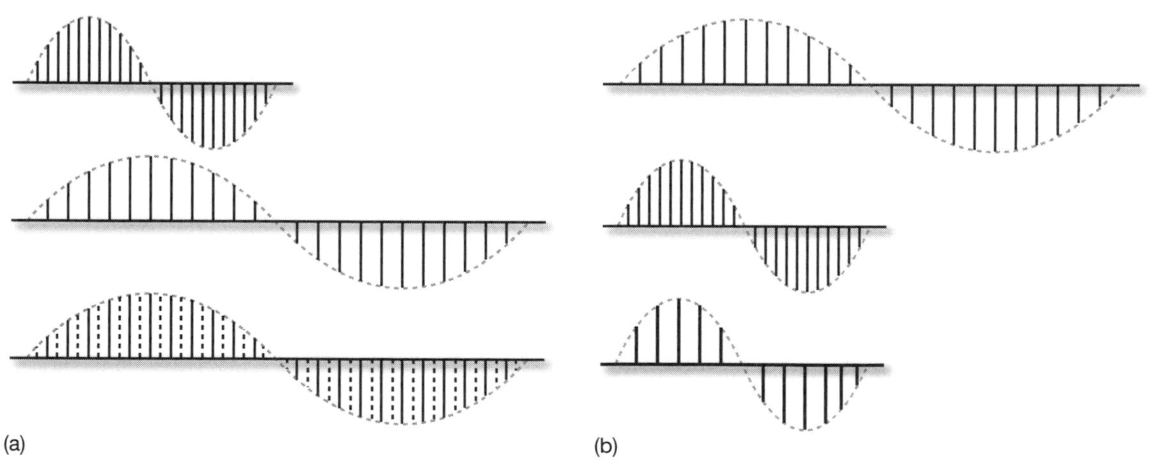

FIGURE 15.38
Two pitch shift examples with an initial 1-kHz digital signal and a sample rate of 44.1 kHz: (a) The signal can be halved in pitch (to 500 Hz) by internally down-sampling to a new rate of 22.05 k. To return the output rate to 44.1 (while retaining the 500-Hz pitch), new sample points must be added into each dropped position. (b) The signal can be doubled in pitch (to 2 kHz) by internally upsampling to a new rate of 88.2 k. To return the output rate to 44.1 (while retaining the 2-kHz pitch), every other sample point must be dropped.

A degree of caution should be used when changing the pitch of a program or audio segment. Whenever uneven or minute interval changes are made, the interpolation of samples doesn't always fall perfectly into place. This can lead to digital artifacts that add unacceptable amounts of harmonic distortion. If the track is in the background, there shouldn't be a problem; however, care should be taken with upfront instruments and vocals. It's important to keep in mind that large pitch changes might be more noticeable. As always, your ears are the best judge.

Time and pitch changes

By combining variable sample rates and pitch shifting techniques, it's possible to create three different variations:

- *Time change:* A program's length can be altered, without affecting its pitch, by raising or lowering its playback sample rate.
- *Pitch change:* A program's length can remain the same while pitch is shifted either up or down.
- *Both:* Both a program's pitch and length can be altered by means of simple resampling techniques.

These functions have become an important part of the signal processing and music production arsenals that are used by the audio-for-video, film and broadcast industries. These tools help give producers control over the running time of film video and audio soundtracks while maintaining the original, natural pitch of voice, music and effects. For example, using a DAW, we could add a 5-second trailer onto the end of an existing 30-second public service radio spot simply by time compressing the 30-second spot to 25 seconds (while keeping the pitch intact) and then adding the trailer.

In addition to the basic time/pitch techniques that are commonly used in music production (most often by electronic musicians), this technology has allowed for the huge explosion in loop-based music composition and production. These popular programs and music plug-ins involve the use of recorded soundfiles that are encoded with headers that include information on their native tempo and length (in both samples and beats). When you set the loop program to a master tempo (or a specific tempo at that point in the score), a loop segment, once imported, can go about the process of recalculating its pitch and tempo to match the current session tempo and … voilà! The file's in sync with the song! Further info on loop-based production tools can be found in Chapter 8 (Groove Tools and Techniques).

It's important to realize that the algorithms that control how time and pitch are to be re-circulated differ between programs (as well as having various algorithm options within the same program. It would be wise to read up on how your favorite programs deal with these shifting techniques and to familiarize yourself with them. The quality of your soundfiles will thank you.

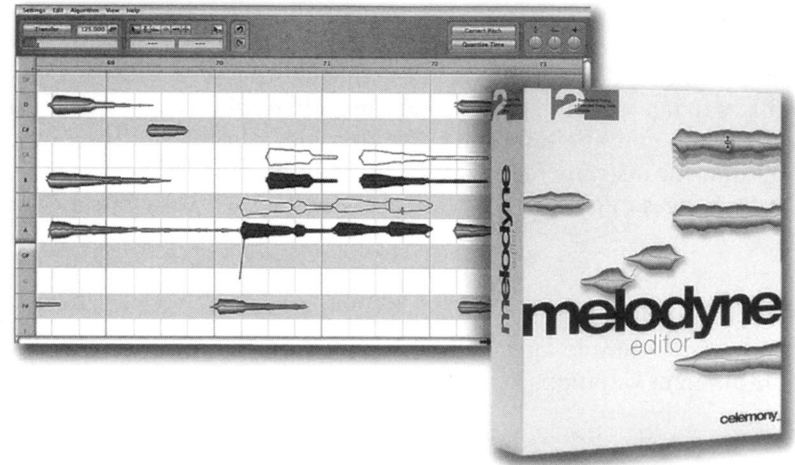

FIGURE 15.39
Melodyne Editor 2 auto pitch correction system. (Courtesy of Celemony Software GmbH, www.celemony.com).

Automatic pitch correction

One other category of pitch correction system makes use of auto-tuning software or plug-ins that is able to detect the off-tune or slight pitch-bend nature of a track and then correct these inaccuracies (according to user parameters). These devices (Figure 15.39) are commonly used in modern music production to compensate for tuning problems in a performance, or (when over-used) to create a tuning effect that exaggerates the various tuning steps in pitch that gives a non-human, robotic-like effect to the vocal or instrumental performance.

Often, these pitch editors allow us to change the pitch in a straightforward, graphical environment, according to musical and/or user-defined parameters. Single notes or entire chords can be built up, altered or changed, making it possible to correct a wide range of material. It's even possible to alter individual notes within a polyphonic recording. For example, you could fix a single wrong note in a piano recording or change the notes within a guitar backing track … all after the recording has been made.

Of course, the use of automatic tuning to fix a track isn't without its critics, as many feel that such a device takes away from the artistry and humanity of a performance, while others have built careers upon the use of this effect both in the studio and on-stage. As with most devices, the automatic use of pitch correction can be greatly over-used, or it can be used by an experienced operator to make minor and necessary corrections to fix a track or live performance.

MULTIPLE-EFFECTS DEVICES

Since most digital signal processors are by nature multifunctional chameleons, it follows that most hardware and certain plug-in processors can be easily programmed to perform various functions. For this reason, many digital systems

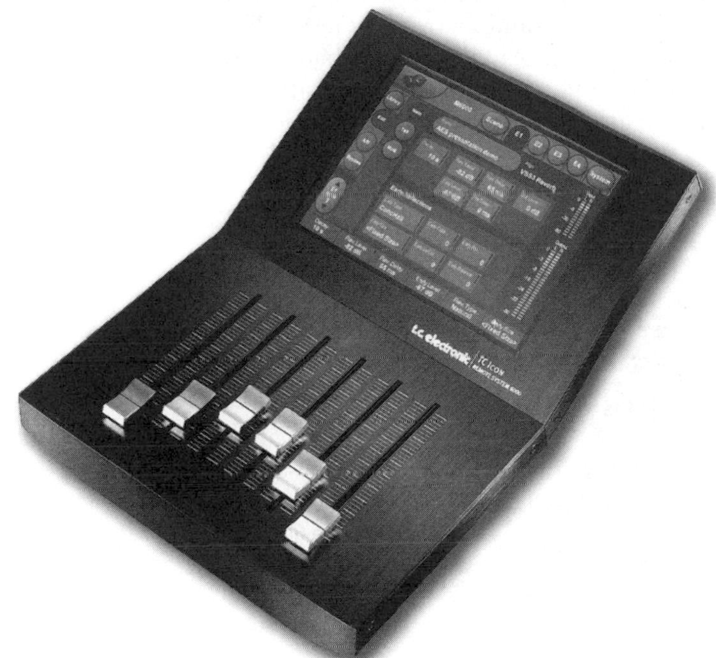

FIGURE 15.40
Controller for tc electronic System 6000 digital effects processor. (Courtesy of tc electronic, www.tcelectronic.com).

have been designed to perform as multiple-effects devices (Figure 15.40). Multiple effects, in this case, can have several basic meanings:

- A single device might offer a wide range of processing functions but allow only one effect to be called up at a time.
- A single device might offer a range of processing functions that can be "stacked" to perform a number of simultaneous effects.
- An effects device might have multiple ins and outs, each of which can perform several processing functions (effectively giving you multiple processors that can be used in a multichannel mixdown environment).

DYNAMIC EFFECTS AUTOMATION AND EDITING

One of the joys of working with effects is the ability to manipulate and vary effects parameters in real time over the duration of a song or audio program. By altering parameters, changing settings and mixing effects levels … the subtle variations in expression can add a great deal of interest to a project.

The ability to dynamically automate effects settings can be accomplished in any number of ways, including:

- Via MIDI control and parameter change messages
- Via external hardware controller
- Via DAW or other form of automation control

In closing, an almost unlimited degree of effects control is available to us through most high-level (and many entry-level) digital audio workstations. Through the use of any of the readily available hardware controllers or on-screen automation controls, it's possible to manipulate and automate effects within a DAW session with an amazing degree of sophistication and ease.

Tutorial: DAW Effects Automation

- The vast majority of plug-in effects can be directly and dynamically automated within the computer's DAW program. It's my greatest hope that you'll:
- Take a look at your favorite DAW manual and start reading!
- Open up a tutorial session or better yet, make your own.
- Call up some effects on various tracks.
- Call up some automation control parameters and start grabbing controls.
- Learn how to edit these automation functions, so as to be able to finesse your effects in new and interesting ways.

Go ahead, get hold of these fun and effective tools … and experiment your heart out!

CHAPTER 16
Noise Reduction

With the advent of newer and better digital audio technologies, low-noise systems, hi-res audio and surround-sound home theaters, an increase in dynamic range and a demand for better quality sound has steadily been on the rise. Because of this, it's more important than ever that those in audio production pay close attention to the background noise levels that are produced by pre-amp and amplifier self-noise, synths, analog magnetic tape and the like. Although the overall dynamic range of human hearing roughly encompasses a full 140 dB and well-designed digital systems can deal with such dynamics … in practice, such dynamics won't be fully captured, played-back or appreciated for several reasons:

- An acoustic or electronic weak link in the chain might be introducing noise and/or restricting the program's dynamic range.
- The medium itself might be incapable of encoding a wide dynamic range.
- Background noises in the environment might mask the subtleties of the sound.

Not all the blame for added noise can be placed on our older technology friends. Even though a 16-bit digital recording has a theoretical dynamic range of 96 dB and a 24-bit system can actually encode 144 dB, noises can (and often will) crop up from such modern-day gremlins as mic preamps, effects and outboard gear, analog communication lines and poorly designed digital audio converters. Honestly, though, one of the biggest noise problems that you'll often encounter is the need for restoring tracks that were poorly recorded or were made under adverse and/or noisy acoustic conditions.

DIGITAL NOISE REDUCTION

As you might expect, in modern music production, digital signal processing (DSP) is most commonly used to reduce noise levels within a recorded soundfile. These noises might include artifacts such as tape hiss, hum, obtrusive background ambience, needle ticks, pops and even certain types of distortion that are present in the original recording. Although stand-alone digital noise reduction processors do exist, most of these processors exist as plug-ins that can be introduced at multiple points into the signal chain of a DAW. For example, an "NR"

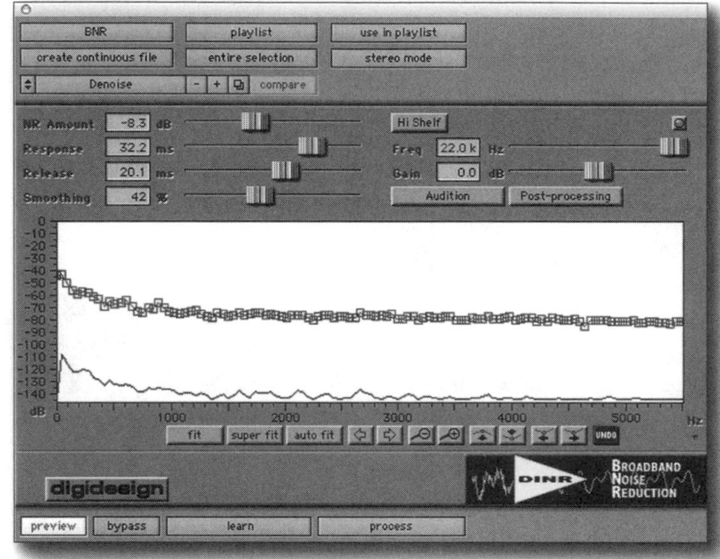

FIGURE 16.1
Digidesign DINR noise-reduction plug-in. (Courtesy of Digidesign, a division of Avid Technology, Inc., www.digidesign.com).

plug-in can be inserted into a single track to reduce the amp noise on a guitar, synth track, etc. to instantly clean up a mix that might otherwise be problematic. Likewise, an NR plug-in can be inserted into the final output mix stage to clean-up a mix that's overly noisy.

Fast Fourier transform

The most commonly used noise reduction programs make use of a mathematically intense algorithm known as *Fast Fourier Transform* (FFT). These applications and plug-ins (Figure 16.1) are able to analyze the amplitude/frequency domain of an audio signal in order to reduce hum, tape hiss and other extraneous noises from your recordings. This digital analysis generally begins by taking a digital "snapshot" of a short snippet of the offending noise (a brief section that contains only the noise will yield the best results). This noise template can then be digitally subtracted from the original soundfile or segment in varying amounts (and under the control of various program parameters), such that the footprint noise is reduced, while (under the best of conditions) the original program material is left intact and unaffected.

In addition, certain FFT noise reduction systems are able to display frequencies in an easy to detect graphic form that lets the user view the overall spectral analysis of a recorded section (Figure 16.2). These various frequencies, noises, pops, etc. can then be easily seen and "drawn out," and then mathematically removed from the audio signal. Such a useful tool can also be used to remove coughs, squeaky chairs, sirens and any unfortunate noises that might make their way into a recording.

Although FFT algorithms for reducing noise have greatly improved over the years, it's still important that we briefly discuss a few of the unfortunate artifacts that

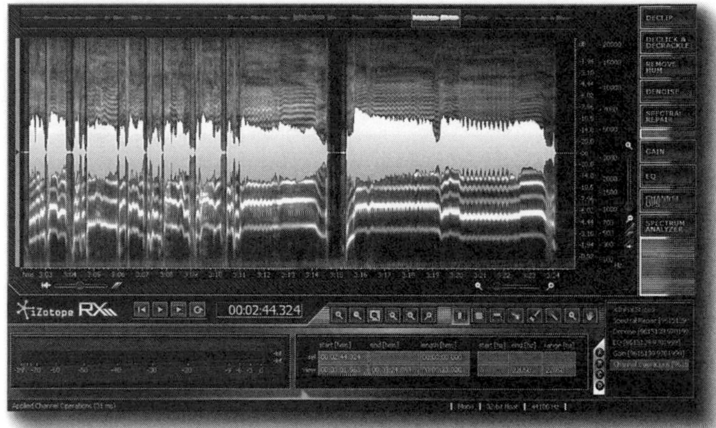

FIGURE 16.2
iZotope's RX 2 audio repair toolkit is capable of displaying the spectral (frequency) content of a passage, allowing the user to "draw out" any unwanted noises from a program. (Courtesy of iZotope Inc., www.izotope.com)

can occur when using (and over using) FFT-based noise reduction. The most notable of these is "chirping." This audible artifact most often occurs when too much FFT processing is applied. It literally sounds like a flock of small chirping birds that can either be heard in the background or in an obnoxious way that sounds like a bad Alfred Hitchcock movie. If you find yourself running for cover, it's best to pull back on the FFT settings (and/or increase the processing quality level) until the artifacts are less noticeable.

Should chirping and/or bandwidth limitations become a problem, you might consider using equalization or a single-ended noise reduction device/plug-in instead. Because single-ended noise reduction (to be discussed later) uses an adaptive filter to intelligently change the program's bandwidth, no chirping artifacts will be introduced.

Finally, it's a misconception that an FFT-based noise reduction application can only be used for reducing noise. Literally, ANY sound can be used as a sonic removal footprint, and as a result, vocal formants, snare hits or any sound that you can imagine, can be pulled from a soundfile to create unique and interesting effects. The sky's literally the limit!

Tutorial: FFT-Based Noise Reduction

1. Load a track containing an excessive amount of noise into the DAW of your choice.
2. Call up an FFT-based noise-reduction plug-in, select a short segment of noise and follow the application's user directions.
3. Apply varying amounts of noise reduction in real time and listen to the results. Can you make it chirp? Were you able to achieve acceptable results?

From the above, it's easy to tell that the noise-reduction process isn't an exact science, but more of a balancing act. Because of these and other band-limiting artifacts, you might consider writing the processed signal to a new track ... while keeping your original soundfile intact ... thereby keeping your options open for future changes, decisions and/or technological advances.

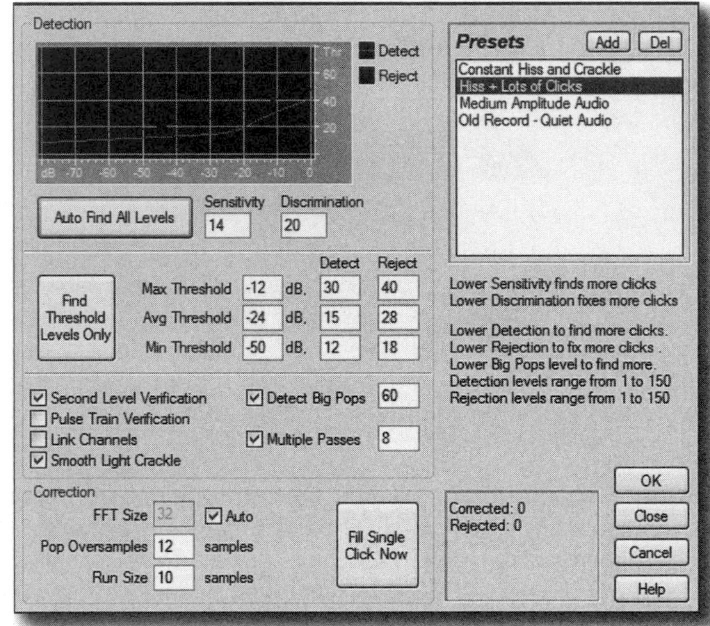

FIGURE 16.3
Click/pop eliminator application within Adobe's Audition 3. (Courtesy of Adobe Systems, Inc., www.adobe.com)

RESTORATION

In addition to removing noise, programs and DAW plug-ins also exist for removing *clicks* and *pops* from vinyl and older recordings (Figure 16.3). Although FFT analysis is often involved in the process, click removal differs slightly from FFT noise reduction. This multistep process begins by detecting high-level clicks (or those exceeding a user-defined threshold) that exist within a soundfile or defined segment. Once the offending noises are detected, the program performs a frequency analysis (both before and after the click) and then goes about the business of making a best plausible "guess" as to what the damaged amplitude/frequency content should sound like. Finally, the calculated sound is pasted over the nasty offender—ideally rendering it less noticeable or gone—and then moves onto the next click and restarts the process.

Because click and pop noises can be different in nature from each other (both in duration and frequency makeup), noise-reduction plug-ins might offer applications that are specifically suited to reducing each one. In addition, certain programs also offer FFT-based applications that allow offending sounds (such as a cough, noisy music stand, etc.) to be visually erased in an on-screen drawing environment (as can be seen in Figure 16.2). Such tools allow us to visually tackle sonic offenders in new and interesting ways.

SINGLE-ENDED NOISE-REDUCTION PROCESS

The *adaptive filter* or *single-ended noise-reduction* process differs from the above digital FFT systems, in that it extracts noise from an audio source by combining a downward dynamic-range expander with a variable low-pass filter. These devices (which can be analog, digital or plug-in-based in nature) can be used to dynamically analyze, process and EQ an existing program to reduce the unwanted noise content with little or no audible effects (or giving us a best possible compromise, in extreme cases).

Single-ended noise reduction systems (Figures 16.4 and 16.5) work by breaking up the audio spectrum into a number of frequency bands, such that whenever the signal level within each band falls below a user-defined threshold, the

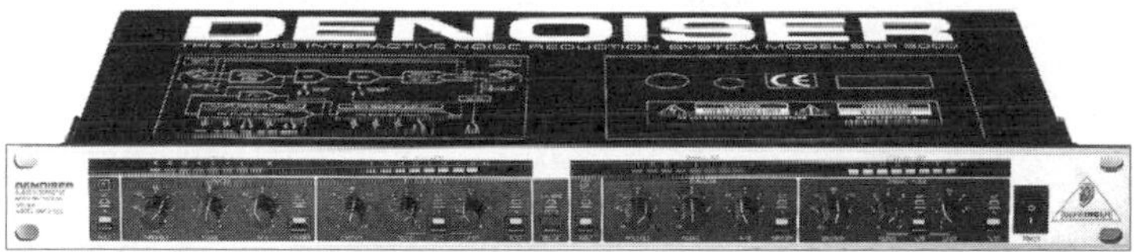

FIGURE 16.4
Behringer's Denoiser system Model SNR 2000. (Courtesy of Behringer Int'l GmbH, www.behringer.de).

FIGURE 16.5
DeNoiser Noise app within Cubase/Nuendo. (Courtesy of Steinberg Media Technologies GmbH, a division of Yamaha Corporation, www.steinberg.net).

signal will be attenuated. This downward expansion/filtering process accomplishes noise reduction by taking advantage of two basic psycho-acoustical principles:

- Music is capable of masking lower level noise that exists within the same bandwidth.
- Reducing the bandwidth of an audio signal will reduce the perceived noise.

It's a psycho-acoustical fact that our ears are more sensitive to noises that contain a greater number of frequencies than to those containing fewer frequencies. Thus, whenever the program's high-frequency content is reduced or restricted to a certain bandwidth, the dynamic filtering process will sense this and reduce the frequency bandwidth accordingly (thereby reducing the noise content). When the program's high-frequency signal returns, the filter will again pass the frequency bandwidth up as far as necessary to pass the signal (allowing the increased program content will mask the background noise).

FIGURE 16.6
Universal Audio SSL E Series Channel Strip with the noise gate section highlighted. (Courtesy of Universal Audio, www.uaudio.com. © 2013 Universal Audio, Inc. All rights reserved. Used with permission).

The noise gate

A noise gate (Figure 16.6) can also be a very effective tool for eliminating noise from a track or tracks within a mix. This device allows a signal above a selected threshold to pass through to the output at unity gain (no gain change) and without dynamic processing. Once the input signal falls below this threshold level, however, the gate acts as an infinite expander and effectively mutes background noises and other unwanted sounds by attenuating them.

When "gating" a specific track, it's often necessary to take time out to fine-tune the device's attack and release controls. This is done to reduce or eliminate any unwanted "pumping" or "breathing" as the noise or leakage signal falls and rises around the threshold point. The general rules for these settings are the same as those that apply to gain-change processors (see the compressor, limiter and expander settings section in Chapter 15). Fortunately, these settings are often audibly more obvious than with any other dynamic tool. Setting the attack and release times either too fast or too slow will be immediately audible because the sound will cut in and out at inappropriate times. For this reason, care should be taken when adjusting the settings by both listening to the track on its own (solo track) and by listening to it within the context of the full mix.

CHAPTER 17

Monitoring

Within the recording process, our ability to judge and adjust sound is primarily based on what's heard through the monitor speakers in a project studio or control room environment (Figure 17.1). In fact, within the audio and video industries, the word *monitor* refers to a device that acts as a subjective professional standard or reference by which program material can be critically evaluated.

Despite steady advances in design, speakers are still one of the weakest links in the audio chain. This weakness is generally due to potential nonlinearities that can exist in a speaker system's frequency response. In addition, interactions with a room's frequency response can often lead to peaks and dips that affect a speaker's sonic character in ways that are difficult to predict. Add to this the factors of personal "tastes" in the sound, size and design types, and you'll quickly find that speakers are also one of the most subjective tools in a production environment.

SPEAKER AND ROOM CONSIDERATIONS

Unless you have several rooms with precisely matching dimensions, materials and furnishings (an unlikely scenario), you can bet your bottom buck that the

FIGURE 17.1 Example of a professional monitoring environment: Hit Factory-Criteria Miami, FL. (Courtesy of Solid State Logic LTD, www.solid-state-logic.com).

same speaker system will sound different in different room environments. That's to say, it will interact with the various complex environmental factors to exhibit a unique frequency response curve and delay characteristic.

Although variations between production rooms often play a big part in giving a facility its own particular sound, extreme variations in a room's frequency response can lead to difficulties that can be heard in the final product. For this reason, certain basic principles (which are covered in the Symmetry in Control Room Design section in Chapter 3) have become common knowledge to many who attempt the art of control room design. A few examples include:

- Reducing standing waves to help reduce erratic frequency response characteristics within a room
- Reducing excessive bass buildup in room corners through the use of bass traps
- Keeping the room/equipment layout symmetrical throughout a room so the left/right and front/rear imaging is consistent
- Using a careful balance of absorptive and reflective surfaces to help "shape" a room's sonic character

Because of the untold number of acoustic variables that are involved, a project that's been recorded in one facility will often sound quite different when played and/or mixed in another, even when high acoustical construction standards are followed. Fortunately for us, production and mixdown room designs have greatly improved over the last several decades. This is largely due to an added awareness of careful room design and the increased availability of acoustical products that can help shape a room's sound.

Beyond careful acoustic design and construction, a professional facility might choose to further reduce variations in frequency response by *tuning* (equalizing) its speakers to the room's acoustics so that the adjusted frequency response curve will be reasonably flat and, therefore, reasonably compatible with most other control rooms. Tuning a speaker system to a room can be carried out in either of two ways:

- Altering settings on the speaker itself
- Equalizing the monitor output lines

One of the simplest ways to alter the acoustic and frequency response of a speaker system is through the careful control of the basic EQ and system setting controls that are found on most actively powered speaker systems (Figure 17.2). These simple controls allow the user to roughly match level and EQ settings to best fit their application or placement layout. Often these settings can be used to:

- Finely match audio balance levels within a stereo and surround system.
- Allow for basic high- and low-end tuning.
- Compensate for bass buildup (whenever speakers are placed in or near a corner or other large boundary).
- Offer various speaker "emulation" modes.

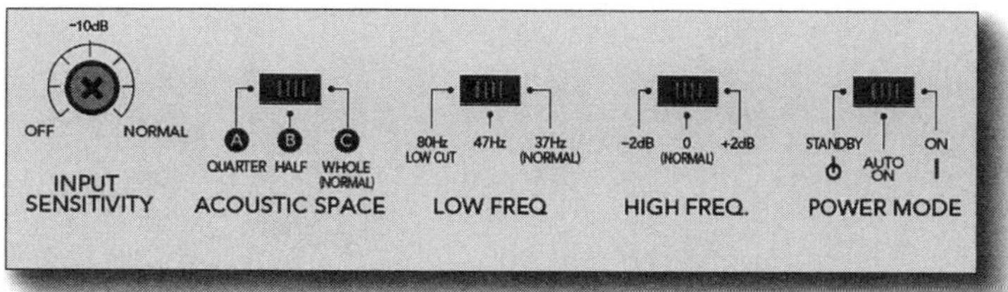

FIGURE 17.2
Rear controls for the Mackie HR824mk2 active monitor speaker. (Courtesy of Loud Technologies, Inc., www.mackie.com)

Larger, passive monitors (often a farfield pair) can be tuned by placing a 1/3-octave bandwidth graphic equalizer between each of the console's control room monitor outputs and the power amplifier. Of course, there are various ways to fine-tune a speaker system and room response to improve a studio's overall monitoring conditions. The simplest approach is to place a high-quality omnidirectional mic at the center listening position and insert it into a channel strip on your DAW. By recording a loop of pink noise (search Wikipedia for the "colors of noise") and playing it back equally to each speaker in the system, basic level matching and frequency measurements can be taken.

It should be noted that the above measurement method is often inaccurate (due to time-based variables and other acoustical factors). Fortunately, standalone software can accurately measure such variables as level, EQ and time delay reflections, to help with the accurate tuning within a control room, studio, performance hall or auditorium. Note also that these measurements are often best interpreted by those who are well versed in acoustics, studio design and the fine art of common sense … in other words; careful fine-tuning "might" be best left to a competent professional.

MONITOR SPEAKER TYPES

Since the buying public will be listening to your mixes over a very wide range of speaker types, under an infinite number of listening conditions – it's often wise to listen to a mix over several standardized speaker types during a record and/or mixdown session. Quite often, a console or monitor control system will let you select between speaker/monitor types, with each set commonly having its own associated amplifier for power and level matching flexibility. These types include far-field, near-field, small-speaker and headphone monitoring.

Far-field monitoring

Far-field monitors often involve large, multi-driver loudspeakers that are capable of delivering relatively accurate sound at moderate to high volume levels. Because of their large size and basic design, the enclosures are generally soffit mounted (built into the control room wall to reduce reflections around and behind the enclosure and to increase overall speaker efficiency). This monitor type has lost popularity to the easier-to-use and less expensive near-field bookshelf system. Further reading on soffit design and construction can be found in Chapter 3.

FIGURE 17.3
Genelec 1238CF main reference monitor. (Courtesy of Genelec Inc., www.genelec.com)

FIGURE 17.4
Yamaha HS-80M studio monitor speakers. (Courtesy of Yamaha Corporation of America, www.yamaha.com)

These large-driver systems (Figure 17.3) are often used during the recording phase because of their ability to safely handle high sound levels (which can come in handy should a microphone drop or a vocalist decide to be cute and scream into an open mic). They're also great for listening to a mix at loud levels in order to hear the impact that it'll have on the dance floor or in a souped-up car system. In fact, certain types of music rely on bass levels that can only be supplied by such a system at moderate-to-high SPLs. (*Note:* It's important to be aware of the danger of long-term exposure to such sound levels.)

Near-field monitoring

Although far-field monitors are generally the best reference at high listening levels, few systems are equipped with speakers that can deliver "clean" sound at such high SPLs. For this reason, most professional and project studios use *near-field monitors* that more realistically represent the type of listening environment that John and Jill Q. Public will most likely have.

The term *near-field* refers to the placement of small to medium-sized bookshelf speakers on each side of a desktop working environment or on (or slightly behind) the metering bridge of a production console. These speakers (Figures 17.4

through 17.6) are generally placed at closer working distances, allowing us to hear more of the direct sound and less of the room's overall acoustics.

FIGURE 17.5
Mackie HR824mk2 active monitor speaker. (Courtesy of Loud Technologies, Inc., www.mackie.com)

In recent times, near-fields have become an accepted standard for monitoring in almost all areas that relate to audio production for the following reasons:

- Quality near-field monitors more accurately represent the sound that would be reproduced by the average home speaker system.
- The placement of these speakers at a position closer to the listening position reduces unwanted room reflections and resonances. In the case of an untuned room, this helps to create a more accurate monitoring environment.
- These moderate-sized speaker systems cost significantly less than their larger studio reference counterparts (not to mention the reduced amplifier cost because less wattage is needed).

One other consideration that should be taken into account when using near-field speakers on a table, desk or console top, are the natural resonances that can be transmitted from the speaker onto the desktop. These resonances can add a distinct coloration and added resonances to the sound. For this reason, several companies have begun to develop foam isolation pads (Figure 17.7) that act to decouple the speakers from the surface, thereby reducing resonances.

As with any type of speaker system, near-fields vary widely in both construction and fundamental design philosophy. It almost goes without saying that care should be taken when choosing the speaker system that best fits your production needs and personal tastes.

FIGURE 17.6
PMC AML2 active monitor speakers. (Courtesy of PMC Ltd., www.pmc-speakers.com)

Small speakers

Because radio, television and web (computer and tablet) airplay are major forces in audio production and help in the

(a)

(c)

(b)

FIGURE 17.7
Speaker isolation pads can help to reduce speaker/stand resonances. (a) isoAcoustics ISO-L8R155 stands (courtesy of Isoacoustics Inc, www.isoacoustics.com); (b) Auralex MoPAD™ speaker isolation pad. (Courtesy of Auralex Acoustics, www.auralex.com). (c) Primacoustic Recoil Stabilizer pad (courtesy of Primacoustic, www.primacoustic.com)

distribution and sales of recordings, it's often good to monitor your final mix through a small, inexpensive speaker set that mimics the nonlinearities, distortion and poor bass response of those media. Such speakers can be either bought or easily made.

Before listening to a mix over such small speakers or over laptop speakers (Figure 17.8), it's often a good idea to take a break in order to allow your ears and your brain to recover from the prolonged exposure of listening to higher sound levels over larger speakers.

FIGURE 17.8
AVANTONE Active MixCubes Full-Range Mini Reference Monitors. (Courtesy of Avantone Pro, www.avantonepro.com)

FIGURE 17.9
Sony MDR-7506 professional dynamic stereo headphones. (Courtesy of Sony Electronics, Inc., www.sony.com/proaudio.)

Headphones

Headphones (Figure 17.9) are also an important monitoring tool, as they remove you from the room's acoustic environment. Headphones also offer excellent spatial positioning in that they let the artist, engineer or producer place a sound source at critical positions within the stereo field without reflections or other environmental interference from the room. Because they're portable, you can

FIGURE 17.10
Kickin' it in the ride ... preferably on Sunset Boulevard with the top down!

take your favorite headphones with you to quickly and easily check out a mix in an unfamiliar environment.

It should be noted that while headphones eliminate the acoustics of a room from the monitoring situation, they don't always give a true representation of how sounds will behave through loudspeakers (especially with regard to imaging). Monitoring through headphones will also often emphasize low-level sounds like reverb and other effects more than loudspeakers in a room. As a result, listening to a mix over both monitor types will often be beneficial.

Your car

Last, but not least, your (or any other) car can be a big help in determining how a mix will sound in one of modern society's most popular listening environments. You might take your mix out for a spin on a basic car system, as well as a supped-up, window-shakin' bass bomb (Figure 17.10).

SPEAKER DESIGN

Just as the sound of a speaker system will vary when heard in different acoustic environments, speakers of different design and operating type will usually sound very different from one another, even when heard in the same room. Enclosure size, number of components and driver size, crossover frequencies and design philosophy contribute greatly to these differences in sound quality.

Usually professional speaker enclosures are one of two design types: air suspension and bass reflex. An air-suspension speaker enclosure (Figure 17.11a) is an airtight system that seals the air in its interior from the outside environment. This system type (which is often used in smaller "bookshelf" designs) generally provides a strong, "tight" bass response, while often being rolled off at the extreme low end. The more commonly-found bass-reflex or vented-box design (Figure 17.11b), uses a tuned bass porthole that's designed into the front or rear of the speaker enclosure. This allows the air mass inside the enclosure to mix freely with the outside air in such a way as to act as a tuned

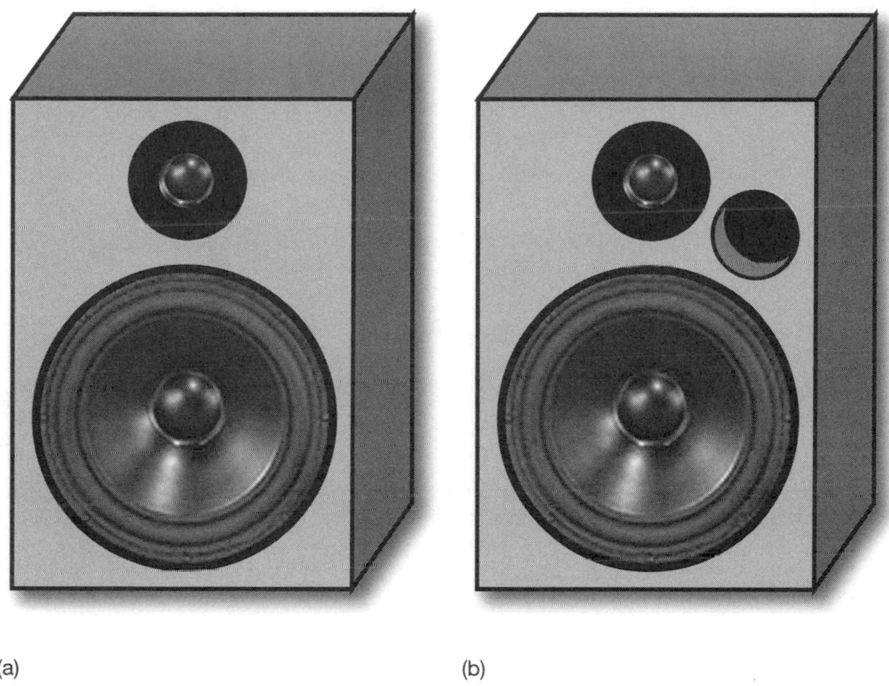

FIGURE 17.11
Speaker enclosure designs: (a) air suspension; (b) bass reflex.

resonator (which serves to acoustically boost the speaker's output at the extreme lower octaves).

Crossover networks

Because individual speaker elements (drivers) are more efficient in some frequency ranges than others, different driver sizes and types are often used in combination to give the desired frequency response and level output. For example, large-diameter bass drivers (such as 10- and 12-inch units) produce low-frequency information more efficiently than at high frequencies; medium-sized speakers (such as 4- and 5-inch units) operate best in the midrange frequencies; and small speakers (such as ½- to 1 ½-inch diaphragm sizes) reproduce highs more effectively.

These speakers are usually connected to a series of passive crossover networks, which are used to divide the frequency spectrum into two or more bands. These bands split the frequency range in such a way that the speakers are driven in the most efficient manner (i.e., signals below the crossover frequency are routed to the bass driver or drivers, while mid and/or high frequencies are routed to the high drivers, as shown in Figure 17.12).

If a speaker system has only one crossover frequency, it is called a two-way system, because the signals are divided into two frequency bands. Likewise, if the signal has two crossover frequencies, it's called a three-way system. The Westlake Audio

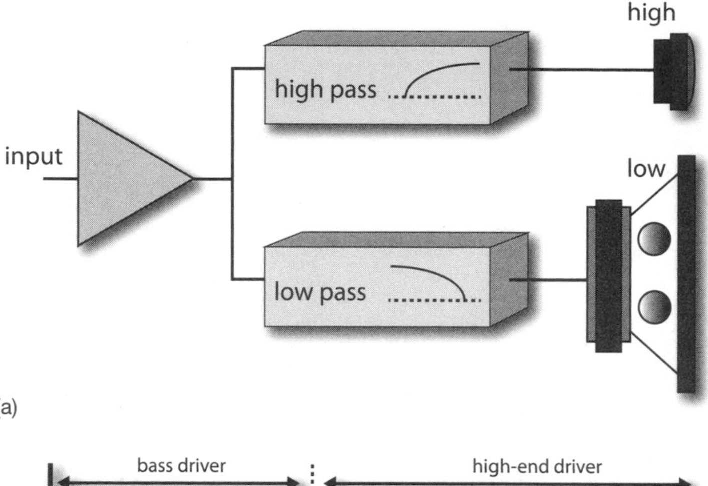

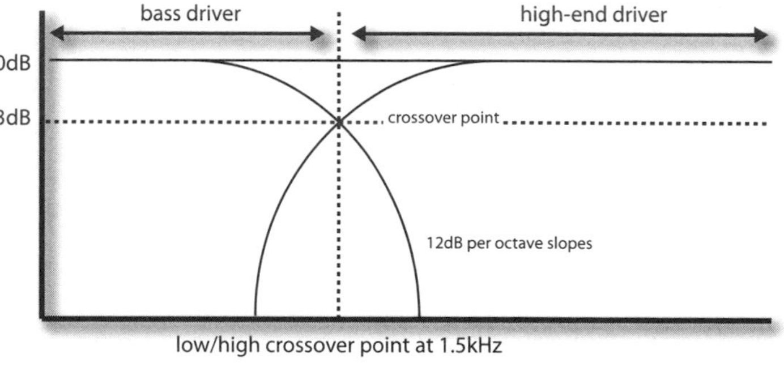

FIGURE 17.12
Example of a passive two-way crossover system: (a) crossover/ amp layout; (b) frequency response curves showing crossover frequencies of 1500 Hz.

FIGURE 17.13
Westlake BBSM-15 Reference Series monitor. (Courtesy of Westlake Audio, www.westlakeaudio.com)

BBSM-15 monitor speaker (Figure 17.13), for example, is a ported three-way system that includes two 10-inch woofers for the bass, a 6.5-inch driver for the midrange frequencies, and a 1.5-inch diaphragm dome tweeter for the highs with the crossover frequencies being tuned at 600 Hz and 4 kHz, respectively.

Actively powered vs. passive speaker design

As you might expect, many of the more popular monitor types that are in use today incorporate an actively powered amplifier into their design. These cost-effective systems have become widely accepted by the professionals and project communities due to their:

- Compact design
- High-quality sound (often these systems are bi- or tri-amplified)
- Expandability (additional speakers can be cost effectively added for surround-sound monitoring)
- Lack of a need for an external power amplifier

For these reasons, these systems are often ideal for project- and DAW-based facilities and are steadily increasing in popularity.

Electronic crossover networks (Figure 17.14), called *active crossovers*, differ from conventional passive crossover systems in that the line-level audio signal is split into various frequency bands that use more complex analog and digital circuitry. Each equalized line is fed to its own power amp, which in turn is used to drive the respective bass, mid-, and/or high-driver elements. Such a system is generally referred to as being bi-amplified or tri-amplified, depending on the number of crossovers and power amps that are required per channel. These systems have several advantages:

- The crossover signals are low in level, meaning that inductors (which can introduce audible ringing and intermodulation distortion) can be eliminated from the design.
- Power losses (due to the inductive resistance within the passive crossover network) can be eliminated.
- Each frequency range has its own power amp, so the full power of each amplifier in the respective speaker efficiency range will be available (meaning that the drawing of excessive current in one range won't affect the sound in another frequency range).

Because there are so many variables in speaker and amplifier design, it quickly becomes clear that there's no such thing as the "ideal" monitor system. The final choice is often more of a matter of personal taste and current marketing trends

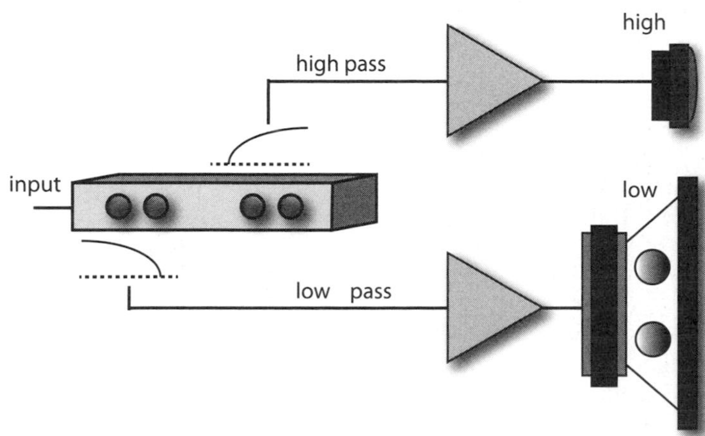

FIGURE 17.14
Example of an active two-way active crossover system.

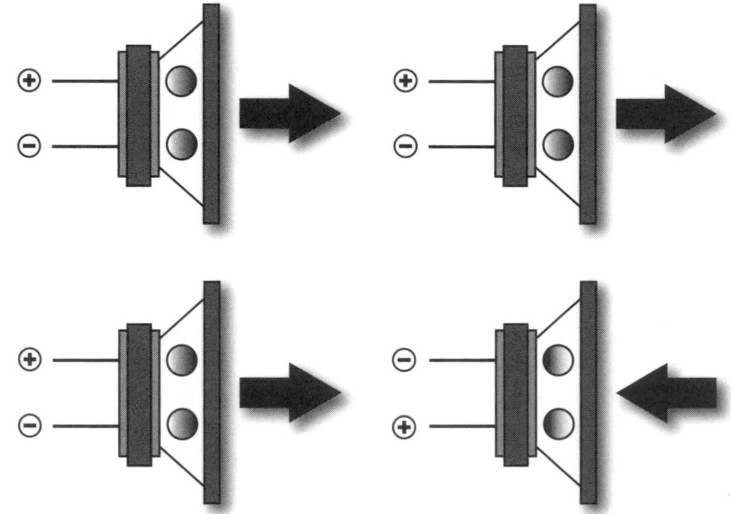

FIGURE 17.15
Relative in-phase (left) and out-of-phase (right) cone motions.

than one of subjective measurements. Monitors that are widely favored over a long period of time tend to become regarded as the industry standard; however, this can easily change as preferences vary. Again, the best judge of what works best for you should be your own ears and personal sense of style.

SPEAKER POLARITY

A common oversight that can drastically affect the sound of a passive multi-speaker system is to wire them out-of-phase with respect to each other (and let's not forget that home-made powered speaker cables can be improperly wired, too). *Speaker polarity* is said to be electrically in-phase (Figure 17.15a) whenever one signal that's equally applied to both speakers causes their cones to move in the same direction (either positively or negatively). When the speakers are wired out-of-phase (Figure 17.15b), one speaker cone will move in one direction while the other will move in the opposite direction.

Speaker polarities can be easily tested by applying a mono signal to both or all of the speakers at the same level. If the signal's image appears to originate from a localized point directly between the speakers, they have been properly wired (in-phase). If the image is hard to locate and appears to originate beyond the outer boundaries of a stereo speaker pair, or shifts as the listener moves his or her head, it's a good bet that the speakers have been improperly wired (out-of-phase).

An out-of-phase speaker condition can be easily corrected by checking the speaker wire polarities. On a passive speaker design, the "hot" lead (+ or red post) leading from each amp channel should be secured to the same lead on its respective speaker (Figure 17.16). Likewise, the negative lead (– or black post) should be connected to its respective lead for each speaker in the system.

FIGURE 17.16
Color-coded banana plug binding posts are often used to help ensure a proper speaker connection.

Speaker cable is generally color coded, with white or red being positive (+) and black being negative (–). If no color coding is present, speaker cables will often have a notched ridge (or set of ridges) or have a printed white band that's generally connected to the negative lead post.

Passive speaker wire gauges should always be as heavy duty as is possible or practical. The #18 wire is considered to be the minimum for lengths of less than 25 to 50 feet, while #14 is considered the minimal length that should be used for 50- and 100-foot runs. (*Note:* The smaller the gauge number, the thicker the wire; therefore, #14 is thicker than #18.) Two reasons for increasing the thickness of the conductor as cable length increases are:

- All cable has resistance, which will increase with length. Thinner cables generally have greater resistance values, meaning that more power will be dissipated in the cable and will therefore be unavailable to drive the speaker.
- The higher the cable resistance, the lower the amplifier's effective damping factor. Damping factor is related to how well the amplifier is able to control the motion of the speaker cone. A lowered damping will often result in a loss of tightness, definition, and clarity in the low end. Again, thicker conductors will have less resistance and thus help to minimize damping problems.

BALANCING SPEAKER LEVELS

Another important factor that should be taken into consideration when setting up speakers within a properly-designed, symmetrical room and speaker setup,

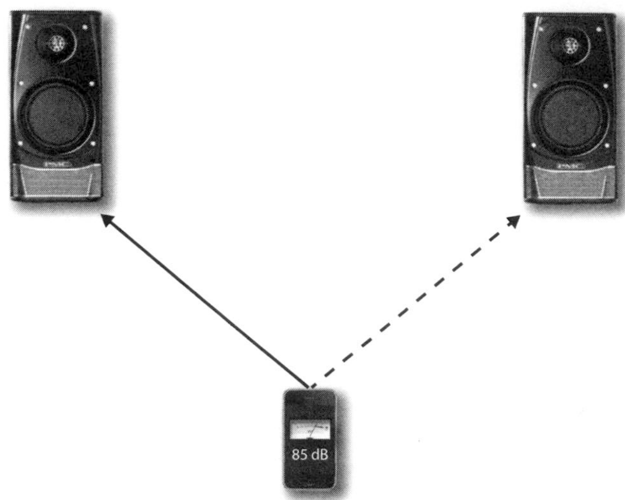

FIGURE 17.17
General setup for balancing your speaker levels.

is the need for balancing your speaker levels, so monitoring and playback occurs at equal and matched levels. If the speaker level settings (at the passive or active power amp) are improperly set, your final mix levels could be off-balance … meaning that your center and overall imaging might be wrong.

In order to set your system properly, simply record a soundfile that contains a stretch of pink noise, import the file into your DAW on as many tracks as there are speakers (2 for stereo or 6 for surround). Set the channels to playback at equal level and place the level meter at the center listening position (you can buy one at Radio Shack, Ebay or simply download a level meter app into your smart phone). Begin playback through one speaker and set the monitor level so that the meter reads at "0" or at any chosen level. While being careful not to be in the way of the mic or meter (with respect to the room), switch to the next speaker, and repeat the process until all speaker output levels are matched (Figure 17.17). Once done, you can finally be sure that the speaker levels are properly matched. … and whenever changes are made at a later time or it you're in doubt, just call up the noise "session" and repeat the setup test. Better to be safe than sorry.

MONITORING

When mixing, it's important that the engineer be seated as closely as possible to the center of the soundfield (making allowances for the producer, musicians and others who are also doing their best to be in the "sweet spot") and that all the speaker volumes are adjusted equally. For example, if the engineer is closer to one speaker than another, that speaker will sound louder and the engineer may be tempted either to pan the instruments toward the far speaker or boost that entire side of the mix to equalize the volumes. The resulting mix would sound properly centered when played in that room, but in another environment, the mix might be off-center.

Here are a few additional pointers that can help you get the best sound from your control room monitors:

- Make sure that the room's reverb time is both low and smooth over the audible range (absorptive and diffusive materials can help).
- Keep large reflections to a minimum within the room (again, absorptive and diffusive materials can help reduce reflections to a level that's at least 20 dB down from the direct signal).

- Keep all room boundaries and reflections as symmetrical as possible along the L/R and front/back axis of the mixing soundfield.
- If diffusers are used, place them at the rear part of the room.
- If the speakers are mounted in soffits, make sure the front wall has a hard, smooth surface and that the speakers are acoustically isolated from the surface (as much as possible; see Chapter 3 for more info).
- Angle the monitors symmetrically toward the listening position in both the horizontal and vertical planes.
- If nearfield monitors are used, place them on medium-density foam blocks to reduce console- and desk-borne vibrations.

In Chapter 14 (mixing), we talked about creating a high-quality desert island compilation disc or music directory that's made up of a number of your favorite or most respected songs that can be listened to as a reference for evaluating a studio or listening room and its speaker monitors. It's always a good idea to take these songs for a spin when entering a new environment … actually; it's a good idea to take a listen from time to time in your own "trusted" room, just to keep a sense of perspective about how it sounds. Trust me, you'll never come across a sound device that's as subjective and variable as your hearing perception.

Monitor volume

Before continuing, I'd like to revisit another important factor—volume. During the record and mixdown stage, it's important to keep in mind that the Fletcher–Munson curves will always have a direct effect on the frequency balance of a mix. Because our ears perceive recorded sound differently at various monitoring levels, our ears will easily perceive the extreme high and low frequencies in the mix when monitoring at loud levels (sounds good doesn't it?). However, when the mix is played back at lower levels (such as over the radio, TV or computer), our ears will be much less sensitive to these frequencies and the bass and extreme highs will probably be deficient (leaving the mix sounding distant and lifeless).

Unlike earlier years, when excruciatingly high SPLs tended to be the rule in most studios, recent decades have seen the reduction of monitor levels to a more moderate 75 to 90 dB SPL. (A good rule of thumb is that if you have to shout to communicate in a room, you're probably monitoring too loud.) These moderate levels offer a good compromise for mixing, as they more accurately represent listening levels that are likely to be encountered in the average home (i.e., the Fletcher–Munson curves will be more closely matched). Ear fatigue and potential ear damage due to prolonged exposure to high SPLs by industry professionals can also be avoided at these levels. For more information on safe monitor levels and hearing conservation, contact the House Ear Institute at www.hei.org.

The big bad bottom end

Before we continue, it's important that we spend a bit of time talking about the bottom end of the audible spectrum. … That's to say, the big-bang-boom of the lows!

THE SUB

A subwoofer, in its truest sense, is an additional low-end driver that's added to a system for the purpose of assisting with low-end bass reproduction … and, hopefully, for improving low-end definition. Although many expect a "sub" to add a VERY BIG BOOM to the low end, this use will almost always end up as an overly accentuated "big bottom" that could easily lead to a false response in the low end. The improper overuse of a sub could actually lead to a final master that's bass shy, due to excessive bass levels.

The proper use and setup of a subwoofer can help extend the low end, while adding a tight definition to the bass that can actually bring a monitor playback system to life. As you might expect, the response and overall sound of a sub is often heavily influenced by the acoustics of the playback environment. In addition to using a software frequency and acoustics analyzer (that's capable of displaying phase relationships as well as frequency response). One of the best ways to set up and tune a sub is by careful experimentation. When placing the subwoofer, the following concepts should be kept in mind:

- Because low-frequency audio is basically nondirectional, some feel that a sub can be placed in any spot in the front of the room. However, a centrally placed position is most likely preferable, as a difference can be heard with upper-bass definition and image localization.
- Although a sub's output will be greater when placed near a wall boundary, a "muddy" or "boomy" sound might exist due to unfavorable room and corner reflections.
- Most active subs will include level and crossover frequency adjustments. These settings should be used to closely match your sub to the chosen speaker set or sets.
- Active subs will also include a phase switch that can match the phase (driver motion) to your main speaker set. Using an analyzer or your ears, set the phase to a position that sounds best.

As with all Big-Bad-Boom-Boxes (of both sub and LFE types), extreme care should be taken when setting up the system's overall levels, placement and sound. Too much low-end might make your system sound rough-n-tough in your room, but will make the mix bass shy when played back in another room. All of this simply leads to the fact that sub placement and setup are super-critical in getting the best sound possible… and an imbalanced system can cause really big problems for you, your room and your reputation.

THE LFE

Within a surround system, the *LFE* channel stands for *low frequency effect* and was developed by the film industry to give an extra "ummmffff" in the low-bass end to add an extra degree of impact … especially during explosions and low rumbles. Contrary to popular misconceptions, the ".1" aspect of an LFE has little or nothing to do with the musical bass line of a music program or film score. Instead, it was originally designed as a band-limited channel (120 Hz for Dolby Digital or 80 Hz for DTS) for adding 10 dB of headroom to a film. (I recall that it was first used in the film *Earthquake* to add extra realism to the earth-shaking score.)

There is an ongoing debate among top music mixers as to whether to use the LFE channel at all in music mixing. Almost everyone agrees that all of the speakers within a music mixing scenario should be full range … however, there are those who don't believe in using the LFE channel at all, while others will put some degree of low content in this channel (stating that customers feel cheated out of their well-earned .1 when they don't get some bottom in that channel). Personally, I fall in the camp of putting a light-to-moderate degree of low-end content in the LFE channel (while being very careful that al calibrations are correct). It's no fun to work hard on a surround project, only to have the bass levels playback at too high a level, muddying up your carefully-crafted mix. When in doubt, it's always a good idea to calibrate your surround system according to the Grammy P&E surround setup guidelines from http://www.grammy.com/recording_academy/producers_and_engineers/guidelines.

On a final note: Critical musical material should never be sent to the LFE channel, because Dolby decoders will completely drop it when folding 5.1 material down to stereo.

BASS MANAGEMENT

At first glance, bass management might look like an LFE channel or a simple subwoofer system (which is a good reason why bottom-end issues can get so confusing). In reality, it is entirely different in that a bass management system (which is used in most home theatre systems and high-end auto systems) use filters to extract low-frequency information from the main channels and then routes the bass to the bass speaker, while the highs are sent to the system tweeters. In short, this method has the advantage of allowing for one bass speaker and smaller, easier-to-build multiple (stereo or surround) tweeters that can be placed in confined spaces, and helps reduce intermodulation distortion.

Although these systems are in wide use in home playback and theater systems, it's widely held that these speaker systems aren't suitable for studio playback due to irregularities in midrange response, image localization and overall sound.

Monitoring configurations

In addition to getting the best overall sound out of a mix, another monitoring concern that has gained importance is the need to tailor the mix to the intended room/speaker configuration (i.e., mono-stereo, mono-surround and stereo-surround).

It's important to remember that a large percentage of your potential customers may first hear your mix over a TV, computer or AM/FM radio in mono. Therefore, if a recording sounds good in stereo but poor in mono, it might not sell as well because it failed to take these media into account. The same might go for a surround-sound mix of a music video or feature release film in which proper attention wasn't paid to phase cancellation problems in mono and/or stereo (or vice versa). The moral of this story is simply this:

To prevent potential problems, a mix should be carefully checked in all its release formats in order to ensure that it sounds good and that no out-of-phase components are included that would cancel out instruments and potentially degrade the balance.

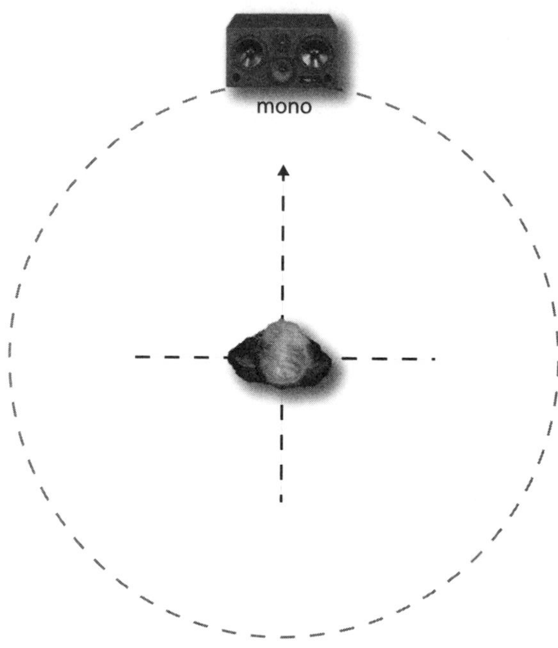

FIGURE 17.18 Example of a monaural monitoring configuration.

The most commonly accepted speaker configurations are mono, stereo and surround sound.

1.0 Mono

Even in this day and age, a number of people will first experience a mix in *monaural* (mono) sound (Figure 17.18). That's to say, they'll hear your song in an elevator, over a small radio or an iPad, etc. For this reason, it's often a good idea to listen to a mix with the stereo channels combined into mono for overall sound, phase and compatibility. The days of doing a special mono mix are pretty much behind us (given the fact that most people experience music over headphones or computer-related devices), but it's always wise to check for mono compatibility. Phase cancellations can cause instruments or frequencies in the spectrum to simply disappear whenever a mix is summed to mono. The best tools for reducing phase errors are good mic technique, a phase plug-in or display and, of course, your ears.

2.0 Stereo

Ever since the practical development of the 45°/45° record cutting process, *stereophonic* (stereo) sound (Figure 17.19) has ruled the turntable. Of course, over the years, stereo has also grown to rule the computer, internet, FM radio, the CD player, TV and the auto. For these reasons, the creation of a quality stereo mix is extremely important with relation to L/R balance, overall frequency balance, dynamics, depth and effects.

Some of the basic rules for monitoring in stereo are:

- Try to make your mixing environment as acoustically and physically symmetrical (within reason) in order to ensure that the L/R balance, effects balance and overall imagery are accurate within the stereo soundfield.

- Balance your speaker levels carefully, so that they match in level.
- Use your ears and have fun!

2+1 Stereo + sub

In actuality, 2+1 isn't really a standard, but it's a relatively easy title to remember. The "+1" represents the addition of a powered sub to a set of stereo speakers (not a .1 LFE channel). Instead of providing an additional sonic big-bang-boom, the subwoofer is actually used to extend and to help define the music's low end. **Extreme care** should be taken when setting up such a system, as a loud and improperly setup sub and crossover combination can create a low end that can sound great, but be very inaccurate. If your system creates a big, bad (but false) bass sound … your mixes might be bass shy and boxy (not what you had in mind for your special mix).

Although it's not commonly known, there are two ways to set up a 2.1 system. Through careful adjustment and care, either method can work:

- Connect the stereo signal to the sub inputs and use the sub's internal crossover outputs to feed a bass-limited signal to the main stereo speakers (Figure 17.20). This traditional system will band-limit the stereo speaker's low end and should be tuned carefully.
- The active sub (which usually has two inputs) can be combined with the stereo signal by using a simple "y" connection or summing network. This method will NOT band-limit the stereo speaker's low end and should be tuned with VERY special care, because interference and interactions between the bass drivers can definitely cause problems.

4.0 Quad

During the 1960s and 1970s, quad was used to introduce music enthusiasts to the joys of surround. Although 5.1 has dominated the field in

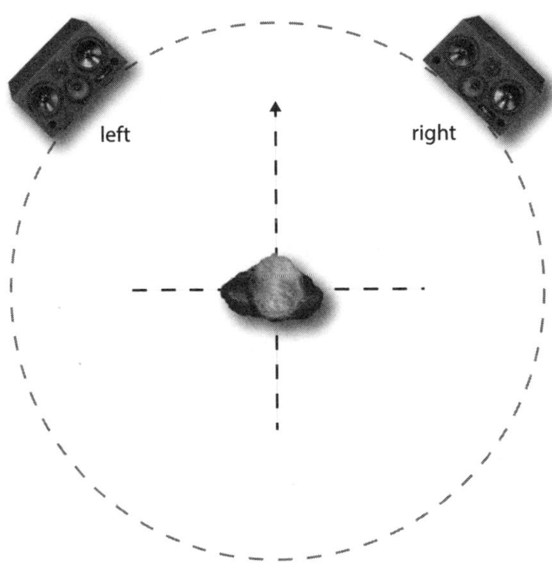

FIGURE 17.19
Example of a stereo monitoring configuration.

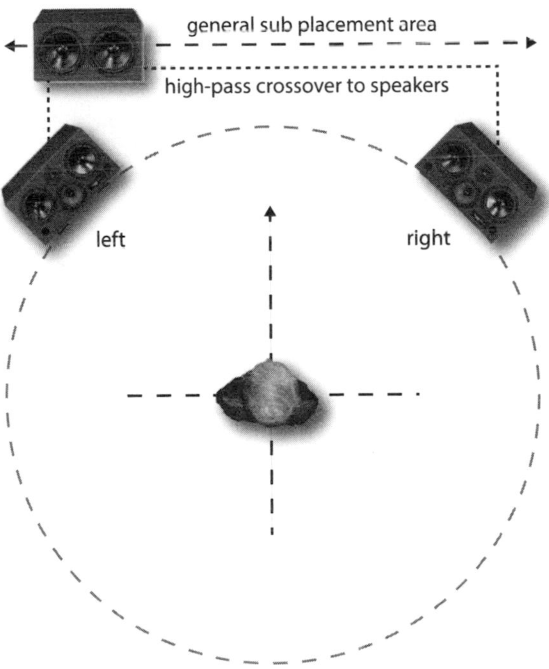

FIGURE 17.20
Example of a 2.1 stereo monitoring configuration, showing the crossover connections through the sub.

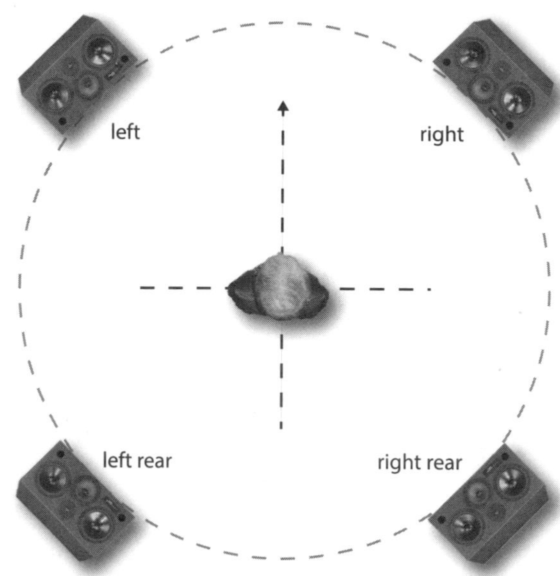

FIGURE 17.21
Example of a quad monitoring configuration, showing a width placement range in the rear channels.

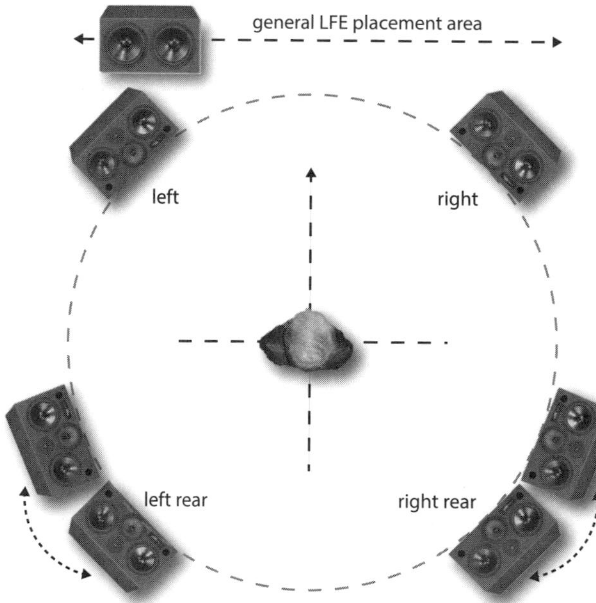

FIGURE 17.22
Example of a 5.1 surround-sound monitoring configuration, showing a width placement range in the surround channels.

recent decades, the use of a four-channel (left/right/rear-left/rear-right) soundfield is still relevant as a surround recording, playback and live performance tool (**Figure 17.21**).

5.0 Surround minus an LFE

Simply put, 5.0 is 5.1 without an LFE big-bang-boom channel. Since LFE is band-limited to the extreme bass frequencies and is intended to provide an added EFFECT!!!!, many surround professionals argue that it's unnecessary to add the LFE channel to a surround music project at all. As with much of recording, surround is an open frontier that's totally open to personal interpretation.

5.1 Surround with an LFE

With the advent of 5.1 playback in home and audio "theaters," surround sound (**Figure 17.22**) has grown into a major professional and consumer entertainment market. The name refers to the five full-range channels that offer up a left/right/center/LFE/ left-surround/right-surround soundfield. DVD videos discs commonly use 5.1 encoding in the form of Dolby Digital (a scheme that encodes the discrete 5.1 information into a single bitstream, known as AC-3) as well as the DTS format.

7.1 Theater plus an LFE

Within large-scale theater soundtracks, the eight-channel 7.1 format is often used to provide two extra (left-side and right-side) speakers for additional spatial placement and localization.

9.1 Height channels

In recent years, a number of systems have begun to crop up that add height channels to an already existing surround sound setup. Having applications for film, game and music, this type of playback makes it possible for jets to fly literally over your head, reverb can float above your head in a mix … adding a dimension to the sound that truly has to be heard to be believed.

FIGURE 17.23
Motu 828Mk3 Hybrid surround audio interface control panel. (Courtesy of MOTU, Inc., www.motu.com)

Although an increasing number of theaters are embracing this technology, time will tell whether 3D audio will be generally accepted by the masses.

Monitor level control

As monitor systems grow to accommodate various production and playback formats, controlling the monitor level, adjustments and switching can become problematic. Many multiple-outout consoles and high-end DAW systems are capable of handling the monitoring requirements of the various formats (including surround sound); however, even these can fall short when multiple sources, level trims and straightforward level control are taken into consideration (although newer interface designs are finally offering multichannel level control). For this reason, many have turned to using a high-quality interface or dedicated studio monitor management system (Figure 17.23) in order to handle the monitoring needs of a professional or project studio.

Spectral reference

Even if your monitor speakers have perfect time/frequency response readings, few people who buy recordings will have a perfectly accurate reproduction system (if there is such a thing). As a result, they won't hear the exact mix that was heard in the control room. The buying public will often hear different frequency balances, due to response variances between the almost limitless types of speaker/listening room combinations. Faced with this fact, the best we can do as professionals is to rely on our judgment, our experience and our ears to create a balanced mix that'll do the best possible justice to a project under a wide range of listening conditions.

In addition to our best set of tools—our experience, judgment and ears—a visual tool known as a *spectral analyzer* is often used to give visual cues as to an audio program's overall frequency balance at any point in time. These applications (which can be either stand-alone or found in a DAW program or suite of plug-ins) give a

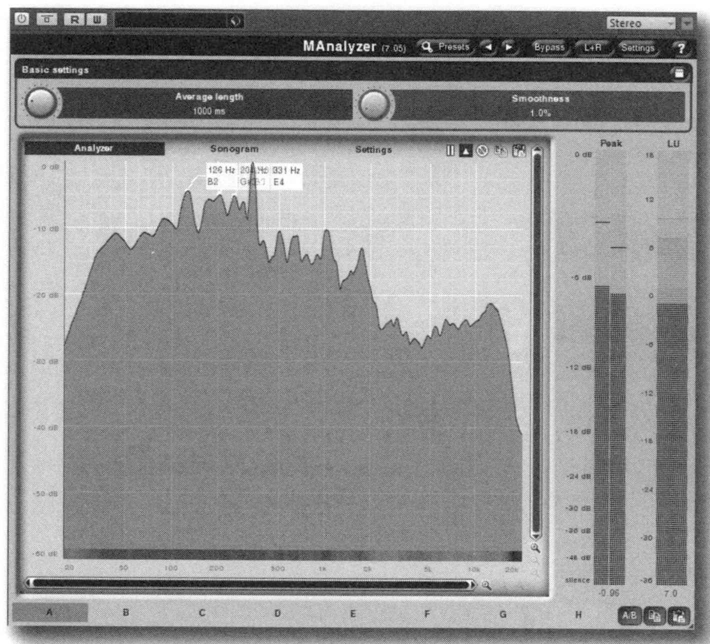

FIGURE 17.24
MAnalyser advanced spectral analyser. (Courtesy of MeldaProduction, www.meldaproduction.com)

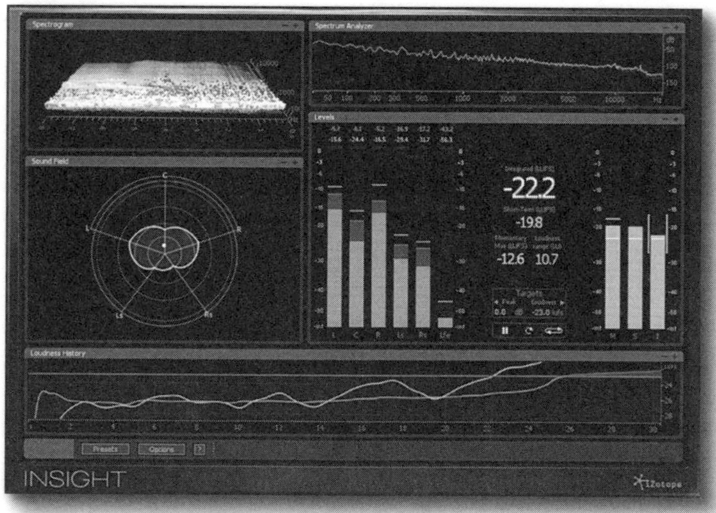

FIGURE 17.25
iZotope's Insight Essential Metering Suite offers level and loudness meters, spectrogram, analyzer, surround scope and a loudness history graph. (Courtesy of iZotope Inc., www.izotope.com)

visual bar readout of a signal's level at various frequencies throughout the audible band (**Figures 17.24 and 17.25**). Obviously, such a tool can help an engineer or producer to zero in on an offending or deficient frequency and/or bandwidth simply by looking at the display over time. During both the record and mix phases, these tools can help point out and avoid potential phase and spectral problems.

MONITORING IN THE STUDIO

In addition to the need for an accurate monitoring environment in the control or production studio, musicians often have special needs for playing back or monitoring the sound that's being recorded during a session.

Headphones in the studio

Monitoring over headphones in the studio is by far the most common way to monitor sound during a recording session.

When recording, it's generally best to use sealed headphones to prevent or minimize the monitor feed from leaking back into the newly recorded track. The number of headphones will vary with the particular session requirements. An orchestral film overdub, for example, could easily use upward to 60 pair during a single session and then, might only need one pair on the next overdub.

Such variations also place demands on the distribution of headphone power and the number of required feeds. As you might imagine, the power that would be needed to run 60 headphones is quite considerable, requiring that a power amplifier be used to drive any number of headphone distribution boxes throughout the room. On the other hand, the power that would be required to drive one or two headphones in a project studio might be so small that they could be driven from the console/mixer's internal headphone or a basic headphone distribution amp (Figure 17.26).

Likewise, the need for separate monitor mix feeds will vary from studio to studio, as well as with varying session requirements. For example, a straightforward session involving only a few musicians will most likely require a single headphone monitor mix. During a tracking or a more complicated session, two or more separate mixes could be made available to the musicians by sending separate "cue" feeds from the mix outputs of multiple auxiliary sends.

For those who wish to allow the musicians to create their own mix, headphone systems (Figures 17.27 and 17.28) exist that can draw separate instrument or submix feeds from various bus or cue feed outputs. This gives the musician complete control over volume, pan, and instrument mix within his or her headphones.

Playback speakers in the studio

Often, there's not enough time for the musicians to leave their instruments and walk into the studio to hear a playback between takes. For this reason, studio

FIGURE 17.26
Powerplay Pro-8 8-channel headphone distribution amp. (Courtesy of Behringer Int'l GmbH, www.behringer.com.)

FIGURE 17.27
HRM-16 16 channel personal headphone mixing station. (Courtesy of Furman Sound, Inc., www.furmansound.com).

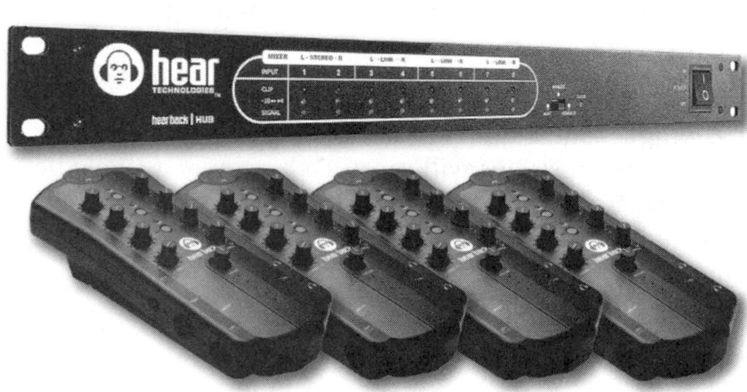

FIGURE 17.28
Hear Technologies Hear Back Personal Monitor Mixer System. (Courtesy of Hear Technologies, www.heartechnologies.com)

monitors are commonly mounted in the studio for immediate playback. These are usually larger monitors, because they will need to be driven to levels that can adequately fill the studio and will often need to withstand unintentional abuse (from loud talkback levels, feedback, etc.).

CHAPTER 18
Surround Sound

At a special media event that was attended by about 500 people, my good friend George Massenburg asked the crowd, "How many of you truly know the joys of surround-sound playback or production?" In that crowd, about 20 of us raised their hands. He then went about the business of playing back several world-class projects that most of the crowd was familiar with, and talking about the benefits of surround-sound music and audio-for-visual production (Figure 18.1). Right-on, George!

Truth is, whether you're an advocate or an adversary of the concept of surround sound, one thing is for sure … it exists in the here and now, and it is certain to play an ever-growing role in the media technologies of tomorrow.

I think that at this point I have to break from my role as a neutral author and state flat out that I'm a HUGE surround fan. For me, the ability to compose and mix in surround has been an uplifting and hugely beneficial experience. I clearly remember as a kid, placing two album covers behind my ears and listening as the music came to life around my head (go ahead, try it)! The ability to augment music and visual media by placing sounds within a 360° circle has literally opened up new dimensions in mixing and effects-placement technologies in a way that keeps me creatively young.

Most of the people that I know who are ideologically closed to the idea of surround haven't worked with or listened to the medium. I urge that you keep an open mind to the process, watch movies and listen to music in surround (in any available format) and, if at all possible, take the time and effort to familiarize yourself with the process of producing sound in surround.

Before I put my neutral writer's cap back on, I'd like to present the strongest argument for becoming familiar with the production techniques of surround: enhanced job opportunities. I have friends living in the technological heart of numerous city centers who are completely unfamiliar with any and all forms of surround. It never occurred to them that they could increase their opportunities, client base and perceived prestige in the fields of mixing music, soundtracks for movies and gaming … by investing in a surround monitoring system and learning

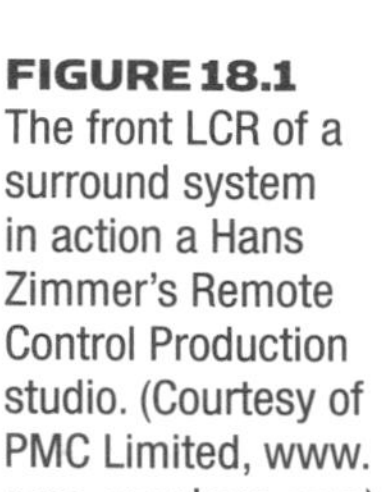

FIGURE 18.1
The front LCR of a surround system in action a Hans Zimmer's Remote Control Production studio. (Courtesy of PMC Limited, www.pmc-speakers.com).

the basic tools and techniques of mixing and mastering media for multichannel sound. If for absolutely no other reason, the ability to understand and work in new and upcoming technologies can help give your career a marketing edge.

SURROUND SOUND: PAST TO THE PRESENT

Of course, it all started with the movies. In the pre "talky" days movie theaters were anything but silent … organs, chimes and all sorts of percussion clanged from all parts of the room behind ornate wall features. With the introduction of movie sound and the use of the musical score in the late 1920s, all of the soundtracks were played back by the delivery format of the time: mono.

On November 13, 1940, Walt Disney's *Fantasia* opened up the soundfield to stereo when it premiered at New York's Broadway Theater. Although it wasn't the first film that was produced using a "multiple channel recording" process, *Fantasia* was the first to introduce multichannel sound to the public.

The final mix of *Fantasia* was printed onto four master optical tracks for playback using a special RCA system called "Fantasound." (Unlike the two-channel format that was adopted for home playback, film "stereo" sound started out with, and continues to use, a minimum of four discrete channels.) This multispeaker setup placed three horns behind the screen and 65 smaller speakers around the walls of the theater. Due to the outlandish setup costs (estimated at about $85,000 for each theater at the time), RCA stopped making this fantastic system after setting another one up at the Carthay Circle Theater in Los Angeles.

In the early 1950s, the first commercially successful multichannel sound formats came onto the scene with the development of CinemaScope (four-track 35 mm) and Todd-AO (six-track 70 mm). Both of these formats made use of magnetic tracks that existed alongside the release print picture, and required that the projector be fitted with special playback heads, amps and speakers.

In the early 1970s, the home consumer stereo market was gaining in popularity and audio quality. With the development of higher quality amps, speakers and record turntables came new experimentations in systems design that eventually led to the development of Quadraphonic Sound (Quad). This playback system made use of four speakers (Figure 18.2) that were placed in the four corners of a room, which enveloped the home listener in a L/R/LS/RS listening experience.

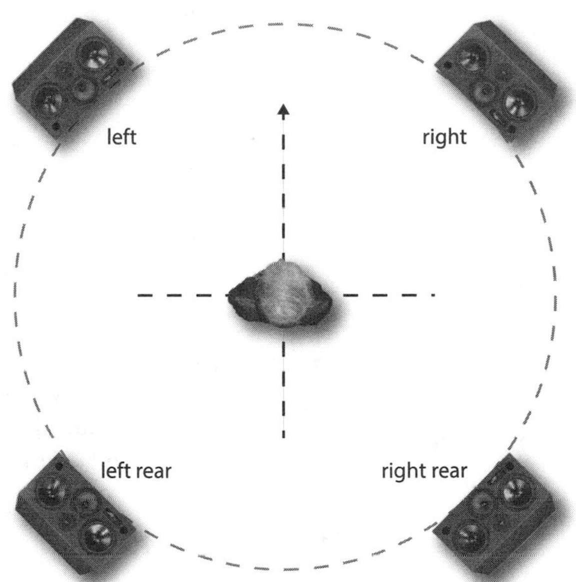

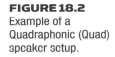

FIGURE 18.2
Example of a Quadraphonic (Quad) speaker setup.

Although analog reel-to-reel and cassette tape machines were used in homes, they were still relatively expensive. By elimination, this meant that playback would have to be carried out by the most popular medium of the day—the LP record.

Reproducing four channels from a record wasn't easy. Often, the task of encoding four channels onto the two walls of a vinyl record was done with relative phase or by using a complex, high-frequency carrier tone that was used to modulate the sum and difference channels. However, the real difficulties lay in the wide assortment of incompatible encode/decode formats that were offered by various manufacturers. Given the fact that your system might not play back the latest and greatest release, and that discs were both expensive and prone to deterioration over a short period of time (the high-frequency signals on modulated records would literally wear away) … the Quad revolution died away.

Stereo comes to television

Since its inception, surround sound had been used in motion picture soundtrack production with great success. With the introduction of Dolby noise reduction and multichannel audio in the theater, good sound was not only appreciated, it was expected! On the other side of the media tracks, television sound was strictly a lo-fi, mono experience up until the early 80s. Sound was strictly an afterthought to the visual image. However, with the adoption of the video cassette recorder (VCR) and later hi-fi stereo sound from a VCR, discriminating audiences began to appreciate the higher-quality sound that accompanied the almighty image. With the dawning of the music video (I want my MTV!), stereo broadcast television and the stereo VCR, TV was finally forced into an era of the higher-quality, multichannel, visual experience.

Theaters hit home

In 1982, Dolby Labs introduced "Dolby Surround," an extension of their professional Dolby Film Sound Project. By 1987, millions of homes were beginning to be fitted with consumer receivers and high-end audio systems that were integrated with video. With the introduction of Dolby Pro Logic, a simple system was put into place that allowed phase information to be extracted from the two tracks of a stereo program to reproduce the L/R/C/Surround field. In 1992, with the introduction of Dolby's AC3 surround encode/decode system (Dolby Digital), it became possible for discrete 5.1 surround sound to be encoded directly with the new visual entertainment medium of the early 21st century—the DVD … and now to the newer Bluray disc.

SURROUND IN THE NOT-TOO-DISTANT FUTURE

With the recent introduction of the Acura TL car with an Elliott Sheiner (ELS) surround-sound DVD-A system, many in the audio world await the open acceptance of surround sound into the domain of the open highways. In fact, many feel that the automobile, when joined with the home theater system, will greatly help to propel surround into the open market mainstream. Meanwhile, video games and audio codecs continue to be developed that allow surround-sound music to be encoded and distributed over the web for personal use with the millions of surround systems that are now installed on home personal computers.

MONITORING IN 5.1 SURROUND

With the recent proliferation of surround-sound speaker options for both the home and studio, options exist at all levels of quality, functionality and cost effectiveness for installing a surround monitoring environment. As with most new technologies, it's important that your existing facility be taken into account, so as to maximize control over monitor levels and monitor format choices (discrete surround, stereo and mono), as well as its integration with your current console and/or DAW system. Before choosing a 5.1 speaker system/setup, it would be wise to consider the following:

- What are the commercial advantages to producing audio in surround? Are there any new clients or business ventures that can make use of this technology?
- What is the budget for such a system?
- Can your existing speakers be integrated into the surround system? Many powered monitor systems can be upgraded to surround by adding matching or similar speakers from the same product line.
- Can your console produce audio in surround sound? If your console has six or more output buses (8 bus +), your system can output surround in some manner; however, true surround panning and surround monitor control are difficult to pull off when using such systems.
- Can your DAW produce audio in surround sound? As above, certain DAWs are capable of routing audio to multiple output buses (6 bus +); however, most surround-capable workstations are able to integrate tracks, effects and track exporting in surround with an amazing degree of versatility.

- How do I plan to monitor in surround? If the console, interface or DAW offers true surround monitor capabilities, you're in luck (Figure 18.3). If not, a hardware surround monitor control system or a surround preamp will be necessary.
- In what types of surround mastering tools should I invest? Creating a surround-sound mix is only part of the battle. Often the real challenge comes in the mastering of the six tracks into a final format that can be played on a commercial playback system … when it comes to the intricacies of surround mastering of the various formats (dealt with later in this chapter), knowledge, attention to detail and patience definitely lead to power.

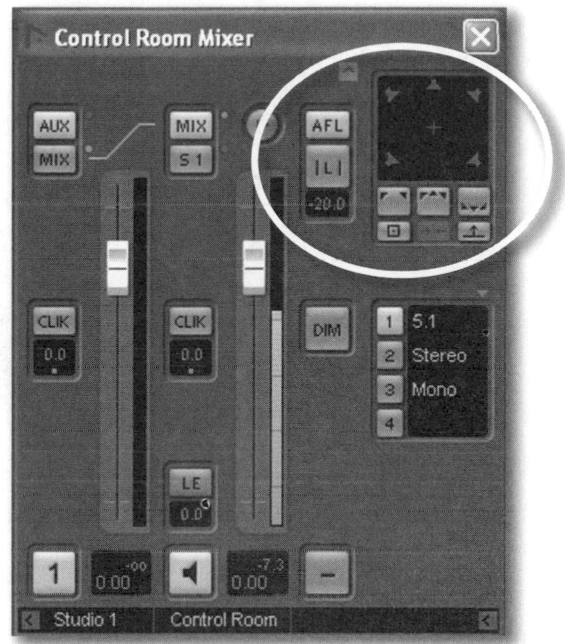

FIGURE 18.3
Nuendo's Control Room Mixer highlighting the 5.1 speaker selector. (Courtesy of Steinberg Media Technologies GmbH, www.steinberg.net)

Once these and other considerations have been taken into account, the task of choosing and installing a 5.1 surround system into the production control room can get under way. This can be a daunting task, requiring technical expertise and acoustical knowledge, or it can be a straightforward undertaking that requires only basic placement and system setup.

5.1 Speaker placement and setup

As defined by the International Telecommunications Union (ITU), the "official" 5.1 speaker setup is made up of five full-range monitors that are positioned in a circular arc, with the speakers being placed at equal distances to the listener (at the center position). Three of the speakers are placed to the front, with the center speaker being placed dead center (0°) and the left/right speakers being placed at 30° arcs to the center point. The surround speakers are then placed behind the listener at 110° arcs to the center point (Figures 18.4 and 18.5).

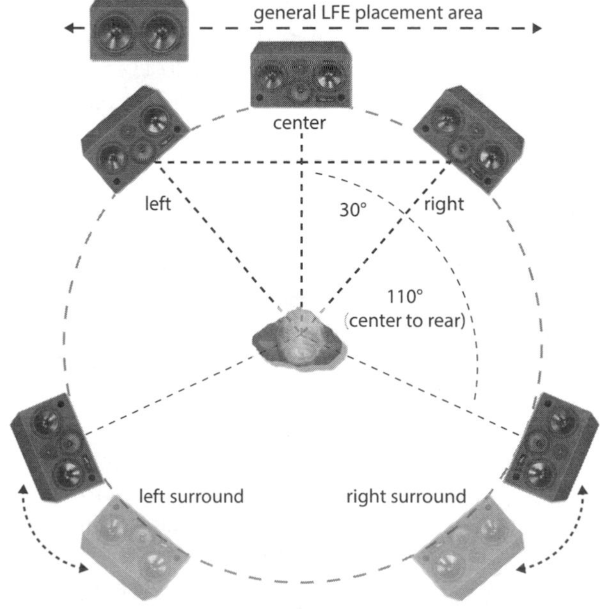

FIGURE 18.4
The ITU 5.1 speaker setup diagram.

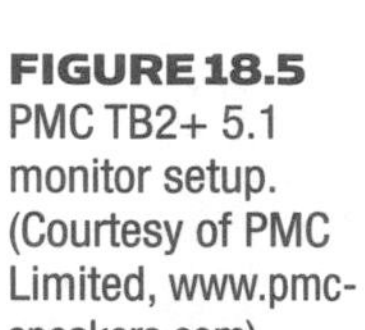

FIGURE 18.5
PMC TB2+ 5.1 monitor setup. (Courtesy of PMC Limited, www.pmc-speakers.com)

It should be noted that the 110° arc in the surround field is a debatable point. During a special playback event, a rather noteworthy group of surround producers and engineers (including myself) compared a 110° arc with a wider 130° arc. We decided on the latter, with popular opinion being that the head and ears didn't provide as much of a barrier to the wide-angle sound. You might want to perform a DIY and check it out for yourself.

The full-range monitors are augmented in the bass end through the use of a low-pass low-frequency effects (LFE) channel, which is then sent to the system's sub-woofer. In film mixing, the low-pass crossover value for Dolby's AC3 has been defined to be 120 Hz (although lower values, such as 80 Hz, are often chosen for use with the DTS system).

Not all music producers and engineers agree that the LFE should be used for surround music production, because it's intended as a sub "effects" channel and not a music bass extension channel … definitely a subject for debate over a couple of pints of lager. However, I would definitely argue that LFE levels are critical … as the overuse of an improperly setup LFE is not a pretty thing. Too much extended low-bass can ruin an otherwise excellent mix.

The LFE's placement and setup within the room should also be taken seriously, as placing it in a corner or in a position that's affected by a harmonic node in the room could greatly affect its response. If possible, the sub should be placed on the floor, near dead center, at a reasonable distance from the front wall (to prevent excessive bass buildup). Finally, most active subs will offer full control over gain and crossover frequency in order to best match the response to the room and main speaker set.

From a level standpoint, all five of the surround speakers should be gain-adjusted to provide the same sound output level. This level matching can be done in many ways. For example, most surround preamps and certain surround monitor systems are able to output white or pink noise to the speakers at equal levels. These signals are sequentially switched from speaker to speaker throughout the room. This allows the levels to be individually matched (either by ear or by using a level meter). Setting LFE levels properly is a bit more complicated. If you're serious about setting up a system properly, I'd strongly recommend that you consult the surround technical setup guidelines from the Producers and Engineers Wing of NARAS (the Grammy folks). It's well worth downloading this and other freely-distributed guidelines from **http://www.grammy.com/recording_academy/producers_and_engineers/guidelines**.

Remember that trusty high-quality desert island compilation disc or music directory that I've been talking about over the course of the book? Well, for those who are serious about surround, you might consider putting together a series of favorite discs that can be played as a reference for your and other 5.1 rooms. As always, you'll probably be amazed at the difference in sound and mixing styles between the discs.

Practical placement

From a practical standpoint, I have often found surround to be a bit more forgiving in placement than the above "spec" suggests. For example, it's often not practical to place the three front speakers in an equidistant arc on most consoles or DAW desks because there simply isn't room. Usually, this means placing the speakers in a straight line or smaller arc (while angling the speakers for the best overall soundfield coverage).

Placing three matched speakers on the front bridge of a console, on floor/ceiling mounts or flush in a soffit generally isn't difficult. However, I found that placing the center speaker on a DAW desk is sometimes a challenge. This relates to the fact that the computer monitor (or monitors) is commonly placed at the center position. If matched speakers are used, where can the center speaker be placed? You might be able to place the center speaker on its side. In addition, a few forward-thinking manufacturers make a dedicated, low-profile center speaker that allows the computer monitor to be placed on top or under a shelf that can hold the monitor(s).

In cases where space, budget and time are a consideration, it might not be possible to exactly match the rear speakers to those in the front. In such cases, your best judgment should be used to match the general characteristics of the speaker sets. For example, if a company makes a speaker series that comes in various sizes, you might try putting a pair of their smaller monitors in the rear. If this isn't an option, intuition and ingenuity are your next best friends.

ACTIVE/PASSIVE MONITORS IN SURROUND

As you might recall from Chapter 17, active monitors include a powered amplifier(s) within their design, whereas passive monitors require that an external power amplifier be used to power their drivers.

One of benefits of a powered speaker system is the ability to upgrade a stereo system to a surround environment by simply adding three monitors to a stereo production system.

Regarding surround monitor control, it's a fortunate fact that many of the newer high-end audio interfaces now include a master volume that can be programmed to control all of the channels in a surround-sound system … a lucky fact, since dedicated hardware monitor controls can cost in the thousands. For project studios on a budget that would also like the added benefit of having extensive control over surround monitoring, a logical choice might be to invest in a high-quality surround receiver system (which integrates preamp, monitoring and surround amp features into a single unit). Such a unit would offer monitoring features that are actually difficult to attain in most professional studios (i.e., mono, stereo, surround, ProLogic, multiple input switching, record/play, dub switching, etc.).

With regard to the LFE channel, almost all surround sub speakers are actively self-powered. Although many of these systems provide active crossover outputs for diverting the low-frequency energy to the sub speakers, while sending the highs out to all five speakers (or more likely in a pro system, only the front stereo channels) … extreme care should be taken when considering the use of this crossover network. The following should only be taken as a recommendation when setting up your system. There are a number of powered/passive variables, and it's best that you fully research your options for obtaining the best sound:

- If the system is a bass managed "satellite" system (meaning that the speakers are not full range, but only contain mid-high drivers, while bass is directed to the sub), the crossover should be installed per the manufacturer's instructions.
- If the surround speakers are full range, you might consider not routing the signals through the active sub's crossover, because this might affect the natural low end of your monitors and could adversely affect the overall sound.
- You could experiment with routing the front stereo pair through the sub and see which way best suits your ears—as with much of audio, it's a very subjective subject that should be considered carefully.

SURROUND INTERFACING

Because surround requires that you have at least six output channels, it logically follows that your audio interface should be either:

- A dedicated audio interface with at least two inputs and six outs
- A multichannel audio interface (for example, having 8 ins × 8 outs)

Over the years, a number of USB/FireWire interfaces have come onto the market that fully support surround sound. These devices allow audio to be easily and cost effectively routed from your host DAW or application to an external amp or active monitor system. In certain cases, these audio cards might include a driver that has extensive monitor controls for varying setup levels, switching between modes (mono, stereo, surround), inserting surround processing and generating setup tones—often well worth the investment.

The various output buses of a DAW are designed to route to the first six outputs in a L/R/C/S/LS/RS configuration (although this scheme might change between DAW platforms and can be changed to fit the program and mastering requirements).

SURROUND FINAL MASTERING AND DELIVERY FORMATS

One of the underlying fallacies of surround-sound production is that you can take a console (or even a DAW for that matter) with at least six output buses, connect 5.1 speakers to their outputs, and then believe that you're ready to fully produce and distribute audio in surround. In fact, there are many hardware and software considerations that must be taken into account, including some that relate to the mastering of these six channels into a final format that can be played by both the average and the not-so-average consumer.

This process often involves the conversion of a set of master recorded tracks or DAW session into a final format that can be distributed and reproduced using existing surround technology. Several of the production formats that are in current use include:

- Dolby Digital (AC3)
- DTS
- Dolby Pro Logic
- FLAC
- MP4
- WMA

Dolby Digital (AC3)

Dolby Digital (a technical spec that's also known as *AC3*) is a popular codec that's used to encode digital audio into a multichannel (mono through 5.1) bitstream through the use of perceptual coding techniques. Because this code eliminates unused audio that is masked or imperceptible, the amount of data that's encoded within the bitstream can be greatly reduced (often by a factor of 10), when compared to its uncompressed counterpart.

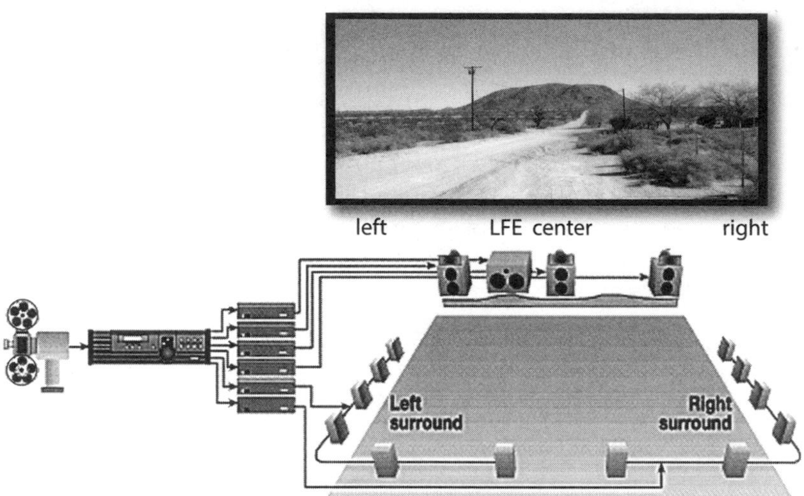

FIGURE 18.6
Dolby Digital Cinema System setup. (Courtesy of Dolby Laboratories, Inc., www.dolby.com)

Dolby Digital is used in the encoding of surround audio on the ever-present DVD, and has also been adopted for use in high-definition Blu-ray discs, HDTV television production, digital cable and satellite transmissions. The traditional flavor of Dolby Digital provides for five full-bandwidth channels in the L/R/C/LFE/LS/RS format, while the LFE channel provides a low-pass channel (usually cutting off frequencies above 120 Hz) for special effects and action sequences in movies (Figure 18.6).

Not all members of the home audience are able to listen to Dolby Digital in discrete 5.1, so a down-mixing feature was implemented into the codec to ensure compatibility with any mono or stereo playback device (meaning that additional tracks in other channel formats aren't necessary). In addition, the compressed nature of this format allows for various alternative track options (including soundtracks in various languages and bonus director/actor/background tracks).

The encoding of a six-channel audio master into a Dolby Digital encoded bitstream can take place through the use of a Dolby licensed hardware encoder or through the use of an encoding software program or plug-in.

Digital Theater System (DTS) (Figure 18.7) is an audio encoding scheme that supports up to 6.1 channels of discrete audio from a single datastream for use in cinema sound, DVD/home theater and multimedia. The DTS codec exists in several format flavors (depending on its intended application) and, as of this writing, these include:

- *DTS Digital Surround:* Provides 5.1 channels of discrete digital audio for commercial movie soundtracks, consumer electronics products and software content.
- *DTS-ES:* A digital audio format for delivering 6.1 channels of discrete audio in the consumer electronics market. DTS-ES is also fully backward-compatible with DTS decoders that aren't "Extended Surround" equipped.

(a)

(b)

FIGURE 18.7
DTS-HD Master Audio Suite encoder plug-in: (a) DTS encoder™. (b) DTS-HD StreamPlayer™ (Courtesy of DTS, Inc., www.dts.com)

- *DTS Neo:6:* Provides up to six channels of matrix decoding from stereo matrix material.
- *DTS 96/24:* Offers 24-bit/96-kHz encoding for multichannel sound on DVD.
- *DTS Interactive:* Delivers real-time, discrete multichannel interactive playback for PlayStation2 games when connected to a DTS-equipped device (such as an AV receiver).
- *DTS Virtual*: Down-converts DTS 5.1- or 6.1-channel soundtracks to stereo while processing a realistic simulation of surround sound for two-channel equipment.

DTS is used to encode audio data at rates of 1.5 Mbit/sec or 754 kbit/sec (compared to the basic rate of 448 kbit/sec, or the 384 kbit/sec rate of Dolby Digital). Although far fewer titles have been released in DTS than in its Dolby Digital counterpart, the new 754 kbit/sec rate has allowed a number of movie production studios to offer soundtracks in both Dolby Digital 5.1 and DTS 5.1 formats.

FLAC

One of the most hopeful codecs for delivering lossless audio within a multi-channel bistream (up to 8 channels) is FLAC or Free Lossless Audio Codec. This codec, which is completely open-source in nature and doesn't make use of any copy protection schemes, is now included in an increasing number of media players.

MP4

MPEG-4 (commonly known as MP4) is essentially based on Apple's Quicktime MOV format, and is capable of containing audio, video and subtitle datastreams in various bitstream formats including surround sound. This popular format allows for the distribution of audio and video content over download and other data transmission media in a way that can be protected using an optional digital rights management scheme.

Initially, MPEG-4 was aimed at being a low-bit-rate video medium; however, its scope was later expanded to be efficient across a variety of bit rates ranging from a few kilobits per second to tens of megabits per second.

WMA

Through the use of plug-in encoders from various DAW manufacturers, it's possible to encode two-channel, as well as discrete 5.1 and 7.1 surround-sound audio into a bitstream that can be played back using Microsoft's Windows Media player (as well as others). The *Windows Media Audio (WMA)* codec, which was designed for digital download and multimedia formats, has been optimized for streaming or download-and-play delivery at bit rates ranging from 128 kbits/sec to 768 kbits/sec, at depths of 16 or 24 bits using sample rates of 44.1, 48, 88.2 or 96 kHz.

When set up properly, true surround sound can be played from a standard Windows XP computer, using an off-the-shelf surround sound card/speaker system or multichannel audio interface. It's also possible to play back surround-encoded files from a stereo computer or media player through the use of a down-mix codec that proportionately mixes the center and rear channels into the L-R channels.

Up-mix to 5.1

One of the more recent developments in surround technology is the ability to make use of processing and spatial technology to "up-mix" an existing stereo soundfile into a fully discrete surround set of files. With such a tool (Figure 18.8), it's possible to take an existing stereo soundfile and (using various width, timbre, spaciousness and other parameters) create and control a wide range of variables to create a surround mix for music, television or film.

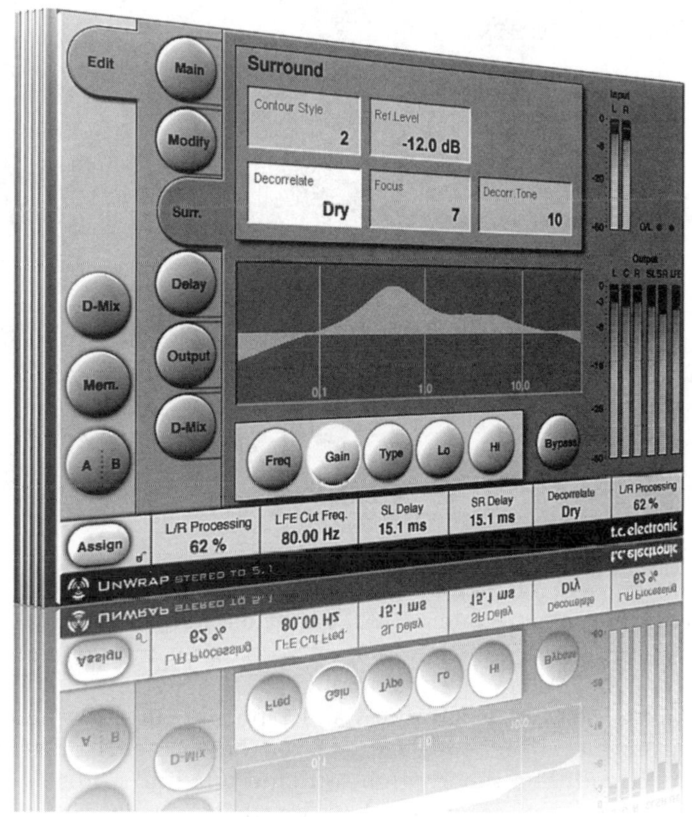

FIGURE 18.8
TC Electronic UnWrap Plug-in for Stereo to 5.1 Up- and Down-mixing. (Courtesy of TC Electronic A/S, www.tcelectronic.com)

MIXING IN SURROUND

Any device having a number of multiple bus consoles (typically having eight or more buses) can be used to create a surround-sound mix, but the important question of the day is this: How easily can signals be routed, panned and affected in a surround environment to create a 5.1 mix without going nuts with frustration?

Whether you're working in an analog hardware, digital hardware or DAW "in-the-box" mixing environment, the ability to pan mono or stereo sources into a surround soundscape, place effects in the 5.1 scape and monitor multiple output formats without difficulty can make the difference between a difficult, compromised mix and one that lifts your spirits.

Surround mixers

As you might imagine, designing an analog console for use with full-scale surround features is almost always a massive and costly undertaking. Digital

FIGURE 18.9
Surround-sound panners within the Nuendo mixer. (Courtesy of Steinberg Media Technologies GmbH, a division of Yamaha Corporation, www.steinberg.net)

consoles, on the other hand, are better suited to controlling the spatial and effects panning functions of numerous inputs through the use of a central control panel. Alternatively, a number of DAW systems offer "in-the-box" software mixers that can mix, process and export in surround in a powerful and cost-effective environment (Figure 18.9). These systems are able to mix, automate, effect and then export a full-surround mix into a final format in a way that's often a thing of beauty—and in a way that's often hard to match in the hardware domain … regardless of the price.

After having mixed many projects in surround myself and in talking with many top engineers and producers, I've come to realize that the first rule of recording—"There are no rules, only guidelines"—definitely applies to surround-sound mixing.

The ways in which sound can be placed in the soundscape are as varied as those who place them. For example:

- There are those who believe that the sound should be primarily placed in the front (L/C/R), while the reverb, natural ambient and/or audience should be placed in the rear.
- Others feel that there literally are no rules, and that sounds can be placed anywhere that sound good, interesting or just plain "right" for the material.
- Others will place monophonic sounds at pinpoint positions throughout the field, while others will place multiple stereo fields throughout the soundscape (i.e., along any of the side or corner axes).
- Many will not use the LFE (sub) channel at all, believing that it adds a degree of unnaturalness to the low bass that's better handled by the full-range speakers.

- Others won't use the center speaker, believing that it messes with the center image along the L/R axis in a way that's reminiscent of the quadraphonic era.

It's my strong belief that how you choose to convey your music through the art of surround-sound mixing is up to you and the muses. It's a fun medium that should be experimented with.

REISSUING BACK CATALOG MATERIAL

On a final note, one of the unintended by-products of older, classic projects that are being reissued into surround sound by the record companies is the resurrection and the rescue of their older analog (and digital) masters that are aging (and/or are so badly documented) to the point that restoration becomes a monumental task.

Regarding efforts to transfer these tracks to 24/96 digital archive soundfiles—it's not uncommon to hear horror stories of 16- and 24-track masters that have to be reconditioned and carefully baked, then played onto an analog machine … only for the iron oxide to shed off the tape and separate from its backing as it plays onto the floor during the actual transfer.

Rescue stories like these are varied and awesome (in the truest sense of the word). Sometimes the masters have already deteriorated past the point of playability and the safety backup must be used. Others must be tracked down in order to find that "right take" that wasn't properly documented. Suffice it to say that surround reissues are definitely doing their part toward helping to keep the original master tracks alive and kickin' into the 21st century, not to mention the fact that they are breathing new life into the tacks in the form of a killer, new surround-sound version.

In a nutshell, listening to a well-crafted surround project over killer speakers in the studio, home playback or a home theater system can rank way up there with chocolate, motorcycle ridin' and sex—absolutely the best!

CHAPTER 19
Mastering

The *mastering* process is an art form that uses specialized, high-quality audio gear in conjunction with one or more sets of critical ears to help the artist, producer and/or record label attain a particular sound and feel before the recording is made into a finished manufactured product. Working with such tools, a mastering engineer or experienced user can go about the task of shaping and arranging the various cuts of a project into a final form that can be replicated into a salable product (Figures 19.1 and 19.2).

In past decades, when vinyl records ruled the airwaves and spun on everyone's sound system, the art of transferring high-quality sound from a master tape to a vinyl record was as much of a carefully guarded art form as it was technology. Because this field of expertise was (and still is) well beyond the abilities of most engineers and producers, the field of vinyl mastering was left to a very select few.

FIGURE 19.1
Emily Lazar, chief mastering engineer, The Lodge, New York City. (Courtesy of The Lodge, www.thelodge.com)

FIGURE 19.2 Darcy Proper's mastering room at Wisseloord Studios, Hilversum, Netherlands (Courtesy of Wisseloord Studios, www.wisseloord.nl, acoustics/photo by Jochen Veith, jv-acoustics.de)

The art of transferring sound to a CD, DVD, Bluray or downloadable media is also still very much an art form that's often best left to those who are familiar with the tools and trade of getting the best sound out of a project. However, recent advances in computer and effects processing technology have also made it much easier for producers, engineers and musicians to own high-quality hardware and software tools that are fully capable of creating a professional-sounding final product in the studio or on a desk/laptop computer.

THE MASTERING PROCESS

In addition to the concept of capturing the pure artistry of a production onto tape, hard disk or other medium, one of the primary goals during the course of recording a project is the overriding concept that the final product should have a certain "sound." This sound might be "clean," "punchy," "gutsy" or any other sonic adjective that you, the artist, the producer or the label might be striving for. Of course, once all of the cuts have been mixed, you'll hopefully be able to sit back and say, "Yeah, that's it!" If this isn't the case, having an experienced mastering engineer help you to "shape" the project's sonic character through the careful use of level balancing, dynamics and EQ could help save your sonic Technicolor day.

Another factor that can affect a project's sound is the reality that recordings may have been recorded and/or mixed in several studios, living rooms, bedrooms and/or basements over the course of several months or years. This could mean that the cuts would actually sound different from each other. In situations like this, where a unified, smooth sound might be hard to attain, it's even more important that someone who's experienced at the art of mastering be sought out.

In essence, the process of successfully mastering can help a project:

- *Sound "right":* This is often accomplished through the use of careful EQ matching and dynamics processing. As was previously mentioned, this process not only takes the right set of processing gear, but also requires experienced ears that intuitively know how the project will most likely sound under a wide range of playing conditions.
- Be in the right sequential order. Choosing a project's song order is an art form that's best done by the artist and/or producer to convey the overall "feel" of a project.
- Have the proper timings between songs. The intuitive process of setting the gap times between songs can also make the difference between having awkward pauses and a project that "flows" smoothly from one cut to the next. Just remember, automatically setting the "gaps" to their default 2 second settings will probably not be in the project's best interest.
- *Playback at optimum levels:* Traditionally, the industry as a whole tends to set the average level of a project at the highest possible value. This is often due to the fact that record companies will always want their music to "stand out" above the rest when played on the TV, radio, MP3 player or the Web. This is usually accomplished by applying compression to the track or overall project. Again, this is an artistic technique that often requires experience. Over-compression can lead to audible artifacts or a sound that can "squash" the life out of your hard-earned sound. In fact, light compression or even no compression at all is also an alternative. Classical music lovers, for example, often spend big bucks to hear a project's full dynamic range.
- *Match levels throughout the project:* In addition to getting the best overall level, it's often important that levels be properly matched when transitioning from one song to the next. The goal is to improve the flow and professionalism of a project by keeping songs from sticking out like a sore thumb.

The equipment that deals with the art and technology of creating a finished "master" is available in many different guises. Often, top-level mastering engineers will use specially designed EQ, dynamics and level matching gear that often won't commonly be found in the recording studio environment. Having said this, with the advent of CD and DVD burning software, dedicated hardware and software processing systems are now on the market that give musicians, producers and engineers a greater degree of control over the final mix and/or finished master than ever before.

To master or not to master—was that the question?

As mentioned, the process of mastering often requires specialized technical skills, audiophile equipment, a carefully tuned listening environment and talented ears in order to pull a sonic rabbit out of a problematic hat or even one that could simply use some dressing up.

A few years back, I had the good fortune of sitting around a big table at a top LA restaurant with some of the best mastering engineers in the US. As you might expect, the general consensus was that it's never a good idea for artists to master their own project … that they're just too close to it to be objective. I'm not sure I agree that this is always the case. However, when approaching the question of whether to master a project yourself or to have a project professionally mastered, it's important that you objectively consider the following questions:

- Are you objective enough to have a critical ear for the sound and general requirements of the project, or are you just too close to it emotionally? (realizing that the "sound" of a particular project might follow you throughout your career)
- Is your equipment and/or listening environment adequate for the task?
- Is the final mix adequate (or more than adequate) for its intended purpose, or would the project benefit most from an outside set of professional ears?
- Does the budget allow for the project to be professionally mastered?

If the services of a professional mastering engineer are sought, the next question to ask is "Who'll do the mastering?" In this instance, it's important that you take a long hard look at the experience level of the person who will be doing the job and make it your job to familiarize yourself with that person's work and personal style. In fact, it's probably best to borrow from the traditional business practice of finding three of the most appropriate mastering house/engineer facilities to bid for the job and follow due diligence in making your decision by considering the following:

- What is their mastering track record?
- Are you familiar with examples of their work? If not, you should definitely ask for a client list and recent examples of their work … then, have a critical listening session with the producer and/or band members.
- Are they familiar with your music genre as well as the type of "sound" that you're hoping to achieve?
- What are their hourly or project rates? Do they fit your budget?
- Are they willing to do a complementary test mastering session on one of your cuts?

Bottom line: Beware of the inexperienced mastering engineer (especially if that person is you). Once the choice of an experienced professional has been made, it's often wise for you/your band and the producer to personally sit in on the mastering session. Make sure that there are several ears around to listen to the project and listen over several types of systems. And above all, be patient, be critical of the project's sound and listen to the opinions of others. Sometimes you get lucky and the mastering process can be quick and painless; at other times it takes the right gear, keen ears and lots of careful attention to detail.

To master or not to master the project yourself—that's the next question!

Obviously, there are different production levels for deciding whether or not to master. Is it a national major release? Is it your first self-produced project? Is there virtually no budget or did the group blow it all on the release party? These and countless other questions come into play when choosing how a project is to be mastered.

Having said this, mastering a project (whether it's your own or not) is increasingly becoming an available option. If you do your own, you should understand a few basic concepts:

- The basic concept of mastering is to present a song or project in its best possible sonic "light." This requires a basic skillset, level of "comfortability" and knowledge of your room's equipment, acoustics and overall downfalls.
- A basic level of detachment is often required during the mastering process. This is a lot harder than it sounds, as sound is a very subjective thing. It's often easy to fool yourself into thinking that your room, setup and sound are absolutely top-notch … and then when you go to play it on another professional system (or any home system) … your hopes can be dashed in an instant. One of the best ways to guard against this is to listen to your mix on as many systems as possible and ask others to critique your work. How does it sound to them on their system?
- As with mixing, mastering is a process of balancing all of the variables that go into making a project sound as good as it can. One of the best ways to hear if you're on the right track or not, is to create a "my favorite songs" directory or CD that includes mixes by artists and producers that you love and admire. By listening to your project alongside of your favorite "best of" recordings, you can often get quick insights into how your work sounds compared to the work of others.

If you're willing to put in the time and effort, it's my belief that mastering your own project is definitely an option.

"Pre"paration

Just as one of the best ways to make the mixdown process go more smoothly is to fix most of your technical problems and issues BEFORE you start the mixdown process, one of the best ways to ensure that a mastering session has as few problems as possible is to ask the right questions and deal with the technical issues BEFORE the mastering engineer even receives your soundfiles. By far, the best way to avoid problems during this phase is to ask questions ahead of time. The mastering engineer should be willing to sit down with you or your team to discuss your needs and preplanning issues (or at least direct you to a document checklist that can help you through the preparation process). During this introductory "getting to know you and your technical requirements" session, here are just a few questions that you might ask:

- What should the final master sample rate and bit rate be?
- What should the master's maximum level be?
- Should all master compression be turned off? Would you like for us to supply you a copy with the bus compression turned on, so you can hear our intended ideas?
- Would you like separate instrument/vocal stem tracks, so they can be treated separately?
- Are there any special requirements that we should be aware of?

Important Notes to Remember

- It's ALWAYS wise to consult with the project's mastering engineer about the general specifications of the final product BEFORE beginning the mixdown (or possibly even the recording) process. For example, that person might prefer that the files be recorded and/or mixed at a certain bit rate and bit depth, as well as in a specific format.
- A mastering engineer might prefer that the final master be changed or processed as little as possible with regard to normalization, fade changes, overall dynamic changes (compression) and applied dither. These processing functions are best done in the final mastering process by a qualified engineer.

Mastering the details of a project

From the mastering standpoint (as well as those of the artists and producer who are overseeing the project), a wide range of artistic decisions need to be carefully finessed in order to master a recording into its final, approved form. Just a few of the decision-making steps include:

- Choosing the proper song order
- The use of level changes to balance the relative track levels and improve the overall "feel" of the project
- The application of EQ to improve the sound and overall "tone" of the project
- The judicial use of dynamics to balance out the sound and increase the project's overall level.

Sequencing: the natural order of things

Whether the master is to be assembled using analog tape, or on a DAW/editor, the running order in which the songs of a project will be played often affects the overall flow and tone of a project. The considerations for which song follows which is infinitely varied, and can only be garnered from experience and having an artistic "feel" for how their order and interactions will affect the listening experience. Of course, sequence decisions are probably best made by the artist and/or producer, as they have the best feel for the project. A number of variables that'll directly affect the sequenced order of a project include:

- *Total length:* How many songs will be included on the disc or album? If you've recorded extra songs, could we include the Bonus Tracks on the disc? Is it worth adding a weaker song, just to fill up the CD?
- *Running order:* Which song should start? Which should close? What order feels best and supports the overall mood and intention of the project?
- *Transitions:* Altering the transition times between songs can actually make the difference between an awkward silence that jostles the mood and a transition that keeps up with the pace and feel of the project. The Red Book CD standard calls for 2 seconds of silence as a default setting between tracks. Although this is necessary before the beginning of the first track, it isn't at all the law for spacings that fall between the first and second or later songs. Most editors will allow you to alter the index space timings between tracks from 00 seconds (butt splice) to longer gaps that help maintain the appropriate transitional mood.
- *Cross-fades:* In certain situations, the transition from one song to the next is best served by cross-fading from one track directly to the next. Such a fade could seamlessly intertwine the two pieces, providing the glue that can help convey any number of emotional ties.

Digital sequence editing

With the advent of the DAW, the relatively cumbersome process of sequencing music tracks in the analog domain using magnetic tape has given way to the faster, easier and more flexible process of editing the final masters from hard disk. In this process, all of the songs can be loaded into a workstation, and then adjusted with respect to volume, equalization, dynamics, fades and timing (song start and endpoints) for assembly into a final, edited form.

Although the length of time between the end of one song and the beginning of the next can be constant, this isn't always the wisest choice, as the timing between songs can have a profound effect upon the "feel" of a transition from one song to the next. Decreasing the time between them can make a song seem to blend into the next (if they're similar in mood) or could create a sharp contrast with the preceding song (if the moods are dissimilar). Longer times between songs help the listeners get out of the mood of the previous song and prepare them for hearing something that might be quite different.

Another way to vary the gaps between songs is to do the process manually. This can be done in the DAW by placing the gap times directly within the session itself. In this way, the song can be exported (including blank silence at the end), so that project songs can be entered into any regular CD burning program and burned to a final CD disc (remembering to enter in a "0 sec" timing into the global settings).

It's always a good idea to make at least one master copy of the final mix soundfiles, session data and master files as a backup (just in case the record company, producer or artist wants to make changes at a later date). This simple precaution could save you lots of time and frustration.

Analog sequence editing

Although the process of assembling a final master in the analog domain occurs far less frequently than in the digital realm, it's still done. During this process, the engineer edits the original mixes out from their respective reels and begins the process of splicing them together into a final sequence on a master reel set. At this time, the level test tones (which were laid down at the beginning of the mixdown session) should be placed at the beginning of side one. Once this is done, the mix master in/out edits should be tightened (to eliminate any noise and silence gaps) by listening to the intro and outro at high volume levels, while the heads are in contact with the tape (this might require that you place the transport into the edit mode). The tape can then be moved back and forth (a process known as "jogging" or "rocking" the tape) to the exact point where the music begins (intro) and after it ends (outro). Once the in (or out) point is positioned over the playback head, the exact position is marked with a grease pencil. If there's no noise directly in front of this spot, it's a good practice to cut the tape half an inch before the grease pencil mark as a safety precaution against editing out part of the first sound. If there is noise ahead of the first sound, the tape should be cut at the mark and a leader inserted at that point. Paper (rather than plastic) leader tape is used because plastic will often cause static electricity pops.

The tail of the song might need to be monitored at even higher volume levels because it's usually a fade-out or an overhang from the last note and is, therefore, much softer than the beginning of the song. The tape is marked and cut just after the last sound dies out to eliminate any low-level noises and tape hiss. Of course, the length of time between the end of the song and the beginning of the next can be constant in a sequenced project, or the timings can vary according to the musical relationship between the songs.

Relative volumes

In addition to addressing the overall volume levels of a project, one of the tasks of the mastering process is to smooth out the relative volume differences between songs over the course of the disc or album. These differences could occur from a number of sources, including general variations in mixdown and program content levels, as well as levels between projects that have been mixed at different studios.

Cues as to smoothing out the relative rms and peak differences can be obtained by:

- Using your ears to fine-tune the volume levels from song to song
- Looking at the general attributes of a soundfile from within a digital audio editor

- Carefully watching the master output meters on a recorder or editor
- Watching the graphic levels of the songs as they line up in a digital audio editor

Contrary to popular belief, the use of a standard DAW normalization tool can't smooth out these level differences, because this process only detects the peak level within a soundfile and raises the overall level to a determined value. Since the average (rms) and peak levels will often vary widely between the songs of a project, this toll isn't always useful, although certain editors provide normalization tools that have more variables that are more useful and in depth.

EQ and mastering control

As is the case in the mixdown process, *equalization* is often an extremely important tool for boosting, cutting or tightening up the low end, adding presence to the midrange and tailoring the high end of a song or overall project. Again, EQ can be used as a tool to smooth out differences between cuts or for making changes that affect the overall character of the entire project. Of course, wide ranges of hardware and software plug-in EQ systems are available for applying the final touches both within a studio and project-based setting (Figures 19.3 and 19.4).

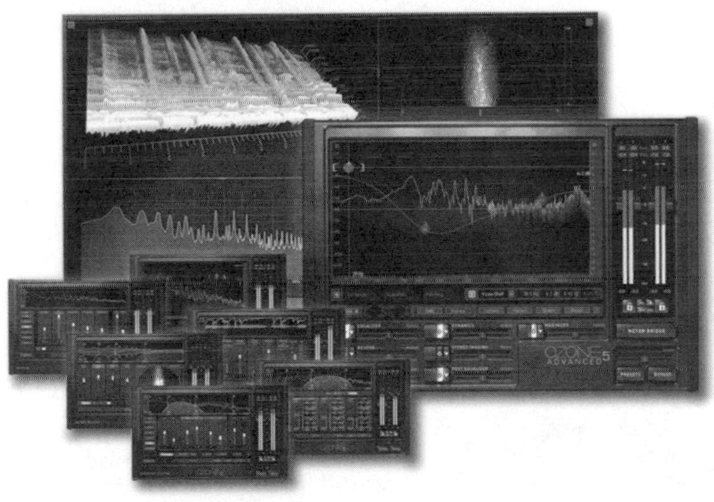

FIGURE 19.3
iZotope's Ozone 5 Mastering Plug-In System with maximizer, equalizer, dynamics, imaging, exciter, reverb, and dithering. (Courtesy of iZotope Inc., www.izotope.com)

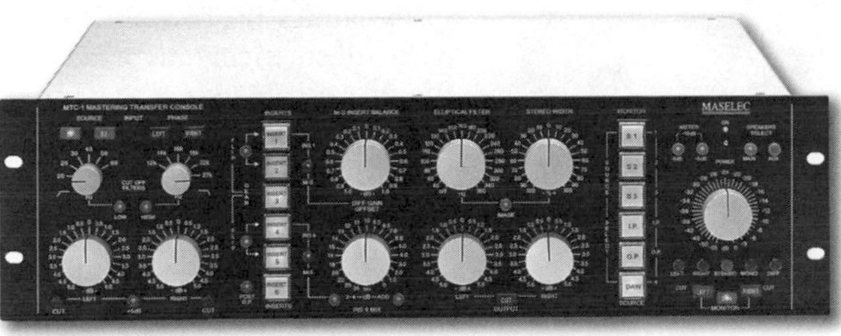

FIGURE 19.4
Maselec MTC-1 Mastering Transfer Console. (Courtesy of Mases Electronics Ltd., www.maselec.com)

FIGURE 19.5 Manley Stereo Variable Mu® mastering outboard compressor. (Courtesy of Manley Laboratories, Inc., www.manleylabs.com)

FIGURE 19.6 The Shadow Hills Mastering Compressor plug-in for the Apollo and the UAD effects processing card. (Courtesy of Universal Audio, www.uaudio.com © 2013 Universal Audio, Inc. All rights reserved. Used with permission).

FIGURE 19.7 EMI TG12412 mastering outboard EQ plug-in. (Courtesy of Digidesign, a division of Avid, www.digidesign.com)

Dynamics

One of the most commonly used (and overused) tools within the mastering process relates to *dynamics* processing, or specifically, compression (Figures 19.5 through 19.7).

Although the general name-of-the-game is to achieve the highest overall average level within a song or complete project, care must be taken so as not to apply so much compression that the life gets dynamically sucked out of the sound (Figure 19.8). As with the first rule in recording – "There are no rules" – the amount of dynamics processing is entirely up to those who are creatively involved in the final mastering process. However, it's important to keep in mind the following guidelines:

- Depending on the program content and genre, the general dynamic trend is toward raising the overall levels to as high a point as possible.

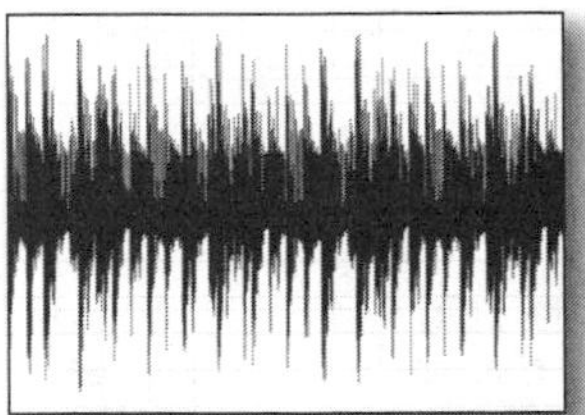

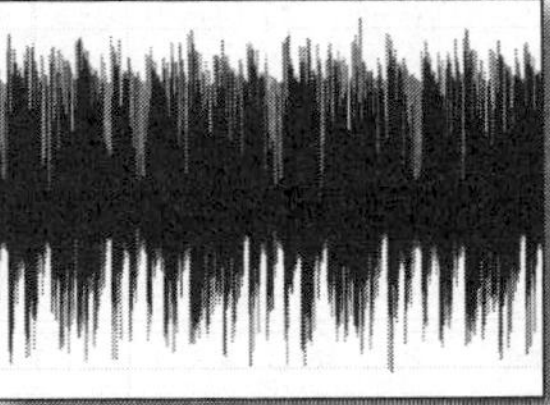

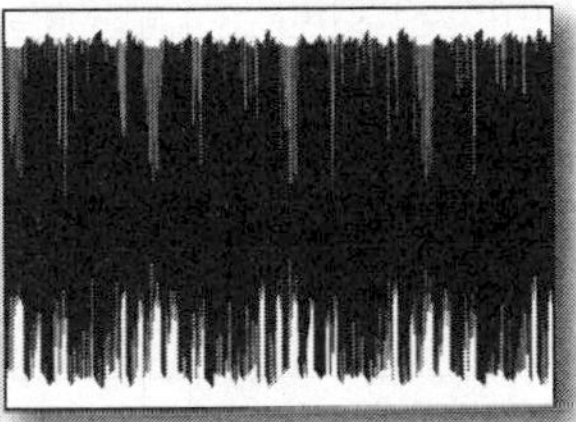

FIGURE 19.8
Figure showing the same passage with varying degrees of compression.

- When pushed to an extreme, compression will often have an intended or unintended side effect of creating a sound that has been "squashed," giving a "wall of sound" character that's thick (a good thing) and/or one that's sonically lifeless (a bad thing).
- When compression is not applied (or little is used), the sound levels will often be lower, thinner (that might be bad) or full of dynamic life (that's good).

From all of this, you'd be correct if you said that the process is entirely subjective and relative! The use of compression can help add a strong presence to a recording, while overuse can actually kill its dynamic life ... so use it wisely!

Multiband dynamic processing

The modern-day mastering process often makes use of multiband dynamic processing (Figure 19.9) in order to break the frequencies of the audio spectrum into bands that can be individually processed. Depending on the system, for example, up to five distinct bands might be available for processing the final signal. Such a hardware or software system could be used to strongly compress

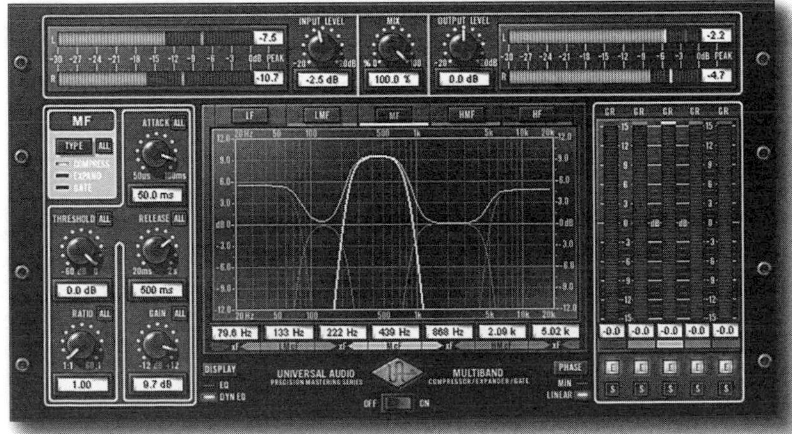

FIGURE 19.9
Precision Multiband Plug-in for the Apollo and the UAD effects processing card. (Courtesy of Universal Audio, www.uaudio.com © 2013 Universal Audio, Inc. All rights reserved. Used with permission).

the low frequencies of a song using a specific set of parameters, while applying only a small amount of compression to the sibilance at its upper end.

Soundfile volume

While dynamics is a hotly debated topic among mastering and recording engineers, most agree that it's never a good idea deliver a soundfile to a mastering house that has been compressed and raised in level. This, of course, effectively limits the dynamics of the song and takes control over the final dynamics and sound of the song away from the mastering engineer. For this reason, most mastering engineers will ask that the project be delivered without any compression or dynamics of any kind … and that overall session gain be reduced so that it peaks at a lower level (-12dB for example). Again, it's always a good idea to check the facts beforehand.

Soundfile resolution

The sample rate and bit rate resolution that a project is recorded at is a personal matter (as with many things in sound recording). Given that high-quality converters and a DAW with a high-quality bit-processing structure are used, many believe that recordings made with 44.1-kHz/16-bit file resolutions are sufficient to capture the full nuances of sound (for both the recorded tracks and the final mixdown). Others believe that rates upward to 96 kHz/24 bits or higher are necessary to capture the extended high frequencies and increased resolution of music.

One thing is for sure, if higher bit rates are chosen, it's almost always best to deliver the final master recording to the mastering lab at the original native rate (i.e., if the session was recorded and mixed at 24/96, the final mixdown resolution should delivered to the mastering engineer at that rate and bit resolution). I should mention, though, that some mastering engineers will prefer that the final track be exported at a higher bit-depth than the native session depth. For example, if the session was recorded at 44.1/16, many engineers will ask that the session be exported (bounced) to a soundfile at 44.1, but with a 24-bit depth. The reasoning behind this is that the mastering engineer will be able to process the mix at the higher, more detailed depth resolution and will have better tools by which to dither the soundfile down to the necessary 16/44.1 rate for final transfer to CD or the intended medium.

Dither

As was stated in Chapter 6, the addition of small amounts of randomly generated noise to an existing bitstream can actually increase the overall bit resolution (and therefore low-level noise and signal clarity) of a recorded signal. Through the careful addition of dither, it's actually possible for signals to be encoded at levels that are less than the data's least significant bit level. You heard that right … by adding a small amount of random noise into the A/D path, the resolution of the conversion process can actually be improved below the least significant bit level and reduce a soundfile's harmonic distortion.

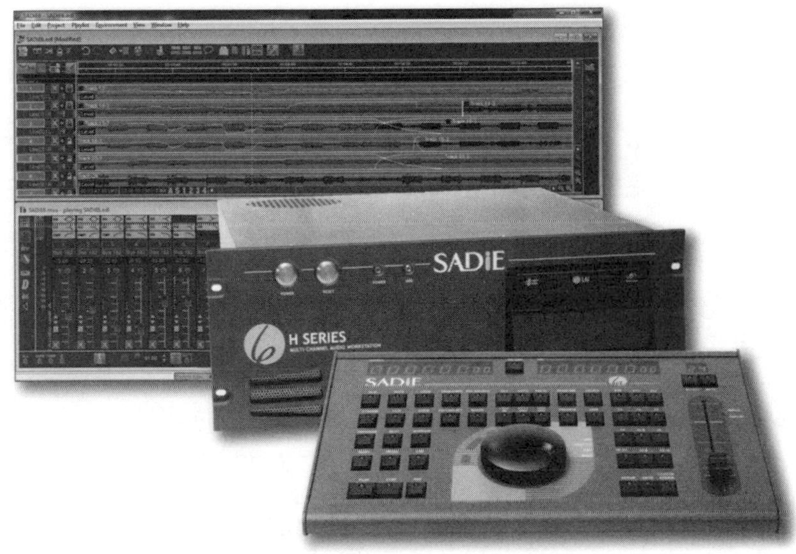

FIGURE 19.10
Sadie PCM4 Editing System. (Courtesy of Prism Media Products Ltd, www.sadie.com)

Within mastering, dither is often manually applied to soundfiles that have been recorded at 20- and 24-bit depths. DAW plug-ins can be used to apply dither to a soundfile or master mix, so as to reduce the effects of lost resolution due to the truncation of least significant bits. For example, mastering engineers might carefully experiment with applying dither to a high-resolution file before saving or exporting it as a 16-bit final master. In this way, noise is reduced and the soundfile's overall clarity is increased.

THE DIGITAL AUDIO EDITOR IN THE MASTERING PROCESS

By far, the most commonly used system for modern-day mastering is the digital audio editor (Figure 19.10). These two-channel and multichannel workstations make use of the personal computer's existing processing, disk storage and data I/O hardware to perform a wide range of audio editing, processing and mastering production tasks. These programs allow each song to be loaded onto its own track and be independently processed according to its own needs, while possibly adding overall effects processing to the master output section (allowing EQ, dynamics, dither, etc., to be applied to the entire project mix). In such a graphic, on-screen environment, individual songs can be imported, processed and exported into a final form that can then be burned to disc—either directly from within the editing program or by using a CD burning software application.

Of course, a DAW doesn't have to be especially designed for the mastering phase. In fact, most mastering is done using many of the workstations that are commonly available on the market (Figure 19.11).

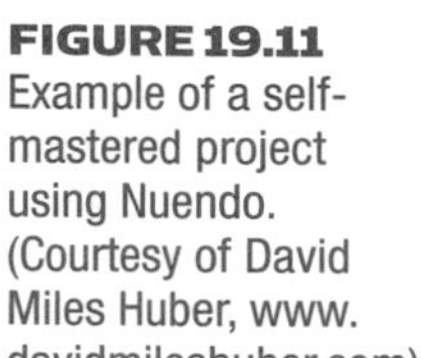

FIGURE 19.11
Example of a self-mastered project using Nuendo. (Courtesy of David Miles Huber, www.davidmileshuber.com)

ON A FINAL NOTE

In closing, it's always a wise idea to budget in some time with the final mastered recording before committing it to the final product. If at all possible, take a week and listen to it in your car, on your boom box, home theater system, in another studio—virtually everywhere! As a musician, producer or record label, it will be your calling card for quite some time to come. Once you're satisfied with the sound of the finished product, then you can move from the mastering phase to making the project into a finished, salable product.

CHAPTER 20

Product Manufacture

One of greatest misconceptions surrounding the music, visual and other media-related industries is the idea that once you walk out the door of a studio with your final master in hand, the creative process of producing a project is finally over. All that you have left to do is upload a file to your favorite online store or hand the CD over to a duplication facility and ... Ta-Dah!—the adoring public will be clamoring for your product, website and merchandise. Obviously, this scenario is almost always far from the truth. Even before you have your blood, sweat and tears physically in hand, it's vitally important to think through and implement your master plan, if your product is to make it into the hands (or iDevice) of the consumer.

Early in this book, I told you about the first rule of recording ... that "there are no rules, only guidelines." This actually isn't true. There is one rule:

If you don't pre-plan, and follow through with these plans once the project is recorded, you can be fairly sure that your project will sit on a shelf. Or, worse, you'll have a 1000 CDs sitting in your basement that'll never be heard—a huge shame given all the hard work that went into making it.

Given the huge number of changes that have occurred within the business of music marketing and distribution, it stands to reason that an equally large number of changes have occurred when it comes to planning and considering how you are going to get your new-born project out to the masses. In short, many of the rules have changed ... and it's the wise person who takes the time to put their best foot (and plan) forward to make it all shine. So, having said this ... what are our options in this day and age? Well there's:

- Online (download) distribution
- Hi-res download distribution
- Music Streaming
- CD
- DVD and Bluray
- Vinyl

Just as each recording facility has its own unique personality and particular "sound," the right mastering and physical duplication and even online download service might also have a profound effect on the outcome of a project. If a project is being underwritten and distributed by an independent or major record label, they will generally be fully aware of their production needs and will certainly have an established production and product manufacturing network in place. If, however, you're distributing the project yourself, the duty of choosing the best facility or manufacturing organization that'll fit your budget and quality needs will fall on your or your band's shoulders.

On the subject of online distribution, besides the important fact that your music has the potential to reach a large fanbase, the best part about downloadable media is that there's no expensive physical media that needs to be manufactured and distributed. Gone are the days when you have to give up a closet or area in your basement for storing CDs and LPs. However, the flipside to this is that the artist will need to research and understand their online distribution options, needs and legal responsibilities before signing up to a download service.

DOWNLOADABLE MEDIA

In this day of surfing and streaming media off the Web, it almost goes without saying that the WWW has become the most important and effective marketing tool for the musician and labels alike. It allows for us to cost-effectively upload, distribute, and promote our songs, projects, promotional materials, touring info and liner notes out to mass audiences. As with other media, mastering for the Internet can either be complicated, requiring professional knowledge and experience, or it can be a straightforward process that can be carried out from a laptop computer. It's a matter of meeting the level of professionalism and development that's required by you and your audience's needs.

Once the recording and mix phases of a project have been completed (assuming that you're also doing your business and promotion homework as to your audience, distribution methods, live and Web marketing presence, production budgeting, etc.), the next step toward getting the product out to the people is to transform the completed song or project into a form that can be mass produced, distributed, marketed and SOLD. Over the web, this will often mean direct sales to the public via popular download sites, such as iTunes, CDBaby, Amazon, etc. Although most sites will have their own system for having the artist upload their music, it's often wise to be aware of the final codec, bitrate and overall look/feel of the final product. The definitions and details for the most popular distribution formats and bitrates can best be found in Chapter 10. Considerations for these final media formats might include the concept of mixing for your intended audience, taking care to take final levels, mastering dynamics and EQ, as well as the overall look and attitude of the final product.

Hi-res downloads

Over the last decade, the concept of quality vs. data compression size has had some time to shake out a few cobwebs. With improvements in data download transfer rates and increased device memory storage, the problems, constraints and need for smaller, more "compressed" data soundfiles have greatly reduced. Given the recent advances in bandwidth, storage and memory storage, the buying public is now asking for a return to uncompressed (or less-compressed), higher-quality downloads, allowing them to enjoy their favorite music in its originally-intended fidelity. For example, such higher-sample-bitrate files might be encoded and distributed as 320kps MP3, AAC, FLAC or even WAV.

Music streaming

Although music downloads are one of the strongest-growing markets in the music industry, another market that's also showing huge growth potential is the online music and media streaming market. Unlike downloads (where the data is physically stored onto the buyer's computer and/or music playback device), media streaming actually stays in the "clouds" ... that's to say, stays on the provider's server and is "streamed" in real-time to the authorized "pay-per-month" buyer or free service listener. With the advent of "smart phones, TVs and media centers, data streams can be enjoyed from any number of media devices ... not just from a computer.

PHYSICAL MEDIA

Although downloadable and streaming media are well on their way to dominating the music and media industries, the ability to buy, own and gift a physical media object still warms the hearts of many a media product purchaser. Often, containing uncompressed, hi-quality music, these physical discs and records are held in regard as something that can be held onto, looked at and packaged up with a big, bright gift bow. Their market-share days may be numbered ... but don't count them out any time soon.

The CD

Beyond the process of distributing audio over the Internet (using an online service or from your own site), as of this writing, the compact disc (CD) is still a strong and viable medium for distributing music. These 120-mm silvery discs (Figure 20.1) contain digitally encoded information (in the form of microscopic pits) that's capable of yielding playing times of up to 74 or 80 minutes at a standard sampling rate of 44.1 kHz.

The pit of a CD is approximately half a micrometer wide, and a standard manufactured disc can hold about 2 billion pits. These pits are encoded onto the disc's surface in a spiraling fashion, similar to that of a record, except that 60 CD spirals can fit into the groove of a single long-playing record. These spirals also differ from a record in that they travel outward from the center of the disc, are impressed into the plastic substrate, and are then covered with a thin coating of aluminum (or occasionally gold) so that the laser light can be reflected back to a

FIGURE 20.1
The compact disc. (Courtesy of www.davidmileshuber.com).

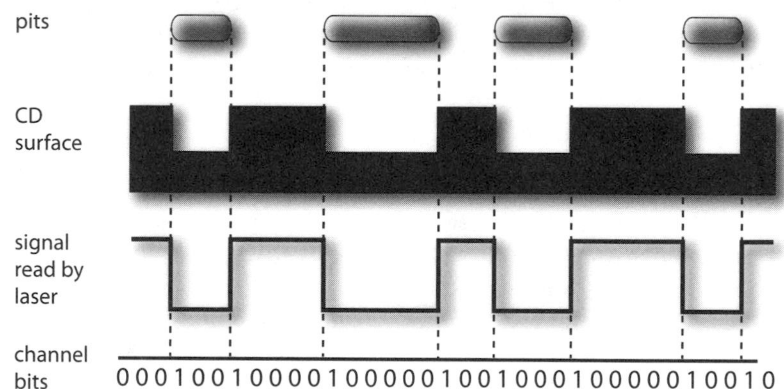

FIGURE 20.2
Transitions between a pit edge (binary 1) and the absence of a pit edge (binary 0).

receiver. When the disc is placed in a CD player, a low-level infrared laser is alternately reflected and not reflected back to a photosensitive pickup. In this way, the reflected data is modulated so that each pit edge represents a binary 1, and the absence of a pit edge represents a binary 0 (Figure 20.2). Upon playback, the data is then demodulated and converted back into an analog form.

Songs or other types of audio material can be grouped on a CD into tracks known as "indexes" This is done via a subcode channel lookup table, which makes it possible for the player to identify and quickly locate tracks with frame accuracy.

Subcodes are event pointers that tell the player how many selections are on the disc and where their beginning address points are located. At present, eight subcode channels are available on the CD format, although only two (the P and Q subcodes) are used.

Functionally, the CD encoding system splits the 16 bits of information into two 8-bit words with error correction (that's applied in order to correct for lost or erroneous signals). In fact, without error correction, the CD playback process would be so fragile and prone to dropouts that it's doubtful it would've become a viable medium. The system then goes about translating this data (using a process known as eight-to-fourteen modulation or EFM) into a special code, known as a data frame. Each data frame contains a frame-synchronization pattern (27 bits) that tells the laser pickup beam where it is on the disc. This is then followed by a 17-bit subcode word, 12 words of audio data (17 bits each), 8 parity words (17 bits each), 12 more words of audio, and a final 8 words of parity data.

The process

In order to translate the raw PCM of a music or audio project into a format that can be understood by a CD player, a compact disc burning system must be used. These come in two flavors:

- Specialized hardware/software that's used by professional mastering facilities
- Disc burning hardware/software systems that allow a personal computer to easily and cost effectively burn CDs.

Both system types allow audio to be entered into the system, after which the tracks can be assembled into the proper order and the appropriate gap times can be entered between tracks (in the form of index timings). Depending on the system, cuts might also be processed using cross-fades, volume, EQ and other parameters. Once assembled, the project can be "finalized" into a media form that can be directly accepted by a CD manufacturing facility. By far, the most common media that's received by CD pressing plants for making the final master disc are user-burned CD-Recordable (CD-R) discs, although some professional systems will still make use of a special Exabyte-type data tape system.

Note that not all CD-R media are manufactured using high-quality standards. … In fact, some are so low in quality, that the project's data integrity could be jeopardized. As a general rule:

- It's always good to use high-quality "master-grade" CD-Rs to burn the final master (you can sometimes see the difference in pit quality with the naked eye).
- It's always best to send two identical copies to the manufacturer (just in case one fails).
- Speaking of failing, you should check that the manufacturer has run a data integrity check on the final master to ensure that there are few to no errors.

Once the manufacturing plant has received the recorded media, the next stage in the process is to cut the original CD master disc. The heart of such a CD cutting system is an optical transport assembly that contains all the optics necessary to write the digital data onto a reusable glass master disc that has been prepared with a photosensitive material.

After the glass master has been exposed using a special recording laser, it's placed in a developing machine that etches away the exposed areas to create a finished master. An alternative process, known as nonphotoresist, etches directly into the photosensitive substrate of the glass master without the need for a development process.

After the glass or CD master disc has been cut, the compact disc manufacturing process can begin (Figure 20.3). Under extreme clean-room conditions, the glass

(a)

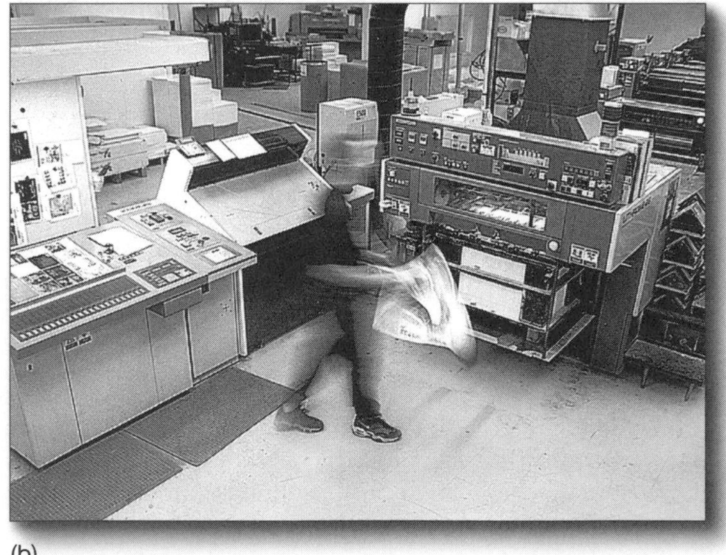

(b)

FIGURE 20.3
Various phases of the CD manufacturing process: (a) the lab, where the CD mastering process begins; (b) once the graphics are approved, the project's packaging can move onto the printing phase; (c) while the packaging is being printed, the approved master can be burned onto a glass master disc; (d) next, the master stamper (or stampers) is placed onto the production line for CD pressing; (e) the freshly stamped discs are cooled and checked for data integrity; (f) labels are then silk-screen printed onto the CDs; and (g) finally, the printed CDs are checked before being inserted into their finished packaging. (Courtesy of Disc Makers, Inc., www.discmakers.com)

(c)

(d)

FIGURE 20.3
Continued

disc is electroplated with a thin layer of electroconductive metal. From this, the negative metal master is used to create a "metal mother," which is used to replicate a number of metal "stampers" (metal plates which contain a negative image of the CD's data surface). The resulting stampers make it possible for machines to replicate clear plastic discs that contain the positive encoded pits, which are then coated with a thin layer of foil (for increased reflectivity) and encased in

(e)

FIGURE 20.3
Continued

(f)

clear resin for stability and protection. Once this is done, all that remains is the screen-printing process and final packaging. The rest is in the hands of the record company, the distributors, marketing and you.

As with any part of the production process, it's always wise to do a full background check on a production facility and even compare prices and services from at least three manufacturing houses. Give the company a call, ask a few

(g)

FIGURE 20.3
Continued

questions and try to get a "feel" for their customer service abilities, their willingness to help with layout questions, etc. You'd be surprised about how much you can learn in a short time. Once you've found a few companies that seem to fit your needs, ask them for a promotional pack (which includes product and art samples, service options and a price sheet). You might also want to ask about former customers and their contact information (so you can email them about their experiences).

Once you've settled on a manufacturer, it's always a good idea to research what their product and graphic arts needs and specs are before delving into this production phase. When in doubt about anything, give them a call and ask; they are there to help you get the best possible product (besides, asking questions helps to get the job done right the first time, with less mess, fuss, time and $$$).

> Whenever possible, it's always wise and extremely important that you be given art proofs and test pressings BEFORE the final products are mass duplicated.

The absolute last thing that you or the artist wants is to have several thousand discs arrive on your doorstep that are … WRONG! Receiving a test pressing and graphic "proof" is almost always well worth the time and money. It's never wise to assume that a manufacturing or duplication process is perfect and doesn't make mistakes. Remember, Murphy's Law can pop up at any time!

CD burning

Software for burning CDs/DVDs/Blu-ray media is available in various forms for both the Mac and PC. These include the simple burning applications that are

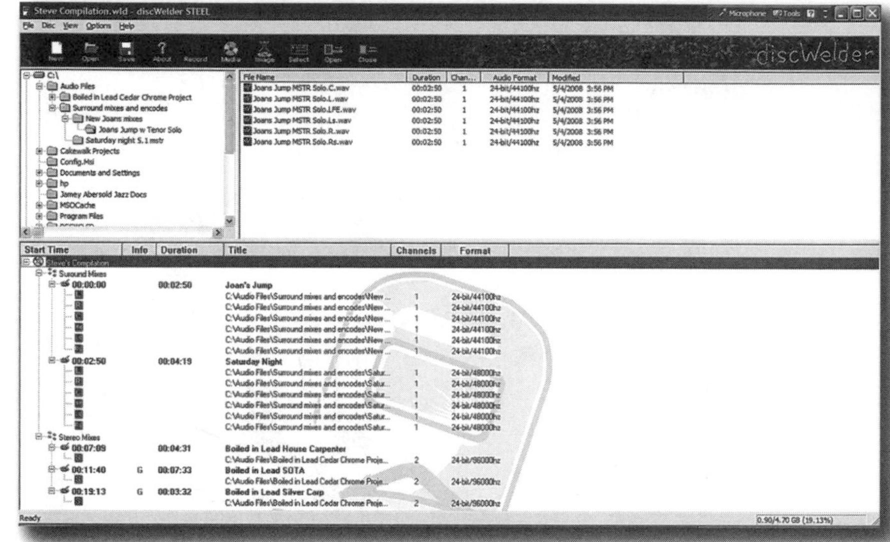

FIGURE 20.4
DiscWelder Steel CD/DVD audio authoring tool for the PC. (Courtesy of Minnetonka Audio Software, Inc., www.minnetonkasoftware.com)

included with the popular operating systems, popular third-party burning applications and more complex authoring programs that are capable of editing, mastering and assembling individual cuts into a final burned master (Figure 20.4).

There are numerous ways in which a CD project can be prepared and burned. For starters, it's a fairly simple matter to prepare and master individual songs within a project and then load them into a program for burning. Using this method, any program should be able to burn the audiofiles in a straightforward manner from beginning to end. Keep in mind that the Red Book CD standard specifies a beginning header silence (pause length) that's 2 seconds long. After this initial lead-in, any pause length can be user specified; the default setting for silence between cuts is 2 seconds, however, any musician/producer will tell you that these lengths will vary, as one song might want to flow directly into another, while the next might want a longer pause to help set the proper mood (it's an artistic "feel" kinda thing).

Most CD burning programs will allow you to enter "CD Text" information (such as title, artist name/copyright and track name field code info) that is then written directly into the CD's subcode area. This can be a helpful tool, because important artist, copyright and track identifiers can be directly embedded within the CD itself and will be automatically displayed on most modern CD players. As a result, illegal copies will still contain the proper copyright and artist info, and discs that are loaded into a computer or media player will often display these fields. While you're at it, you should check with the manufacturer to see if they're able to enter your track and project information into iTunes and Gracenote. These databases (which should not be confused with CD Text) allow CD

FIGURE 20.5
CD Baby is a distribution service that helps sell physical CDs and downloadable music over the Internet. (Courtesy of CD Baby, www.cdbaby.com).

titles, artist info, song titles and graphics to appear on most media-based music players … an important feature for any artist and label.

As always, careful attention to the details should always be made, making sure that:

- High-quality media is used
- The media is burned using a stable, high-quality drive
- The media is carefully labeled using a recommended marking pen
- Two copies are delivered to the manufacturer, just in case there are data problems on one of the discs

Rolling your own

With the rise of Internet music distribution and the steady breakdown of the traditional record company distribution system, bands and individual artists have begun to produce, market and sell their own music on an ever-increasing scale (Figure 20.5). This concept of the "grower" selling directly to the consumer is as old as the town square produce market. By using the global Internet economy, independent distribution, fanzines, live concert sales, etc., savvy independent artists are taking matters into their own hands by learning the inner workings of the music business. In short, artists are taking business matters more seriously in order to reap the fruits of their labor and craft … something that has never been and never will be an easy task.

Beyond the huge tasks of marketing, gigging and general business practices, many musicians are also taking on the task of burning, printing, packaging and distributing their own physical products from the home or business workplace.

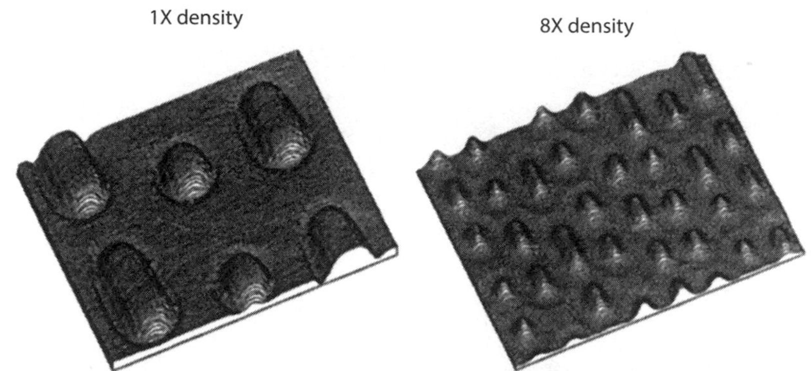

FIGURE 20.6
Detailed relief showing standard CD and DVD pit densities.

This homespun strategy allows for small runs to be made in an "on-demand" basis, without tying up financial resources and storage space with CD inventories.

Creating a system for burning CD-Rs for distribution can range from being a simple home computer setup that creates discs on an individual basis to sophisticated replication systems that can print and burn stacks of CD-Rs or DVD-Rs under robot control at the simple touch of a button.

DVD and Blu-ray burning

Of course, on a basic level, DVD burning technology has matured enough to be available and affordable to the general Mac and PC public. From a technical standpoint, these CD-drive compatible discs differ from the standard CD format in several ways. The most basic of these are:

- An increased data density due to a reduction in pit size (Figure 20.6)
- Double-layer capabilities (due to the laser's ability to focus on two layers of a single side)
- Double-side capabilities (which again doubles the available data size)

In addition to the obvious benefits that can be gained from increasing the data density of a standard CD from 650 Mbytes to a maximum of 17 Gbytes, DVD discs allow for much higher data transfer rates, making DVD the ideal medium for the following applications:

- The simultaneous decoding of digital video and surround-sound audio
- Multichannel surround sound
- Data- and access-intensive video games
- High-density data storage

As of this writing, Blue-ray disc (BD) technology has established itself as the video delivery media du jour. (Of course, this might be old hat by the time you read this.) Because Bluray uses a shorter wavelength red laser (650 nm) and smaller pit densities than a DVD, the capacity has risen to 25 Gbytes on a single disc (or 50 on a dual-layer)!

As DVD and BD-R/RW drives have become commonplace, affordable data backup and mastering software has come onto the market that brought the art of video and Hi-Def production to the masses. Even high-level DVD media mastering is now possible in a desktop environment, although creating a finished product for the mass markets is often an art that's best left to professionals who are familiar with the finer points of these complex technologies. More information on the finer points of codec data compression and media technologies relating these technologies can be found in Chapter 10.

DISC HANDLING AND CARE

Here are a few basic handling tips for optical media (including the recordable versions) from the National Institute of Standards and Technology:

DO:

- Handle the disc by the outer edge or center hole (your fingerprints may be acidic enough to damage the disc).
- Use a felt-tip permanent marker to mark the label side of the disc. The marker should be water or alcohol based. In general, these will be labeled as a non-toxic CD/DVD pen. Stronger solvents may eat though the thin protective layer to the data.
- Keep discs clean. Wipe with a cotton fabric in a straight line from the center of the disc toward the outer edge. If you wipe in a circle, any scratches may follow the disc tracks, rendering them unreadable. Use a disc-cleaning or light detergent to remove stubborn dirt.
- Return discs to their cases immediately after use.
- Store discs upright (book-style) in their cases.
- Open a recordable disc package only when you are ready to record.
- Check the disc surface before recording.

DON'T:

- Touch the surface of a disc.
- Bend the disc (because this may cause the layers to separate).
- Use adhesive labels (because they can unbalance or warp the disc).
- Expose discs to extreme heat or high humidity; for example, don't leave them in direct sunlight or in a car.
- Expose discs to extreme rapid temperature or humidity changes.
- Expose recordable discs to prolonged sunlight or other sources of ultraviolet light.

ESPECIALLY DON'T:

- Scratch the label side of the disc (it's often more sensitive than the transparent side).
- Use a pen, pencil or fine-tipped marker to write on the disc.
- Try to peel off or reposition a label (it could destroy the reflective layer or unbalance the disc).

Disc labeling

Once you've burned your own CD-R/RW or DVD-R/RW, there are a number of options for printing labels onto the newly burned discs (burning the disc first will often reduce data errors that can be introduced by dust, fingerprints or scratches due to handling):

- *Use a felt-tip pen:* This is the easiest and fastest way to label a disc. However, water-based ink pens should be used, because most permanent markers use a solvent that can permeate the disc surface and cause damage to either the reflective or dye layer. When properly done, this is an excellent option for archiving discs.
- *Use a disc printer:* Specially designed ink-jet or laser printers are able to print high-quality, full-color layouts onto the face of a printable (white or silver-faced) disc. This is a cost-effective option for those who burn discs in small batch runs, but still want a professional look and feel.
- Use a disc manufacturing company to professionally print a limited run of CD-Rs that can be burned on-demand. These CDs are often printed using the same high-quality screen printing techniques that are used to print mass-duplicated discs.

VINYL DISC MANUFACTURE

Although the popularity of vinyl has waned in recent years (as a result, of course, of the increased marketing, distribution and public acceptance of the CD and online distribution), the vinyl record is far from dead. In fact, for consumers that range from Dance DJ trip-hopsters to die-hard classical buffs, the record is making somewhat of a popular comeback. However, the truth remains that many record pressing facilities have gone out of business over the years, and there are far fewer mastering labs that are capable of cutting "master lacquers." It may take a bit longer to find a facility that fits your needs, budget and quality standards, but it's definitely not a futile venture.

Disc cutting

The first stage of production is the disc-cutting process. As the master is played from a digital source or on a specially designed tape playback machine, its signal output is fed through a disc-mastering console to a disc-cutting lathe. Here the electrical signals are converted into the mechanical motions of a stylus and are cut into the surface of a lacquer-coated recording disc.

Unlike the compact disc, a record rotates at a constant angular velocity, such as 33 1/3 or 45 revolutions per minute (rpm), and has a continuous spiral that gradually moves from the disc's outer edge to its center. The recorded time relationship can be reconstructed by playing the disc on a turntable that has the same constant angular velocity as the original disc cutter.

The system that's used for recording a stereo disc is the 45/45 system. The recording stylus cuts a groove into the disc surface at a 90° angle, so that each wall

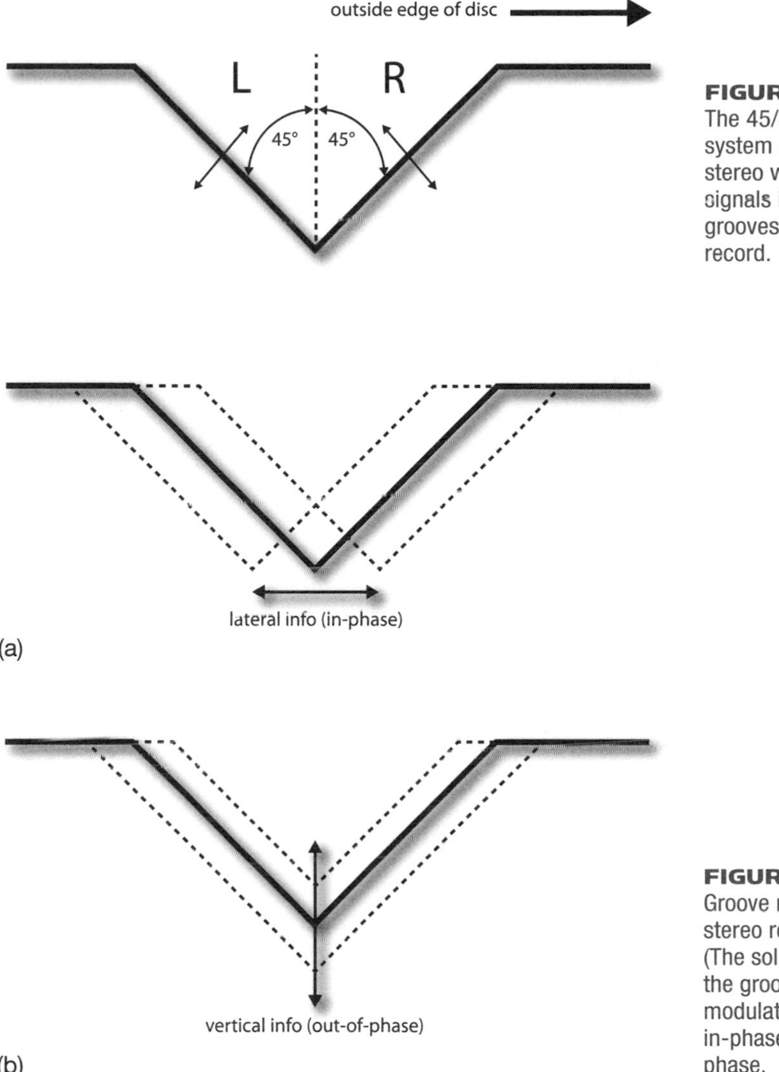

FIGURE 20.7 The 45/45 cutting system encodes stereo waveform signals into the grooves of a vinyl record.

FIGURE 20.8 Groove motion in stereo recording. (The solid line is the groove with no modulation.) (a) in-phase; (b) out-of-phase.

of the groove forms a 45° angle with respect to the vertical axis. Left-channel signals are cut into the inner wall of the groove and right-channel signals are cut into the outer wall, as shown in Figure 20.7. The stylus motion is phased so that L/R channels that are in-phase (a mono signal or a signal that's centered between the two channels) will produce a lateral groove motion (Figure 20.8a), while out-of-phase signals (containing channel difference information) will produce a vertical motion that changes the groove's depth (Figure 20.8b). Because mono information relies only on lateral groove modulation, an older disc that has been recorded in mono can be accurately reproduced with a stereo playback cartridge.

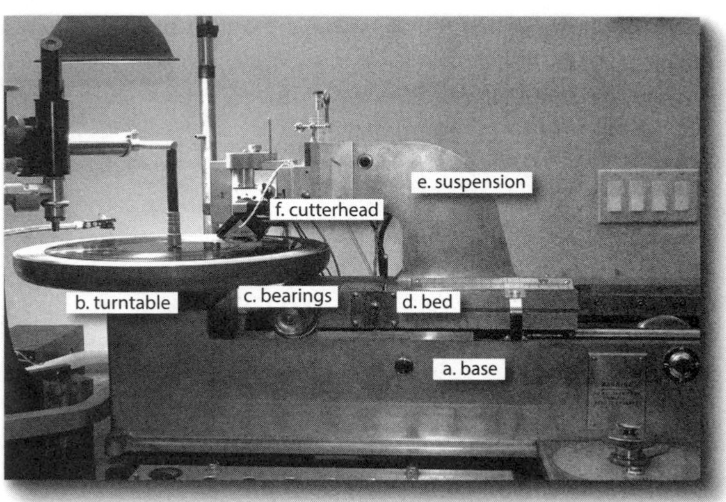

FIGURE 20.9
A disc-cutting lathe with automatic pitch and depth control. (Courtesy of Paul Stubblebine Mastering and Michael Romanowski Mastering, San Francisco, CA, www.paulstubblebine.com and www.michaelromanowski.com).

DISC-CUTTING LATHE

The main components of a vinyl disc-cutting lathe are the turntable, lathe bed and sled, pitch/depth control computer and cutting head. Basically, the lathe (Figure 20.9) consists of a heavy, shock-mounted steel base (a). A weighted turntable (b) is isolated from the base by an oil-filled coupling (c), which reduces wow and flutter to extremely low levels. The lathe bed (d) allows the cutter suspension (e) and the cutter head (f) to be driven by a screw feed that slowly moves the record mechanism along a sled in a motion that's perpendicular to the turntable.

CUTTING HEAD

The cutting head translates the electrical signals that are applied to it into mechanical motion at the recording stylus. The stylus gradually moves in a straight line toward the disc's center hole as the turntable rotates, creating a spiral groove on the record's surface. This spiral motion is achieved by attaching the cutting head to a sled that runs on a spiral gear (known as the lead screw), which drives the sled in a straight track.

The stereo cutting head (Figure 20.10) consists of a stylus that's mechanically connected to two drive coils and two feedback coils (which are mounted in a permanent magnetic field) and a stylus heating coil that's wrapped around the tip of the stylus. When a signal is applied to the drive coils, an alternating current flows through them creating a changing magnetic field that alternately attracts and repels the permanent magnet. Because the permanent magnet is fixed, the coils move in proportion to a field strength that causes the stylus to move in a plane that's 45° to the left or right of vertical (depending on which coil is being driven).

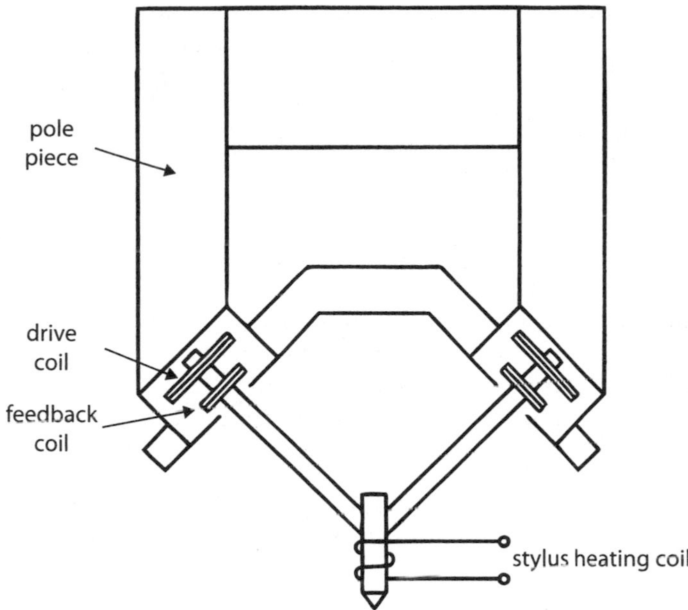

FIGURE 20.10
Simplified drawing of a stereo cutting head.

PITCH CONTROL

The head speed determines the "pitch" of the recording and is measured by the number of grooves, or lines per inch (lpi), that are cut into the disc. As the head speed increases, the number of lpi will decrease, resulting in a corresponding decrease in playing time. Groove pitch can be changed by:

- Replacing the lead screw with one that has a finer or coarser spiral
- Changing the gears that turn the lead screw, so as to alter the lead screw's rotation speed
- Vary the lead screw's rotation by changing the motor's speed (a common way to vary the program's pitch in real-time)

The space between grooves is called the *land*. Modulated grooves produce a lateral motion that's proportional to the in-phase signals between the stereo channels. If the cutting pitch is too high (causing too many lines per inch, which closely spaces the grooves) and high-level signals are cut, it's possible for the groove to break through the wall into an adjacent groove (causing overcut) or for the grooves to overlap (twinning). The former is likely to cause the record to skip when played, while the latter causes either distortion or a signal echo from the adjacent groove (due to wall deformations). Groove echo can occur even if the walls don't touch and is directly related to groove width, pitch and level.

These cutting problems can be eliminated either by reducing the cutting level or by reducing the lines per inch. A conflict can arise here as a louder record will have a reduced playing time, but will also sound brighter, punchier, and more present (due to the Fletcher–Munson curve effect). Because record companies and producers are always concerned about the competitive levels of their discs relative to those that are cut by others, they're reluctant to reduce the overall cutting level.

The solution to these level problems is to continuously vary the pitch so as to cut more lines per inch during soft passages and fewer lines per inch during loud passages. This is done by splitting the program material into two paths: undelayed and delayed. The undelayed signal is routed to the lathe's pitch/depth control computer (which determines the pitch needed for each program portion and varies the lathe's screw motor speed). The delayed signal (which is usually achieved by using a high-quality digital delay line) is fed to the cutter head, thereby giving the pitch/depth control computer enough time to change the lpi to the appropriate pitch.

Pitch is divided into two categories: coarse (which refers to spacing between 96 and 150 lpi) and microgroove (which is between 200 and 300 lpi, or more). Microgroove records have less surface noise, wider frequency range, less distortion and greater dynamic range than do coarse-pitched recordings. They can also be tracked with lower stylus pressure, resulting in a longer life; however, the stylus is more likely to skate across the record if the turntable isn't level. The playback stylus for a stereo microgroove record must have a tip radius of 0.7 mil or less (compared to 2.5 mils ± 0.1 for coarse-groove records). Older 78-rpm and early 33 1/3-rpm records were recorded with a coarse pitch; however, virtually all current records are microgroove (having an average pitch of 265 lpi). At maximum pitch, the playing time of one side of a 12-inch disc, with no modulation in the grooves, is about 23 to 26 minutes, while the duration of a variable-pitch 12-inch disc cut at average levels is about 45 minutes per side.

Recording discs

The recording medium used on the lathe is a fiat aluminum disc that's coated with a film of lacquer, which is dried under controlled temperatures, coated with a second film, and then dried again. The quality of these discs (called lacquers) is determined by the flatness and smoothness of the aluminum base. Any irregularities in this surface (such as holes or bumps) will cause similar defects in the lacquer coating. Lacquers are always larger in diameter than the final record, which makes it easy to handle them without damaging the grooves. For example, a 12-inch album is cut on a 15-inch lacquer and a 7-inch single is cut on a 10- or 12-inch lacquer. As always, it's wise to cut a reference test lacquer in order to hear how the recording will sound after being transferred to disc.

The mastering process

Once the mastering engineer sets a basic pitch on the lathe, a lacquer is placed on the turntable and compressed air is used to blow any accumulated dust off the lacquer surface. A chip suction vacuum is started and a test cut is made on

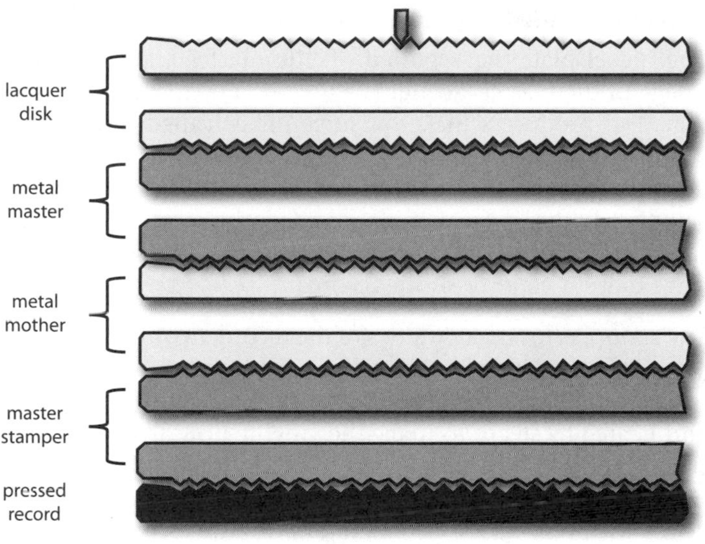

FIGURE 20.11
The various stages in the plating and pressing process.

the outside of the disc to check for groove depth and stylus heat. Once the start button is pressed, the lathe moves into the starting diameter, lowers the cutting head onto the disc, starts the spiral and lead-in cuts, and begins playing the master production tape. As the side is cut, the engineer can fine-tune any changes to the previously determined console settings. Whenever an analog tape machine is used, a photocell mounted on the deck senses white leader tape between the selections on the master tape and signals the lathe to automatically expand the grooves to produce track bands. After the last selection on the side, the lathe cuts the lead-out groove and lifts the cutter head off the lacquer.

This master lacquer is never played, because the pressure of the playback stylus would damage the recorded soundtrack (in the form of high-frequency losses and increased noise). Reference lacquers (also called reference acetates or simply acetates) are cut to hear how the master lacquer will sound.

After the reference is approved, the record company assigns each side of the disc a master (or matrix) number that the cutting room engineer scribes between the grooves of the lacquer's ending spiral. This number identifies the lacquer in order to eliminate any need to play the record, and often carries the mastering engineer's personal identity mark. If a disc is remastered for any reason, some record companies retain the same master numbers; others add a suffix to the new master to differentiate it from the previous "cut."

When the final master arrives at the plating plant, it is washed to remove any dust particles and then electroplated with nickel. Once the electroplating is complete, the nickel plate is pulled away from the lacquer. If something goes wrong at this point, the master will be damaged, and the master lacquer must be recut.

Vinyl disc plating and pressing

The nickel plate that's pulled off the master (called the matrix) is a negative image of the master lacquer (Figure 20.11). This negative image is then electroplated to produce a nickel positive image called a mother. Because the nickel is stronger than the lacquer disc, several mothers can be made from a single matrix. Since the mother is a positive image, it can be played as a test for noise, skips, and other defects. If it's accepted, the mother can be electroplated several times, producing stampers that are negative images of the disc (a final plating stage that's used to press the record).

The stampers for the two sides of the record are mounted on the top and bottom plates of a hydraulic press. A lump of vinylite compound (called a biscuit) is placed in the press between the labels for the two sides. The press is then closed and heated by steam to make the vinylite flow around the raised grooves of the stampers. The resulting pressed record is too soft to handle when hot, so cold water is circulated through the press to cool it before the pressure is released. When the press opens, the operator pulls the record off the mold and the excess (called flash) is trimmed off after the disc is removed from the press. Once done, the disc's edge is buffed smooth and the product is ready for packaging, distribution and sales.

CHAPTER 21

Studio Tips and Tricks

As we near the end of this book, I'd like to take some time to offer some tips that can help make a session go more smoothly, both in the project and in the professional studio environment. Of these, one of the most important insights to be gained (beyond an understanding of the technology and tools of the trade) is the fact that there are no rules for the process of recording. This rule holds true insofar as inventiveness and freshness tend to play a major role in keeping the creative process of making music alive and exciting. However, there are guidelines, equipment setup and procedures that can help you have a smoother-flowing, professional-sounding recording session, or at the very least, help you solve potential problems when used in conjunction with four of the best tools for guiding you toward a successful project:

- Preparation
- Attention to detail
- Creative insight
- Common sense

PREPARATION AND THE DETAILS

You've all heard the old adage "The devil's in the details"? It's important to remember that the hallmark of both a good production and a good production facility rests with the nitty-gritty small stuff that will help you to rise above the crowd. Basically, the glory goes not so much to those who simply do the job, but to those who take the time to get the details right. OK, let's take some time to look at some of the details that can help your projects shine. Probably the most important step toward ensuring that a recording project will become successful and marketable is *careful preparation and planning* (Figure 21.1).

By far, the biggest mistake that a musician or group can make is to go into the studio, spend a lot of money and time, press a few thousand CDs, make a template website … and then sit back and expect an adoring audience to spring out of thin air! It ain't gonna happen! Beyond a good dose of business reality and added experience, the artist(s) will have the dubious distinction of joining the throngs

FIGURE 21.1
Workin' out the kinks beforehand in the practice space. (Courtesy of Yamaha Corporation of America. www.yamaha.com)

that have thousands of unheard CDs sittings in their closet or basement. Getting your "product" out there takes hard work, persistence, marketing and luck.

What's a producer and when do you need one?

One of the first steps that can help ensure the success of a project is to seek the advice and expertise of those who have extensive experience in their chosen fields. This might include seeking legal counsel (for help and advice with legal matters, business contacts or both). Another important "advisor" can come in the form of that all-important title, *producer*. Basically, the producer of a project can fill one of two roles:

- The first type can be likened to a film director, in that his or her role is to be an artistic, psychological and technical guide that can help the band or artists reach their intended goals of obtaining the best possible song, album, remix, film score, etc. It's this type of producer's job to stand back and objectively look at the big picture, to offer up suggestions as to how to shape and guide the performance and to direct the artist or group in directions that would result in the best possible final product.
- The second type also encompasses the directorial role, but also has the added responsibilities of being an executive producer. He or she will be charged with many of the business responsibilities of overall session budgeting, making arrangements for all studio and session activities, contracting (should additional musicians and arrangers be needed on the project), etc. This type of producer may even join with a music lawyer to help find the best sales avenue for the artist and negotiate contact relations with potential record companies or distributors. If you find such a person that fits with your artistic vision … count yourself as being very fortunate!

From this, you can see that a producer's role can be either limited or broad in scope. His or her roles should be carefully discussed and agreed on long before any record button is pressed. The importance of finding a producer that can work best with your particular personalities, musical style and business/marketing needs can't be stressed enough, and finding the right match can be a rewarding experience. Here are a few tips to prepare you for the hunt:

- Check out the liner notes of groups or musicians that you love and admire. You never know—their producer just might be interested in taking you on!
- Find a local up-and-coming producer that might be right for your music. This could help fast-track your reputation.
- Talk with other groups or musicians. They might be able to recommend someone.

Here are just a few of the questions to ask when searching for a producer:

- Does he or she openly discuss ideas and alternate paths that contribute to growth and better artistic expression?
- Is he or she a team player, or are the rules laid out in a dictator-like fashion?
- Does the producer know the difference between a creative endeavor and one that wastes time in the studio?
- Does he or she say "Why not?" a lot more often than "Why?"

Although many engineers have spent most of their lives with their ears wide open and have gained a great deal of musical, production and in-studio experience, it's generally not a good idea to always assume that the engineer can fill the role of a producer. For starters, he or she will probably be unfamiliar with the group and their sound, or may not even appreciate or like them! For these and other reasons, it's always best to seek out a producer that is familiar with you, your goals and your style (or is contacted early enough in the game that he or she has time to become familiar).

Do you need a music lawyer?

It's important to realize that music in the modern world is a business. Once you get to the phase of getting your band or your client's band out to the buying public, you'll quickly realize just how true this is. Building and maintaining an audience with an appetite for your product can easily be a full-time business—one where you'll encounter well-intentioned people as well as those who would think nothing of taking advantage of you or your client.

Whether you're selling your products on the street, at gigs, on MySpace or in the stores through a traditional music label, it's often wise to retain the counsel of a trusted music lawyer. The music industry is fraught with its own special legal and financial language, and having someone on your side that has insight into the language, quirks and inner workings of this unique business can be an extremely valuable asset. Before we move on, it should be pointed out that many metropolitan areas have "Lawyers for the Arts" organizations that regularly offer seminars and events, as well as one-on-one consultations with artists who are on a tight budget and have need of legal counsel.

1 LONG BEFORE GOING INTO THE STUDIO

As was emphasized earlier in the chapter, one of the most important steps to take when approaching a project that involves a number of creative and business stages, decisions and risks is *preparation*. Without a doubt, the best way to avoid pitfalls and to help get you, your client or your band's project off the ground is to discuss and outline the many factors and decisions that will affect the creation and outcome of that all-important "final product." Just for starters, a number of basic questions need to be asked long before anyone presses the "REC" button:

- How are you going to recoup the production costs?
- How is it to be distributed to the public? Self-distribution? Digital download? Indy? Record company?
- Will other musicians be involved and what will be the cost or financial arrangements?
- Do you need a producer or will you self-produce?
- How much practice will you need? Where and when?
- Should you record it in the drummer's project studio or at a commercial studio?
- If you use the project studio and it works out, should you mix it at the commercial studio?
- Who's going to keep track of the time and budget? Is that the producer's job … or will he or she be strictly in charge of creative and contact decisions?
- Are you going to need a music lawyer to help with contacts and contracts?
- When should the artist or group's artistic and financial goals be discussed and put down on paper? (Of course, it's always best to discuss budget requirements and possible rewards as early as possible in the game!)

These are but a few of questions you should ask before tackling a project. Of course, they'll change from project to project and will depend on the final project's scope and purpose. However, in the final analysis, asking the right questions (or finding someone who can help you ask the right questions) can help keep your career on-track.

Now that you've asked the questions, here's a to-do list of tasks that are often wise to tackle well before going into the studio:

- Create a "mission statement" for yourself/your group and the project. This can help clue your audience into what you are trying to communicate through your art and music, and can greatly benefit your marketing goals. For example, you might want to answer such questions as these: Who am I/who are we? What are my/our musical goals? How should the finished project sound? What emotions should it evoke? What is the budget for this project? How will it be sold? What are the marketing strategies?

- Practice, practice and more practice ... and while you're at it, you might want to record your practices to get used to the process (some of these tracks could be used on your website as bonus tracks and for "making of" music videos).
- Start working on the project's artwork, packaging and website ASAP. Do you want to tackle this yourself, hire a professional or a qualified friend who wants to help or could use some extra $$$?
- Copyright your songs. Government forms are readily available for the copyrighting (identifying and protecting the intellectual property) of your music. The Library of Congress Form PA is used for the registration of "published or unpublished works of the performing arts. This class includes works prepared for the purpose of being "performed" directly before an audience or indirectly "by means of any device or process." Works of the performing arts include (1) musical works, including any accompanying words; (2) dramatic works, including any accompanying music; (3) pantomimes and choreographic works; and (4) motion pictures and other audiovisual works." ... In short, it is used to copyright a work that is intended for public performance or display. Form SR is used for the "registration of published or unpublished sound recordings. It should be used when the copyright claim is limited to the sound recording itself, and it may also be used where the same copyright claimant is seeking simultaneous registration of the underlying musical, dramatic, or literary work embodied in the phonorecord." In other words, it is used to copyright a recording, while also protecting the underlying performance that is recorded onto the media. These and other forms can be found at www.copyright.gov/forms or by searching the Library of Congress at www.loc.gov. Again, this might be a good time to discuss matters with a music lawyer.
- Should you wish to use the services of a professional studio during the recording and/or mixdown phase, it's ALWAYS wise to take the time to check out several studios and available engineers. Take the time to listen to tracks that have come out of the studio, as well as those that have been recorded by the engineer. Finding out which one best fits your style, budget and level of professionalism is an important and potentially expensive decision ... the time taken to find the best match could be the difference between a happy or potentially disastrous experience.

② BEFORE GOING INTO THE STUDIO

Before beginning a recording session (possibly a week or more before), it's always good to mentally prepare yourself for what lies ahead by creating a basic checklist that can help answer what type of equipment will be needed, the number and type of musicians/instruments, their particular miking technique (if any) and where they'll be placed. The best way to do this is for you, your group and the producer (if there is one) to sit down with the engineer and discuss instrumentation, studio layout, musical styles and production techniques. This meeting lets everyone know what to expect during the session and lets everyone become familiar with the engineer, studio and staff. This is time well spent, because it will invariably come in handy during the studio setup and will help get the session off to a good start. The following tips can also be immensely valuable:

- If there's to be no producer on the project, it's often wise to pick (or at least consider picking) a spokesman for the group who has the best production "chops." He or she can then work closely with the engineer to create the best possible recording and/or mix.
- Record your songs during live gigs or rehearsals. It doesn't matter if you record them professionally or not; however, keeping the "always press the record button" adage in mind, if the setup meets basic professional standards, it's always possible to import all or part of a "magical" take into the final project.
- You might want to audition the session's song list before a live audience.
- Work out all of the musical and vocal parts before going into the studio. Unrehearsed music can leave the music standing on shaky ground; however, leave yourself open to exploring new avenues and surprises that can be the lifeblood of a magical session.
- Try to leave plenty of time for laying down the final vocal tracks. Many a project has been compromised by spending too much time on "tracking" the basic instruments and then running short on time to lay down the vocals. This almost always leads to increased tensions and a compromised vocal track.
- Rehearse more songs than you plan to record. You never know which songs will end up sounding best or will have the strongest impact.
- Again, meet with the engineer beforehand. Take time for the producer and/or group to get to know him or her, so you'll both know what to expect on the day of the session.
- Prepare and edit any sequenced, sampled or prerecorded material beforehand. In short, be as prepared as possible.
- Try working to a metronome (click track) if timing is an issue (that, or find a better drummer).
- Make sure that the instruments are in good working condition, bring new strings, etc.
- Create a checklist of all of the small but important details that can make or break a session, for example, extension cords, tuners, instrument cords, drum oil, drum tuning lugs, your favorite good luck charm, comfortable jammies … you name it!
- While we're on the subject of small, but important details, make sure to bring cables that are in good working order … and don't forget extra ones.
- Take care of your body. Try to relax and get enough sleep before and during the session. Eat the foods that are best for you (you might bring some health foods, fruits and plenty of liquids to keep your energy up). Be aware of your energy levels, so that low- or high-blood sugar problems don't become a factor (grrrrrrr ….).
- Don't fatigue your ears before a session; keep them rested and clear.

In short, it's always a good idea to plan out the session's technical and musical arrangement so as to most efficiently budget your time.

Again, it's always wise to confer with the producer and/or engineer about the way in which the musicians are to be tracked. Will the tracks be recorded in a traditional multitrack fashion (with the basic rhythm tracks being laid down first, followed by overdub and sweetening tracks and finishing with vocals) … or is there a layout that would work best for your particular group's taste and organizational way of doing things? Communicating these details to all those involved in the session will help smooth the studio setup process and keep a basic game plan that can help keep the session reasonably on track.

③ SETTING UP

Once the musicians have shown up at the studio, it's extremely important that all of the technical, musical, and emotional preparation be put into practice in order to get the session off to a good start. Here are a few tips that can help:

- Show up at the studio on time or reasonably early. At some studios, the billing clock starts on time (whether you're there or not). Ask about their setup policies: Is there another session before yours? Is there adequate setup time to get prepared? Are there any charges for setup? What is the studio's cancellation policy in case of illness or unforeseen things go wrong?
- Use new strings, chords, drumsticks and heads, and bring spares. It's also a good idea to know the location, phone number and hours of a local music store, just in case.
- Tune up before the session starts, and tune up regularly thereafter.
- Don't use new or unfamiliar equipment (musical, hardware-wise or especially software-wise). Taking the time to troubleshoot or become familiar with new equipment and software can cost you time and money. The frustration could even result in a lost vibe, or worse! If you must use a new toy or tool, it's best to learn it beforehand.
- Take the time to make the studio a comfortable place in which to work. You might want to adjust the lighting to match your mood ring, lay down a favorite rug and/or bean bag, turn on your lava love light or put your favorite stuffed toys on the furniture. Within reason, the place is yours to have fun with!

④ SESSION DOCUMENTATION

There are few things more frustrating than going back to an archived session and finding that no information exists as to what instrument patch, mic type, or outboard effect was used on a session (even one that's DAW-based). The importance of documenting a session in a separate written document or in the notepad apps within a DAW can't be overemphasized. The basic documentation that relates to the who, what, where and when of a recording, mixdown, mastering and duplication session should include such information as:

- Artists, engineers and support staff who were involved with the project
- Session calendar dates and locations
- Session tempo
- Participants in the project and important dates (for future credits)
- Mic choice and placement (for future overdub reference)
- Outboard equipment types and their settings
- Plug-in effects and their settings or general descriptions (you never know if they'll be available at a future time, so a description can help you to duplicate it with another app)

To ease this process, the artist or producer might pull out a camera or camera-phone and start snapping pictures to document the event for prosperity as well as for technical reasons. Also, it's often a wise idea to bring along a high-quality video camera to quietly document the setup or actual session for your fans as extra "behind-the-scenes" online video content such as MySpace or YouTube.

The more information that can be archived with a session (and its backups), the better the chance that you'll be able to duplicate the session in greater detail at some point in the future. Just remember that it's up to us to save and document the music of today for the fans and future playback/mix technologies of tomorrow. Basic session documentation guidelines can be found in the Guidelines & Recommendation section at www.grammy.com/Recording_Academy/Producers_and_Engineers. In addition, check out the Grammy's Give Fans the Credit Initiative for helping artists, producers and labels to better document the who, what, when and where's of those who were involved in the making of recordings and making it easier (or even possible) for them to get their just dues … and even get paid.

5 RECORDING

It's obviously a foregone conclusion that no two recording sessions will ever be exactly alike. In fact, in keeping with the "no rules" rule, they're often radically different from each other. During the recording session, the engineer watches the level indicators and (only if necessary) controls the faders to keep from overloading the media. It's also an engineer's job to act as another set of production ears by listening for both performance and quality factors. If the producer doesn't notice a mistake in the performance, the engineer just might catch it and point it out. The engineer should try to be helpful and remember that the producer and/or the band will have the final say, and that their final judgment of the quality of a performance or recording must be accepted.

From the musician's standpoint, here are a few additional pointers that can help the session go more smoothly:

- It's always best to get the right sound and vibe onto disk or tape during the session. If you need to do another take, do it! If you need to change a mic, change it. Getting the right sound and vibe onto the track will almost always result in less stress and a better final product than trying to "Fix in the mix" at a later time.
- You might want to record "run-throughs", just in case it's magical.
- Know when to quit! If you're pushing too hard or are tired, it'll often show.
- Technology doesn't always make a good track; feeling, emotion and musicality does.
- Beware of adding new parts or tracks onto a piece that doesn't need it. Remember, too much is simply too much! Musicians and techno-geeks alike often don't know when to say "it's done … let's move on."
- Leave plenty of time for the vocal track(s). It's not uncommon for a group to spend most of their time and budget on getting the perfect drum or guitar sound. It takes time and a clear focus to get the vocals right.
- If you mess up on a part, keep going, you might be able to fix the bad part by punching-in. If it's really that bad, the engineer or producer will hopefully stop you.

In his *EQ Magazine* article "The Performance Curve: How Do You Know Which Take Is the One?" my buddy Craig Anderton laid out his experiences of how different musicians will deal with the process of delivering a performance over time. Being in front of a mic isn't always easy, and we all deal with it differently. Here's a basic outline of his findings:

- *Curves up ahead:* With this type of performer, the first couple of takes are pretty good, then start to go downhill before ramping back up again, until they hit their peak before going downhill really fast.
- *The quick starter:* This type starts out strong and doesn't improve over time in later performances. Live performers often fall into this category, because they're conditioned to get it right the first time.
- *The long ramp-up:* These musicians often take a while to warm up to a performance. After they hit their stride, you might get a killer take or a few great ones that can be composited together into the perfect take.
- *Anything goes:* This category can vary widely within a performance. Often, snippets can be taken from several takes into a single composite. You want to record everything with this type of performer, because you just never know what gem (or bad take) you'll end up with.
- *Rock steady:* This one represents the consummate pro who is fully practiced and delivers a performance that doesn't waver from one take to the next; however, you might record several takes to see which one has the most feeling.

From the above examples, we can quickly draw the obvious conclusion that there are all types of performers, and that it takes a qualified and experienced producer and/or engineer to intuit just which type is in front of the mic … and to draw the best possible performance from him or her.

⑥ MIXDOWN

Many of the same rules of preparation and taking good care of your ears and yourself apply during the session's mixdown phase. Here are a few tips:

- Regarding monitoring, it's often a good idea to use reference monitors that you trust. As such, it's common practice for a mix engineer and/or artist to request their favorite speakers or to bring their own into the studio for the final mix.
- Unlike during the 1970s, when excruciatingly high SPLs tended to be the rule in most studios, recent decades have seen the reduction of monitor levels to a more moderate 75- to 90-dB SPL. A good rule of thumb is that if you have to shout to communicate in a room you're probably monitoring too loudly. From time to time, you can always jack it up to 11 to check the mix at higher volumes and then turn it back down to a moderate level. Taking occasional breaks isn't a bad idea either.
- Listen on several speaker types … at home, in your car, on your iPod/phone. Jot down any thoughts and comments that might come in handy, should you need to go back and make adjustments.

⑦ BACKUP AND ARCHIVE STRATEGIES

The phrase "nothing lasts forever" is especially true in the digital domain of lost 1's and 0's, damaged media and dead hard drives … you know, the "Oh @#$%! factor". It's a basic fact that you never quite know what lies around the techno bend – and it's extremely important that you protect yourself as much as is humanly possible against the inevitable. Of course, the answer to this digital dilemma is to back up your data in the most reliable (or redundant) way possible. Hardware and program software can (usually) be replaced; on the other hand, when un-backed valuable session soundfiles are lost, they're lost!

Backing up a session can be done in several ways, depending on the level of certainty that the files can be played back in the future with a minimum of hassle. Here are a few tips that can help you avoid the loss of data:

- As you might expect, the most straightforward backup system is to copy the session data, in its entirety, to the most appropriate media (most commonly onto one or more hard drives).
- In the longer run (5 or more years), the most ironclad way to back up the track data of a session is to print each track as its own .wav or .aif file. Each track should always be recorded or exported as a contiguous file that flows from the beginning of the session (00:00:00:00 or appropriate beginning point) to the end of that particular track. In this way, the individual track files can be loaded into any type of DAW, at the beginning point, for processing and future mixdown.
- In such a track-by-track safety restoration situation, you might want to save two copies of a track that has a particular effect—one that contains the original, unaltered sound and one that contains the effected signal.

- Those who want additional protection against the degradation of unproven digital media may also want to back each track (or group of tracks) to the individual tracks of a multitrack analog recorder.
- For those sessions that contain MIDI tracks, you should always keep these tracks within the session (i.e., don't delete them). These tracks might come in very handy during a remix or future mixdown.
- Whenever possible, make multiple backups and store them in separate locations. Having a backup copy in your home as well as in the studio can save your proverbial butt in case of a fire or any other unforeseen situation. Remember the general backup rule of thumb: Data is never truly backed up unless it's saved on three drives (and I would argue, in two places)!

8 HOUSEHOLD TIPS

Producers, musicians, audio professionals and engineers spend a great deal of time in the control room and studio. It only makes sense that this environment should be laid out in a manner that's aesthetically, functionally and acoustically pleasing from a feng shui point of view. Creating a good working environment that's conducive to making good music is the goal of every professional and project studio owner. Beyond the basics of creating a well-designed facility from an acoustic and electronic standpoint, a number of basic concepts should be kept in mind when building or designing a recording facility, no matter how grand or humble. Here are a few helpful hints:

- Given the fact that an engineer spends a huge amount of time sitting on his or her bum, it's always wise to invest in both your and your clients' posture and creature comforts by having comfortable, high-quality chairs around for both the production team and the musicians (Figure 21.2).
- Velcro™ or tie-straps can be used to organize studio wiring bundles into groups that can be laid out in a way that reduces clutter, improves organization (color-coded straps can be used) and makes the studio look more professional.
- Most of us are guilty of cluttering up our workspace with unused gear, papers … you know, junk! I know it's hard, but a clean, uncluttered working environment tells your clients a lot about you, your facility and your work habits.
- Unused cables, adapters and miscellaneous stuff can be sorted into storage boxes and stacked for easy storage.
- Important tools and items that are used every day (such as screwdrivers, masking tape or markers) can be stored in a rack-mounted drawer that can be easily accessed without cluttering up your space … don't forget to pack a reliable LED flashlight (your phone's screen display might also work in a pinch).
- Portable label printers can be used to identify cable runs within the studio, identify patch points, I/O strip instrumentation … you name it.

FIGURE 21.2
The venerable Herman Miller Aeron® chair. (Courtesy of Herman Miller, Inc.; www.hermanmiller.com).

9 MUSICIANS' TOOLS

By now it's probably painfully obvious to most musicians that producing the music is only the first step toward building a career in the business. It takes hard work, perseverance, blood, sweat, tears and laughter. For every person who makes it, a large number don't. There are a lot of people waiting in line to get into what is perceived by many to be a glamorous biz. So, how do you get to the front of the line? Well, folks, here are some keys to help you through the golden gate.

- A ton of self-motivation
- Good networking skills
- A good, healthy attitude
- The realization that "showing up is always huge!"

The business of art (the techno-arts of recording and music production being no exception) is one that's generally reserved for self-starters. Even if you get a degree from XYZ College or recording school, there's absolutely no guarantee that a label will be knocking on the door with a contract in hand (if they do, get a lawyer, quick!). It takes a large dose of perseverance, talent and personality to make it. In fact, one of the best ways to get into the biz is to get down on your knees and knight yourself on the shoulder with a sword (figuratively or literally—it doesn't matter) and say: "I am now a ___________!" Whatever it is you want to be, simply become it ... Shazammm! Make up a business card, start a business, begin contacting artists to work with or make the first step toward becoming the artist you want to be.

There are many ways to get to the top of your own personal mountain. You could get a diploma from a school of education or the school of hard knocks (it usually ends up being from both), but the goals and the paths are up to you. Like a mentor of mine says: "Failure isn't a bad thing … not trying is!"

Another part of the success equation lies in your ability to network with other people. Like the venerable expression says: "It's not [only] what you know … it's WHO you know." Maybe you have an uncle or a friend in the business, or a friend who has an uncle—you just never know where help might come from next. This idea of getting to know someone who knows someone else is what makes the business world go around. Don't be afraid to put your best face forward and start meeting people. If you want to play gigs around your region (or beyond), get to know a promoter or venue manager and hang out without being in the way. You never know—the music maven down the street might know someone who can help get you in the proverbial door. The longer you stick with it, the more people you'll meet, thus making a bigger and stronger network than you thought could be possible.

Like a close buddy of mine always says, "Showing up is huge!" It's the wise person who realizes that being in the right place at the right time means being at the wrong place hundreds and hundreds of times. You just never know when lightning is going to strike—just try to be standing in the right field when it does.

Here are some more practical and immediate tips for musicians:

- *Build a personal and/or band website:* Making a site on www.MySpace.com or creating your own personal site helps to keep the world informed of your gigs, projects, bio and general goings-on.
- *Build a relationship with a music lawyer:* Many music lawyers are open to building relations that can be kicked into gear at a future time. Take the time to find a solicitor who is right for you. Does he or she understand your personal music style? If you don't have the bucks, is this person willing to work with you and your budget, as your career grows?
- The same questions might be asked of a potential manager. This symbiotic relationship should be built with care, honesty and safeguards (which is just one of the many reasons you want to know a music lawyer).
- *Copyright your music:* Always protect your music by registering it with the Library of Congress. It's easy and inexpensive and can give you peace of mind about knowing that the artistic property that you're sending out into the world is protected. Go to www.copyright.gov for more information (www.copyright.gov/forms).
- On a personal note as a musician, I've come to realize that making music is about the journey—not the goal of being a star, or being the big man on campus. It's about the friendships, the collaborations, good and bad times at gigs…. To me, it's about making music.

10 RECORD YOUR OWN CONCERTS AND PRACTICES

As a musician, I've found that recording my concerts and putting them up on the Web can help in many ways.

- Freely distributing concerts on your site helps to promote your music and provides a degree of goodwill that can go a long way with your fans.
- These recordings can be really helpful as a business card promo tool, in that a link can be sent to potential venues, allowing booking agents to hear and appreciate your music firsthand.
- These recordings can also be helpful as learning tools for the band and yourself. In general, they don't lie, and can help to point out shortcomings in your performance.

This process doesn't have to be involved. For example, connecting a portable recorder to your audio interface or house mixer and hitting the record button before going on stage will often do the trick.

11 PROTECT YOUR INVESTMENT

When you've spent several years amassing your studio through hard-earned sweat-equity and bucks, it's only natural that you'll want to take the necessary precautions to protect your investment.

Obviously, the best way to protect your data is through a rigorous and straightforward backup scheme (the general rule is that something isn't backed up unless it's saved in three places—preferably with one of the backups being stored off-site). However, you'll also want to take extra steps to protect your hardware and software investments as well, by making sure that they're properly insured.

The best way to start the process of properly insuring your studio is to contact your trusted insurance agent or broker. … If you don't have one, now's the time to get one. You might get some referrals from friends or people in your area and give them a call, set up some appointments and get several quotes.

If you haven't already done so, sit down and begin listing your equipment, their serial numbers and replacement values. Next, you might consider taking pictures or a home movie of your listed studio assets. These steps will help your agent come up with an adequate replacement plan and will come in handy when filing a claim, should an unfortunate event occur. Being prepared isn't just for the Boy or Girl Scouts.

12 PROTECT YOUR HARDWARE

One of the best ways to ensure against harmful line voltage fluctuations (both above and below their nominal power levels) is to use an adequately powered uninterruptible power supply (UPS). In short, a quality UPS works by using a regulated power supply to constantly charge a rechargeable battery or bank of batteries. This battery supply is again regulated and used to feed sensitive studio equipment (such as a computer, bank of effects devices, etc.) with a clean and constant voltage supply.

Just as the "uninterruptible" part of the name implies, should the power be momentarily interrupted or give out altogether, a good UPS can draw on its battery supply to see you through a momentary power loss or to give you enough time to safely shut your system down without losing data during a total power failure. The biggest concern here is to make sure that you buy a UPS that has enough power reserves to adequately and continuously power the equipment during normal operation, and will give you enough time to save your session and/or file data and then safely shut the system down during an outage. In short, make sure the UPS has a continuous power rating that's high enough for your supply needs.

13 UPDATE YOUR SOFTWARE

Periodic software updates might help to solve some of those annoying problems that you've been dealing with in the studio. Often, the software that's been pressed onto a commercially available CD or DVD will be out of date by the time it reaches you. For this reason, it's good to check the Web regularly to see if there's a newer version that can be loaded at the outset or periodically over the course of the software's life.

I will say, however, sometimes it's wise to research an update (particularly if it's an important one) … it's a rare update that will rise up and bite you and your system in the butt … but it does happen.

14 READ YOUR MANUALS

Unfortunately, I really am not a big reader, but taking an occasional glance at my software manuals helps me to get a better understanding of the finer points of my studio system.

15 A WORD ON PROFESSIONALISM

Before we close this chapter, there's one more subject that I'd like to touch on—perhaps the most important one of all—*professional demeanor*. Without a doubt, the life and job of a typical engineer, producer or musician isn't always an easy one. It often involves long hours and extended concentration with people who, more often than not, are new acquaintances. In short, it can be a high-pressure job. On the flip side, it's one that's often full of new experiences, with demands that change on almost a daily basis, and often involves you with exciting people who feel passionately about their art and chosen profession.

It's been my observation (and that of many I've known) that the best qualities that can be exhibited by anyone in "The Biz" are:

- Having an innate willingness to experiment
- Being open to new ideas (flexibility)

- Having a sense of humor
- Having an even temperament (this often translates as patience)
- Being open to communicating with others
- Being able to convey and understand the basic nuances of people from all walks of life and with many different temperaments.

The best advice I can possibly give is to *be open, be patient and above all, be yourself*. Also, be extra patient with yourself. If you don't know something, ask. If you made a mistake (trust me, you will; we all do), admit it and don't be hard on yourself. It's all part of the process of learning and gaining experience.

This last piece of advice might not be as popular as the others, but it might come in handy someday: It's important to be open to the fact that there are many, many aspects to music and sound production, and you may find that your career calling might be better served in another branch of the biz. That's totally OK! Change is an important part of any creative process—that and taxes are the only constants you can count on!

Tutorial: Handling a Successful Session

Your DIY: you can totally do whatever you want, whatever sounds good to you and what's best for the session, the artist and the music ... and that's a very liberating thing!

IN CONCLUSION

Obviously, the above tips are just part of an ever-changing list. The process of producing, recording and mixing in any type of studio environment is an ongoing, lifelong pursuit. Just when you think you've gotten it down, the technology or the nature of the project changes under your feet—hopefully, you'll be the better for it and will learn from the process. Far more than just the technology, the process of coming up with your own production style and your own way of applying the tools, toys and techniques to a production is what makes an artist—whether you're in front of the proverbial glass or behind it. Over time, your own list of studio tips and tricks will grow. Take the time to write them down and pass them on to others, and be open to the advice of your friends and colleagues. Use the trade mags, conventions and the Web to lead you to new ideas. This way, you're opening yourself up to new insights to using the tools of your profession and to finding new ways of doing stuff. Learning is an ongoing process ... have fun along the way!

CHAPTER 22
Yesterday, Today and Tomorrow

I'm sure you've heard the phrase "Those were the good old days." I've usually found it to be a catch-all that refers to a time in one's life that had a sense of great meaning, relevance and all-around fun. Personally, I've never met a group of people who seem to bring that sense of relevance and fun with them into the present more than music and audio professionals, enthusiasts and students. The fact that many of us refer to the tools of our profession as "toys" says a lot about the way we view our work. Fortunately, I was born into that clan and have reaped the benefits all my life … the here and now are always, my "good ol' days"!

Music and audio industry professionals, by necessity, tend to keep their noses to the workaday grindstone. But market forces and personal visions often cause them to keep one eye focused on future technologies, whether these are new developments (such as advances in digital, touch and audio technologies), rediscovering retro trends that are decades old (such as the reemergence of tube technology and the reconditioning of older devices that sound far too good to put out to pasture), or future technologies that excite the imagination. Such is the time paradox of a music and audio professional, which leads me to the book's final task: addressing the people and technologies in the business of sound recording in the past, present and future.

YESTERDAY

I've always looked at the history of music and sound technology with a sense of awe and wonder, although I really can't explain why. Like so many in this industry, I tend to get shivers when I see a wonderful old tape machine or an original tube compressor. For no reason whatsoever, I get all giggly and woozy when I read about some of the earlier consoles that were used to record industry greats, see pictures from past recording sessions (Figure 22.1) or see an original Ampex 200 (the first commercially available professional tape machine). I experience the same sense of awe when I read about my personal historical heroes such as Alan Dower Blumlein (Figure 22.2), who was instrumental in developing stereo mic techniques, the 45°/45° stereo disc-cutting process, the TV camera and radar. To many, his list of accomplishments is second only to those of Nikola Tesla (who

FIGURE 22.1
Nat King Cole clowning around with engineer Bill Putnam (out of shot, in the control room). (Courtesy of Universal Audio, www.uaudio.com)

deserves *at least* as much credit than his nemesis Edison). Mary C. Bell (Figure 22.3) who was probably the first woman sound engineer also comes to mind, along with another unsung hero, the late John (Jack) T. Mullin (Figure 22.4), who stumbled across a couple of German magnetophones at the end of WWII and was smart enough to send them back to San Francisco. With the help of Alexander M. Poniatoff (founder of "AMP"ex) and Bing Crosby, Jack and his machines played a crucial role in bringing the magnetic tape recorder into commercial existence.

Every once in a techno-blue moon, major milestones come along that affect almost every facet of information and entertainment technology. Such milestones have ushered us from the Edison and Berliner era of acoustic recordings, into the era of broadcasting, electrical recording and tape, to the environment of the multitrack recording studio (Figures 22.5 and 22.6), and finally into the age of the computer, digital media and the Web. With the introduction of personal home recording equipment (Figure 22.7), the cassette-based portastudio and affordable mixer/console designs, recording technology was to change forever into being a medium that can be afforded by the general masses.

FIGURE 22.2
The life of Alan Dower Blumlein (truly, one of my long-time heroes) has been published by Focal Press. (Courtesy of Focal Press, www.focalpress.com)

When you get right down to it, the foundation of the modern information and digital age was laid with the invention of the integrated circuit (IC). The IC has likewise drastically changed the technology and techniques of present-day recording by allowing circuitry to be easily designed and mass produced at a fraction of the size and cost of equipment that was made with tubes or discrete transistors.

FIGURE 22.3
Mary C. Bell in NBC's dubbing room #1 (April, 1948) inspecting broadcast lacquer discs for on-air programs. (Courtesy of Mary C. Bell)

FIGURE 22.4
John (Jack) T. Mullin (pictured left) with the two original German magnetophones. (Courtesy of the late John [Jack] T. Mullin)

Advances in digital hardware and software have brought about new developments in equipment and production styles that have affected the ways in which music is created. Integrating cost-effective yet powerful production computers with digital mixing systems, modular digital multitracks, MIDI synths/samplers,

FIGURE 22.5
Gilfoy Sound Studios, Inc., circa 1972. Notice that the room is set up for quad surround! (Courtesy of Jack Gilfoy, www.jackgilfoy.com)

FIGURE 22.6
Little Richard at the legendary LA Record Plant's Studio A (circa 1985) recording "It's a Matter of Time" for the Disney film *Down and Out in Beverly Hills*. (Courtesy of the Record Plant Recording Studios, photo by Neil Ricklen)

plug-in effects and instruments, digital signal processors, etc., gives us the recipe for having a powerful production studio in our homes, bedrooms, or on the bus. Such laptop and desktop music studios have made it possible for more and more people to create and distribute their own music with an unprecedented degree of ease, quality and cost effectiveness.

Peter Gotcher (Digidesign cofounder) was one of the first to envision the creation of a cost-effective "studio-in-a-box" (Figures 22.8 and 22.9). This conceptual spark, which started a present-day goliath, helped to create a system that

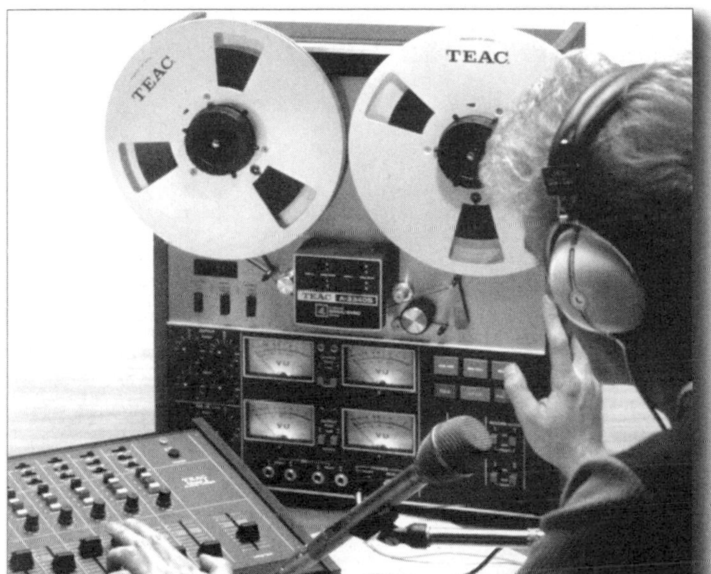

FIGURE 22.7
Early multitrack home recording system. (Courtesy of Tascam, a division of TEAC America, www.tascam.com)

FIGURE 22.8
"Sound Designer," the first cost-effective digital audio workstation, was released by Digidesign in the late 1980s. (Courtesy of Avid Technology, Inc., www.avid.com)

would offer the power of professional hard-disk-based audio at a price that most music, audio and media producers could afford. (It's important to realize that previous to this, digital systems started at over $100,000!) His goal (and those of countless others since) has been to create an integrated system that would link together the many facets that go into audio and audio-for-visual production, via the personal computer. Years later, this dream has totally transformed music and audio production.

FIGURE 22.9 Digidesign's ICON console/Pro Tools-based recording system. (Courtesy of Avid Technology, Inc., www.avid.com)

One of the cooler by-products of this digital age is an upsurge in "retro-future" trends in music and technology that blend together older and newer technologies. With this has come an increased interest and awareness in museums and websites that help us to learn about our musical and technological roots throughout music production history. A few of the physical and cyber sites that can help give us a better understanding of our roots include:

Museums:

- Computer History Museum
  1401 N. Shoreline Boulevard
  Mountain View, CA 94043
  (650)810-1010
  info@computerhistory.org
- Experience Music Project
  325 5th Avenue N. (in the Seattle Center, near the Space Needle)
  Seattle, WA 98109
  877.EMPLIVE; (877)367-5483
  www.empsfm.org
- Grammy Museum
  800 W. Olympic Boulevard, Suite A245
  Los Angeles, CA 90015
  www.grammymuseum.com
- Rock and Roll Hall of Fame and Museum
  1100 Rock and Roll Boulevard
  Cleveland, Ohio 44114
  (216)781-ROCK
  www.rockhall.com

- Stax Museum of Soul Music
  926 E. McLemore Avenue
  Memphis, TN 38106
  (901)946-2535
  www.soulsvilleusa.com

Sites:

- Audio Engineering Society Historical Committee (www.aes.org/aeshc)
- Synthmuseum.com (www.synthmuseum.com)
- Tinfoil.com (www.tinfoil.com)

When it comes to understanding the tools, toys and techniques of our trade, I've always felt that there are a lot of benefits to be gained from looking back into the past … as well as by gazing into the future. A wealth of experience in design and application has been laid out for us. It's simply there for the taking; all we have to do is search it out and put it to good use.

On a final note about the past, I'd like to invite you to visit my personal site and browse through the free online booket "The History of Recorded Sound" (www.davidmileshuber.com/hrs). Here you can follow the timeline of recording from its beginning to recent time, with pictures and linked videos (Figure 22.10) … I hope you enjoy and benefit from it!

TODAY

"Today" is a really difficult subject to talk about, since new equipment literally comes out on a monthly basis. However, it safely goes without saying that the two buzz terms that best sum up the dawn of the 21st century are *digital audio* and the *Web*.

FIGURE 22.10
The History of Recorded Sound … a free online booklet. (www.davidmileshuber.com/hrs).

1's and 0's

The grouping of digital 1's and 0's into words that represent alphanumeric values, sampled voltage levels over time (digital audio) or pixilated color and brightness values on a screen (digital graphics) have changed communications and creative production forever. Literally, even the most diehard analog fanatic can't escape its far-reaching grasp.

With the advent of the personal computer, DAW, cost-effective peripherals, application and effects plug-ins and digital downloads, music creation isn't only more cost effective than ever before—it also offers a degree of power, flexibility and portability that boggles the mind!

Fortunately, this newfound digital technology has spawned a primal urge, almost a frenzied lust to marry cutting-edge hardware/software with vintage gear or newly reissued toys that are based on decades-old technologies, particularly tube devices (such as tube condenser mics, preamps, and signal processing gear). So why embrace this older technology? Well, for starters, tube electronics inherently have a sound that's very different than their IC- or transistor-based counterparts. When a tube is overdriven to the point of clipping, the square edge of the distorted signal is actually rounded off. This tends to yield a smoother, fuller, more "rounded" distortion, when compared to a sharply distorted, harsh edge that's usually exhibited by transistor and IC circuit designs. In addition, tubes generate even-order harmonics, which are far more musical sounding in nature than the harsher odd-order distortions that are generated by its modern solid-state counterparts.

The Web

Beyond the overall concept of digital media, another huge mover and shaker of everyday life is the Web. Cyberspace made the creation of this book much easier for me as a writer (search engines and company sites make research a relative breeze, and photos can be quickly emailed) and, of course, social media is now a major force in communication and self-expression and music (Figure 22.11). Although many of the music share sites that contain ripped downloads of major releases have been shut down, legitimate pay-per-download sites have sprung into megabuck action. Personal and indy music sites have allowed upstart and established artists to directly sell their music, inform their fans about upcoming tours and publish fanzine info to keep their public begging for more. In fact, from a philosophical standpoint, I feel that the trend toward the breakdown of the traditional music industry isn't all bad news. I'm definitely not alone in thinking that this shift has taken the overall power out of the hands of a few … and has given it back to the indies and to the individual "many." You might even say that the shift toward downloadable distribution has taken music production back into the home where it began over a hundred years ago with the popularity of having the family piano or other instruments in the parlor.

FIGURE 22.11 Social websites allow people to connect and share their everyday lives. One such site, www.allsoundsmatter.com was designed to connect professionals and musicians who are driven by their passion for sound. (Courtesy of allsoundsmatter, www.allsoundsmatter.com)

In short, although we're still at the dawning of the Web distribution age, cyberspace has long proven itself to be a source of increased visibility, viability and hope for the budding artist, as well as for the established corporate world of the music biz.

TOMORROW

Usually, I tend to have a decent handle on the forces that might shape the sounds and toys of tomorrow … but it's getting increasingly harder to make specific predictions in this fast-paced world. Today, there are far more choices for gathering information and entertainment than reading a book or watching TV. Now, we can interact with others in a high-speed, networked environment (Figure 22.12) that does far more than let us be just spectators; it lets us participate and share our thoughts with others, which leads to increased knowledge, creative discourse and personal growth.

This idea of intercommunication through the web has changed the face of business, communication, and recreation towards an e-based, on-demand commerce.

As I write this, digital audio has and will continue to become more portable, more virtual and more powerful. For example, I have a killer laptop production system that fits snuggly into my M-Audio backpack, an iPad that lets me produce and perform my music wirelessly and a smart-phone that'll practically cook eggs for breakfast. Although my main surround-sound studio involves keyboards, synths and music controllers of various types, I've definitely welcomed the continued march toward quality virtual instruments and useful plug-ins that seamlessly integrate into the DAW. I think I'll always marvel at a computer's ability to be a chameleon (in its many incarnations)—one moment it's a music

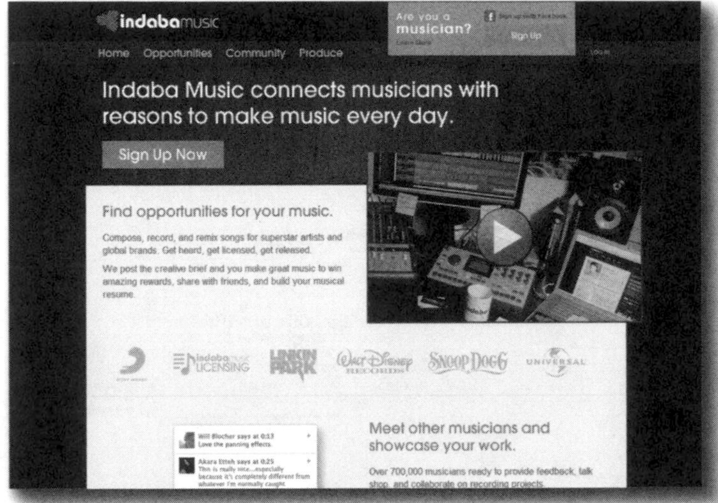

FIGURE 22.12
There are numerous sites that allow musicians to network, collaborate and remix other folks' work on the Web. (Courtesy of indaba music, www.indabamusic.com)

production system, next it's a video editor, then a word processor, then a graphic workstation, then a partridge in a neon pear tree that'll let you talk to a buddy across the world. This aspect of technology literally frees us to be creative in an amazing number of ways that's truly an astonishing joy.

HAPPY TRAILS

Before we wrap up the 8th edition, I'd like to take a moment to honor one of the greatest forces driving humanity today (besides sex) … the dissemination and digestion of information. Through the existence of quality books, trade magazines, university programs, workshops and the Web, a huge base of information on almost any imaginable subject is now being distributed to and understood by a greater number of aspiring artists and technicians than ever before. These resources often provide a strong foundation for those who are attending accredited schools as well as those who are attending the school of hard knocks. No matter what your goals are in life (or in the business of music), I urge you to jump in and read, surf, skip through pages and keep your eyes and ears open for new sounds, ideas, technologies and experiences. The knowledge and skills you gain will always be well worth the expended time and effort.

On a final note, I'd like to paraphrase Max Ehrmann's "Desiderata," when he urges us to keep interested in our own career, however humble, as it's an important possession in the changing fortunes of time. Through my work as a producer, musician, writer and educator, I've been fortunate enough to know many fascinating, talented and fun people. For some strange reason, I was born with a strong drive to have music and production technology in my life. By "keeping interested in my own career" and working my butt off, while having several brushes with extreme luck, I've been able to turn this passion into a successful career.

To me, all of this comes from following your bliss (as some might call it), listening to reason (both your own and that of others you trust) and doing the best work that you can (whatever it might be). As you know, thousands of able and aspiring bodies are waiting in line to make it as an engineer, a successful musician, a producer, etc. So how does that one person make it? By following the same directions that it takes to get to Carnegie Hall—practice! Or as the t-shirt says, "Just do it!" Through perseverance, a good attitude and sheer luck, you'll be led through paths and adventures that you never thought were possible. Remember, being in the right place at the right time, simply means being in the wrong place a thousand times … "Showing up in life is huge!"

Have fun!

David Miles Huber
3x Grammy nominated artist/producer
www.davidmileshuber.com

APPENDIX A
Continued Education

As the equipment used by engineers, producers and musicians places an ever-increasing emphasis on technology, education must play a greater role in the understanding of basic industry skills. Education can take many forms, ranging from a formal education to simply keeping abreast of industry directions with the help of the many industry magazines.

Beyond getting the necessary hands-on education by playing with all of the tools and toys, one of the best ways to get a better handle on the pulse of this industry is to read. Read all the magazines, books and articles that you can get your hot little hands on. Of course, there's a ton of schools that offer courses in recording, media and music production. If this approach works for you, I'd recommend that you carefully check the institution and department out before potentially laying down your hard-earned cash to further your education towards getting into the music biz.

THE WEB

I obviously encourage you to be creative about jumping onto an internet connection. Go to your local library, bug a friend, hang out at your favorite wifi coffee shop with your laptop, phone or pad… whatever it takes to surf the worldwide info-stream.

Most media-related magazines have a web presence that lets you read and download the latest issue and even dig deeper by watching the latest technology videos, audio examples and the like. Users can also use the web to communicate using public and private forums (allowing the users to voice their comments and ideas to other users and manufacturers) in addition to contacting the company directly using electronic mail services…

The most important lesson to be learned from this short appendix is to "Never Stop Learning"! Put a magazine rack in the bathroom, check out new technologies, see who's doing what, surf the mediawaves and soak it all in… and then put that knowledge into practice. The art of expanding your mind will soon translate into an expanded set of skills that will help keep your skills and your outlook on life keen and sharp!

GETTING OUT THERE!

One of the less obvious—but most important—tools for getting yourself out into the world happens through the joining industry-related organizations and going to industry conventions. On a personal note, most of my best friends and industry contacts have come to me through the attending of conventions such as NAMM and the AES… while personal contacts and priceless opportunities have come to me through joining and active participating in "The Recording Academy" (better known to most as the Grammy organization).

Using the all-important adage "Showing up is huge!"… going to special industry events, conventions and music happenings will put you in the smack-dab in the middle of the action, putting all the people that are in the best position to help you out with information, contacts—and even jobs—right there in the same room. Literally stepping up to the plate and helping with these organizations and events can help connect you to the industry in ways that are amazingly huge.

For all of you students (in the US)… my first suggestion would be to immediately join Grammy U (www.grammy.com/grammyu). For a very nominal price, this membership would allow students living near one of the many local chapters across the country to attend special Grammy events, lectures and parties… it's WELL WORTH the time and effort! Here's what the site has to say about "The U"…

> GRAMMY U is a unique and fast-growing community of college students, primarily between the ages of 17 and 25, who are pursuing a career in the recording industry. The Recording Academy created GRAMMY U to help prepare college students for their careers in the music industry through networking, educational programs and performance opportunities. GRAMMY U is designed to enhance students' current academic curriculum with access to recording industry professionals to give an "out of classroom" perspective on the recording industry.

A small listing of events and organizations include:

Conventions:

- AES (Audio Engineering Society) – www.aes.org
- APRS (Association of Professional Recording Services, UK) – www.aprs.co.uk
- Musikmesse (Europe) – musik.messefrankfurt.com
- NAMM (National Association of Music Merchants) – www.namm.org

Organizations:

- AES (Audio Engineering Society) – www.aes.org
- APRS (Association of Professional Recording Services, UK) – www.aprs.co.uk
- The Recording Academy (Grammy) – www.grammy.com
- SPARS (Society of Professional Audio Recording Services – www.spars.com

Happy reading, surfing and learning!

APPENDIX B
Tax Tips for Musicians

by Jeffrey P. Fisher

Author's Note—Some might wonder why I included this excellent article on taxes in this book. Well, it's a well-known fact that artists (and the general public at large) aren't always money conscious... and with the burden of student loans, equipment costs and the decisions that go with day-to-day living, it's more important than ever to learn the value of careful financial decisions that could affect your credit rating, your ability to save and your financial security. Understanding the value of money, the power of a personal budget and the responsibilities that go with incurring a debt load are just as important as anything you might learn in this book. When I started in the biz, I had very little money and was living on my own on the West Coast. Since I've never been one for living on the edge, I quickly learned the necessity of saving up for times when the going might get rough. As a result, I've always been able to set aside enough to live comfortably, without the fear that goes with incurring too much debt. Over the years, this simple idea has given me freedoms that I never would've thought possible... all through the simple concept of saving for a rainy day. Underneath all of the technicalities of Jeffrey's article is the implied moral of saving, living within your personal means and setting back enough bucks to enjoy the fruits of your artistic labors.—DMH

Everybody complains about taxes. But few do anything about it. Well, there are several actions you can take to improve your tax situation right now. If you're making even the tiniest amount of music-related scratch, there's no reason to pay more taxes than you have to. To reap the most tax benefits, start running your music career as a legal small business. The IRS loves small business. According to the Small Business Administration, there are 25 million small businesses in the US today. And a large percentage of them are sole proprietorships—one person shops. As a sole proprietor, you report your music business income as part of your personal income using Schedule C (and a few other forms).

A BUSINESS OF YOUR OWN

It makes real financial sense to run your project studio as legal business. Follow these basic steps:

- Set up your business by choosing its legal structure (sole-proprietorship, partnership, corporation, etc.)—Consult with a tax adviser for details about the financial aspects of each form. Contact a legal adviser for answers to liability issues.
- Get legal—Make sure you meet any regulations that pertain to running a business in your town. For example, you may have to get a business license from your local clerk's office.
- File a doing-business-as (dba) with your local government if you call your business something other than your legal name—You may need a separate tax ID for your business and some states require a sales tax ID number.
- Open a business checking account—Deposit your project studio income into it and pay your business expenses using checks drawn on it. Also, use a credit card only for business purchases and pay it off on time from your business checking account.
- Use bookkeeping software to track all your business income and expenses – This makes tax preparation and monitoring your financial situation easier. Understand the various tax consequences of your business, too. You'll probably need to make quarterly tax payments in addition to yearly tax preparation.
- Protect yourself through health and property insurance—Also, consider life, disability, and liability insurance if it makes sense for your situation.

It all comes down to income and expenses—the money you make and the money you spend. The more you make, the more you pay in taxes. Simple, right? Even the most convoluted of IRS instructions make that point painfully clear. That means the inverse is also true. Since the IRS only taxes your business profits, cut back on the profit and pay less taxes.

"BUT, DUDE, I GOTTA EAT"

Whoa there. I'm not saying that you should earn less. Instead, look for all the possible ways to convert some of your everyday expenses into legitimate business deductions. Even some personal expenses may be deductible against the business. The more expenses you have, the more you reduce your taxable income. And since you were going to spend the money anyway, you might as well realize some tax benefits, too.

WRITE-OFFS

Basically, all the expenses you incur to run your little music business are deductible. To be fully deductible, however, these business expenses must be "ordinary and necessary" according to the IRS. That's just fuzzy enough to be dangerous. Ordinary means the expenses must be typical for the business. Buying a new guitar could apply; buying a dishwasher wouldn't. Necessary just means the expense is vital to the success of your business. Office equipment, postage, phone charges, graphic design charges, recording studio fees, duplication, dues, magazine subscriptions, and other such related items are definitely necessary for the success of the typical music business.

Here is a basic list of business deductions for musicians:

- Advertising and promotion costs
- Car and truck expenses
- Commissions and fees you pay to other people/businesses
- Depreciation and section 179 deduction (see article)
- Insurance (except health insurance—see article)
- Interest (on business loans)
- Legal and professional fees (hire a lawyer; deduct the charge)
- Office expenses
- Rent/lease payments
- Repairs and maintenance
- Supplies
- Taxes and licenses such as a business license
- Travel costs
- Meals and Entertainment (see article)
- Utilities (see home-office deduction in article)
- Wages, salaries paid

For every $100 you earn, you pay approximately $45.30 in taxes (if you're in the 27% tax bracket, pay the 15.3% self-employment tax, and send an additional 3% to your state). Of course, that also implies that for every legitimate $100 business expense you incur, you also save the $45.30 you would otherwise pay in taxes. Hey, that's like getting everything you buy at a discount.

Why does the IRS let you deduct all these expenses? They want you to succeed. So, they let you invest money (spend it) in your business as incentive for you to earn more. And the more money you make, the more you'll pay in taxes. You see they have an ulterior motive; they ain't jus' bein' neighborly.

But there's a caveat (isn't there always?). You need to be gainfully engaged in making a buck. You must turn a profit in your business three out of every five years or your business will be classified as a hobby, and you forfeit the expense deductions. Bottom line: very bad news and very high tax bill. (And one dollar in profit those three years ain't gonna make you popular down at the Treasury Department neither).

Since the burden of proof falls solely to you, it's vital that you record all your music business income and expenses diligently. A shoebox full of receipts does not a bookkeeping system make. Get help setting up your books or look for a software solution to help document your business financial transactions.

Another important gotcha: If you're just launching your music business, startup expenses can't be deducted all at once. You must amortize them over five years by taking 20% portions of the total expenses and deducting them over five consecutive years.

GEAR LUST = TAX SAVINGS

Did you know that the gear you buy for making your music magic could be a sweet tax deduction? Under section 179 of the tax code you can deduct or "expense" up to $100,000 of tangible property and write it all off when you prepare your taxes. For tangible property think expensive, long-lasting items, such as a new computer. This amount can be above and beyond many other normal business expenses you might incur.

If you've had a particularly strong earnings year, you might want to offset some of that gain by deducting all the cost of large purchases in one year (up to the limit). Alternately, you can choose to depreciate what you buy and deduct a portion of those costs over the next several years.

HOME SWEET HOME

If you do the majority of your music work in your home office, you can deduct a portion of the same expenses that now do little or nothing to lessen your tax burden. You can write off rent or mortgage interest, property taxes, utilities (gas, electricity, water/sewer), insurance, repairs, and depreciation. First, dedicate a portion of your home entirely for your music business. Keep it free of personal items and make it your primary business location. Beware that if you do most of your work elsewhere (gigging, for instance), and only use this home office occasionally, your deduction may be limited or entirely verboten.

Here's how to figure your deductions. Total up the square footage of this exclusive and principal place and compare it to the total square footage of your crib. Say your math works out to 10% (100 square feet of a 1000 square foot home – you need a bigger place!). You can then deduct 10% of the aforementioned expenses using Form 8829: Expenses for Business Use of Your Home. The total deduction then flows through to your Schedule C reducing your income, and therefore your taxes, considerably.

There is a recapture clause (which doesn't apply to renters). If you sell your home and make a profit, those profit dollars become taxable business income at the same percentage rate as your deduction. Score a $50,000 gain (good for you!), and, following the above example, $5,000 of it belongs to the business (subject to self-employment tax and regular income tax, of course). It's important to note that the personal income you make from a house sale is generally not taxed, though! Stop taking the home-office deduction for two tax years prior to the home sale and this recapture clause doesn't apply.

SELF-EMPLOYMENT TAX

Yes, we self-employed have a special tax just for us. Actually every worker pays the same tax - funding for Social Security and Medicare - it's just a little different when you're on your own. You must contribute both the employee and employer contributions which total up to a whopping 15.3%. Yep, just over fifteen pennies on every buck you earn goes right into the Social Security kitty. This is, of course, before you start paying any regular income taxes. Ouch! And you have to pay the self-employment taxes (along with income taxes) quarterly. You need to predict what you are going to earn this year, and the taxes that would be due on that dollar amount. Then, you send in 25% of that money on April 15, June 15, September 15, and January 15 of the next year. These estimated tax payments are important, because if you don't pay enough, there's a penalty due the next April 15.

HEALTH INSURANCE

You can deduct the premiums paid for health insurance, too. This doesn't come off the Schedule C, but is actually a front page deduction on your personal 1040. Self-employed individuals can deduct 100% of the premiums they pay (unless Congress changes its collective mind). Other typical medical costs are deductible on Schedule A (if you qualify).

EAT, DRINK, AND BE MERRY

When you entertain your clients, the money you spend is another write-off. However, these meals and entertainment are subject to a 50% limitation. Spend a $100 on a pizza party, and take $50 off on Schedule C. Give clients gifts, up to $25 per client, and you can take that as a full deduction, though.

When you travel as part of your music business, those expenses are deductible including airfare, lodging, and meals. You must support your travel and lodging deductions with receipts. However, instead of keeping track of your meals, you can take the government's standard per diem allowance of $30 for Meals and Incidentals. Other cities may have higher amounts so check the official Web site (www.policyworks.gov/perdiem). Meals on the road are, of course, still subject to the 50% limit. The IRS figures you gotta eat anyway, so they limit the expense. Bummer.

VEHICULAR DEDUCTIONS

Yes, that old beater is worth money! Keep track of actual vehicle expenses (gas, repairs, etc.) or take the standard mileage rate (which changes every year; check with the IRS). In either case, you must document the miles you drive for business, date, and purpose of trips, along with expenses incurred. A dedicated notebook/diary earns a gold star from the IRS.

Even if you use your ride for business and personal use, the business portion of your expenses is still deductible. Determine your business percentage by dividing your business miles by the total miles driven (2500 business/10,000 total=25%). If you just use the standard mileage rate, multiply your business miles driven by that rate (2500 × $/mile=deduction) to arrive at your deductible expense amount. You can also deduct the full cost of tolls and parking fees incurred while on business. And the loan interest on the car is deductible (subject to the business use percentage, if it applies).

FEED THE NEST EGG AND SAVE, TOO

You also save money by contributing to a qualified retirement plan. The IRS makes it easy to sock away some cash for a rainy day and rewards you with a nice, fat deduction each year. This is another 1040 deduction, not Schedule C. IRAs are the first method that pop up.

However, they're limited to $2250 (which changes regularly). With a SEP (Simplified Employee Pension) you can deduct as much as 15% of your business income topping out at $30,000 total per year. The more you put away, the more you save. And since you're really helping yourself down the road, it's a smart way to manage your taxes and your retirement. For some of us, a Roth IRA may be more prudent. Roths give you no up front deduction, but the earnings are tax-free. Talk to a financial planner.

EOY TAX TIPS

At the end of each year, you have another opportunity to reduce your tax burden: accelerate expenses and decelerate income. First, spend some cash on business expenses. Don't just blow the wad. Make sensible purchases this year that will reduce your taxable income. Ideal last minute purchases include postage, equipment, general office supplies, and promotions. You can also pay your mortgage and health insurance premium before the year-end to realize some other tax savings on the personal side. Second, put off collecting money this December to January by billing your clients a little later. Though you'll have to pay taxes on the money eventually, you defer that payment for a whole year.

GET HELP FROM THE IRS

Surf on over to the always exciting IRS Web site (www.irs.gov) and download the free guides that explain the specific tax benefits for small business owners.

- #334, Tax guide for small business
- #463, Travel, entertainment, gift, and car expenses
- #533, Self-employment tax
- #535, Business expenses
- #583, Starting a business and keeping records
- #587, Business use of your home

FINAL WORD

While we all have to pay taxes, we are only required to pay our fair share. Make sure you are not throwing money out the window. Take advantage of these and all the other tax breaks available to you. And put more music money in your pocket… where it belongs!

AUTHOR BIO

Jeffrey P. Fisher provides audio, video, music, writing, training, and media production services. He also writes about music, sound, and video for print and the Web. He's published six books including: Instant Sound Forge (CMP, 2004), Moneymaking Music (Artistpro, 2003), and Profiting From Your Music and Sound Project Studio (Allworth Press, 2001). He teaches at the College of DuPage in Glen Ellyn, IL. Also, Jeffrey co-hosts the Acid, Sound Forge, and Vegas forums on Digital Media Net (www.dmnforums.com). For more information visit his Web site at www.jeffreypfisher.com or contact him at jpf@jeffreypfisher.com.

Index